IoT for Defense and National Security

IEEE Press
445 Hoes Lane
Piscataway, NJ 08854

IoT for Defense and National Security

Edited by

Robert Douglass
Alta Montes, Inc., USA

Keith Gremban
University of Colorado School of Law, Boulder, CO, USA

Ananthram Swami
Army Research Laboratory, Adelphi, MD, USA

Stephan Gerali
Lockheed Martin, Inc., Bethesda, MD, USA

IEEE PRESS
WILEY

Published by John Wiley & Sons, Inc., Hoboken, New Jersey.
Published simultaneously in Canada.

For general information on our other products and services or for technical support, please contact our Customer Care Department within the United States at (800) 762-2974, outside the United States at (317) 572-3993 or fax (317) 572-4002.

Wiley also publishes its books in a variety of electronic formats. Some content that appears in print may not be available in electronic formats. For more information about Wiley products, visit our web site at www.wiley.com.

Library of Congress Cataloging-in-Publication Data Applied for:

Hardback ISBN: 9781119892144

Cover Design: Wiley
Cover Image: © Lines forming abstract shapes – stock illustration

Set in 9.5/12.5pt STIXTwoText by Straive, Chennai, India

Contents

List of Contributors

Tarek Abdelzaher
Department of Electrical and Computer Engineering
University of Illinois at Urbana-Champaign
Urbana
IL
USA

Jae-wook Ahn
IBM Thomas J. Watson Research Center
Yorktown Heights
New York
USA

Jonathan Ashdown
Air Force Research Laboratory
Rome
NY
USA

Michael Atighetchi
Raytheon BBN
Cambridge
MA
USA

Nathaniel D. Bastian
Army Cyber Institute
United States Military Academy
West Point
NY
USA

Luis Bathen
IBM Research – Almaden
IBM
San Jose
CA
USA

Christina Baxter
Emergency Response TIPS
LLC
Melbourne Beach
FL
USA

Gisele Bennett
MEPSS LLC
Indian Harbour Beach
FL
USA

Elisa Bertino
Department of Computer Science
Purdue University
West Lafayette
IN
USA

Ramesh Bharadwaj
Information Technology Division
U.S. Naval Research Laboratory
Washington
DC
USA

Stephane Blais
Raytheon BBN
Cambridge
MA
USA

Kyle Broadway
Chief Technology Officer
Armaments Research Company
University of Missouri
Columbia
MO
USA

and

Johns Hopkins University
Baltimore
USA

Jonathan Bunton
Department of Electrical and Computer Engineering
University of California
Los Angeles
CA
USA

Seraphin Calo
IBM Thomas J. Watson Research Center
Yorktown Heights
New York
USA

Armando Caro
Raytheon BBN
Cambridge
MA
USA

Samrat Chatterjee
Data Sciences & Machine Intelligence Group
Pacific Northwest National Laboratory
Richland
WA
USA

Satish Chikkagoudar
Information Technology Division
U.S. Naval Research Laboratory
Washington
DC
USA

Shyama Prosad Chowdhury
IBM GBS
IBM India
Kolkata
WB
India

Dan Coffin
Raytheon BBN
Cambridge
MA
USA

William Crowder
Logistics Management Institute
Tyson
VA
USA

Nirmit Desai
IBM Thomas J. Watson Research Center
Yorktown Heights
New York
USA

Vicente Diaz-Casas
Grupo Integrado de Ingeniería
CITENI
Campus Industrial de Ferrol
Universidade da Coruña
Ferrol
Spain

Robert Douglass
Alta Montes
Sandy
UT
USA

M. Douglas Williams
Seed Innovations
Colorado Springs
CO
USA

Zheng Fang
Department of Computer Science
University of California
Davis
CA
USA

Tiago M. Fernández-Caramés
Department of Computer Engineering
CITIC Research Center
Universidade da Coruña
A Coruña
Spain

Paula Fraga-Lamas
Department of Computer Engineering
CITIC Research Center
Universidade da Coruña
A Coruña
Spain

Auroop Ganguly
Department of Civil & Environmental Engineering
Northeastern University
Boston
MA
USA

Luis A. Garcia
Information Sciences Institute
University of Southern California
Marina Del Rey
CA
USA

Stephan Gerali
Enterprise Operations
Lockheed Martin Corporation
Bethesda
MD
USA

Keith Gremban
Ann and H.J. Smead Aerospace Engineering Sciences and Silicon Flatirons Center
University of Colorado Boulder
Boulder
CO
USA

Keith Grueneberg
IBM Thomas J. Watson Research Center
Yorktown Heights
New York
USA

Marcos Míguez González
Grupo Integrado de IngenierÍa
CITENI
Campus Industrial de Ferrol
Universidade da Coruña
Ferrol
Spain

Sara Ferreno-Gonzalez
Grupo Integrado de IngenierÍa
CITENI
Campus Industrial de Ferrol
Universidade da Coruña
Ferrol
Spain

Nancy Greco
IBM Thomas J. Watson Research Center
Yorktown Heights
New York
USA

Tadanobu Inoue
IBM Research
IBM Japan
Chuo-ku
Tokyo
Japan

Susmit Jha
Neuro-symbolic Computing and Intelligence
CSL
SRI International
Menlo Park
CA
USA

Dhiraj Joshi
IBM Thomas J. Watson Research Center
Yorktown Heights
New York
USA

Joud Khoury
Raytheon BBN
Cambridge
MA
USA

Paul J. Kolodzy
Kolodzy Consulting, LLC
Falls Church
VA
USA

Sastry Kompella
Information Technology Division
U.S. Naval Research Laboratory
Washington
DC
USA

Bhaskar Krishnamachari
Department of Electrical and Computer Engineering
University of Southern California
Los Angeles
CA
USA

Hyunwoo Lee
Department of Computer Science
Purdue University
West Lafayette
IN
USA

Wei-Han Lee
IBM Thomas J. Watson Research Center
Yorktown Heights
New York
USA

Ninghui Li
Department of Computer Science
Purdue University
West Lafayette
IN
USA

Tommaso Melodia
Department of Electrical and Computer Engineering
Northeastern University
Boston
MA
USA

Steve Morgan
Chief Technology Office
Raft LLC
Herndon
VA
USA

Prasant Mohapatra
Department of Computer Science
University of California
Davis
CA
USA

Anand Mudgerikar
Department of Computer Science
Purdue University
West Lafayette
IN
USA

Alicia Munin-Doce
Grupo Integrado de Ingeniería
CITENI
Campus Industrial de Ferrol
Universidade da Coruña
Ferrol
Spain

Sam Nelson
Raytheon BBN
Cambridge
MA
USA

Konstantinos Poularakis
Department of Electrical Engineering &
Institute for Network Science
Yale University
New Haven
CT
USA

Francesco Restuccia
Department of Electrical and Computer
Engineering
Northeastern University
Boston
MA
USA

Stephen Russell
Department of Research Opportunities and
Innovation
Jackson Health System
Miami
FL
USA

Qiaofeng Qin
Department of Electrical Engineering &
Institute for Network Science
Yale University
New Haven
CT
USA

Gowri Sankar Ramachandran
School of Computer Science
Queensland University of Technology
Brisbane
Queensland
Australia

Lucía Santiago Caamaño
Grupo Integrado de Ingeniería
CITENI
Campus Industrial de Ferrol
Universidade da Coruña
Ferrol
Spain

Tim Strayer
Raytheon BBN
Cambridge
MA
USA

Ananthram Swami
U.S. Army DEVCOM Army Research
Laboratory
U.S. Army Futures Command
Adelphi
MD
USA

Paulo Tabuada
Department of Electrical and Computer Engineering
University of California
Los Angeles
CA
USA

Leandros Tassiulas
Department of Electrical Engineering & Institute for Network Science
Yale University
New Haven
CT
USA

Bishal Thapa
Raytheon BBN
Cambridge
MA
USA

Darlene Thorsen
Data Sciences & Machine Intelligence Group
Pacific Northwest National Laboratory
Richland
WA
USA

Venugopal Veeravalli
ECE Department
University of Illinois at Urbana-Champaign
Champaign
IL
USA

Dinesh Verma
IBM Thomas J. Watson Research Center
Yorktown Heights
New York
USA

Gunjan Verma
U.S. Army DEVCOM Army Research Laboratory
U.S. Army Futures Command
Austin
TX
USA

Shiqiang Wang
IBM Thomas J. Watson Research Center
Yorktown Heights
New York
USA

Jaime Wightman
Chief Data and Analytics Office
Lockheed Martin Corporation
Bethesda
MD
USA

Maggie Wigness
U.S. Army DEVCOM Army Research Laboratory
U.S. Army Futures Command
Adelphi
MD
USA

David Wood
IBM Thomas J. Watson Research Center
Yorktown Heights
New York
USA

Introduction: IoT for Defense and National Security

Robert Douglass

Alta Montes, Inc., Sandy, Utah, USA

1 An Introduction to the Topic of IoT for Defense and National Security

The Internet of Things, IoT, connects physical objects through digital networks. It fuses sensors, processors, data storage, smart algorithms, and actuators to observe and physically alter the world and the people in it. In the winter of 2022, IoT weapons wielded by badly outnumbered but determined and courageous Ukrainian light infantry won the battle of Kyiv, destroying hundreds of Russian tanks and perhaps thousands of military vehicles along with the Russian soldiers in them. Javelin anti-tank missiles and Switchblade loitering missiles were instrumental in securing Ukraine's victory. Javelins are nascent IoT weapons while Switchblade missiles are quintessential IoT systems. They move their human operators back from the most hazardous combat zones. They operate with a diverse collection of sensors and actuators, wired and wireless tied together. They coordinate and share their attack across a network connecting reconnaissance drones, command nodes, and other soldiers. They function both as weapons receiving off-board intelligence and as a source of intelligence. They put soldiers in top-level supervisory control while moving them out of the real-time control loop, allowing their weapons to find, track, attack, and destroy their targets in a highly autonomous manner. They turbocharge the tempo of combat. They dispel the fog of war. They save the lives of their operators. In the battle of Kyiv, they help win the fight against long odds and an overwhelming weight of armor.

IoT is not new. More than half a century ago, the internet protocols were conceived as the glue of an IoT system, one that would connect early warning sensors detecting incoming intercontinental missiles with command nodes and systems of response. While the first internet was not ultimately used for this application, its rationale was clearly the creation of a network of systems to sense and respond in a nation's defense – it was conceived as an IoT system of systems for war. Realizing the power of IoT required decades of maturation in five core technologies: sensors, wireless networking and communication, cloud or distributed computing, intelligent algorithms, and digitally controlled actuators. Like the Internet itself, defense investment largely invented these technologies and their underlying concepts while the much greater commercial market powered their expanded performance and reduced their size, weight, power, and especially cost. These technologies continue to advance, but they already enable emerging IoT systems to realize a vision anticipated for decades. As described below, IoT will revolutionize warfare. It will revolutionize how nations secure themselves and maintain peace. The technology is already doing so, although it is not often referred to as IoT. Despite their impact to date and their still greater future promise, IoT systems used for military affairs still face many challenges – challenges not faced or faced to the same degree by commercial IoT.

In the commercial sector, hundreds of books have been written on IoT. Arguably IoT technology impacts defense and national security more than any commercial domain; however, not a single book exists on the topic. The subject is shrouded in secrecy and government restrictions, isolating it from the public realm. This Book provides a first look at what can be released. Leading scientists and technologists describe their latest research results. They address a wide range of topics including IoT security in hostile environments, artificial intelligence (AI) in IoT for defense, and tactical networks in a disrupted, intermittent, and limited-bandwidth battlefield. For example, this Book explains that in such a world IoT can be sustained in the cauldron of combat using content-based routing, configured with mobile ad-hoc networks that ride jam-defiant and intercept-resistant electronically formed beams.

This book offers solutions to special challenges of IoT for defense that set it apart from the commercial environment. It enumerates some of the outstanding and unsolved technical problems. It looks at several different visions for the future of IoT ranging from IoT-enabled rifles to entire logistics systems powered by IoT. It provides several case studies by practitioners in the field from defense manufacturing to the design of warships. It provides a roadmap for policy and regulation of government use of IoT. This Book in no way purports to be a comprehensive review of IoT for defense, for two reasons. First, no one volume can span such an extensive domain. Second, governments sequester much of the material. This book presents an introduction to the subject – a sampling of its many aspects. The public remains largely unaware of these issues, but they are critically important to governments and individuals who are protected by IoT systems and in places oppressed by them.

Sections 1 to 4 of this Book assume some prior knowledge of IoT. For IoT neophytes or those who want more background and context on the elements of IoT for defense and national security, the first part of Chapter 21 provides that introduction, and the beginner might want to start there. Section 1 of this Book presents a sampling of challenges, applications, and opportunities for IoT used for defense. Section 2 reviews the role of AI in IoT for defense and addresses selected case studies and key challenges associated with AI-based IoT. Section 3 discusses security issues and solutions for operating distributed IoT networks in adversarial environments. Section 4 addresses the key challenge of providing IoT systems with reliable communications and networking in mobile, dynamic, and hostile environments. It highlights differing requirements between defense and commercial IoT and suggest how one might build on the other. The final Chapter, 21, addresses issues of regulating IoT to advance its use while blocking its misuse and abuse.

2 What Is IoT?

As a revolution in technology, IoT rivals the Internet on which it depends. It is not just another application riding on the Internet, but a fundamental advance in technology. It can automate our world. IoT senses the world, analyzes the data in the light of mission requirements, and then takes actions that affect the physical world. This is unique – closing the loop automatically in the physical world. The only similar technology consists of control systems. In some sense, IoT is a control system for the world. IoT combines three elements – sensing and information extraction, processing, and action. These elements ride on top of digital communication and networking, an infrastructure that has become cheaper, smaller, and ever greater in capacity, especially with the advent of 5G wireless technology. IoT goes beyond sensing and processing information. It fuses the technology that can take control and modify our physical world. Society-altering consequences, both good and bad, will flow as IoT advances. IoT can leverage human senses, thought, and power, putting people on top of the control loop rather than in the middle of it. Alternatively, IoT can displace human control. Perceiving, formulating a response, and acting to achieve a desired goal are some of the hallmarks of sentient beings. But IoT capabilities extend beyond human senses and exceed our manual powers. Its ability to plan actions already surpasses human performance in some domains.

IoT uses distributed sensors, databases, digital documents, and other software applications to extract information from multiple sources across time and space. Extracted information can be processed locally or by processors geographically dispersed across a cloud of computing resources. Based on what its sensors perceive, IoT processors make decisions and plan actions. The processing can be as simple as assessing a single sensor value, such as a temperature on a thermostat, or it may consist of sophisticated AI algorithms that recognize people, places, and events. IoT actuators carry out plans immediately or alternately synchronized them over time using remote devices that physically alter the environment. IoT's actuators may be as trivial as turning on a remote smoke alarm or they may aim a weapon to track and kill a person that IoT has identified. The actuators can operate on scales both far smaller and far larger than human actions. Human operators may be in the loop of IoT actions but need not be. We will want to retain supervisory control and oversight.

IoT endows us with countless beneficial advances, many not yet envisioned. But when misused, IoT becomes the ultimate tool of authoritative regimes for pervasive surveillance and automatic control of their citizens – a tool which can easily destroy free societies. Some nations are already using the power of IoT to help suppress terrorism, protect their security, and enforce the rule of law. Other nations are using it to suppress all opposition to the ruling government. The public and policy makers must understand both the potential and the dangers that extensive IoT networks pose for democracies. They must find a way to regulate IoT to advance the benefits for defense while protecting against misuse and abuse.

3 Why Is IoT of Great Importance Now?

All three of IoT's key elements exist today: sensors/info-extraction, intelligent processing, and network-enabled actuators. The infrastructure that supports these elements also already exists, such as high-bandwidth wireless networking and cloud and edge computing. The concept of integrating these elements into a an IoT system is not new, as noted above. What is new now is the maturity of the core components that make up each of these elements. Many of these elements were invented by the US Department of Defense and other governments but matured in the commercial domain. Their maturation creates an inflection point in IoT. The key components now powering IoT networks, can be summarized in the following list. Many of these topics are addressed in the Chapters of this book.

- **Communications**: Wireless communication provides high bandwidth, low-latency, and near universal availability via 5G, WiFi, Bluetooth, ZigBee, and other standards. Performance surpasses what was available from wired networks just a decade or two ago.
- **Networking**: The continuing reductions in cost, size, weight, and power of network interfaces embedded in devices creates an explosion of devices and sensors that can interact with one another. Widely used network standards supports interoperability across different manufactures of IoT devices. For example, China announced plans in 2021 to network together different sensor systems observing all public spaces in the nation. In private residences Amazon's Echo ties together many devices and appliances made by many different companies.

- **Processing Power and Storage**: The ongoing operation of Moore's law creates processor performance capable of digesting "big data" and training and executing intelligent algorithms, such as deep neural networks. As an example, a single mobile phone contains as much storage and processing power as any supercomputer just a few decades ago.
- **Distributed Processing**: Cloud, fog, and edge computing makes massive computing power available to IoT devices without having to embed it at every node of an IoT network.
- **Security:** The understanding of the digital vulnerabilities of IoT and how to protect against them has expanded in both commercial and defense enterprises. However, this understanding is occasionally ignored in defense practice and routinely ignored in commercial applications. Anti-tamper techniques developed and applied in both domains of defense (increasingly) and commercial (selectively) add protection from physical attempts to coopt devices and software.
- **Localization and Common Timing:** Proliferation of geolocating satellite systems and small, low-cost receivers, and techniques such as Bluetooth beacons make it possible for most devices to locate themselves precisely and continuously. They also allow IoT devices to coordinate on precise timelines by sharing a common timing framework. People who carry devices like cell phones or fitness trackers or who drive a recent model of car can be tracked continuously and precisely. For example, one commercial fitness-tracker aggregates individuals' locations, revealing the location of individual soldiers as well as secret intelligence facilities.
- **Big Data and AI Algorithms**: Machine-learning/AI algorithms demonstrate increasing sophistication and human-like intelligence in such activities as pattern recognition, tracking, language understanding, and autonomous control. As an example, a simulated aircraft controlled by an AI algorithm recently defeated a top human fighter pilot in an air-to-air simulated dog fight, six out of six times. *Science Magazine*, the journal of the American Association for the Advancement of Science, called AI's ability to predict protein folding the most important scientific innovation of 2021.
- **Digital Sensors**: Video cameras, infra-red cameras, accelerometers, health status monitors, lidars, and other sensors have dropped dramatically in size, weight, and power as well as in cost while at the same time improving sensor quality. Many sensors now have integrated position, timing, and network interfaces. For example, one of the first lidars (a type of 3D imaging sensor) cost the US Defense Department approximately $690,000 in 1985 in inflation-adjusted dollars and weighed about 50 pounds. A lidar costs less than one 10,000th of that amount today and can fit inside a smart phone. Today's version provides both higher resolution and faster framerates.
- **Digital Information Extraction**: The past two decades have seen an explosion of information exfiltrated from software applications. As an example, one information aggregator combines the detailed smart phone location data reported from over 80,000 phone apps residing on millions of phones. By combining information from different data bases, supposedly anonymous tracks can easily be associated with named individuals. Purchasing a week's worth of such aggregated data allowed the New York Times to locate and track the US President through his daily movements.
- **Actuators**: The size, weight, power, and cost of many actuators have declined, sometimes dramatically, while the types of action and control of actuation have expanded. The scale of commercial markets supports this advance by funding the necessary non-recurring engineering. Specific advances include newly available materials or cost reductions in specialized materials (e.g. carbon-fiber composites, titanium, printable metals). New possibilities for IoT arise from 3D printers, micro-electro-mechanical systems (MEMS), and actuators with embedded Bluetooth or other wireless network interfaces. MEMS-based accelerometers tied to micro electro-mechanical actuators control the flight of many unmanned aerial vehicles as well as enable motion tracking and haptic interaction in smart watches and fitness devices.

- **Widely Adopted Standards**: Shared standards for communications and networking, widely adopted by commercial product and service providers, expand their markets globally and support the interaction of thousands of devices and applications. Examples of standards driving IoT include 5G and Bluetooth in the communications realm, Amazon's Ring and Alphabet's Nest device-interface standards, and Apple's IoS and Google's Android operating systems and application environments.
- **Smart Phones as a Common Human-machine Interface:** Several billion people use one of two types of smart phones. These smart phones provide a common hardware and software platform hosting thousands of applications. The two platforms provide human-interface environments and infrastructure, rich in hardware and media for human interaction. They form ubiquitous human-supervisory control nodes for IoT devices and networks. The use of common human-interface conventions, although they fall short of standards, helps users move quickly from a known application to a new one. They reduce training and learning time and expand market sizes. When combined with MEMS sensors and new displays, they make virtual-reality a reality through IoT systems such as Occulus.
- **Successful IoT Applications:** IoT systems are rapidly expanding in popularity and use across the developed world, demonstrating the power of IoT and increasing trust in new applications and automation that amplifies human potential. For example, Tesla cars and Ring security networks are two such IoT systems gaining popularity and acceptance. Automotive IoT now provides warnings, detects moving obstacles, and in constrained situations even becomes the driver. Digital home assistants, such as Amazon's Alexa, tie microphones to cloud processors running AI algorithms to understand spoken language and control many home devices. In defense and national security, emerging weapons and intelligence systems include IoT technology, such as tracking supplies, semi-autonomous missiles, drones, and surveillance networks. Existing IoT applications demonstrate their power, gain public acceptance, and attract the notice of defense ministries.

These advances in core technology mean that IoT can today automate more tasks. IoT can now reduce the cost of many operations and increase efficiency, performance, and tempo of actions. Increasingly, IoT networks will begin to alter and control our commercial and private environments. For defense and national security, IoT's promise is just beginning to unfold. However, IoT cannot yet meet its full potential for defense. The special challenges IoT faces in this domain still limit it. This Book addresses many of these challenges, presenting emerging R&D results and case studies. These advances let IoT alter, control, and automate our affairs both for defense and in peaceful society. It is already beginning to do so.

4 IoT Will Change How War Is Waged and How Peace Is Maintained

National defense and security are founded on situation awareness that provides insights into hostile intentions and actions. Accurate situation awareness serves as a basis for planning operations and then controlling them by observing and adjusting their effects. Excellent situation awareness depends on continuous surveillance provided by diverse sensors. To be effective for defense, sensors must be coupled together, coordinated, and controlled along with thousands of other devices and entities. The entities are distributed, mobile and heterogeneous. They include packages, weapons, vehicles, aircraft, soldiers, and thousands of other types of equipment and supplies. They must operate effectively together whether waging war, maintaining peace, assisting in emergency relief, ensuring effective logistics, or protecting the homeland. For many of the world's militaries, IoT is already altering logistics, command and control, tactics, and surveillance. Coordination and control require planning and monitoring a plan's execution. Yet, as Helmuth von Moltke reputedly observed, no plan survives contact with the enemy. The reason is not because parts of the plan fail – good commanders develop contingencies in their plans. Plans fall apart because of what Clausewitz called the "fog of war" – a breakdown of visibility and communication in the battlespace. The fog of war prevents a commander from knowing what parts of his plan are succeeding or failing and from deploying the means to alter it. Lacking insight into a battle's progression, a commander may not know where the enemy is or even the disposition of his own forces. Individual units lose track of the location and status of their supporting units, supplies, and opposing forces. In the future, IoT promises more than just improved efficiency in envisioning the battlefield and coordinating and executing plans. IoT can change the course of battles, wars, and peace. It can do so by physically altering the world by dynamically joining vast numbers of intelligent, heterogenous sensors, processors, and actuators over robust and secure networks. IoT dispels the fog of war. It extends the eyes and ears of the intelligence analyst. It multiplies and amplifies the arms of the warfighter.

Some of the main benefits of IoT for defense and national security can be summarized as follows:

- **Autonomous/Automated Weapons:** IoT increases automation and enables autonomous weapons and systems.
- **Increased the Tempo of War:** More autonomy in defense systems moves humans from inside the control loop to supervisory, top-level control. Actions can go from sensors to actuators in milliseconds instead of seconds, minutes, or hours.
- **Reducing Your Force's Casualties**: In a supervisory role, humans use IoT remotely for sensing, control, and action. Warfighters move back from the most dangerous environments. Javelin anti-tank missiles and Switchblade loitering missiles keep soldiers miles back from their targets, for example, by using a Javelin in a "fire-and-forget" mode and controlling a Switchblade with remote supervisory oversight.
- **Increasing the Probability of Destroying Targets While Reducing civilian Casualties:** Onboard and off-board sensors in an IoT network can guide weapons more accurately to their intended target. Automated control loops update flight paths more rapidly, more often, and more accurately than weapons that are directly sighted, aimed, and controlled by soldiers. By tying multiple types of dispersed weapons together, IoT increases the attack surface on the adversary.
- **Expands the View for Military Operations and Intelligence:** IoT networks can join large numbers of unattended and dispersed sensors covering wide areas. IoT also gleans information from data bases, by sniffing networks, and by monitoring software applications. While traditional standalone sensors can provide rudiments of this sort of information, large numbers of sensors and information sources become infeasible to manage and integrate without the network structure IoT provides. IoT observes an area, event, or activity in a more pervasive, multi-view, and continuous manner. IoT sensor networks support improved human interpretation of sensor data and deliver training data through the cloud for intelligent processing algorithms.
- **Dispels the Fog of War:** A fog envelopes a commander's view of an unfolding situation when sensors are destroyed, or data and reports become unavailable. IoT's creates dynamically self-healing, adaptable, and resilient networks of information sources with multiple secure communications paths. These features allow a commander to continue to observe the situation even in the chaos of combat. The fog is lifted. When information sources are overlapping or closely adjacent, data can be fused into a high-quality operational picture even when using poor quality sensors or when individual sensors or reports go missing. IoT's ability to flexibly place

and replace and move sensors provides further resiliency to uninterrupted situation awareness. When IoT includes specialized sensors in the IoT network, such as radar, thermal imagers, and radio-frequency imagers, it can penetrate literal atmospheric fog as well as smoke, clothing, and even the walls of buildings. Advances in tactical communications and networking, such as those described in Section 4, keep information flowing from and into the battlefield.

- **Intelligent Processing (AI and Big-data Techniques) to Convert Data to Information, Information to Awareness, Awareness to Plans, and Plans to Action:** IoT provides intelligent processing to expand human ability to observe. It makes sense of vast amounts of data and information. It can automatically develop plans and autonomously control responses. While IoT may dispel the fog of war, it can create a new fog by overloading decision makers with data. Smart processing provides the essential element to digest, filter, prioritize, and abstract information so commanders are not buried by the flow. AI algorithms are beginning to demonstrate human-level intelligence in aspects of military operations, such as air combat.
- **Automation of Logistics:** Logistic activities, such as resupply, deployment, and maintenance, become much more efficient and experience an increased tempo. For example, as described in Chapter 2, if a soldier's weapon always knows what ammunition is remaining and can automatically order more and can monitor the rifle's use and wear, then both resupply and maintenance become streamlined. Logistics becomes supercharged when IoT can stay abreast in real-time of what, where, when, and how much is needed. IoT knows where supplies currently reside and what resources are available to move them. Just-in-time and just-in-case supply chains can add efficiency, lower costs, and improve reliability. When supply chain data is collected and entered manually, advanced supply concepts become infeasible. By networking sensors and actuators to monitoring algorithms, new logistics approaches become possible, such as conditioned-based maintenance.
- **Encourage the Use of Standards:** The advantages of having flexible and dynamic networks of sensors, processors, and actuators will encourage the development and adoption of standards allowing future defense systems to interoperate as part of larger IoT systems. Expanding standards also allows common IoT components to be used across stovepipe systems, simplifying logistics.
- **Breaking and Connecting Existing Stovepipes and Silos:** IoT can join legacy stovepipe systems (and eventually make them obsolete) by using automatically compiled software to create interfaces between legacy systems. This eliminates human fingers, keyboards, and manual delays crossing the gaps between legacy systems.
- **Rapid Evolution and Turn-over of Technology**: The nature of IoT supports rapid and low-cost evolution of technology by allowing new, improved components to be easily inserted as nodes in the network. For example, IoT LTE communication nodes support incremental replacement with 5G as it becomes available. This does not free an IoT systems designer from considering the impact on the entire system from upgrading a node. But it does make it technically easier and therefore quicker and cheaper to effect the upgrade. It also makes it possible to do it incrementally, spreading the cost of upgrades and allowing them to be continuous, keeping closer pace with rapidly advancing commercial technology.
- **Forcing Function for Advances in Policy, Doctrine, Logistics, and Acquisition:** The power to be gained by adapting commercial IoT technology to defense and continually upgrading defense capability may incentivize or force defense departments and defense industry to develop the policy, doctrine, logistics, and procurement processes necessary to quickly leverage commercial technology. The advantages of adapting commercial IoT to military requirements on an ongoing basis may entice defense ministries to make necessary organizational and process changes.

Advancing and mastering IoT technology and its application is essential for any government attempting to maintain parity with its peers. At present, commercial IoT far outstrips defense IoT. The challenge for defense ministries and industry is to leverage commercial investment in IoT while augmenting it to meet the domain's special requirements. Many of the challenges are technical and many technical solutions are the topic of the Chapters of this Book. But nations must also adapt their processes and policies to accomplish an efficient flow of commercial IoT advances into defense applications. Those who do so will likely dominate military power in the rest of twenty-first century. To better understand how commercial and defense IoT have and can intertwine, a brief history of the development of IoT for national defense follows.

5 A Short History of IoT for National Defense

As noted, the concept of IoT originated in defense organizations many decades ago. Only with the commercial maturation of component technology has its impact become significant. Defense departments are traditionally conservative in adopting new technology into practice, although, ironically, they are often in the forefront of inventing new technology. Commonly, the much larger commercial marketplace is required to move inventions to innovations in practice. Once mature the technology moves back into defense systems. That is the case with IoT. A brief review of the development and use of IoT in the US Defense Department provides a case study in understanding how IoT is adopted for defense applications. It also illustrates the interplay between defense IoT development and commercial technology maturation. Understanding the history of IoT in the US defense enterprise provides a context for the discussion of the interplay between commercial and defense IoT going forward, as described in part 6 of this Introduction.

It should be noted when studying the history of defense technology and systems that the term "IoT" is relatively new, even though the concept is not. Many military systems are becoming IoT systems, even though they still are not typically described using the term "IoT". This is unfortunate. A greater awareness of IoT provides systems designers with the insight to build new systems using the best IoT concepts, practices, components, and standards.

5.1 IoT Predecessors and Emergence of "Internets" and "the Internet"

The concept of networking sensors and control systems and actuators is likely to be as old as organized militaries and defense. The Romans built Hadrian's Wall across the breadth of England beginning in AD 122 to stop incursions by the Celtic tribes of Scotland into Roman Britain. The Wall functioned as a kind of human powered IoT defensive network where sensors, processing of threats, and execution of actions in response were all carried out by people. People functioned as the underlying communications network. By 1936, 18 centuries later, IoT systems for defense had evolved to electronic devices and radio communications. One of the earliest examples was the British Chain Home coastal radar installations designed to detect approaching German bombers. Detections were passed by landlines and radios to command headquarters where Royal Air Force fighters were scrambled to intercept the bombers. Warnings were issued to civilian populations through sirens and broadcast media. Although an advance over entirely human-powered IoT systems of Hadrian's time, early defense systems used dedicated communications channels that were expensive and provided little flexibility for new types of sensors or new locations. Data was usually moved by manual action between sensors, command centers, and human operated response systems. They were often unintentionally constructed with single points of failure.

The concept of networking sensors and control systems and actuators using a flexible and survivable digital network originally emerged in the 1960s. The US Department of Defense through its Advanced Research Projects Agency (now the Defense Advanced Research Projects Agency or DARPA) conceived the idea of an internet of ballistic-missile surveillance systems coupled to missile control systems for rapid detection and response to nuclear attack. The Internet Protocol (IP) and Transmission Control Protocol (TCP) that are the basis for today's Internet were developed for the ARPANET, a precursor to the Internet, to explore distributed, redundant, adaptable communication that would be robust and survivable. Although the early applications of the ARPANET were for data exchange between computer science researchers, the funders envisioned a survivable network linking control stations for intercontinental ballistic missiles with early warning sensors, such as the Defense Early Warning (DEW) Line across arctic North America.

The TCP/IP architecture of the ARPANET and later the Internet, underlying IoT, creates a communication system having ideal properties for defense and national security – not surprisingly considering they were invented by a defense agency. The IP protocol established packet-based communication between a sender and a receiver without the need for a fixed, dedicated communication path. If one node in a network is destroyed or overwhelmed with traffic, the IP packets automatically flow through any available alternate route. One great advantage of the IP packet scheme is that packets can be communicated over any type or combination of transmission medium from wires to optical fiber to WIFI radio waves to cell phone signals.

Because large messages are broken into smaller IP packets, they may be delivered out of order or lost. As a result, an additional protocol is needed that allows a receiver to reassemble the original message in order and request re-transmission of any lost packets. The TCP protocol accomplishes this task. It establishes a virtual connection between a sender and receiver built on top of the robust and adaptable IP packet system. Higher-level protocols ride on top of TCP to provide special services such as secure file transfer. Overtime alternatives to TCP and IP protocols have been developed with specialized usages. When capitalized, the word "Internet" refers to a specific network of routers that share common routing tables for IP addresses that have been registered on the global Internet. No one entity owns all the routers or controls the Internet, but it was initially established when ARPA/DARPA transferred the ARPANET from the US Defense Department to the US National Science Foundation (NSF). NSF created an open, distributed, global network. NSF moved away from the Defense Department as the owner of all Internet hardware and addressing tables to a model with distributed ownership and distributed control. Senator Al Gore sponsored the bill in the US Senate that converted the ARPANET to the "Internet" prompting him to take credit once, infamously, for "creating the Internet". This distributed ownership and distributed routing are possible because of the nature of the TCP/IP protocols. They provide one unifying and underlying source of IoT's power.

The utility of the Internet was further strengthened in 1994 with the introduction of Uniform Resource Locators (URLs) to uniquely name resources on the Internet. From the late 1980s to the early 1990s the Hypertext Transfer Protocol (HTTP) was developed. HTTP information rides on top of TCP/IP protocols. URLs integrated with HTTP developed in parallel with web browsers. They render in human-readable form HTTP data retrieved from a URL. Taken together these developments created the World Wide Web. Research leading to URLs, HTTP, and web browsers was funded initially by governments in Europe, the US, and other nations. Defense departments and ministries made significant contributions, for example, Carnegie Mellon University's ZOG hierarchical-paging concept, funded by DARPA, is incorporated in evolved forms by almost all commercial web browsers. The core of the world-wide-web concept was funded by

EU governments. The World Wide Web and supporting tools allow even simple IoT devices to interface intuitively with people.

When not capitalized, "internet" typically refers to a digital network that uses the TCP/IP and related standard protocols even though it may not be tied to the common global "Internet". The distinction is important for the "Internet of Things" for defense applications because an IoT system built for national security may not be tied to the global "Internet", even though it uses the TCP/IP internet protocols to form a closed "internet" of things. As used in this Book, IoT will be used for networks of devices that may use either the global Internet or a closed private internet or both. Today, an internet allows specialized sensor systems, processors, and actuators to cheaply interconnect and be easily modified or extended in a standard way. IoT internets make them dynamically adaptable, which in turn makes them resilient and survivable – properties of special utility for defense and national security.

5.2 IoT Begins Overtaking Legacy Systems for National Defense

For the next forty years after the creation of the Chain Home Radar Systems, defense IoT systems typically were custom built and did not interact or directly communicate with any other systems. These standalone systems are commonly referred to in the defense industry as "stovepipes" or as "silos" in commercial parlance. They are built to address specific missions and functions, such as air operations, artillery command and control, or signals surveillance. They consist of fixed types of sensors and controllers with dedicated firmware or software and dedicated communication systems. Stovepipe systems, configured in fixed networks serving specific applications with unique protocols, usually require manual intervention to communicate with other stovepipe systems.

Even after internet protocols were invented, defense systems made little use of them because in the early years TCP/IP protocols demanded higher bandwidth, increased transmission latency, and required expensive additional hardware. When DARPA funded the development of the first autonomous land vehicle (ALV) in 1985, it connected many of the vehicle's sensors, processors, and actuators with an internet; however, no deployed military systems or networks at the time relied on internets. The widespread advent of personal computers and office workstations in mid to late 1980s through the early 1990s, drove down the cost of hardware to process TCP/IP protocols while at the same time digital bandwidth expanded along with the number and capacity of routers. IP addresses on the global Internet exploded. Internets began to move into military systems. Commercial advances moved TCP/IP processing directly onto the computer's main chip with the central processing unit (CPU).

By the early 1990s, IoT-style applications were under development in specialized areas, such as the US Army and DARPA program called Personnel Status Monitor (PSM). PSM instrumented soldiers with sensors for vital signs such as temperature, heartrate, and blood oxygen. The sensors networked together to continually monitor the health status of individual soldiers and their units. PSM was the conceptual forerunner of IoT systems for health monitoring, such as Fitbit, Strava, and the Apple Watch. PSM required a backpack's worth of electronics and computers. The size of the commercial marketplace generated the funding for the non-recurring engineering needed to reduce PSM's backpack to a watch-sized object.

By 1998, another DARPA program call Battlefield Awareness and Dissemination (BADD) demonstrated an IoT network that communicated from Eastern Europe to the US using land lines, wireless communication, and space-based relays to route TCP/IP packets. BADD demonstrated an IoT network that for the first time linked a robotic surveillance aircraft over Bosnia to local tactical command posts on the ground in Bosnia and Germany manned by the US Army, to a NATO air operations center in Italy, to a US Navy command-ship in the Mediterranean, to an airborne command center operated by the US Air Force, to the Pentagon. By the early 2000s, the widespread development of small, cheap RFID tags (radio frequency identification) made it practical for the world's militaries to begin on a large scale to instrument packages and automate aspects of logistics – a central pillar of warfare. At the same time, smaller, but more heterogenous and dynamic demonstrations of early IoT technology were underway in defense research establishments. By the middle of the first decade of the twenty-first century, the defense contractor SAIC and the US Space and Naval Warfare Center at Point Loma, CA demonstrated an advanced infantry combat system that combined an instrumented solider with a rifle outfitted with sensors and a small combat-support robot. These elements were networked together and linked to commanders and logistical support.

This concept of integrating soldiers, weapons, sensors, and robots was expanded around 2010 on programs such as US Army and DARPA's Small Unit Operations (SUO) and Squad X. DAPRA through SUO introduced the notion of content-based networking, a potential mainstay of future combat IoT communications, described in Chapter 18. Content-based networking was demonstrated with US Army units on the CBMEN program in the mid 2010s. In 2016, the US Army initiated the Internet of Battlefield Things (IoBT), a leading collaboration of universities, industry and the government aimed at advancing basic IoT technology as it relates to defense. Several Chapters of this Book present the results of the IoBT collaboration. Other IoT-focused efforts started in the US and abroad at the same time or shortly afterward, such as the DARPA/US Navy's Internet of Ocean Things. Beyond the US, countries and alliances are investing in and exploring IoT for defense, including NATO, China, India, South Korea, Spain, and the EU, among many others.

Concurrently with US Defense Department experiments that integrated soldiers and tactical gear into IoT networks, other groups developed analytical software using artificial intelligence and big data analysis. They showed how to derive new types of defense and intelligence information from sensor networks. As an example, DAPRA's Combat Zones That See program in the mid 2000s networked together video cameras mounted across military bases. The program used video analytics software, a type of AI, to track all vehicles across a zone of interest. Efforts in other countries have connected video cameras in networks across entire sections of cities to detect and track individuals as well as vehicles. These systems use video analytics for tracking, face recognition, and license-plate reading. For example, the City of London began detecting, identifying, and tracking all vehicles entering the City with a focus on counterterrorism. Other types of sensing and tracking, such as location reports from cell phones, were being integrated into networks that could provide comprehensive surveillance across a region. Most recently, the government of the People's Republic of China developed the most comprehensive and ambitious set of national security programs aimed at using IoT networks to monitor and control its citizens, for example the Skynet and Sharp-Eyes initiatives, as described in Chapter 21.

On the battlefield, IoT weapons systems are now impacting the way wars are fought. The Javelin anti-tank missile system is a tightly integrated predecessor to IoT weapons. It consists of several sensors, actuators, and processors running intelligent pattern tracking and control algorithms. Its design grew out of the US Army/DARPA Tank Breaker program in the 1970s and therefore predated the use of internet protocols. It functions as a shoulder-fired missile that allows a soldier to designate a target using several different sensors. Once the soldier releases the missile, it tracks the target autonomously for more than a mile to impact. The sophisticated visual and thermal sensors can be used by themselves for surveillance. Javelin's images can be shared over battlefield networks with other soldiers and commanders. Because it watches the target and controls its own flight path, it can take a non-line-of-sight path to strike the more vulnerable top of a targeted tank.

The soldier firing the weapon can release the missile and seek cover. The closed nature of Javelin makes it difficult to integrate it digitally with other sensors. Tying in off-board information to improve or coordinate its operation is also more manual than it need be.

The much more recently developed Switchblade loitering missile is a more modern IoT weapon. The somewhat miss-named "missile" is actually a fixed-wing drone (or unmanned aerial vehicle or UAV), that can be launched, flown several miles beyond the line of engagement, and used to find a target or wait for one to appear. The concept grew out of the DARPA/US Army program called Net-Fires begun around 2003. From the beginning the loitering missile was conceived as an IoT system that networked with other missiles, off board sensors, vehicles, and both autonomous and human command-and-control systems and weapons. Switchblade can function as a standalone weapon or coordinate with other drones to share targets and target locations. Both Javelin and Switchblade showcase many of the advantages of IoT weapons. They move the soldier back from the most dangerous zones. They can operate with a high degree of autonomy under top-level supervisory control. They share information and coordinate with other surveillance systems and weapons.

Recently, IoT technology demonstrated the automatic integration of legacy systems. While new systems are likely to be designed around an IoT framework, the bulk of existing defense systems and processes are not. Legacy systems most often are developed as stovepipe or silo systems. In contrast to stovepipe systems, IoT technology allows systems to be open, flexible, adaptive, and easily extended. Their distributed nature allows for redundancy and dynamic reconfiguration, which in turn makes them more resilient. However, given the time and money expended to develop legacy systems, few are likely to be redesigned as flexible IoT systems. Consequently, the US Department of Defense has begun to research ways to integrate entire legacy systems together into a larger IoT network. While the individual legacy systems do not receive the benefits of IoT design, they can function with other legacy systems in an IoT network of systems. They thereby create a system of systems with greater power and flexibility, and much greater speed of interaction. Such systems interact at electronic speeds rather than slowly through a human intermediary as currently occurs.

An example of this approach is the US Air Force and DARPA's System of Systems Integration Technology and Experimentation (SoSITE) program. It brought existing, custom-built surveillance systems, weapons, and command-and-control systems into an IoT paradigm. SoSITE developed tools that use formal message definitions and interface protocols to automatically compile software to translate between existing, custom-built systems. SoSITE sponsored a combined ground, air, and sea combat exercise showing, for example, that an artillery unit could directly interact with a strike fighter in a "call for fire". An IoT network can incorporate new components as they are developed elsewhere for other applications. SoSITE proved that an IoT paradigm can also effectively include existing, "closed" defense systems as "devices" in the network. It should be noted, however, that technically integrating legacy systems is not sufficient. They must be integrated in terms of mission requirements and constraints – a challenge beyond merely exchanging commands and information. While in the future it may become increasingly easy to technically integrate IoT networks with legacy systems developed for different missions, care must be taken to integrate them at a mission level as well.

6 Special Challenges for IoT Used for Defense and National Security

The nature of war and national security lead to some profound differences between commercial IoT requirements and those of defense and national security. Adapting commercial technology successfully imposes formidable technical challenges but also process and organizational challenges. This Book focuses on technical solutions. The equally important, but non-technical problem of meeting organizational and process challenges requires changes to defense organization, policy, funding, acquisition, doctrine, logistics, and military operation. Needed process and organizational changes may well be more difficult to institute than overcoming the engineering and scientific hurdles. This section briefly summarized both types of challenges. However, process and organizational changes largely remain a deserving topic of future books.

Along with the benefits for national defense come special challenges. IoT networks and devices, by design, must operate in hostile environments where misinformation, deception, and both physical and cyber-attacks are the rule. An adversary will attack IoT in military and national security applications on all three fronts of confidentiality, integrity, and availability. In the commercial world the adversaries are relatively rare, rogue individuals and criminal organizations. In defense applications, hostile nations are the adversary, and they are well funded, sophisticated, and persistent. Nation states mount large-scale, coordinated attacks on IoT networks. IoT in the commercial world should be designed to fend off occasional specific cyber-attacks, some of large scale. Military and national security applications will certainly be attacked in the cyber domain, but they must also anticipate physical attacks and physical compromise of IoT nodes and infrastructure. Because military operations occur unexpectedly in unforeseen locations, subject to ongoing devastation, IoT systems cannot rely solely on in-place commercial communications, networks, and cloud infrastructure. Commercial 5G wireless networks support civilian IoT, but commercial networks are not designed to face likely jamming, destruction, interception, physical capture, or compromise in battle and the lead-up to war. Existing commercial IoT security developments concentrate more on processes and procedures rather than innovative technology for security solutions, such as those described in this Book.,

While IoT amplifies defense operations, it can also amplify damage caused by an adversary if an IoT system is compromised. Today, if you capture an enemy's rifle, you've captured one weapon. But if that rifle is networked through IoT to all other rifles and to its entire logistics chain, then capturing one rifle with a cyber vulnerability may change the course of a war. By their very nature, IoT elements are networked together and the compromise of one element may threaten them all. System dependencies and emergent behaviors mean that ensuring the integrity of individual IoT nodes does not ensure integrity of the system. While this feature of IoT threatens commercial IoT networks, the consequences of attacks replicating across an IoT network are far greater for defense and national security applications.

Another defense-specific security concern is that some components of the IoT devices, on which every military relies, are likely manufactured by an adversary, and compromised before they leave the factory. Finding and protecting against embedded threats in foreign supplied devices presents a significant technical challenge. An added challenge is that it is frequently difficult to determine what components are foreign-made and even deciding what constitutes a foreign source. Many high-tech firms are globalized. For example, Apple is an Irish firm that designs its devices largely in the US but manufactures them largely in China. The components come from several countries, such as Taiwan. The foreign suppliers may be based in one country but use manufacturing facilities in other countries. For example, Apple uses Foxconn, a Taiwanese company, to manufacture their phones, but Foxconn has manufacturing plants in China and other countries. The software comes in part from the US, but foreign firms participate in the development. Software developed nominally in the US may in fact involve Apple employees working remotely in another country or foreign nationals working in the US. An iPhone is clearly not a US product; it is clearly a global product of multiple, uncertain origins. An iPhone's complex nationality is representative of many IoT devices and components.

In summary, IoT security and protection are essential and different for national security than for commercial applications. Adding security and protection to commercial products and services is an economic cost-benefit trade-off, where security often loses the trade. In defense and security applications, security and protection are essential, not a profit-based decision. IoT security requires changes to organizations, processes, and administration as well as technical solutions. However, this Book concentrates on technical solutions with new research results for IoT security. The Chapters of Section 3 and some in Sections 1 and 4 of this Book discuss IoT security; they are authored by some of the leading researchers in the area.

Beyond security concerns, in contrast to many commercial applications of IoT, military operations often involve continuing changes in tempo, scale, location, and objectives. IoT systems for national security must be flexible, dynamic, and adaptive. Today, multi-national coalitions are the norm, and IoT systems must be capable of quickly incorporating diverse devices, networks, protocols, standards, and regulations. IoT for the defense industry, from design to manufacturing to logistics, must reflect the same dynamism as the military operations it supports. Military operations can and do occur in areas with little to no existing infrastructure. Where infrastructure does exist, it may quickly cease to. This Book includes chapters on optimizing performance in such changing environments, both for military operations and industrial IoT. Many of these approaches can be used to adapt commercial components to defense use.

In no area does defense IoT technology lag commercial industry more than in communications and networking. In the past, defense ministries and industry followed their own path in radios and networks to meet their special needs. As the commercial communications market for IoT accelerates away from the limited-sized defense market, the gap between military and commercial capabilities widens. Defense organizations must build effectively on commercial advances in communication and adapt them to special needs of military and intelligence operations. Section 4 addresses issues of communications and networking and issues related to leveraging commercial products and technology.

Defense applications also differ from commercial IoT in that the consequences of IoT failures are much more likely to lead to loss of life and destruction of property. As a result, safety, reliability, and predictability of IoT behavior are of the utmost importance for military operations. Defense and security activities also necessarily require the collection and control of information from not only willing but also unwilling and unaware people. Therefore, the use of IoT technology must often function with stealth and covertness at a level not found in commercial applications. Because they may be used covertly and on the unwilling and unaware, IoT used for defense and intelligence outside of war must overcome a high bar of ethical, privacy, and security obligations that exceeds those of commercial applications. IoT systems increasingly incorporate artificial intelligence, a technology that can be opaque with little ability to explain its decisions. As a result, testing and trusting IoT systems' behavior becomes problematical. These challenges weigh especially heavily on IoT when integrated with the destructive capabilities required for defense and security. Section 2 and Chapter 21 address some of these issues.

As nations build IoT systems capable of surveilling, confronting, defeating, and controlling opposing nations, they will have built systems capable of surveilling and controlling their own populations. When misused by individuals, organizations, or governments, IoT can unjustly threaten the rights, freedoms, property, and lives of individuals and society. Awareness, regulation, and vigilance can ensure that emerging IoT technology realizes its benefits while preventing its abuse. The final Chapter, 21, reviews potential challenges to society presented by IoT, especially the misuse of IoT employed by governments for defense and intelligence. Chapter 21 provides an initial framework for regulating IoT to prevent abuse and misuse.

As mentioned above, IoT challenges for defense and national security consist of both technical and organizational and process challenges. This Book concentrates primarily on the technical challenges. The characteristics of war and national security that give rise to the specific technical challenges addressed in this Book, can be summarized with the following list:

- **Greater Consequences:** Many military systems are designed expressly to destroy lives, structures, infrastructure, economies, and even cities, demanding a great emphasis on safety, reliability, control, predictability, and testing.
- **Adversarial Environments/Enterprises:** Defense and national security operations, by definition, occur in the face of powerful, hostile adversaries. Commercial IoT may be attacked. Defense IoT systems will be attacked on all three fronts of confidentiality, integrity, and availability. Defense IoT will see types of attacks that are rare or non-existent outside military operations, such as jamming, radiation-seeking missiles, nuclear weapons, trojan horses, and physical tampering, among many others.
- **Security, Protection, and Resilience:** Adding protection and security features constitute a cost-benefits trade-off for commercial products. Commercial IoT may be attacked. Defense IoT will be attacked and compromised. IoT networks must continue to function as a network despite hostile attacks. Their safety and protection are not options nor part of an economic trade space. To avoid defeat or retreat, when IoT systems are attacked and partially destroyed or disrupted, they must exhibit resilience and maintain some level of function.
- **Highly Constrained Resource Environments:** In combat environments, supplies and resources are targeted. Resupply is disrupted. New supply chains must be rapidly formed, and they may extend halfway across the globe. Rarely can militaries count on anything more than minimally adequate provisioning of any resource, at least in the middle of a conflict or engagement. National security operations frequently occur short of war or prior to it, but they often occur inside hostile territory where resources are scare and not easily supplied or resupplied.
- **Lack of Existing Infrastructure:** Militaries and national security organizations may operate where little to no infrastructure exists and where existing infrastructure is likely to be soon destroyed or compromised.
- **Dynamic, Uncertain Communications and Networking:** Operating in hostile environments and while on the move, military communications networks are disrupted, intermittent, limited in bandwidth, and dynamic in configuration. Cloud resources may be interrupted or intermittent or non-existent. Emitting electro-magnetic radiation for communication can disclosure a device's location and make it a target.
- **Dynamic in Scope and Mission:** Military IoT systems function in widely varying missions from peace keeping to homeland protection to disaster relief to combat to surveillance and intelligence. Both militaries and national security organizations can be called upon to operate anywhere on the globe with little to no advanced warning. IoT systems must function in all extremes of weather and climate. It is difficult to anticipate where, when, and what the mission will be, or what sort of coalitions will be involved. Missions may require military operations to interface with IoT systems used by police, emergency responders, NGOs, and other civilian operations. The collaboration required with other types of units, services, and nations varies widely. Missions and deployments change and transform as they unfold, often quickly and in unpredictable ways.
- **Dynamic in Scale:** Operations may involve small units to entire armies engaged in combined arms and multinational operations. An operation may unfold across a few acres of land or sky, or it may span a continent or ocean. IoT networks must operate and interoperate across all scales and must be dynamic enough to scale quickly as operations change in size and intensity.

- **Timeliness and Reliability:** Failure to receive sensor data in a timely manner or loss of control of a weapon due to latency in communication or disruptions can lead to unintended loss of life. Timeliness and reliability of many commercial applications have much lower demands and consequences.
- **Sources of Technology:** Commercial IoT products and components may not meet the demands of military operations or may not be capable of being modified or augmented to meet these demands for technical, cost, IP, or schedule reasons. This failure may necessitate custom defense solutions. Additionally, many IoT components are designed and manufactured in whole or in part in countries foreign to the country deploying them. Trojan horse attacks and supply chain vulnerability present potential threats to all nations relying on IoT for defense and national security. Technically protecting against embedded threats is challenging, but so is unraveling the national sources and supply chains of IoT components.

Beyond solving technical challenges, organizational and process changes are also needed if defense IoT is to leverage commercial progress. While this Book largely does not address process and organizational changes, they are briefly listed here. Each item represents a significant challenge and deserves a book of its own. The US defense enterprise is referenced for examples.

- **Organizational Structure**: The organizational structure that controls new technology insertion into US defense practice differs radically from commercial organizations. To succeed in moving commercial technology to defense IoT requires organizational changes if defense is to maintain something close to parity with commercial capability. Different organizations in the decision loop on new technology may have differing or even opposing priorities and objectives. Dismemberment of the defense enterprise into strongly partitioned functional realms leads to duplication and stovepipe systems.
- **Policy and Regulation:** Policies and regulations for IoT appear glaringly absent when one considers that IoT introduces increasing amounts of autonomy into military systems including weapons. Policy and regulation need to evolve to encompass a world in which technology evolves markedly every year. In the commercial arena, products are discarded rapidly to keep pace with advances in performance. Mobile phone users have purchased and discarded their phones three or four or more times over the last two decades as technology has advanced from 1G to 5G. That contrasts with intelligence systems like the U-2 spy plane, which entered service in 1954 and is still flying; or the current mainstay of the US bomber fleet, the B-52, that entered its design phase in 1948 – although both aircraft have undergone numerous upgrades. In a rapidly evolving and transforming field like IoT, a throw-away culture may be regrettable for environmental reasons, but it enables accelerated growth of performance and capabilities. Existing defense and legislative policies do not easily accommodate such practice.
- **Funding:** Funding for defense systems is typically allocated every year and can be as fickle as the most recent election or a single legislator's special interests. Funding for different phases of development is split between separate organizations with interests that are only partially overlapping or even opposing. Difficult problems, such as perfecting more autonomous IoT systems, require sustained funding from inception as basic research through to production and fielding. Development plans should anticipate and plan to fund the full cycle of innovation, anticipating the need for phases of commercially funded maturation where required.
- **Acquisition/Procurement/System-development:** Sharing of IoT components can lower the lifecycle cost of separately developed and procured systems, which are today all too often procured and developed as stovepipes. Incentives need to be created for acquisition organizations to support standards and interoperability between stovepipes to enhance sharing of IoT resources.

Defense enterprises need to better understand the differences in motivations and business process in the commercial domain. They need to do more to understand and find compromises with commercial practice.

- **Doctrine:** Military operations work smoothly when there is a uniformity of equipment, systems, training, and practices. But this efficiency comes at the cost of falling behind advancing technology. In contrast, commercial IoT allows consumers to upgrade technology incrementally and disjointly. If IoT systems are to be refreshed frequently, based on rapidly evolving commercial products that are modified for military operations, then doctrine needs to evolve.
- **Logistics:** Rapid turnover using more powerful, modified commercial components as they emerge presents a host of logistical problems, including retraining, interoperability, documentation, supply, maintenance, and deployment. Compatibility with legacy systems presents logistical challenges as well for rapidly evolving IoT. Organizing logistics around custom, stovepipe systems simplifies the logistics for that system, but it makes the overall logistics more complex and less efficient. IoT enables sharing of components across systems and makes them easier to upgrade, improves inventorying, deployment, resupply, maintenance, and repair.
- **Market Size:** One great disadvantage for national defense and security is that its market size is small compared to commercial markets. In commercial practice, the size of the market determines how much capital is invested in improving performance, adding features, and reducing the costs, size, weight, and power requirements of products. Per-unit costs of maintenance, repair, and training also decrease in proportion to size of the market. Defense organizations can consciously work to maximize their own defense market size for IoT systems. The best approach is to remove barriers to technology flow between defense and commercial enterprise. Some of the specific challenges include the following:
  - **Standards:** More strongly embracing commercial standards allows more components and subsystems to be shared across IoT networks designed for different missions. At present there are few process and organizational incentives to use standards across stovepipes. Adopting more commercial standards would also help unify defense and commercial markets.
  - **Intellectual Property (IP):** In many instances, the US Department of Defense has chosen to combine the worst of two IP practices and thereby limit its market size and impede the flow of technology between the commercial and defense domain. The result is few companies attempt to develop high-tech products for both commercial and defense purposes. With a smaller market, disconnected from the commercial world, defense product development falls behind commercial industry.
  - **Access to Defense Technical Information:** National security and defense establishments need to protect their technology and systems advantages from foreign competitors. However, IoT technology advances with rapidity and depends on a global supply chain. Restrictions on information flow need to be rethought in the light of the current reality of global technology.
- **Source of IoT technology, Products, and Components:** Arguably, no state-of-the-art IoT system of any complexity exists today that does not contain components from more than one country. Trusting the security of foreign IoT devices and systems and protecting international IoT supply chains present formidable process and organizational challenges. Because IoT components will increasingly span numerous functional systems, uniformity in process and common organizational authorities are needed to understand supply chains and sources of IoT technology. Changes to organizations and processes are also needed to apply technical security checks and protective measures in a consistent and comprehensive manner across the defense enterprise.

7 Commercial Vs. Defense IoT

Even though inventions sponsored by defense departments created the initial basis for much of IoT, today's commercial use of IoT dwarfs the defense market and is driving its current and future maturation and proliferation. Defense investment will continue to foster major, long-term IoT inventions as opposed to commercial investment. Businesses focus on new products typically achievable in one to five years. However, commercial investment, given the much larger market, will lead the way in converting defense inventions to products and services and in driving down the cost, size, weight, and power requirements of IoT. Commercial investment will also lead the way in finding new applications and improving performance. However, only the largest and most profitable commercial corporations can afford to engage in basic research and high-risk development that may or may not have payoff for some years. Venture capitalists and private equity focus on moving new technology rapidly into new products or services. Venture funding seeks ideas that have already been invented and shown to work in principle. Commercial ventures are excellent in turning inventions into innovation once the inventions have occurred. In contrast, defense organizations will fund basic research with the speculative hope that advancing science will lead to new inventions that promise advances in defense technology. Some defense agencies, like DARPA, are especially adept at taking breakthrough scientific advances and moving them through a proof-of-principle stage in a military context. Few defense departments or ministries, including DARPA, are effective at moving their own inventions into innovative practice. Defense inventions are more often moved to innovative practice by commercial endeavors that adopt the invention. Once a new technology is part of a mature commercial practice, defense organizations may adopt and translate that commercial technology back into the defense domain.

All too often, defense organizations still develop their own parallel versions of commercial technology customized around the special requirements of military operations without leveraging commercial advances. For example, the prime contractor for the first mainstream Army robotic vehicle program choose to design and manufacture its own custom lidar rather than build upon more advanced commercial products. This practice may maximize profits for the short term in the defense industry, but it will prevent defense systems from even approximating the advances in performance and cost savings realized in the commercial IoT domain. This failing is especially evident in communications for military IoT. For example, commercial wireless networks, key to IoT, are achieving bandwidths and low-latencies unattainable on wired networks just a few decades ago. This advanced largely arises from 5G, Bluetooth, and related technology. 5G and Bluetooth are interoperable across the globe and compatible with billions of mobile devices. In contrast, the US military is stuck with dozens of custom radio systems which do not interoperate, fail even remotely to approach the performance of commercial 5G, and are in many cases almost as susceptible to jamming and compromise as commercial systems.

With rare exceptions, the cycle of defense-invention, to commercial-innovation, back to defense practice occurs without conscious planning or funding. It is rarely recognized or acknowledged. The development of driverless vehicles offers many examples. The first driverless vehicle that could navigate continuously at practical speeds on and off roads was the DARPA-sponsored, Army-run, Autonomous Land Vehicle (ALV) in 1984–1987. The ALV consisted of a combination of commercial and military components that were wired together with internet technology. It was an IoT device as are today's driverless cars. The ALV program was followed by a series of disjoint follow-on efforts sponsored by the Army, DARPA, and the USMC. While each took advantage of improvements in commercial components, little advantage was leveraged from preceding defense department R&D efforts. In terms of defense technology, these programs stood on the feet of their predecessors not their shoulders. Unlike its defense counterpart, commercial progress in components, especially processors and sensors, evolved continuously. But the automotive industry appeared largely uninterested in creating driverless IoT vehicles, and the technology languished making slow incremental progress in both commercial and defense domains. One exception was the Carnegie Mellon University (CMU) Navlab project (originally funded as part of the ALV effort in 1984). Chuck Thorpe and other Navlab principals managed to retain continuous funding by bridging multiple government sources for several decades and consequently generated continuing, accretive progress.

In 2004-2005, DARPA sponsored the Grand Challenge, a race by driverless vehicles across the California desert. It was an unusual effort by a defense establishment to intentionally fund and foster the transition of driverless vehicle technology from the US Defense Department to commercial industry. No new technology was funded by the Defense Department, just infrastructure supporting the race – the Grand Challenge was intended as a media event. The two first place finishers, teams from Stanford University and CMU, used defense-funded technology growing out of the Navlab project at CMU with shared roots in the ALV program. As a media event, rather than an R&D program, the Grand Challenge demonstrated to the commercial world a proof of principal and the maturity of driverless vehicles. Not only did it finally garner attention from the automotive industry, but it also spurred the development of some essential components. For example, while the first lidar for vehicles was funded by the Defense Departments two decades before the Grand Challenge, the first reliable, hardened commercial lidar for driverless vehicles grew out of the Grand Challenge. The commercial product supported experiments at many automotive-related corporations, large and small. It helped progress the commercial auto industry to today's current state-of-the-art in driverless cars. This interplay between defense invention and commercial innovation in practice seems little noticed and rarely acknowledged. The commercial advances in autonomous vehicles have by in large not yet flowed back into defense practice.

There have been several other attempts by intelligence and defense organizations to foster technology transfer between commercial enterprise and national security applications. The US Central Intelligence Agency (CIA), in cooperation with other government agencies, founded a venture capital firm, IN-Q-TEL, in 1999. Its purpose is to fund high-tech startups in the style of venture firms, specifically firms developing commercial innovations that might benefit national security. IN-Q-TEL's founding premise was that the Intelligence Community (IC) was missing out on innovation coming out of Silicon Valley. Because the IC is a closed, classified community, the success of IN-Q-TEL is unknowable by the public. Its funding forms a small piece of the total venture and private equity capital available and it is therefore unlikely to have much impact on commercial innovation. However, if the organizational connections to the IC are adequate, IN-Q-TEL should at the least allow them to remain abreast of emerging commercial products and it may give them less entangled access to the IP that IN-Q-TEL funds. It does not address the problem of adapting commercial technology to national security environments and requirements. Perhaps that adaptation is addressed and funded in secret within the IC.

The US Department of Defense and its Service Departments acknowledge the difficulty in moving commercial technology into defense use. Several organizations have been created to address the problems in recent years. The Defense Innovation Unit (DIU) was established in 2015 with an office in Silicon Valley. DIU works to stay abreast of innovations in Silicon Valley and seek ways to streamline procurement and funding for commercial tech companies that have dual use between defense and commercial applications. AFWERX was created by the US Air Force in 2017 to bridge commercial technology to air force applications. It focuses on forming teams of Air-Force and private innovators as well as streamlined procurement for more rapidly adopting innovative technology. It is not clear that either DIU or AFWERX have understood and systematically address the technical challenges of adapting commercial technology to defense needs and conditions as listed above. Nor is clear that these organizations have the authority to make the needed policy and organizational changes to bridge the gap between advanced development and mainstream defense-system development. They seem unlikely to effect the required adjustments in operational policy, doctrine, system acquisition, and logistics. AFWERX sits within the Air Force Research Laboratory and therefore sits on the R&D side of the of the Valley of Death, the commercial name for the gap between R&D on one side and product development and adoption on the other. Many commercial inventions advance into the Valley of Death never to emerge. In the defense domain, the Valley is even steeper and deeper, and many AFWERX inventions and innovations may end there as well. These challenges are especially formidable for IoT.

In January of 2021 the US Department of Defense announced the creation of the Trusted Capital Digital Initiative (TCDI) to establish trusted sources of funding for small and medium-sized providers of innovative defense-critical capabilities. TCDI may help to de-globalize funding for commercial innovations of critical value for US national security. But the Initiative is unlikely to unravel globe-spanning supply chains for IoT components.

Perhaps the most promising creation is the US Army Futures Command (AFC). Created in 2018, AFC enjoys the status of a top-level Army command, led by a four-star general. Of the US defense organizations aiming to foster commercial-defense transfer, AFC may be the only one of the new organizations with enough authority and span of mission to foster technology solutions as well as organizational and process changes. IoT especially needs such solutions and changes if it is going to evolve in coordination and parity with commercial practice.

The component technologies of IoT are advancing so rapidly at present in the commercial domain, that defense enterprises must find the means-to rapidly adapt commercial technology to meet military requirements. How effectively and quickly a country's defense establishment takes advantage of commercial IoT developments, while still meeting the special requirements of military operations, will determine which nations take the lead in national security.

8 A Guide to This Book

8.1 The Audience for This Book

The Chapters of this Book present new research results and address the potentials and challenges for IoT applied to defense and national security. The authors represent a collection of experts in each area, all engaged in ongoing research, development, or management in their fields. As an anthology of chapters from multiple authors, this Book is aimed at a broad and varied audience. The Chapters cover a range of topics appealing to a diverse range of readers. Some chapters provide detailed, recent research results of special interest to active researchers in the area. They presume technical background in IoT. Other Chapters address decision makers, managers, executives, funders, and policy makers and do not require an extensive IoT background. Because IoT spans many disciplines, readers from many fields will find some or all Chapters of interest. This Book focuses primarily on defense issues and technology in the US – the largest single defense market. However, the topics are relevant to the militaries and national security establishments of all countries. To date, the few book chapters that have been published on IoT for defense come from India, South Korea, Spain, and China. The discussion of IoT for defense in the US and Spain in this Book will expand the picture.

8.2 The Organization of This Book

8.2.1 Section 1: Challenges, Applications, and Opportunities

Section 1 presents a broad view of IoT for defense and national security and describes several applications of IoT to military operations. It provides examples of IoT used in support of ground combat, naval warship design and operation, and defense logistics. Chapter 1 reviews the US Army's Internet of Battle Things (IoBT) collaborative alliance, a premier effort to advance the state-of-the-art in IoT for military affairs. The Chapter identifies key applicational requirements for fielding IoBT technology, including supporting diverse missions, rapid deployment, managing resource-constrained assets, exploiting heterogeneous sensors, supporting varying scale in data, and being deployable within contested environments. Chapter 2 focuses on the application of IoT to ground warfare. It describes how an IoT-instrumented rifle can provide outstanding situational awareness by detecting and tracking such information as when a weapon is fired, how often, in what direction, and when incoming fire is detected. The Chapter also addresses the unique challenges of weapons-based IoT sensors, such as unreliable network connections, prioritization of data to enable critical battle decision making, securing data in contested areas, deploying within extreme environmental conditions, and managing weight restrictions for soldiers on the battlefield.

While IoT may remove the fog of war caused by lack of information about the state of the battlespace, it may instead create its own fog by overwhelming commanders, processors, and communication channels with too much information. Chapter 3 proposes new algorithms to optimize which resources to use and how best to use them, given the constraints of combat. Both submodular and convex minimization algorithms are defined, and their integration is explored with the goal of producing an optimal solution that can execute in time to help battle commanders make fast decisions. Decisions often revolve around how to apply limited resources in the face of adversarial actions. Chapter 4 defines the needs of a data fabric that includes data standards and processes to provide a secure environment for mission execution. Such a fabric can modernize enterprise information management to address gaps, simplifying the security architecture to enable the warfighter. The Chapter also discusses how such a data fabric produces consistent policies to enable interoperability and optimize data management to improve decision making. Chapter 5 turns to the problem of providing edge computing to support IoT, where edge computing refers to provisioning of processing resources on the edge of the network at or near IoT devices. The Chapter reviews common edge connectivity solutions and new visualization approaches for IoT sensor data.

Omar Bradley, the first US Joint Chief of Staff, reportedly said: "Amateurs talk strategy. Professionals talk logistics." Effective logistics involves inventorying, cataloging, transporting, tracking, and delivering thousands, even millions of items to the right place at the right time in the midst of war. Beginning with the advent of RFID tags, militaries understood that IoT could be used to automate logistics making it more efficient and effective. Chapter 6 describes IoT in support of logistics to provide commanders and defense ministries with planning, prepositioning, situation awareness, data collection, and conditions-based maintenance. Also discussed are IoT applications for defense installations, emergency response and disaster recovery.

Digital twins are virtual, software duplicates of a physical IoT system. They can accept sensor inputs from the physical system and monitor responses from the physical system. The final Chapter in Section 1, Chapter 7, provides an example of a digital twin supporting warship systems where a complete virtual environment can reproduce the behavior of a physical ship in support of a nation's navy and its warships.

8.2.2 Section 2: Artificial Intelligence and IoT for Defense and National Security

Expanding compute power and massive digital data supported the most spectacular accomplishments of IoT over the past decade. Massive amounts of digital training data and the processing power to digest it enabled a class of algorithms to display intelligent behavior. These algorithms, commonly referred to as artificial intelligence, can solve many tasks previously thought to require human intelligence. IoT can use intelligent processing to close the loop between sensing and action to accomplish missions currently requiring people in the loop. Closing this loop creates new and greater possibilities for automating defense systems and creating autonomous and semi-autonomous battlefield IoT networks. Such networks have the potential of not only amplifying humans on the battlefield but also moving them back from some of the most dangerous missions and environments to supervisory positions of greater safety. Section 2 consists of three Chapters addressing and IoT.

Chapter 8 addresses some special concerns for defense that arise when inserting AI algorithms into IoT networks. Deep-learning convolution neural networks are among the most powerful and most common forms of intelligent processing or AI. They require significant processing power, and they can be brittle, vulnerable, and untrustworthy. Chapter 8 provides principals and methods to make machine learning robust, resilient to adversarial attacks, and more interpretable for human-on-the-loop decision-making. For those readers who desire more context on the history and mechanisms behind AI algorithms and a discussion of the special challenges for military IoT presented by deep-learning networks, they can find an extended discussion in Chapter 21.

Chapters 9 and 10 review research that uses AI to drive applications of robotics and sound perception, both key emerging defense uses of IoT. Chapter 9 describes an approach to distributed edge processing that can power AI algorithms at the tactical edge while avoiding communications limitations found in battlefield environments. That approach is illustrated with case studies in the use of robotic IoT. In Chapter 10, a team of IBM researchers describe a system for deploying AI based acoustics in real-world environments and review lessons learned from using their system. They show how AI provides an augmented capability for a data-driven understanding of the environment through acoustics – a capability that also comes with several new challenges that the authors address.

8.2.3 Section 3: Security, Resiliency and Technology for Adversarial Environments

Section 3 contains six chapters addressing security, survivability, and operations in adversarial environments. These topics are essential and central to the safe and successful use of IoT for defense. The difficulty and importance of these issues separate commercial IoT from military IoT as much as any other set of topics. Solving these challenges represents the difference between IoT being a battlespace advantage or a strategic liability. Chapters 11 and 16 report research sponsored by the US Department of the Navy; however, they describe approaches to resiliency and protection of IoT network integrity that apply to all defense domains from maritime to air to ground to space. Chapter 11 describes the design of IoT networks that are resilient against physical and cyber threats. The designs can also tolerate misclassification errors resulting from natural and/or adversarial errors within their data-driven components. Chapter 16 describes an automatic code generation approach that compiles formally defined IoT interface specifications into software that is provably free from common software vulnerabilities. Attempts to generate software using neural nets trained on examples have met with limited success. In contrast, auto-code generation successfully generates workable, vulnerability-free software when starting with high-level, formal specifications. Automatically generated interface code can be created, controlled, and distributed by a trusted "IoT Interface Control Authority". When augmented with anti-tamper measures this approach ensures that a compromise of one node in a defense IoT network cannot be propagated through the entire IoT system.

Part of the power of an IoT network comes from its ability to be flexibly and dynamically reconfigured and extended with new devices that may not have existed when the network was designed. This extensibility combined with the complexity of today's international technology supply chain means that protecting an IoT network cannot be guaranteed by ensuring the device or component-level security is adequate once they have been integrated into a network. Chapter 12 provides a detailed view of such vulnerabilities critical for tactical applications. IoT systems used for military operations will be under constant attack, and Chapter 13 introduces the important topic of detection and prevention of intrusions. The chapter discusses research on the design, deployment, and management of intrusion detection systems for IoT. One of the great advantages of IoT is that devices do not have to house all their processing power locally or on each device. Distributed computing allows massive processing power to be applied to specific nodes without co-locating it at the node. In an adversarial environment, the flow of data and results from distributed processing is vulnerable to attack, disruption, interception, and cooption. Chapters 14 and 15 address aspects of protecting data flows in IoT networks. Chapter 14 proposes to overcome some of these challenges by leveraging data-plane programming languages that enable intelligent packet processing. The authors propose two novel data-driven approaches for attack detection. Chapter 15 reviews specific challenges for distrusted IoT processing. The authors present solutions for distributed computing in an adversarial environment where the data being processed may be maliciously manipulated and the computing and networking resources may be compromised. The Chapter also emphasizes the need to protect meta-data about resource availability from attack.

8.2.4 Section 4: Communication and Networking

Few areas of IoT for defense and national security pose challenges as significant as those of communications and networking in defense environments, the focus of Section 4. Many of the most exciting IoT applications in the commercial domain depend on wireless communication, especially 5G cellular networks. The highly mobile nature of military operations makes wireless communications essential. However, commercial systems are not built to meet the special requirements of national defense. For example, the cellular infrastructure that underlies the commercial 5G network probably will not survive hostile actions leading up to and occurring during war. The size of the commercial marketplace far outstrips the defense market. Not surprisingly therefore, commercial communication products already outstrip defense IoT with both higher performance and lower cost. Military and national security IoT systems must find a way to leverage commercial developments. Section 4 addresses some of these challenges. Chapter 17 describes the differing demands of commercial and defense communications. It pays special attention to the relationship between defense and commercial advances in communications and networking. Chapter 18 discusses the use of cloud technology at the tactical edge to share content across IoT networks to reduce bandwidth demands and increase survivability of information flows. Content-based networking may be the single most important new concept in how to move information around the battlefield, both to support IoT networks and in general. As commercial IoT devices demand ever more bandwidth and move higher up in the spectrum, they compete with spectrum needed for military IoT systems. Chapter 19 discusses spectrum challenges for IoT for air operations. It outlines the state-of-the-art and next steps. A general design of a tactical edge IoT communications architecture is presented in Chapter 20. It describes how edge-computing approaches address some limitations of IoT in military operations. This chapter identifies scenarios in which defense and national security can leverage commercial off-the-shelf (COTS) Edge IoT to deliver greater survivability to warfighters or first responders, while lowering costs and increasing operational efficiency and effectiveness

8.2.5 Final Considerations

Section 3 addressed the problem of protecting IoT systems in hostile defense and national security environments. The final Chapter of this Book, Chapter 21, asks the complementary question: if governments build and command vast IoT systems for national security, how does the public protect itself from a government that then uses those IoT systems to further its own power at the expense of those governed? This question is not idle speculation. Some governments already use IoT to impose their will on their citizens to benefit of the rulers at the expense of the ruled. Chapter 21 reviews the nature of the elements of IoT in some detail, assessing how they could abuse individuals and organizations if not constrained. IoT cannot advance national security if it becomes a tool used by a nation's government to oppress a free society. Chapter 21 presents a preliminary framework for regulating IoT. It serves as a basis for discussions among the public and policy makers on regulating IoT to enhance its benefits while protecting against possible abuse and misuse of its power.

9 This Book Is a Beginning

IoT's power and automation will pervade all aspects of our life. The changes in our relationship to technology are fundamental. IoT will automate many tasks in our life, replacing human networks with increasingly autonomous networks of things. Nowhere will that power have more impact than in securing nations, enabling their defense, empowering their military, and maintaining peace. Its benefits need to be extoled but balanced by caution. If misused or abused by governments, IoT can erode and eventually steal the freedom of individuals and societies. Thousands of books, articles, and conference proceedings discuss commercial IoT. Almost none address IoT in the service of national defense and security. By necessity, most of the work in the national security domain resides behind walls of classification blocking it from public view. The reader will not find in this Book a comprehensive review of IoT for defense and national security; instead, they will find an initial peek inside this important world. The Editors hope this peek will be of use to defense establishments and arouse public attention and discussion. It provides researchers with a view of the state-of-the-art in many areas and a view of what's possible. With luck, it will stimulate policy makers and decision makers to bring more information on IoT forward to the public and to take an active role in establishing policy and expanding the use of IoT.

Section 1

Introduction: Vision, Applications, and Opportunities

Stephan Gerali

Enterprise Operations, Lockheed Martin Corporation, Bethesda, MD, USA

Future battlefields engagements will increasingly utilize myriads of multimodal intelligent sensors, seemingly unlimited computational power available using the cloud and a fast always-on networking all relying on the power of 5G. This will fundamentally change the way nation states will engage in conflict and resolution for the foreseeable future. To better understand the future battlefield, it is important to understand the vision, applications, and opportunities for Internet of Things (IoT) technologies within the defense and national security domain.

This section presents a broad view of IoT for defense and national security with several applications of IoT specific to military operations in support of ground combat, naval warship operations and defense logistics. The section starts with a review of the U.S. Army's Internet of Battlefield Things (IoBT) to include key fielding requirements (supporting diverse missions, rapid deployments, managing resource constrained assets, exploiting heterogenous sensors, supporting varying scale in data, and deployable within contested environments) along with enabling faster sensor-to-effect decision loops. In building the IoBT vision, all warfighters will need to be sensorized in order to digitize the battlefield and understand who is friend or foe during engagements. With lots of IoBT data becoming operationalized, it will be critical to have decision support tools that can help commanders quickly optimize their decisions for when and how to place resources to win the battle.

In order for joint operations between military forces to coordinate efforts across the battlefield, common data standards will need to be developed to enable interoperability of all the systems used in battle. Effective edge-based monitoring solutions will need to be developed to help collect real-time information of the battlefield and analytics will need to be created to constantly improve execution. Having all of this IoBT data, senior leaders within the military will know the past, current and future performance of military assets to help win the engagement. Digital twins of the battlefield will be developed to support simulation and analysis of mission objectives along with real-time monitoring as the battle occurs. This section will cover all aspects of digitizing the battlefield and warfighters, collecting operational data, optimizing decisions, and understanding the behavior of complex engagements by using IoT technologies. In addition, this section discusses the special challenges, applications, and opportunities for leveraging IoT in the defense and national security domain and presents a variety of approaches for dealing with them.

In Chapter 1, *Internet of Battlefield Things: Challenges, Opportunities, and Emerging Directions*, the concept for the IoBT is developed which is a convergence of technologies and devices that

IoT for Defense and National Security, First Edition. Edited by Robert Douglass, Keith Gremban, Ananthram Swami, and Stephan Gerali.

are pervasively networked, interconnected, intelligent and often dynamically composed to serve the purpose of delivering effects that can occur in both cyber and physical space. The success of IoBT, as a common operating environment and control framework, hinges on the development of scientific foundations of performant and resilient computational, sensing and triggering services for the battlefield. Its goal is to ensure the execution of sensor-to-effect decision loops (that involve multiple intelligent devices and systems) that meets the challenges arising from spatial distribution, accelerated mission-tempo, heterogeneity, transient resource availability, environmental dynamics and the presence of adversarial activity.

Within Chapter 1, IoBT identifies key applicational requirements to fielding the technology that include: supporting diverse missions, rapid deployment, managing resource constrained assets, exploiting heterogeneous sensors, supporting varying scale in data, and being deployable within contested environments. Further the chapter covers the complexity and challenges of deploying IoBT: synthesis of joint all-domain operational data, timeliness of connecting sensors to shooters to understand intents, robustness to adversarial disruption when using deceptive data, and being deployable at the point of need when computing resources might be constrained.

In Chapter 2, *Sensorized Warfighter Weapon Platforms: IoT Making the Fog of War Obsolete*, there is no platform or tool closer to ground-truth engagement data than the warfighter's weapon. Weapons-based IoT sensors provide a path to harness individual engagement data into actionable intelligence insights across echelons and enable turning each weapon into a new information node. Every warfighter information node can provide engagement data that includes weapons metrics such as discharge detection alerts, shot counting, ammunition remaining, weapon rates of fire, weapon orientation and directionality, and overall weapon readiness. This information represents a critical window into multiple use cases that include predictive maintenance, real time situational awareness across echelons, ammunition and weapon supply chain optimization, warfighter and squad performance metrics and ground truth historical reporting.

Within Chapter 2, weapons based IoT sensors also carry unique challenges such as dealing with unreliable network connections, prioritization of data to enable critical battle decision making, securing data over contested areas, being deployable within extreme environments and conditions, and managing weight restrictions for soldiers on the battlefield. If the unique challenges for weapons based IoT sensors can be addressed, thus immediately and automatically knowing friendly positions, maneuvering, and status, the threat would be efficiently located and neutralized with improved enemy kill chains and little to no risk of fratricide. With real time weapons data available, timely engagement and support decisions can be made to take advantage of a treasure trove of insights, intelligence and usage data that can enable better warfighters.

Chapter 3, *IoT Resource Allocation via Mixed Discrete and Continuous Optimization*, analyzes how the deluge of data from IoT sensors makes it way into battlefield systems. Battle commanders will need to be able to constantly monitor existing resources and allocate them in response to adversarial actions. To augment the battle commanders, Chapter 3 proposes new algorithms that will support which resources to use and how best to use them especially when fractions of existing resources can be deployed. To make use of available budget, the battle commanders must reason on which sets of tasks to allocate resources to and the best way to allocate resources to them simultaneously.

Within Chapter 3, resource allocation problems with both mixed continuous and discrete costs of optimization are explored and defined over two fundamentally connected lattices. Both submodular and convex minimization algorithms are defined. Methods are explored to integrate such algorithms to produce a guaranteed optimal and efficient solution to help the battle commanders with faster decision making in applying existing resources based on adversarial actions.

In Chapter 4, *Operationalizing IoT Data for Defense and National Security*, discusses how important sensors are in interpreting the battlefield conditions, being able to quickly acquire data and operationalize the data in decision making is critical to winning the battle. Due to information protection policies across government agencies and the defense industrial base, leveraging technology across domains has become excessively complex. Foundational data architectures that can enable digital threads from product design to implementation to sustainment have not been effectively implemented. Many of the battlefield management systems today were designed as point solutions without consideration of the needs of joint military operations and integrated missions, thereby making the solutions suboptimal for winning the battle. Further, since common threads were not identified across battlefield systems, supportive data models and master data management systems could not support the level of integration required to fully enable the warfighter.

In order to win with data, Chapter 4 documents the needs for data standards on the representation, format, definition, structuring, tagging, transmission, use, and management of such data. Further, business process standards need to be created on the input, output and outcomes from the executing business processes. A data fabric must be established that can adapt quickly to new data requirements, connect data from disparate data sources, and enable a semantic data layer to remove complexity from the process of finding, sharing, and analyzing data. Finally, Chapter 4 defines the needs of the data fabric to provide a secure environment for mission execution by modernizing enterprise information management to address gaps, simplifying the security architecture to enable the warfighter, producing consistent policies to enable interoperability, and optimizing data management operations to improve data interoperability and decision making.

In Chapter 5, *Real Time Monitoring of Industrial Machines using AWS IoT*, IoT sensors for acquiring data from the battlefield and leveraging common data standards/processes to enable interoperability are critical to competing in future warfare. In order to enable all of this, an edge computing framework must be established that can connect the data, transform the data, store the data, and perform high level analytics off of the data to enable the warfighter. Chapter 5 provides a real-world example of leveraging edge computing to enable collection and analysis of all of this data to enable the warfighter.

Chapter 5 goes through common edge-connectivity solutions like KepServerEx for connecting to IoT equipment using common protocols, AWS IoT Core for scalable IoT data processing, Amazon EMR (Elastic Map Reduce) for sensemaking of the data and Tableau for visualization of the data. Finally, Chapter 5 talks about next generation visualization capabilities with Augmented Reality and Virtual Reality to help visualize the IoT sensor data to streamline operations and ultimately help the warfighter in future engagements on the battlefield.

Chapter 6, *Challenges and Opportunities for IoT for Defense and National Security Logistics*, discusses how once an IoT system has been deployed, senior leaders within the military are enabled to increase readiness through understanding of the condition of their things (people, equipment, facilities, operations and battlefield). The ability to know the past, current, and future performance of military assets is a powerful tool for tailoring the force to meet the mission need. Chapter 6 lays out the challenges (policies and legal implications) and opportunities (situation awareness, prognostics, health monitoring, data fusion, analysis, and distribution of information for combat, emergency response, humanitarian, and disaster relief) that IoT presents to the nation in defense and national security.

Within Chapter 6, IoT systems for military applications can provide platforms for operational and strategic advantage with the ability to sense and analyze the world around them to meet the mission with the right equipment and the right personnel at the right time. Further, the chapter covers the issues with avoiding decision paralysis through the development of decision support tools to help the warfighter to make the best decisions under challenging conditions.

In Chapter 7, *Digital Twin for Warship Systems: Technologies, Applications and Challenges*, discusses how once all the IoT sensor data has been acquired and integrated, the warfighter can begin seeing a digital representation of the battlefield for either real-time monitoring or simulation and analysis purposes. Chapter 7 provides an example of a digital twin supporting warship systems where a complete virtual environment can reproduce the behavior of a physical system in support of the U.S. Navy and its warships.

Within Chapter 7, the purpose of the digital twin is to acquire experience and knowledge about the operation of the system that is being simulated, identify problems within a complex process, propose alternative scenarios to understand the behavior of complex systems and to apply the digital twin through any phase of the lifecycle design of the system. Further, with the recent advancements in artificial intelligence, digital twins can benefit from algorithms that can optimize operations, perform predictive maintenance versus preventative maintenance, detect anomalies, and support diagnosis and correction of such anomalies. Focusing on the simulation models that are applied in the development of digital twins in ships and in ship's systems, these will be in high proportion multiphysics models to help the warfighter better understand the physical behavior of different components in ship's systems and interactions between them.

1

Internet of Battlefield Things: Challenges, Opportunities, and Emerging Directions

Maggie Wigness[1], Tarek Abdelzaher[2], Stephen Russell[3], and Ananthram Swami[1]

[1] *U.S. Army DEVCOM Army Research Laboratory, U.S. Army Futures Command, Adelphi, MD, USA*
[2] *Department of Electrical and Computer Engineering, University of Illinois at Urbana-Champaign, Urbana, IL, USA*
[3] *Department of Research Opportunities and Innovation, Jackson Health System, Miami, FL, USA*

Abstract

The internet of battlefield things (IoBT) is expected to be a major feature of future tactical wireless networks. Multiple challenges arise from the expected scale, heterogeneity, information sharing (in a joint or coalition environment), dynamics, and actions of sophisticated adversaries. The chapter will discuss these challenges in detail, and explore emerging directions which provide opportunities to enable a scalable, secure, and performant IoBT.

1.1 IoBT Vision

The rapid emergence and acceptance of the internet of things (IoT) has been fueled by advances in *machine intelligence* and *networked communications*. When appropriately managed, each of these two aspects makes the collection of "things" more practical, capable, and coordinated. We refer to an IoT operating with military "things" and under military constraints as the *internet of battlefield things* (IoBT). The IoBT is already becoming a reality [1, 2], and will likely become a dominant component of the future battlespace [3, p54], [4].

IoBT research challenges arise from the increasing complexity of networked computational and physical resources in future battlefields, as well as the challenging environment in which they will operate. Future battlefields will increasingly utilize myriads of multimodal intelligent sensors and heterogeneous computing resources that are deeply embedded in an adversarial environment, networked together to enhance real-time decision making, support local autonomy, penetrate areas of denial, and improve insight into enemy actions. Intelligent computation and physical assets must work together to achieve mission goals. Mission success will depend as much on the capabilities of computational artifacts as it will on the capabilities of the physical devices in the loop. It is in this context that one must advance the performance and resilience of the networked resource substrate in future cyber-physical battlefields.

In general computing contexts, streamlining application development and operation necessitates the introduction of *operating systems* to address common challenges such as resource-efficiency, resource-sharing, deconfliction, execution robustness, scalability, and responsiveness that applications require. In a modern battlefield, where mission success depends in large part on

IoT for Defense and National Security, First Edition. Edited by Robert Douglass, Keith Gremban, Ananthram Swami, and Stephan Gerali.

computational artifacts, a new operating-system-like construct is in order. Its goal is to ensure that the execution of sensor-to-effect decision loops (that involve multiple intelligent devices and systems) meets the challenges arising from spatial distribution, accelerated mission-tempo, heterogeneity, transient resource availability, environmental dynamics, and the presence of adversarial activity [5]. IoBT is intended to be such an operating environment. The focus of IoBT is on the *tactical edge*, where deployed military assets must perform their combat functions.

Understanding the concept of the IoBT needs further elaboration. Many want to think of the IoBT as a singular technology, much like a new networking protocol such as 5G, or new methods for computation such as machine learning, or a new sensing modality or capability like LiDAR. The IoBT encompasses these technologies and many more. Critically, the IoBT should not be viewed as a singular technology. Rather, it is a convergence of technologies and devices that are pervasively networked, interconnected, intelligent, often dynamically composed to serve the purpose of delivering effects that can occur in both cyber and physical space. Within this context, the *vision for IoBT* is to manage intelligent complex systems-of-systems, pervasively hosting intelligent sensing and actuating devices, driven by learning processes that adapt and achieve the Army's macro and micro goals.

The foundation of such an intelligent system-of-systems already exists as the IoT, but the IoBT introduces new complexities resulting from military-centric technological advancements, provenance, and operations. The success of IoBT, as a common operating environment and control framework, hinges on the development of scientific foundations of *performant* and *resilient* computational, sensing, and triggering services for the battlefield. By *performant*, we refer to the capability to perform functions under unevenly resourced and likely contested environments, while meeting battlefield mission dynamics, demands, and time constraints. By *resilient*, we refer to the ability to withstand or effectively recover from a wide array of adversarial and environmental threats. In the rest of this chapter, we explore the underlying challenges and possibilities of the IoBT, and exemplify recent progress on meeting them. We start with some background: How are IoBTs different from IoTs? What are the main demands imposed on their operational characteristics? What technical design requirements arise by virtue of the characteristics demanded by Army applications? And how does the need for innovation in IoBT create research challenges and opportunities? Below, we elaborate.

1.2 IoBT vs. IoT

The explosive growth of technologies in the commercial sector that leverage the convergence of cloud computing, ubiquitous mobile communications, networks of data-gathering sensors, and artificial intelligence (AI) presents an imposing challenge and opportunities for defense applications. These IoT technologies have the potential to offer increasing capabilities that may become the decisive factor in winning future conflicts. However, commercial development is not focused on the unique challenges of battlefield environments, such as those described in the US Army's Operating Concept: "Win in a Complex World" [6].

Battlefield environments give rise to wickedly difficult problems. Goals, objectives, and constraints evolve in unpredictable ways and are frequently challenged by adversarial intention to impede agendas. Complexity of supporting systems grows from the increasing heterogeneity, connectivity, scale, dynamics, functionality and interdependence of networked elements, and from the increasing velocity and momentum of human interactions and information. Events unfold in internet-time (as noted by the Defense Science Board 2014 Study on Decisive Army Strategic and Expeditionary Maneuver [7]). Novel mathematical approaches and techniques are needed to represent and reason about current state, understand emerging behaviors, and deliver predictive analytics in diverse and dynamic environments. Rapid asset deployment and adaptation must occur in the presence of high mobility, resource constraints, and extreme heterogeneity, in both very dense and sparse settings. Capabilities that enable dominance through technological adaptability, continuous speed, and a system-whole that is stronger than its individual parts become critical to battlefield success.

The above battlefield challenges impose unique requirements on the characteristics of IoBT systems that are different from their civilian counterparts. We review such IoBT characteristics and related definitions next.

1.3 IoBT Operational Requirements

In recent work [8–10], an IoBT has been defined as a set of interdependent and interconnected entities or "things" that can include:

- Sensors
- Actuators
- Devices – computers, weapons, vehicles, robots, human-wearables
- Infrastructure – networks, storage, processing
- Analytics – on-node, in-network, centralized
- Information sources and Open Source Intelligence
- Humans

The unique characteristics of the operating domain, discussed earlier, suggest that an IoBT must be dynamically composed to meet multiple mission goals, capable of adapting to acquire and analyze data necessary to predict behaviors/activities and effectuate the physical environment, self-aware to operate autonomously and autonomically, and have the capacity to interact with networks, humans, and the environment in order to enable predictive decision augmentation that delivers intelligent command and control (C2), and battlefield services. More specifically, IoBTs must offer support for several application requirements:

- **Diverse missions, tasks, and goals.** An IoBT will be specifically created (and adapted) to meet a mission, complete a task, or accomplish a goal. Tasks may vary from wide area persistent surveillance, tracking dispersed groups of targets moving throughout a cluttered environment, to local tasks, such as monitoring the physiological and psychological state of soldiers. Tasks are not expected to start or end simultaneously, and new tasks may emerge throughout a mission.
- **Rapid composition, deployment, and adaptation to changing missions/tasks.** To maintain operation tempo, goal-driven IoBTs must be composed and deployed rapidly, e.g. at a quickly-formed forward operating base. Once deployed, the dynamics of the battlefield will generate changes to missions/tasks, cause new tasks to emerge, and provoke analysts/commanders to change tactics. These changes make it necessary for the IoBT to adapt to maintain synchronization with the current mission needs and specifications.
- **Highly dynamic, mobile, and resource-constrained assets.** Many IoBTs will be forward-deployed and consist of disadvantaged assets with limitations on energy, power, storage, bandwidth, infrastructure (fixed infrastructure may not be available), computing, and communications. Additionally, there may be many IoBTs operating simultaneously, possibly competing for resources. IoBTs must be able to operate under these severe constraints without hindering their support for tasks with stringent time deadlines.

Characteristic	Resilient IoBTs	Performant IoBTs
Diverse missions, tasks, and goals Rapid composition, deployment, and adaptation	• Provide quantifiable/bounded/thresholded learning and domain adaptation in variable environmental conditions and changing mission contexts	• Optimally or near optimally (re)allocate resources such that external dynamics minimally impact functional objectives/goals
Highly dynamic, mobile, and resource-constrained assets	• Adapt to disruptions and latency in power and communications efficiency and handle channel variabilities • Provide temporal guarantees for learning and inference, given non-stationary environmental latencies and/or observational rates	• Prevent leakage of blue information, minimize data loss, and has awareness of probability-of-detection • Have acquisition and ML/AI that pays attention to performance degradation to "keep-up" with real-time/rapid change
Extreme heterogeneity	• Exploit heterogeneity to avoid excessive reliance on any one sensor/modality type, i.e. exploit sensor diversity, given availability, accessibility, or capability constraints	• Optimally or near optimally selects/utilizes sensors, across disparate sensor types, given complex tradeoffs such as different compute, power, and networking capabilities
Varying scale	• Rapidly detect, with low overhead, and mitigate aberrant behaviours/conditions resulting from the number of network nodes/assets/conditions/data	• Rapidly or selectively make sense of data generated from a vast number of devices, inputs, or observations
Contested and adversarial environments	• Condition outputs, given devices' provenance, configuration, and/or trustworthiness, dependability, or reliability • Detect and/or adapt to threats intended to generate incorrect inferences/outputs	• Safely and securely exploit assets that may not be blue, may be spoofed, or may be red in disguise • Mitigate attacks designed to delay inference or deplete blue resources (e.g. jamming, power consumption, disconnection)

Figure 1.1 Examples of resilient and performant capabilities that are desirable in an IoBT for each of the operational environment characteristics.

- **Extreme heterogeneity.** Heterogeneity is seen across multiple dimensions of an IoBT. Variety of sensing devices creates multi-modal data, and variety of compute devices creates diverse processing capacities. Further, IoBTs will contain a mixture of entities: blue – devices controlled by friendly forces, red – devices controlled (or partially controlled) by the adversary, and gray – commercial and open source devices. Coexistence and co-deployment of commercial IoT devices and networks with purposefully built, certified, and carefully controlled military devices and networks will be required. These things will have a wide variety of security, provenance, and capabilities that must be accommodated and will need to exploit very capable devices and simple disposable devices.
- **Varying scale.** IoBTs will be deployed in a wide variety of places and domains, usually in contested environments. On one extreme is the highly dense and cluttered mega-city environment and on the other extreme sparse terrain with limited entities and gaps in sensor coverage and networks.
- **Contested and adversarial environments.** Many IoBTs will be forward deployed with limited physical security and will include entities in the IoBT that are monitored and controlled by the adversary. It must be assumed that adversaries are already on the IoBT network and that the IoBT must be protected from sophisticated and persistent threats. Cyber/information security measures must be taken to protect IoBTs, and analytics must be able to deal with conflicting and deceptive data to identify adversarial activity.

Figure 1.1 summarizes some of the desirable capabilities that an IoBT will exhibit to remain resilient and performant when faced with the characteristics of a unique Army operating environment. Specific efforts that address these desired capabilities are outlined later in the chapter.

1.4 An Organizing Concept

Meeting the above characteristics is a challenging system-of-systems design problem. Making it manageable requires structuring the problem in a manner that allows meaningful decomposition into simpler subproblems that are easier to address and solve. This divide-and-conquer strategy has been the wisdom behind (and reason for) success of several large systems-of-systems design endeavors, such as the Internet Architecture [11] and the multisensor data fusion model [12]. In both examples, it was the decomposition of the overall architecture into smaller individually more manageable components that enabled innovation and accelerated technological advances. How should one break down the IoBT architecture in a manner that accelerates component innovations and helps solve the underlying research challenges? One possible answer lies, as we show below, in the *MDO Effect Loop*.

1.4.1 The MDO Effect Loop

A key distinction of IoBTs (from, say, the general Internet) is that they are *goal-driven*. The objective is not fairness to all users but rather to carry out the military mission successfully. It therefore behoves us to start our architectural decomposition by analyzing the goal of employing an IoBT. One way of framing the goal of an IoBT is in the context of a decision cycle, where a sequence of steps is executed on a repeated basis to gather information to make informed decisions and learn from the results. Specifically, the multi-domain operations (MDO) effect loop has been described as an organizing principle for military operations and information-driven decision processes [5]. It breaks the end-to-end decision cycle into the stages of (i) assessment/detection, (ii) identification, (iii) localization/tracking, (iv) aggregation/synthesis, (v) distribution, (vi) decision, and (vii) actuation.

These seven stages cover the functions and phases that are found in the Observe-Orient-Decide-Act (OODA) loop, the military decision-making process (MDMP), find-fix-track-target-engage-assess (F2T2EA), and even Simon's decision phases [10], with an emphasis on multi-domain considerations. Contemporary adversaries seek to gain control of contested spaces not only on land, air, and the seas but also in space and cyberspace, as well as in the electromagnetic spectrum. It is this shift to threats that exploit one or more domains, particularly cyber and space, to provide advantages in other domains and at all phases of the battle. Furthermore, contemporary operations must seamlessly utilize assets and capability in all of these domains, often simultaneously. This transparent access to and through multiple domains characterizes MDO and is where technologies such as the IoBT become particularly critical. This is because the underlying systems and capabilities must inter-operate intelligently and fluidly at battlefield op-tempo to efficiently accomplish tactical tasks, mission objectives, and national goals.

An overview of the seven stages of the MDO effect loop is illustrated in Fig. 1.2 (originally shown in [5]). The first three stages are correlated with Simon's [13] "intelligence gathering phase." They are where information is gathered about a decision (effect) opportunity. The next three stages correlate with Simon's "design phase," where the decision problem is structured, understood, and alternatives are generated. From a pragmatic perspective in the MDO effect loop, information is fused, distributed appropriately, and courses-of-action are developed. Other activities such as modeling, simulation, and forecasting would also occur in these phases of the effect loop; ultimately leading to decision-effect pairings and a commander's decision. Simon's model describes the "decide phase" as an atomic instance. The MDO effect loop extends this notion into the military context, where appropriate authoritative or commander approval is obtained. It is followed by the final phase of the MDO effect loop: to actuate an effect, or in other words, "engage."

It is noteworthy that the MDO effect loop may be non-linear and recursive at any individual stage. In other words, a stage can loop back to previous stages to gain additional inputs. Further, given decisions internal to each stage, a nested loop can be executed as necessary for the purpose of stage completion. Elements of the loop may operate at different temporal and spatial scales.

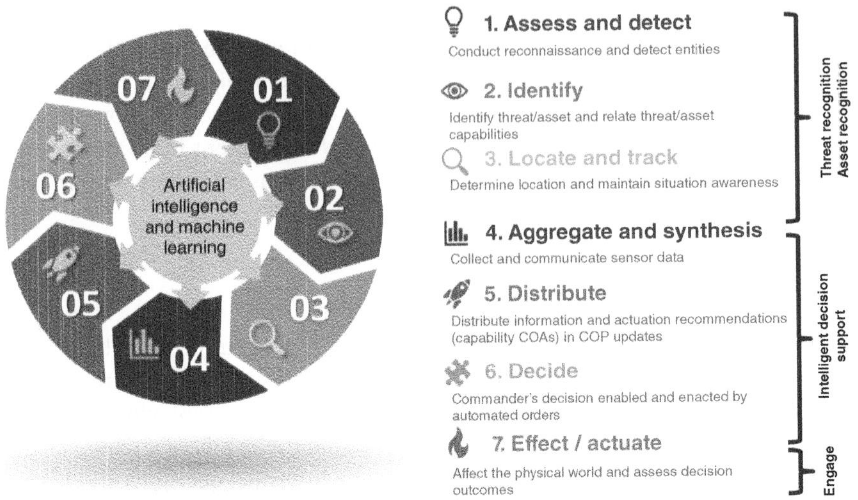

Figure 1.2 A depiction of the seven stages represented in the multi-domain operations (MDO) effect loop. Source: Abdelzaher et al. [2] / SPIE.

Multiple loops may be instantiated in different domains that interact via the aggregation and distribution components. For example, detection components in different domains might operate at the fastest time-granularity, locally. When changes of interest (or of concern) are suspected, the IoBT may be adapted to bring in additional assets to assist with identification, localization, and tracking. While these stages may operate on local objectives, an important capability is to aggregate the information at multiple spatial scales and across domains in order to identify higher-level patterns, such as coordinated movements of troops across large areas. In turn, any conclusions from such larger-scale analysis may need to be shared with the individual loops and decision makers. Decisions can then be made in accordance with mission objectives and commander intent. Effects are ultimately employed in accordance with executed decisions. Assessment follows to understand the impact of employed effects.

Elements involved in the loop may be heterogeneous, and may include assets from the ground, air, space, sea, and cyber domains. While it is common to consider kinetic effects as the desired actuation, the effects may include a broad range of non-kinetic alternatives as well. For example, in an intelligent battlefield, where an adversary's assets (e.g. unmanned combat vehicles, tactical operations centers, military units, etc.) need to communicate to accomplish their joint mission, creating interference in the communication medium may temporarily disable decision-making capabilities across these assets and offer a window of uncontested access to their vulnerabilities. In general, examples of effects include (i) stimulating the adversary to action (e.g. in order to reveal a capability, vulnerability, location, or other information), (ii) revealing the adversary, (iii) striking the adversary, or (iv) assessing outcomes of an interaction with the adversary.

The MDO effect loop offers the desired decomposition of IoBTs for purposes of breaking down challenges into manageable subproblems. Within the MDO effect loop framing, various IoBT research challenges can be mapped to certain stages of the decision cycle, and must come together to support all-domain C2. The IoBT common operating environment is where individual technologies exist, are composed to execute a process, and are employed with a military purpose – to deliver an effect. It is both the aggregated composite technologies and the process in which they are utilized that form the IoBT common operating environment.

1.4.2 Technical Challenges

If IoBT is a disruptive convergence of technologies around the purpose of delivering an effect, and if the MDO effect loop is its decomposition into a set of underlying stages, what are the technical challenges in building IoBTs that streamline the MDO effect loop? Lessons can be learned from the U.S. Department of Defense's (DoD's) Joint All Domain Command and Control (JADC2) systems that are already moving in this direction [14]. For example, United States Air Force (USAF) demonstrated its advanced battle management system (ABMS) and the potential of JADC2 in 2019, connecting aircraft from the Air Force and Navy, a Navy destroyer, an Army air defense sensor and firing unit, a special operations unit, as well as commercial space and ground sensors [15]. Accordingly, the aim of IoBT is to connect sensors from all of the military services into a single network that supports the notion of a dynamically-connected, context-driven, collectively intelligent, self-aware, and self-adaptive interconnected system-of-systems; a system that executes the stages of the MDO effect loop and delivers effects in accordance with mission goals. Several challenges arise:

1.4.2.1 Compositionality and Synthesis

MDO and Joint All Domain Operations are fundamentally challenges of C2 (in both the military and academic sense), where evidence can be seen that the bottleneck lies in interoperability, and the need for an intelligent and adaptive common operating environment. The enduring science and technology challenges of IoBT are thus grounded in the interactions in the underlying multi-domain technologies. IoBT technical issues are most significant in the field of complex systems, where conceptual and (systems) behavioral frameworks may not hold, and where technology gaps exist in our ability to synthesize cohesive interdependent exchanges.

Accordingly, traditional views of technology as isolated functional systems are no longer the paradigm. The *interaction* between devices is a new technological frontier: one where time is a weapon, the boundaries between actors and non-actors are naturally blurred, and intent will autonomously drive outcomes and effects. *Compositionality* in this context refers to our ability to reason about properties of complex systems from properties of their interacting constituent components. Attaining compositionality often entails reducing the complexity of component interactions in order for them to become analyzable for purposes of offering end-to-end assurances and correctness guarantees. *Synthesis* refers to exploiting compositionality properties to put together capabilities that meet mission desired specifications and constraints. The MDO effect loop organizes IoBT interactions into stages with well-defined functions and boundaries, thereby reducing the interactive system complexity and facilitating the systematic investigation of foundations, algorithms, and technological enablers for each of the individual stages, as well as their composite properties, such as end-to-end timing and validity assurances.

When reasoning about synthesis of mission-informed capabilities, challenges of resource federation and information exchange among federated units must also be considered. For example, most services would probably rather retain control of their systems for their purposes, but allow utilization of the information. Dynamic access control, asset recruitment, and top-down composition of desired functions become both a technical and organizational problem, imposing constraints on asset use that need to be considered when synthesizing multi-domain sensing, decision, and actuation solutions.

Embedded in the challenge of dynamic compositionality is the issue of interoperability. Interoperability has been a long standing problem in military systems, where technologies are "stove-piped" for a myriad of reasons. The challenges of interoperability occur at many different operational-military levels: the coalition level, the joint operations level, and even within a single force. Factors that create or contribute to these challenges include syntactic and semantic differences in the data, system update rates, latency requirements, messaging formats, security requirements (including encryption: data at rest and in motion), validation requirements, network protocols, application programming interfaces (APIs), and even certification processes [16]. There is ample literature that frames AI-enabled systems such as IoT, and thus IoBT, as complex systems-of-systems with emergent behaviors. Yet very little attention is given to the relevance of interoperability and technologies that provide the capability to facilitate system-to-system interactions [17]. There have been many prior defense programs centered around interoperability since the introduction of networking in the battlefield. While there are many research issues specifically related to interoperability, this chapter focuses on the next layer up; namely, closing intelligent application loops in novel distributed application contexts (on top of the underlying networking and interoperability "plumbing").

1.4.2.2 Timeliness and Efficiency

In future conflicts, winners will be those who can more quickly connect their sensors and shooters to understand intents, and deliver effects ahead of the adversary. Latency will be a huge technical problem; hypersonic missiles, for example, operate at an op tempo vastly different than tank mobility, all of which impact decisions, such as anti-missile actuation. An end-to-end requirement is therefore to reduce the sensing-to-effects time in the MDO effect loop. Deep technical challenges arise from the need to reason about end-do-end temporal properties of interconnected interacting subsystems.

1.4.2.3 Robustness to Adversarial Disruption

A critical issue embedded throughout all aspects of the MDO effect loop is cyber-physical resilience and robustness to adversarial disruption. The IoBT will face challenges associated with using things it does not have complete control over (e.g. adversary (red) and civilian (gray) assets), communicating in denied, disrupted, intermittent, and limited (DDIL) bandwidth environments, processing and inferring across deceptive data, and countering advanced persistent threats. The technical challenges associated with robustness and resiliency will have overarching implications on the quality and efficiency of IoBT performance.

1.4.2.4 Deployability at the Point of Need

A foundational problem to be addressed through advances in Network and Information Science is the fundamental understanding of how to learn and devise complex models of IoBT goals, networks, information, and dynamics to enable intelligent C2, and battlefield services. Importantly, such learning, adaptation, and inference must occur *at the point of need*; not in a data center with ample computing power and bandwidth, but on the field, where resource-limited components must contend with harsh environmental conditions, adversarial action, large distribution, fast tempo, and possibly strict stealth requirements.

A scientific opportunity for transformative innovation in this space is to re-map the boundaries of feasibility. For example, what design paradigms enable massively reduced (distilled) model sizes for intelligent computation (e.g. deep neural networks) and attain superior accuracy and uncertainty estimation on in- and out-of-distribution data? Scientific advances are needed for altering the fundamental trade-offs. For example, how far can we push the Pareto optimality frontiers of high robustness, accuracy, scale, and efficiency? These challenges call for exploring new dimensions. For example, can models be built with safety as a core principle that offer high-probability assurances of safe behavior?

1.5 Performant and Resilient IoBTs

In this section, we elaborate on research innovations that address the challenges outlined in the preceding section. Four such challenges were identified. These challenges must be addressed in an environment where the large operation scale and high operation tempo potentially exceed human capacity to keep up with the dynamics. Human operators can no longer manually analyze incoming sensor data, aggregate data across multiple sources to confirm gathered intelligence, cross-cue sensors, perform adaptation to adversarial action, and perform weapon-to-threat assignment within a reasonable time scale. IoBTs will help automate these tasks, in accordance with the MDO effect loop architecture, delegating the human to a more supervisory or decision-maker role. As mentioned earlier, several MDO loop challenges get elevated to key research needs in this context: (i) composition and synthesis, (ii) sensor-to-effect timeliness and efficiency, (iii) robustness to adversarial disruption, and (iv) intelligent edge processing at the point of need. Below, we address them in more detail.

1.5.1 Compositionality and Synthesis

Future missions may exploit IoBTs comprised of large numbers of nodes, with a wide range of capabilities, from unattended ground sensors to mobile platforms with Radar and LiDar sensors and from small embedded devices to powerful edge clouds with graphics processing units (GPUs). These missions will need to be exceedingly *agile*. Mission goals and needs may not be known until just before mission execution, and mission planners may not be able to (without the aid of automated tools) recruit and construct, at short timescales, IoBTs with sufficient resources to satisfy mission needs. The large scale of IoBTs will imply continuous churn, so discovery and composition solutions will need to be robust to failure or removal of assets as a normal operating regime.

IoBTs must allow for discovery, recruitment and composition of resources into composite assets with sufficient sensing, compute, and communication capacities to satisfy mission needs and constraints, while at the same time multiplexing individual assets to achieve high return on investment. Recruitment, composition and reconfiguration of such assets must meet several fundamental needs. First, it should be possible to assemble (or re-assemble, for example, upon damage) composite assets quickly on demand, despite high component heterogeneity, large scale, and presence of adversaries. Second, the aggregate properties of the composite, including timeliness, performance/functionality, security, and dependability, must be *formally assured* in an appropriately quantifiable and operationally relevant manner, subject to well-understood assumptions. Finally, component composition should exhibit optimality properties, calling for optimization algorithms that provide synthesis of desired capabilities from specifications of underlying mission requirements and constraints.

A typical network synthesis problem might consist of deciding where to place sensors (or activate sensing), choosing computing resources to which the sensing data should be sent, over a pre-defined communications infrastructure, while meeting complex constraints: quality of sensing, communications latency, communications bandwidth, computing capacity, routing, coverage, placement, visibility, and connectivity constraints. The problem has been shown to be NP hard. Typical sensor utility functions such as mutual information are approximately submodular, and near-optimal greedy algorithms have been developed [18] leveraging the submodularity. Constraints such as those due to visibility and placement can lead to non-convex problems. To cope with this, hierarchical synthesis techniques have been developed in [19], in which the authors compare the performance of two approaches: satisfiability modulo convex optimization (SMC) – a recent approach, and mixed-integer linear programming (MILP). The expressivity of SMC is shown to yield better quality solutions than MINLP at larger problem sizes.

Another approach to dealing with heterogeneity in IoBT draws upon ideas from network science, by modeling the IoBT as a multi-layer network. Then, drawing upon theories of stochastic geometry and mathematical epidemiology, an integrated framework for information dissemination has been developed [20].

As we discussed earlier, an approach to scalability is to a hierarchical one, in which small networks are synthesized and composed to create larger ones. Computational sheaf theory offers a mathematically sound approach to ensuring validity of the composed network [21, 22]. This may be used to compose local views (such as local knowledge of topology and network resources) to a global view (of the network as a whole). We note also the close connection with network slice creation and management problems.

Real-time IoT/IoBT systems must cope with dynamics at both slow and fast scales, thus requiring appropriate reconfiguration of the synthesized network; a framework for this is presented in [23]. A key consideration is scalability beyond computational efficiency (to limit the amount of computations that need to be offloaded to other devices), to include communications overhead (how often a device needs to communicate with other devices, and how much data it needs to send or receive). Some of these scalability issues are addressed in later sections.

1.5.2 Timeliness and Efficiency

Decision dominance is, in part, achieved by reducing the time elapsed between sensing and effect. To this point, on a contemporary battlefield time is a weapon. Aspects of timeliness and efficiency are key in detecting information changes, processing data at scale, and data transmission. If the IoBT is not able to process and share this crucial data at the same rate as an adversary then decision dominance is lost.

Sensing supplies information for decision making, and key changes in sensory measurements may thus require timely detection and response. A fundamental question in this context is the following: how quickly can one detect change from weak indicators in the complexity of modern battlefields, where a variety of factors contribute to measurement errors and detected features are often stochastic and inconclusive? Fundamental feasibility limits on early detection (i.e. earliest change detection results) have recently been developed for classes of dynamic anomalies [24], moving objectives [25], growing anomalies [26], and heterogeneous objectives [27], and extended to the challenging case of distributed detection problems [28, 29].

The scale of data within an IoBT may be massive in part because of the vast number of sensors within the network, but also because of the increased speeds at which sensors are now able to take measurements. To begin to address the need for processing data at scale, a new neural network framework has been designed to prioritize inference execution and reduce latency for mission-critical stimuli [30]. Multimodal information was used to define scheduling of perception-based inference at a subframe level, e.g. parts of images instead of the entire frame, where more critical regions of interest are processed first. Intelligent image resizing based on mission-criticality and resource batch processing has also been designed to improve detection speed [31].

IoBT is by definition a distributed operational environment, and latency induced from data transfer across the network can impact speed of inference, collaborative task execution, and federated learning. Addressing challenges of constrained network bandwidth, a distributed inference approach was developed to partition neural network models across edge and cloud resources by evaluating the dynamic network bandwidth and edge resource capabilities [32]. The combination of progressive model layer slicing and network partitioning enables the above approach to maintain low latency and energy use while ensuring high performance accuracy. For the case where the platform must consult a remote server to complete the full inference, a new brand of compression is introduced, where data over the communication link are compressed in a manner guided by the nature of the inference algorithm to be executed [33, 34]. By taking the algorithm into account, significant improvements are attained in the compression factor. Recent work also has addressed latency optimization in multimodal sensing, where speculative inference is carried out with only a subset of the inputs when some of the fused sensing modalities are inaccessible or delayed [35] resulting in improved end-to-end data fusion and inference latency with only minimum effects on quality. Related to network optimization efficacy, IoBT research has shifted resource allocation protocols to consider the purpose of communication, which opens up additional optimization opportunities that significantly improve the attainable trade-off between resources used and results achieved [36]. Efforts to evaluate the optimal communication patterns for distributed learning [37] or the fundamental storage needs of learning algorithms bound the degree of attainable compression [38, 39]. Federated learning algorithms have employed task-dependent compression schemes to dramatically reduce their communication needs [40]. The ideas behind exploiting purpose from communication have so far resulted in 3–4 orders of magnitude improvement in communication efficiency in distributed learning from complex data samples [41].

1.5.3 Robustness to Adversarial Disruption

Operation in the modern battlefield will necessitate that the IoBT can display characteristics of robustness and resiliency to an active, near-peer adversary. With the advent of machine learning algorithms, the concepts of decoys and camouflage have been extended to exploit properties of machine learning in a manner that produces misclassifications and misprediction [42]. For example, in the visual sensing domain, adversarial inputs have been designed to generate minimal changes in appearance to cause misclassifications in inference algorithms, see e.g. [42]. A key challenge in IoBT systems is therefore to detect such adversarial input manipulation. A recently proposed metric, called attribution-based confidence [43], for assessing the likelihood of adversarial manipulation, has shown significant improvements in adversarial input detection [44], leading to better classification. Several game- and hypergame-theoretic approaches for defense against cyber deception in IoBT are described in [45].

Spoofing attacks also generate challenges for critical sub-tasks in the decision loop, e.g. identification, localization, and tracking. For example, a spoofed sensor may advertise an incorrect target location measurement, leading to a failure in localization. To address this, distributed sensors in the IoBT must jointly decide if a subset of them were compromised. This was shown to be an NP-hard problem [46]. Polynomial solutions were developed for the case, where a "sufficient" number of sensors are present (by a metric of sufficiency defined in the original paper) [46]. Spoofing discovery can inform sensor deployment strategies, where multiplicity is used as a means to combat complexity.

AI-based decision support can be used to analyze the current operational state and recommend actions for commander's decision. A resilient formalism is needed to inform courses of action recommendations in the presence of an adversary. A significant contribution lies in developing such a formalism, called non-cooperative inverse reinforcement learning [47]. In traditional reinforcement learning [48], agents learn to maximize an objective function by executing well-designed actions to "prod" their environment and observing collected rewards from these actions. Conversely, inverse reinforcement learning [49] refers to the ability of an agent to infer an objective function by observing behavior, e.g. apprenticeship learning [50]. Decision-making in IoBT shares with inverse reinforcement learning the need to understand the objectives of another (namely, the adversary) to optimize one's own actions. However, the classical inverse reinforcement learning formalism is not designed for situations where an adversary might purposely obfuscate their intent by executing actions that lead an observer to an incorrect conclusion. Thus, non-cooperative inverse reinforcement learning [47] offers foundations for developing optimal policies in interacting with strategic adversaries who obfuscate their objectives and solutions that increase the ability to tolerate such obfuscation.

1.5.4 Deployability at the Point of Need

The explosive growth in IoT, combined with the continuous advancements in GPU and high performance computing, has produced a large spectrum of heterogeneous compute capabilities. In the battlefield domain, when access to resources may be unattainable due to security concerns or contested communications, it becomes critical to exploit whatever resources are available. In many cases this may mean relying on resource-constrained edge devices. This edge processing allows the IoBT to offer intelligence where it is needed, when it is needed, without the necessity to rely on expensive resources that may not be readily available (or may be hard to contact) in a given dynamic situation. The IoBT must ameliorate the inherent trade-off between resource cost of modern learning algorithms (e.g. deep neural networks) on one hand and their quality of results on another. Attaining both resource economy and quality at the same time is an important feature for IoBT systems, where data transfer latency might be prohibitively high, yet individual edge sensors and processors, have size, weight, and power (SWaP) requirements that make it challenging to run complex tasks locally. Recent work attains a significant reduction in inference model size [51, 52], as well as model execution latency [53], allowing it to execute on a resource-constrained platform with nearly no sacrifice in quality.

As more processing is moved to the edge, the need for uncertainty awareness in the autonomy becomes critical for decision support. That is, the trade-off between latency, energy, and performance continues to exist at the edge and must be encoded in the information produced by automated stages of the decision cycle. Although uncertainty quantification research has made general advances in the field, the ability to provide this uncertainty estimation at the edge still remains a challenge. Recent work has developed sparsity and distillation-based methods for compressing expensive Monte Carlo posterior uncertainty computations into resource constrained neural network models [54].

1.6 Future Directions

Future IoBT challenges arise from scale, heterogeneity, fast op tempo, and operation in an adversarial environment. A common thread is therefore one of taming interactions between the multiplicity of components, algorithms, objectives, uses, or time-scales. Combining considerations of latency and efficiency at the system level, it is important to explore time and resource management challenges that arise when large numbers of heterogeneous intelligent components, algorithms, or objectives interact. New paradigms are needed for managing system-level data acquisition, communication, and processing in the battlefield that break existing performance barriers and trade-offs in order to significantly improve processing speed at the system level. Moreover, considerations of resilience suggest the importance of fast run-time adaptation to dynamic conditions and adversaries. After all, "No plan survives first contact with the enemy." This quote has been attributed (in different variants) to many leaders including Helmuth von Moltke, Carl von Clausewitz, Dwight Eisenhower, and Douglas MacArthur, among others. It reflects the importance of adaptation in adversarial settings. Below, we elaborate on some specific challenges and needs.

1.6.1 Multi-tenancy and Multiplicity of Use

In a high op-tempo distributed battlefield, multiple sensing and computing tasks pose conflicting demands on finite tactical edge resources. The importance of each task varies with context, and the importance of one piece of computation often depends on whether downstream stages can be executed as well. To reduce end-to-end sensing-to-effect latency, a key challenge is to avoid spending scarce local resources processing inputs that do not contribute to global value. Yet information about global value is often not available to local components. Solutions are needed to the above problem that combine advances in information theory, real-time computing, anomaly detection, and hypothesis testing. From an information-theory standpoint, observations that deviate from expectations carry more information, since predicted/expected observations are, in a sense, redundant with past knowledge. Thus, anomaly detection techniques can be used to understand, in a bottom-up fashion, what part of the data landscape deserves further attention. Such techniques will significantly speed up input data processing at both the individual agent level and at system levels. Since individual agents do not have a global view of the battlefield, information exchange algorithms must be designed to share a concise view of battlefield state. Information should be collected to disambiguate lower confidence hypotheses in elements of world state of interest to the application, offering a top-down criterion for prioritizing information pull. Real-time analysis of the system can provide a better understanding of overall capacity constraints, allowing a reconciliation of the bottom-up anomaly-based information push with the top-down confidence-based information pull, guaranteeing a notion of optimality of information throughput, that adapts within the resources available. Protocols for such optimization that minimize end-to-end latency remain to be developed.

1.6.2 Multiplicity of Function

Another approach for accelerating end-to-end timing and improving edge efficiency is to explore opportunities for multi-purpose component use that simultaneously achieve seemingly conflicting system objectives. Several examples of this approach have been reported in recent work. An example is to re-use certain sensing modalities for multiple application purposes simultaneously, such as the use of vibration sensors to detect both motion and sound, or the use of common radios to implement radar-like radio frequency (RF) sensing, as well as communication. At present, solutions that overlay (as opposed to serialize) different uses of the same device are largely one-off and ad hoc. A challenge is to develop a principled architecture that allows simultaneous exploitation of shared resources for different purposes and resolves the conflicting requirements that the different uses might impose on the underlying resource. An advantage, besides improved efficiency, is enhanced resilience. When devices can be repurposed for a different use, substitution, and replacement opportunities are increased, making the system as a whole more inherently adaptable to local failures and disruptions as it increases the ability of other (different) components to substitute for the failed one.

1.6.3 Non-stationarity and Multiplicity of Perturbations

It is important, in an IoBT context, to develop large-scale decentralized optimization algorithms (for real-time adversarial settings) to support adaptation, inference, and decision-making in scenarios that are highly dynamic to the point where the system exists in a never-ending transient state. In such scenarios, the inputs, constraints, and objectives to be optimized are constantly changing and new algorithms are required to support optimization. Standard optimization and control algorithms often assume that system properties and optimization objectives evolve slower than current system state. This time-scale separation makes it possible to optimize state for specific objectives and with specific assumptions on underlying properties. When these objectives and assumptions change rapidly, the system may be in a constant "catch-up" mode. As it converges to one optimal solution, new changes in the system, goals, or environment render the solution inadequate. The research question becomes, how can future changes be anticipated such that optimization is not lagging behind but anticipates future needs and optimizes the system for such conditions, instead of the currently observed ones? This problem becomes more challenging as the system scale and heterogeneity grow, as well as when information sharing is impeded by time and resource constraints. Additionally the speed and memory/storage footprint of the optimization algorithm and its ability to trade off optimality for speed are key concerns.

1.6.4 Multiplicity of Sensing Modalities

How can we seamlessly combine sensing from multiple modalities? Novel techniques in machine learning developed self-supervised variations of an encoder–decoder structure (called auto-encoders) that encode complex signals into a lower-dimensional latent representation, then decode this latent representation again into the original signal. The idea is that if the latent representation captures the essence of the signal, decoding into the original signal is possible. Auto-encoders give rise to an interesting question: can one arrive at a latent representation that is sensing-modality-agnostic? For example, the vibration signature of an object might suggest that it is a Polaris All-Terrain Vehicle. The Polaris brings to mind a specific visual representation thus allowing one to convert from vibration to vision. Can one construct a unified latent representation for a growing multiplicity of concepts with architectures that convert the same latent representation to/from each of the multiple sensing modalities? The approach will allow for more reliable and robust sensing, where any subset of available modalities can be used to produce the same latent representation and thus could be exploited to detect, identify, and track complex objects. The modalities can further corroborate each other, or conversely allow separation of true targets from decoys. The capabilities and limitations of this approach need to be understood using sensors that include vibration, vision, IR, LiDAR, sound, and RF sensing modalities, among others.

1.6.5 Multiplicity of Time-scales

A key candidate for exploration of the multiplicity of time-scales is neuro-symbolic computing. Current neural networks offer superior detection and identification capabilities by recognizing increasingly subtle patterns in input signals. Unfortunately, the approach has severe scalability limitations. As the length of the pattern increases, so does the size and complexity of the underlying neural network. Larger networks require more training data, ultimately rendering this solution ineffective. How can one capture patterns that span a long duration and multiple time-scales? What representation learning techniques can be used to learn and express such patterns? Intuitively, a hierarchical approach is needed, where elementary activities or events are detected at fast time-scales, and can be put together to form patterns spanning much longer intervals. For example, a sensor (with a neural network) might learn to detect a given type of target in the environment. However, to detect gradual accumulation of forces along a border, one will need to remember and aggregate such detections over periods of time that are several orders of magnitude longer than the time it takes to detect a single target. Can one exploit neuro-symbolic computing to jointly learn symbolic larger-scale event patterns, while at the same time learning the representations of individual events?

1.6.6 Architecture

One must also address architectural challenges that improve the efficiency of the intelligent edge. For example, can learning algorithms improve the design of the I/O subsystem that consists of sensors and actuators that are mobile, unreliable, and operate in contested environments to be connected to the computing elements such as edge devices? What hierarchy of caches and storage is needed to significantly reduce the need for execution of complex algorithms (such as neural network inference) and replace it when possible with simple surrogates? This hierarchy will decide which data to cache and store where, how to prioritize it, and in what form to best index it, query it, and visualize it, as well as when to re-compute it to meet freshness, resource capacity, and decision latency bounds. A computational subsystem must be developed that consists of the distributed heterogeneous and mobile processing units executing operational services and applications, such as distributed AI algorithms. On top of this substrate, we could build application platforms that allow exploitation of various IoBT research products.

1.7 Conclusion

We summarize the key takeaways from the IoBT vision, challenges, and research opportunities.

- The IoBT is the ultra-connected convergence of several key technological objects (sensors/actuator, networking/processing, machine intelligence, and workflow enablers/software). Of critical importance to IoBT is the interactions between the technology objects and their utilization in (decision) processes, such as the MDO effect loop.
- From an S&T perspective, the IoBT can be thought of as an intelligent and adaptive common operating environment that manages the MDO effect loop to enhance its performance and resilience in battlefield environments.
- The IoBT is not a theoretical concept or thought exercise. The US Army, the DoD, and coalition partners are already thinking along these lines.
- Military uses of IoBTs are different from civilian and industrial applications, but industry should/will be highly leveraged in terms of underlying technology objects.
- From an architecture perspective, it is helpful to view the IoBT as a multi and all domain operations enabler/multiplier (and threat) that breaks down into well-defined stages aligned with the architecture of decision loops.
- Several key research challenges were articulated for IoBTs, including composition/synthesis, reducing end-to-end decision loop latency, resilience in the presence of battlefield threats, and support for deployable edge intelligence.
- Modernization of the IoBT should be a focal area that emphasizes scientific cross-technology experimentation, underlying technology interactions, and decision process multipliers.

References

1 Kott, A., Swami, A., and West, B.J. (2016). The internet of battle things. *IEEE Computer* 49 (12): 70–75.

2 Seffers, G.I. (2015). Defense department awakens to Internet of Things. *Signal Magazine*. https://www.afcea.org/signal-media/defense-department-awakens-internet-things

3 Army Futures Command (2020). *AFC Pamphlet 71-20-9 - Army Futures Command Concept for Command and Control 2028: Pursuing Decision Dominance*. AFC.

4 Wigness, M., Pham, T., Russell, S., and Abdelzaher, T. (2021). Efficient and resilient edge intelligence for the internet of battlefield things. *NATO IST-190 Research Symposium on Artificial Intelligence, Machine Learning and Big Data for Hybrid Military Operations (AI4HMO)*.

5 Abdelzaher, T., Taliaferro, A., Sullivan, P., and Russell, S. (2020). The multi-domain operations effect loop: from future concepts to research challenges. *Artificial Intelligence and Machine Learning for Multi-Domain Operations Applications II*, Volume 11413, International Society for Optics and Photonics (SPIE).

6 US Army Training and Doctrine Command (2014). The US army operating concept: Win in a Complex World. *US Army Training and Doctrine Command*, Pamphlet 525-3-1.

7 Isaacson, J. (2015). Decisive Army Strategic and Expeditionary Maneuver. *Technical report*. Arlington, United States: Army Science Board.

8 DEVCOM ARL (2017). Program Announcement for the Internet of Battlefield Things (IoBT) Collaborative Research Alliance (CRA), W911NF-17-S-0005. https://www.grants.gov/web/grants/view-opportunity.html?oppId=292214 (accessed 22 January 2022).

9 Russell, S. and Abdelzaher, T. (2018). The internet of battlefield things: the next generation of command, control, communications and intelligence (C3I) decision-making. In *IEEE Military Communications Conference (MILCOM)*, pp. 737–742.

10 Russell, S., Abdelzaher, T., and Suri, N. (2019). Multi-domain effects and the internet of battlefield things. In *IEEE Military Communications Conference (MILCOM)*, pp. 724–730.

11 Cerf, V.G. and Cain, E. (1983). The DoD Internet Architecture model. *Computer Networks (1976)* 7 (5): 307–318.

12 Waltz, E. and Llinas, J. (1990). *Multisensor Data Fusion*. Boston, MA: Artech House.

13 Simon, H.A. (1960). *The New Science of Management Decision*. Harper & Brothers.

14 Congressional Research Service (2021). Joint All-Domain Command and Control: Background and Issues for Congress, March 18, 2021. https://crsreports.congress.gov (accessed 14 October 2022).

15 Bousie, C.C. and Pope, C. (2019). Air Force, Navy, Army conduct first 'Real World' test of Advanced Battle Management System, 23 December 2019. https://www.af.mil/News/Article-Display/Article/2046531/air-force-navy-army-conduct-first-real-world-test-of-advanced-battle-management (accessed 31 January 2022).

16 Russell, S.M., Suri, N., Lenzi, R., and Fouad, H. (2016). Measuring and evaluating interoperability for complex C2 information management system-of-systems. In *21st International Command and Control Research and Technology Symposium*, pp. 1–16.

17 Russell, S., Jalaian, B., and Moskowitz, I.S. (2021). Re-orienting toward the science of the artificial: engineering AI systems. In William F. Lawless, Ranjeev Mittu, Donald A. Sofge, Thomas Shortell, and Thomas A. McDermott (Eds.) *Systems Engineering and Artificial Intelligence*, 149–174. Springer.

18 Bunton, J., Anevlavis, T., Verma, G. et al. (2021). Split to win: near-optimal sensor network synthesis via path-greedy subproblems. In *IEEE Military Communications Conference (MILCOM)*, pp. 789–794.

19 Ghosh, P., Bunton, J., Pylorof, D. et al. (2021). Synthesis of large-scale instant IoT networks. *IEEE Transactions on Mobile Computing*. doi: 10.1109/TMC.2021.3099005.

20 Farooq, M.J. and Zhu, Q. (2018). On the secure and reconfigurable multi-layer network design for critical information dissemination in the internet of battlefield things (IoBT). *IEEE Transactions on Wireless Communications* 17 (4): 2618–2632.

21 Robinson, M. (2013). Understanding networks and their behaviors using sheaf theory. *arXiv:1308.4621*.

22 Joslyn, C.A., Charles, L., DePerno, C. et al. (2020). A sheaf theoretical approach to uncertainty quantification of heterogeneous geolocation information. *Sensors (Basel)* 20: 3418.

23 Chen, T., Barbarossa, S., Wang, X. et al. (2019). Learning and management for Internet of Things: accounting for adaptivity and scalability. *Proceedings of the IEEE* 107 (4): 778–796.

24 Rovatsos, G., Moustakides, G.V., and Veeravalli, V.V. (2019). Quickest detection of a dynamic anomaly in a sensor network. In *53rd Asilomar Conference on Signals, Systems, and Computers*, pp. 98–102. IEEE.

25 Rovatsos, G., Zou, S., and Veeravalli, V.V. (2019). Quickest detection of a moving target in a sensor network. In *IEEE International Symposium on Information Theory (ISIT)*, pp. 2399–2403.

26 Rovatsos, G., Veeravalli, V.V., Towsley, D., and Swami, A. (2020). Quickest detection of growing dynamic anomalies in networks. In *IEEE International Conference on Acoustics, Speech and Signal Processing (ICASSP)*, pp. 8926–8930.

27 Rovatsos, G., Veeravalli, V.V., Towsley, D., and Swami, A. (2021). Quickest detection of anomalies of varying location and size in sensor networks. *IEEE Transactions on Aerospace and Electronic Systems* 57 (4): 2109–2120.

28 Li, J., Towsley, D., Zou, S. et al. (2019). A consensus-based approach for distributed quickest detection of significant events in networks. In *53rd Asilomar Conference on Signals, Systems, and Computers*, pages 1–4. IEEE.

29 Zou, S., Veeravalli, V.V., Li, J. et al. (2019). Distributed quickest detection of significant events in networks. In *IEEE International Conference on Acoustics, Speech and Signal Processing (ICASSP)*, pp. 8454–8458.

30 Liu, S., Yao, S., Fu, X. et al. (2020). On removing algorithmic priority inversion from mission-critical machine inference pipelines. In *IEEE Real-Time Systems Symposium (RTSS)*, pp. 319–332.

31 Hu, Y., Liu, S., Abdelzaher, T. et al. (2021). On exploring image resizing for optimizing criticality-based machine perception. In *27th IEEE International Conference on Embedded and Real-Time Computing Systems and Applications (RTCSA)*, pp. 169–178.

32 Huang, J., Samplawski, C., Ganesan, D. et al. (2020). CLIO: Enabling automatic compilation of deep learning pipelines across IoT and Cloud. In *Proceedings of the 26th Annual International Conference on Mobile Computing and Networking*, pp. 1–12.

33 Deshmukh, A., Liu, J., Veeravalli, V.V., and Verma, G. (2020). Information flow optimization in inference networks. In *IEEE International Conference on Acoustics, Speech and Signal Processing (ICASSP)*, pp. 8289–8293.

34 Yao, S., Li, J., Liu, D. et al. (2021). Deep compressive offloading: speeding up edge offloading for AI services. *ACM GetMobile: Mobile Computing and Communications* 25 (1): 39–42.

35 Li, T., Huang, J., Risinger, E., and Ganesan, D. (2021). Low-latency speculative inference on distributed multi-modal data streams. In *Proceedings of the 19th Annual International Conference on Mobile Systems, Applications, and Services*, pp. 67–80.

36 Lee, J., Marcus, K., Abdelzaher, T. et al. (2018). Athena: Towards decision-centric anticipatory sensor information delivery. *Journal of Sensor and Actuator Networks* 7 (1): 5.

37 Neglia, G., Calbi, G., Towsley, D., and Vardoyan, G. (2019). The role of network topology for distributed machine learning. In *IEEE Conference on Computer Communications (INFOCOM)*, pp. 2350–2358.

38 Bu, Y., Gao, W., Zou, S., and Veeravalli, V.V. (2020). Information-theoretic understanding of population risk improvement with model compression. In *Proceedings of the AAAI Conference on Artificial Intelligence*, volume 34, pp. 3300–3307.

39 Bu, Y., Zou, S., and Veeravalli, V.V. (2020). Tightening mutual information-based bounds on generalization error. *IEEE Journal on Selected Areas in Information Theory* 1: 121–130.

40 Basu, D., Data, D., Karakus, C., and Diggavi, S. (2019). Qsparse-local-SGD: Distributed SGD with quantization, sparsification, and local computations. In *NeurIPS*.

41 Singh, N., Data, D., George, J., and Diggavi, S. (2020). SPARQ-SGD: Event-triggered and compressed communication in decentralized optimization. In *59th IEEE Conference on Decision and Control (CDC)*, pp. 3449–3456.

42 Papernot, N., McDaniel, P., Goodfellow, I. et al. (2017). Practical black-box attacks against machine learning. In *Proceedings of the 2017 ACM on Asia Conference on Computer and Communications Security (AsiaCCS)*, pp. 506–519.

43 Jha, S., Raj, S., Fernandes, S.L. et al. (2019). Attribution-based confidence metric for deep neural networks. In *NeurIPS*.

44 Jha, S., Raj, S., Fernandes, S.L. et al. (2019). Attribution-driven causal analysis for detection of adversarial examples. *arXiv preprint arXiv:1903.05821*.

45 Kamhoua, C.A., Njilla, L., Kott, A., and Sachin, S. (eds) (2020). *Modeling and Design of Secure Internet of Things*. Wiley.

46 Mao, Y., Mitra, A., Sundaram, S., and Tabuada, P. (2019). When is the secure state-reconstruction problem hard? In *58th IEEE Conference on Decision and Control (CDC)*, pp. 5368–5373.

47 Zhang, X., Zhang, K., Miehling, E., and Basar, T. (2019). Non-cooperative inverse reinforcement learning. In *NeurIPS*.

48 Kaelbling, L.P., Littman, M.L., and Moore, A.W. (1996). Reinforcement learning: a survey. *Journal of Artificial Intelligence Research* 4: 237–285.

49 Ng, A.Y. and Russell, S.J. (2000). Algorithms for inverse reinforcement learning. In *International Conference on Machine Learning*.

50 Abbeel, P. and Ng, A.Y. (2004). Apprenticeship learning via inverse reinforcement learning. In *International Conference on Machine Learning*.

51 Yao, S., Zhao, Y., Zhang, A. et al. (2017). DeepIoT: Compressing deep neural network structures for sensing systems with a compressor-critic framework. In *Proceedings of the 15th ACM Conference on Embedded Network Sensor Systems*, pp. 1–14.

52 Yao, S., Zhao, Y., Zhang, A. et al. (2018). Deep learning for the Internet of Things. *IEEE Computer* 51 (5): 32–41.

53 Yao, S., Zhao, Y., Shao, H. et al. (2018). FastDeepIot: Towards understanding and optimizing neural network execution time on mobile and embedded devices. In *Proceedings of the 16th ACM Conference on Embedded Networked Sensor Systems*, pp. 278–291.

54 Vadera, M., Jalaian, B., and Marlin, B. (2020). Generalized Bayesian posterior expectation distillation for deep neural networks. In *Conference on Uncertainty in Artificial Intelligence*, pp. 719–728. PMLR.

2

Sensorized Warfighter Weapon Platforms: IoT Making the Fog of War Obsolete

Kyle Broadway

Chief Technology Officer, Armaments Research Company, University of Missouri, Columbia, MO, USA
Johns Hopkins University, Baltimore, USA

Abstract

There is no platform or tool closer to ground-truth engagement data than the warfighter's weapon. Weapons-based IoT sensors provide a path to harness individual engagement data into actionable intelligence insights across echelons. These IoT platforms work to provide information that will close the data gap between the individual warfighter and their command – relegating the fog of war concept to obsolescence. Weapons-based IoT sensors are uniquely positioned to fill this knowledge gap because their cost effectiveness enables deployment to every warfighter on the battlefield, turning each weapon into a new information node.

Every warfighter information node can provide engagement data that includes weapon metrics such as discharge detection alerts, shot counting and ammunition remaining, weapon rates of fire, weapon orientation and directionality, and overall weapon readiness. This information represents a critical window into multiple use cases that include weapons predictive maintenance, real time situational awareness across echelons, ammunition and weapon supply chain optimization, warfighter and squad performance metrics, and ground truth historical reporting.

Weapon based IoT sensors also carry unique challenges. Many deployed warfighters are burdened with gear and batteries in excess of 100 pounds. While nearly all IoT markets must rectify size, weight, and power (SWaP) constraints, weapons-based IoT sensors must resolve these issues at the extremes while remaining seamlessly integrated into a weapons platform.

Sensorizing warfighter weapons provides the highest fidelity engagement data possible, but also subjects the sensors to the most diverse network conditions of any IoT sensor market segment. Network requirements range from undetectable (no RF profile) to integrating with custom, sometimes multi-national, battlefield networks to provide individual warfighter insights where necessary.

Regardless of network presence, weapon usage data is generated any time a weapon is picked up. Data loss at any time during weapon handling could mean the absence of critical intelligence and ground truth of events. The data path of weapons-based IoT sensors must be designed for resiliency, reliability, and durability in network-agnostic implementations that avoid data loss at all costs without compromising security. In doing so, the data generated from weapons-based IoT sensors ensures that every action taken with the weapon by the warfighter can be aggregated into tactical command decisions that improve individual and small unit performance, optimize supply chains, and, ultimately, save lives.

IoT for Defense and National Security, First Edition. Edited by Robert Douglass, Keith Gremban, Ananthram Swami, and Stephan Gerali.

2.1 Introduction

Ground unit situational awareness between units and command is largely still in the same format as it was in the 1940s – over radio communications. Wireless formats, encryption standards, and transmission techniques have evolved, but the underlying data communicated over these formats has not. Communications have always been an active requirement on the soldier that requires manual intervention to perform. That leaves an immense amount of untapped data on the battlefield, where the fog of war obscures ground truth to command and deployed squads. Every second, critical data is generated through troop movements, ammunition status, engagement line of sight, and threat locations just to name a few. Battlefield internet of things (IoT) sensors provide a unique solution to this problem, enabling every single soldier to be a data node for command insights. More than that, weapons based IoT technology puts the data node in the closest possible location to the data source it is conveying- the engagement itself. Pushing data directly from the source closes the data gap between command and engagements, allowing leaders to utilize absolutely unparalleled insights to tip conflicts in their favor. This paradigm shift represents a major shift in communications responsibility. Rather than a squad leader manually radioing in engagement data, and support requests, the entire end to end engagement support effort can be fully automated. Information between squads becomes free flowing, with details down to knowing the exact location and engagement trajectory of friendly units. Having this information readily available to every single deployed unit translates directly into improved enemy kill chains and reduced fratricide rates within the Department of Defense (DoD) and industries where small arms are utilized. Just by passively communicating the line of fire from a squad, others in the area are immediately alerted to the directionality of a threat, as well as their fellow soldiers line of fire with zero added cognitive burden. No other data source on the battlefield today is positioned to provide this level of data and insight. However, IoT sensors on the battlefield are subject to some of the most restrictive and difficult to manage requirements in the IoT industry. There is no public WiFi connection. Security considerations are far more grave than hackers controlling your garage door. Battlefield IoT sensors are subject to the most extreme environments and conditions on the planet, and are subject to all manner of drops and vibrations. Providing accurate, reliable, and secure data in the midst of these factors is a challenge to say the least. The payoff of doing so though is an information advantage over your enemy. Timely information has been the key deciding factor of many conflicts and whether troops come home or not.

During his deployment to Iraq in 2005, Mike Canty stood staring into a dense palm groves. The enemy was ahead in the vegetation, actively engaged with friendly units. The trees and brush obscured any visibility of the engagement, making it impossible to distinguish where the threat was located. Behind him were nine Iraqi soldiers he was responsible for leading, waiting for his direction. Unable to discern any further details from his position and unable to maneuver due to the risk of cross-fire, Mike was left waiting at the edge for additional information via radio communications. Because of an outright lack of any viable data sources to inform him of any situational awareness, his troops were unable to discern the enemy's location and confirm their engagement trajectory would not put other US soldiers lives at risk. The literal fog of war in the form of thick vegetation restricted his team's ability to make any call other than to stand by and wait for further information. About 20 minutes passed on the battlefield while awaiting more information. By the time radio communications finally came in providing the much needed maneuvering information Mike required, the opportunity for engagement was lost. The enemy had slipped away, and the opportunity for success in this moment of the battle was lost [1].

The battlefields that exist in today's world require more information than before to support a wide variety of engagements in order to prevent these opportunities from slipping away. Small scale engagements, guerilla warfare with ambiguous enemies, or large scale firefights with near peers will all require advances in technology to push the boundaries of what information can be shared. By adding IoT sensors into the equation, more data can flow from the battlefield to command. With IoT, information normally provided through verbal radio communications is now provided automatically across battlefield networks in an instant with the mission critical data needed to leverage wartime assets effectively. Embedded sensors on every soldier relay information about temperature, heart rate, fatigue levels, weapon usage, ammunition remaining – all of the critical data a leader needs to consider when making battle critical decisions. Every decision benefits from the information provided by IoT sensors: who to resupply with limited ammunition reserves, what artillery to call in, airstrike munitions – all need critical information that's only present on the ground with the soldiers generating the data through engagement. Even squad level decisions require information from each individual soldier for maximum effectiveness. Knowing your squad's positioning, line of fire, and the total ammunition across your squad immediately gives you an edge. IoT solves problems by co-locating sensors and controls with the problem space or mechanism to be monitored. Soldier weapon borne IoT sensors are the closest sensors possible to the most important tools to monitor on the battlefield. These simple solutions when applied at scale to every soldier can be aggregated and reported up across echelons for multi-variate insights and battlefield forecasting of conditions such as at-a-glance ammunition remaining for a squad. Suddenly, problems that used to be solved through verbal communication and hand-checking your squad's ammunition are provided without even having to think. Full focus can be given to the real and imminent problems downrange. At the same time, squad leaders are provided ever-present and always accurate information without needing to ask – and wait – for it. Squad leaders, and their commanders in turn, always have actionable data as it changes during an engagement. When a need or a new threat arises, the information ahead of that threat paves the way to defeating it quickly and soundly. All of this information is generated by a soldier's actions with their weapon, with no thought or specific action required. In many cases, it is the absolute most critical information possibly generated on the battlefield. It is both ground truth and historical reference of exactly what is happening or has happened during an engagement, because it is the centerpiece of engagement for the soldier wielding it. There is simply no other place closer to the engagement than a soldier's rifle – which makes it an ideal candidate for hosting IoT sensors (Figure 2.1).

Figure 2.1 ARC concept of the connected battlefield. Source: Cuckoo/Getty Images.

2.2 IoT for Firearms

Modern firearms have been used in combat since the late thirteenth century, with notable use of gunpowder for combat engagements as early as the tenth century [2]. Since then, advances in firearms development have made weapons capable of blasting through hundreds or thousands of rounds per minute. Advances in state of the art firearms since then have nearly all been centered around the accuracy, performance, reliability, and lethality of an individual soldier wielding the weapon with no regard to a connected battlespace. These advances have iterated and improved on the same fundamental mechanics, operation, and usage monitoring systems that have existing since firearms were first used. For a soldier to know if their weapon is loaded or know how many rounds they have left, they still have to count the rounds or look themselves. This is especially shocking when considering the potential linkage to larger battlefield data ecosystems. The lack of advancements in this area is odd for a number of reasons. Of all tools used on the battlefield, a soldier's rifle is one of the most critical – the first and last line of defense and offense of a soldier. It's use is indicative of what the soldier is doing and thinking. Directionality of the weapon indicates a danger zone, the line of fire, and how it's being aimed. It provides the source of truth for how many and how fast rounds are being fired, where they're being fired, the direction they're fired in, and by who. From there, you can deduce how much ammunition each soldier has. Those details cumulatively aggregated can relay critical squad based intelligence that relays engagement trajectories, enemy locations, ammunition supplies, and more. Weapon directionality of multiple soldiers can provide triangulation of their line of fire, making it possible to locate convergence points and deduce a threat's location with zero emissions from lasers or other energy. No sensors or firearms currently exist that include the functionality to recognize, produce, or ingest any of this information, meaning all of it disappears into obscurity as quickly as it is generated.

This is where IoT provides its critical value on the battlefield. IoT is a uniquely suited technology to embed into each and every fielded weapon. By doing so, every piece of information is captured for real time situational awareness and historical profiling of weapon usage. By doing so, suddenly a new trove of information becomes available for ingest into a vast number of larger systems. Real time data displays, weapons maintenance, and logistics dashboards can now include what to this point has been hand-drawn and estimated for post mission analysis or logging, the real data from the weapon. For instance, simply knowing the number of rounds that have been fired on a particular weapon can inform a new field of maintenance. Weapons maintenance is no longer based on a simple monthly or weekly cleaning schedule. Weapons maintenance can be automated to be based on the total number of rounds each component and accessory on the weapon has sustained. Knowing the amount of time between each round adds the ability to provide a rates of fire and timestamps on every round fired. Armed with this information, more accurate maintenance schedules can be created based on data driven insights from IoT. Because this information has never existed until now, it is changing the landscape of how manufacturers and practitioners think of weapons maintenance. Even with this data, it's clear that only knowing the number of rounds and rate of fire is insufficient to completely understand the impact these metrics have on weapons readiness. However, this information can be aggregated with additional IoT data for the creation of a new concept of weapons stress and readiness, further automating the job of an armorer for weapons cleaning, maintenance, and upkeep. By understanding both number of rounds and rate of fire of every one of those rounds, the total amount of stress placed on the weapon can be deduced into a measurement of how impactful every round fired is to the weapon's overall health and readiness. This allows the armorer a more fundamental understanding of the physics and mechanics that have occurred during a particular weapon's use and lifecycle. Barrel changes and

Figure 2.2 Gas overcharging of weapons can lead to degradation of weapon performance faster. It can be visually identified by notable differences in ejection patterns of shells, but can be difficult to observe outside of static scenarios. Gas variability is detectable with weapons-based IoT.

spring changes are now informed directly by the data around the events that should be dictating these actions, rather than manual logging, schedules, and guesswork. At an individual weapon or user level, this information can be greatly helpful in keeping a soldier's rifle ever-ready for combat. At a company level, this information saves billions of dollars and countless hours maintaining weapons that have never even fired a single shot. When data from the individual soldier level is aggregated against data from the rest of the squad, platoon, or company – the data becomes invaluable in real time as well as historical profiles for many use cases. A company's worth of weapons and their maintenance can be fully automated with IoT that tracks the weapons usage and stress. Knowing when to maintain a weapon becomes a simple notification via email, short messaging service (SMS), or peripheral notification. Weapons maintenance becomes as simple as checking your email for your maintenance schedule, which includes the exact parts that need cleaned or replaced, how to replace them, and instructions on how and where to locate the parts or find more. Parts inventory and forecasting can be automated as well – bringing the armory up to the same level as e-commerce giants in the tech industry today (Figure 2.2).

2.3 New Insights into the Battlefield Provided by IoT

IoT gives the firearms industry and DoD new ways to characterize the stress felt on their weapons. By capturing the data around how the weapon experiences various stressors over time, the manufacturer can provide precise maintenance information customized to every weapon in an armory at exactly the right time, preventing failure in the field and unnecessary maintenance costs. Maintenance procedures can be fine-tuned to specific components based on the data harnessed off an embedded IoT platform. With IoT data, armors track weapons usage data all the way down to component level information. Things like the number of rounds through a particular barrel, spring, or bolt group. This allows manufacturers to develop detailed tests that provide thresholds for individual components. Armorers will then be able to swap components out or clean parts well before any failure in the field. Extended further, this IoT information can be used to predict the readiness of a soldier's weapon for combat, all the way up to an entire company's arsenal combat readiness. This data provides another large benefit historically. By collecting IoT information from a weapon over time, any changes in weapon mechanics and performance patterns is noticeable. Malfunctions, jams, stoppages, and altered recoil profiles for a particular weapon can be captured and monitored. This data builds a weapon's historical profile, so it becomes explicitly identifiable when that profile changes due to an imminent mechanical failure. Malfunctions and jams, when detected, are captured and reported back to inform maintenance systems and armorers. Trends of malfunctions, jams, and misfires can be identified on a weapon over time, leading to a predictive capability to inform an armorer to replace faulty components. This data can be used even further by manufacturers to inform warranties, design changes, and other modifications to improve upon failure rates. IoT data aggregated like this lets commanders know and assess their own readiness at any time.

Weapons based IoT also enables real time insights on the battlefield to a data standard that's never been possible before. Streaming real time data from a soldier's weapon conveying the details of how the weapon is being used by the soldier – and changes over time – allow rapid sensor to shooter response times. Knowing when a soldier is engaged with a threat is one of the most basic forms of situational awareness possible. Up to this time, this information has always been passed up to command in a secondary form – typically through voice or visual. While this information provides useful awareness, it isn't conveying the absolute truth of engagement – *exactly* when was the first round fired? How many rounds were fired so far in this engagement? How much ammunition does this soldier (or squad) have remaining, and will they go run out of ammunition before active engagement ends? Am *I* in a friendly's line of fire? Without IoT, this information is either difficult to ascertain in real time or impossible. Sensorizing the weapon itself with IoT provides all of this information organically, which is then made available to the broader ecosystem of battlefield assets. Information from each individual soldier is passed exactly where it needs to go, persistently, and automatically.

Real time weapons data provides the insights across echelons that not only alert commanders to emergent situations on the battlefield – the data provides a persistent source of ground truth information as the situation unfolds. What a soldier is doing with their weapon during battle informs a great deal of the actual workings of the fight. When a soldier is firing, the direction they are firing in, how many rounds they're firing, how quicky, how they are controlling their recoil, and their overall performance in combat scenarios is captured utilizing a simple sensor set for immediate problems. When combined into the larger picture of the entire battlefield of assets, the data takes on it's own new value proposition. For instance, counting the number of rounds a single soldier fires and providing that into a larger ecosystem yields results much larger than a simple round count. Knowing how many rounds are fired by each soldier allows command to ascertain the scale of the conflict the squad is involved in. Assuming it's known how much ammunition each soldier deployed with, it's possible to autonomously count how many rounds a particular squad has remaining. This information is a massive upgrade from manual ammo checks across the squad. The squad leader is informed on ammo utilization, granting him or her the ability to make informed combat decisions in real time. Similarly, platoon and company leadership is informed at scale from every soldier on the ground. This information, when ingested in larger macro battlefield ecosystems, becomes a crucial data point. Resupply decisions between squads can now be prioritized and data driven, all of which happens automatically. Going further, a human-in-the-loop can be removed altogether, allowing for full automation where deployed squads are resupplied with ammunition and assets autonomously through autonomous vehicles and drones. IoT information achieves it's highest value proposition at these levels, when tied into larger ecosystems that enable previously impossible tasks. Similarly, understanding an existing or ongoing engagement is crucial to directing artillery assets effectively. IoT sensors on individual soldiers and weapons on the battlefield make the coordination for their use automatic and driven by data from the field. Threat location and directionality is provided automatically by soldiers during engagement, detailing the direction a discharges were fired in. With multiple friendlies engaged on the same target, a commander viewing the aggregated IoT data is now able to ascertain where the threats are at, as well as how many friendlies are engaged on each target. When ingested into a larger system, the information allows artillery and other support to accurately determine the highest value lowest load targets based on active engagement data. With every soldier providing this information, making these decisions is almost trivialized (Figure 2.3).

Weapons sensors also provide immense benefits to their fellow soldiers within a squad. In close quarters combat, this is especially critical information at all times as the risk of fratricide is high. Fratricide rates are frequently cited around 2% [3]. This rate should be unacceptable to everyone that has an opportunity to change it. Not knowing positioning and line of fire are deadly mistakes that happen far too frequently. The unfortunate part about this situation is that it is preventable with technology. With IoT, all of the information a soldier's rifle is generating is available on friendly networks. Providing a weapon's heading and orientation, and therefore line of fire, provides the key details required to remove this problem entirely. By receiving a squad's worth of data from the network, each squad member is equipped with the knowledge of position and line of fire to keep them safe. If they cross into a friendly line of fire, they're immediately alerted through heads-up display or haptic feedback on a peripheral. Command is notified too, which can also provide preemptive feedback on position and potential for friendly fire scenarios. This does of course, depend upon reliable and secure radio networks extending across the battlefield to link disparate units together. Ultra reliable battlefield networks that provide unfettered communication links have been a priority of the US DoD for some time. The evolution of battlefield IoT sensors will add even more urgency to the discussion [4].

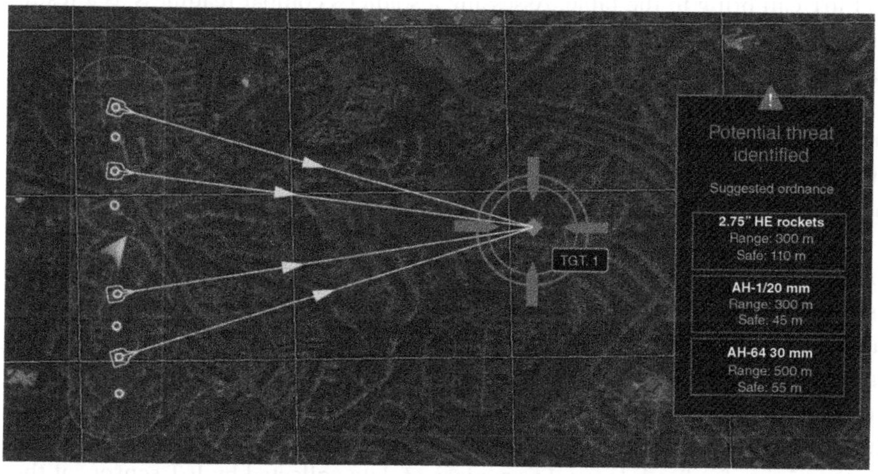

Figure 2.3 ARC IoT enabling top-down battlefield insights and targeting for assets.

Battlefield IoT sensors generate information on a lower level, closer to the ground truth, than any other type of sensors did before. The data generated by these sensors isn't always agnostic to the soldier responsible for creating the data. Every rifle on the field is wielded by a soldier with unique traits, stats, build, performance, and ability. When viewed as situational awareness data, soldiers are treated as if they are all one-in-the same, cut from the same cloth with interchangeable forgettable abilities. IoT provides the opportunity to embed sensors directly into those soldier's weapons, allowing a view into each individual's unique style of mission execution. A soldier's metrics and weapons usage data is all logged, which can be tracked and built into complex modeling around the soldier's skillset and performance. Sensors in the weapon platform provide interesting data on the user's reaction time, fatigue, state, and recoil control to name a few. The soldier's reaction time can change with additional training, fatigue, and alertness. The amount of time it takes for a soldier to turn toward a threat and raise their weapon will vary based on their training and fatigue. When IoT data is captured for multiple soldiers in a squad and compared, a baseline average reaction time to a threat can be produced and soldiers graded according to it. While video and other methods can provide alternative means of collecting this data, none of it is personalized to an individual and tracked over time without painstaking detail and effort. With IoT, it's automatic. A sensor is assigned to a weapon, which is in turn assigned to a user – two anchor points for data profiling.

A soldier's fatigue and readiness level can also be assessed when persistent data streams are provided from IoT. The amount of activity and external factors will influence soldiers differently depending on a wide number of variables. However, the affects of fatigue per soldier can be quite noticeable once baselined for an individual. A soldier's reaction time drops, and their recoil management will become less pronounced resulting in less controlled groupings. This data is fleeting and very difficult to capture without the tools provided by IoT sensors in the weapon. By capturing IoT data in the form of accelerometer and gyroscope sensor measurements, a soldier's baseline performance for fatigued vs. combat ready can be baselined per soldier. This data can be used as a training guide, informing soldiers and their leaders of any shortcomings prior to combat that need to improve. It also allows soldiers to assess their own performance and bring awareness to areas that need improvement. Commanders can use this information to assess at-a-glance the readiness and lethality of their soldiers.

Building a profile around the soldier for additional combat readiness insights is a massive advantage that battlefield IoT can bring to the table. Assessing a soldier's combat readiness as a result of that profile is an insight that can't be provided any other way. A soldier's ability to control the recoil of their weapon is one of the most significant factors in determining a soldier's lethality in controlled scenarios. Recoil control is one of the most significant factors between soldiers. Recoil control (or lack thereof) contributes to a soldier's accuracy or inaccuracy of multiple shots. It has a significant contribution to the data generated by a weapons platform during discharge scenarios, and can convey a great deal of information about a particular individual's experience. Characterizing it as part of a soldier's profile is made possible with IoT. By collecting data on weapon usage by an individual and associating that data with the individual, IoT is providing value at both the individual level as well as higher echelons. IoT data collected from an individual baselined for performance gives more concrete paths forward for improvements in lethality. Instead of qualifying or not qualifying at a range, a soldier is provided more specific metrics based on their individual profile around what needs to improve to qualify, or how to improve their overall accuracy and control profiles. Seeing this progress over time quantifies the training process into tangible results that can be brought back into training feedback loops. Every piece of data collected by IoT sensors at the individual level can feed larger data sets and external integration points.

2.4 Challenges for IoT in Soldier Weapons

The opportunity to gather this information doesn't come without its own challenges. Of all Battlefield IoT sensors, weapons based IoT sensors face some of the most extreme challenges in size, weight, power, and cost (SWAP-C) confined spaces. Weapons platforms and their configurations vary between fielded units, how they are employed in combat varies between individuals, and they are subject to extreme forces of temperature, vibration, and force – none of which bode well for battery operated electrical devices. Sensing and classifying events and actions performed with a rifle also means that the sensor itself needs to experience the same forces applied on the weapon, without damaging the sensor. Great care and attention is required to isolate peripheral motion or degrees of freedom of the sensor with relation to the weapons platform. Any slight misalignments or misconfiguration can result in damaged sensors or inaccurate data. Inaccurate data from weapons platforms can translate to inaccurate ground truths, and potentially cost lives and therefore must be avoided at all costs. Mounting location of the sensor to the weapon is also problematic. Any sensor that characterizes weapons events and actions must be tightly coupled to the platform itself, but at the same time cannot in any way interfere with the normal operation of a weapon. Given that the weapon is essentially an extension of the soldier wielding it, any operational hinderance produced by an IoT sensor on the weapon is not acceptable. While accessory rails on many weapons platforms can provide a common mechanical interface location, the space is already highly contested with a dizzying and ever growing array of scopes, range finders, flashlights, and other equipment. Accessory rail space is also subject to some of the most extreme temperatures and forces the weapon experiences. It's no surprise that battery operated IoT sensors last longer the further away from these stressors they are. The standard augmented reality (AR) grip on a weapon provides for another ideal location for senor placement. Completely isolated from the heat and forces experienced on the barrel, the grip provides additional rigidity as well to absorb shock. Weight is also a significant factor in IoT sensor success or failure. Soldiers have become so accustomed to their weapons and their use, that any weight change over about 40 g is detectable and adversely affects training exercises. Cramming a battery into the packed space with sensors and ruggedized design becomes an immense challenge that, while felt throughout commercial IoT, is heavily emphasized in Battlefield IoT [5].

In addition to the SWAP-C and integration challenges faced by IoT devices in a confined, harsh environment of soldier weapons, deploying IoT sensors in every firearm presents a cognitive challenge as well. Presenting information to individuals or squad leaders that will may overwhelm the casual commercial IoT device user is a non-starter when considering this information may be crucial to understand while under the pressure of live enemy fire. Soldiers deal with an enormous amount of mental pressures including stress and fatigue [6]. These factors can heavily impede a forward deployed decision maker when interpreting multi-source data from a large number of ground sensors. Pre-processing the abundance of information is critical for it to be useful to these users. When true cloud access isn't available, edge processing nodes provide the compute capabilities required to aggregate and summarize the information from disparate IoT senor input across the battlefield. The aggregation and preprocessing of data takes raw sensor data that would be difficult or impossible for a soldier to interpret (even in ideal circumstances), and makes it immediately understandable. Knowing which soldier is currently engaged with a threat, the sector of fire, and estimated enemy position are all immediately recognizable and actionable by a squad leader. However, none of these data points are provided natively by any one sensor without a degree of preprocessing in the cloud or at the edge to interpret the raw acceleration, line of bearing, location data, etc., that are involved in producing useful (actionable) data. When this information is provided to a leader from every node (weapon) on the battlefield, it becomes critical to identify which nodes are producing information that is pertinent and actionable, and which is not. For firearms IoT, the distinguishing factor in useful data vs. passive, non-actionable data is the soldier action occurring. Soldiers actively engaged in a firefight are obviously more pertinent to pay attention to than soldiers at a mess hall with their firearms. Machine learning algorithms lend themselves very well to identifying the patterns and trends common to critical engagements vs. non-critical information. By utilizing a pattern recognition algorithm on the edge, data can easily be flagged when certain criteria and thresholds are met that identify information that is critical to decision makers, thereby filtering incoming data and reducing the overall cognitive load.

2.5 Battlefield Challenges to Aggregating and Exfiltrating Data

While real time communications and situational awareness does require reliable, consistent battlefield networks, IoT data doesn't need to be collected and streamed immediately in real time to be valuable. The data generated from sensors in direct line with actual combat are always generating data, regardless of network availability. That means losing data just because a network isn't available means mission failure. This is a major difference between weapons-based Battlefield IoT sensors and many commercial IoT sensors. The data generated on the battlefield needs to be preserved for numerous different uses including debriefs and future combat operations training. The ability to pull all the data gathered by weapons based IoT sensors after a mission allows command to review precise actions of a soldier, reaction time and response to threats, and a soldier's weapons and ammunition status during an engagement. All of this data can be aggregated with cross-echelon data after missions, meaning leadership can review and replay the exact timeline and scenario around a conflict on a map. Command is able to incorporate a tactics feedback loop based on real world and current engagements with weapons usage and engagement data that has never existed before. Tacticians are able to know exactly how long an engagement lasted, exactly how many shots were fired, where they were fired, the direction they were fired in, and the orientation of the weapon as the soldier absorbs the recoil of each discharge. This data sheds light on enemy tactics and force responses with far greater transparency than any tool or method today. Precision debriefing like this has never existed before, largely because the data just isn't recorded. It's not recorded because no other sensor or system can provide this data except IoT because of the extreme requirements placed on solution providers for this information. Without this data, units are left with sometimes choppy and unreliable helmet camera footage and inconclusive data, translating to a painstaking effort by analysts to achieve any level of detail around ground truth.

Every IoT sensor deployed to battlefield environments needs to deliver capabilities like this whenever data is captured. Being able to review all captured data after a mission is, in many cases, one of the most critical tools and assets to any debrief. Unfortunately, the battlefield is not IoT friendly. There are no WiFi networks or reliable cellular connectivity in many cases. When engaging in conflicts in third world or developing countries, the supporting network infrastructure simply does not exist and must be created by the forces wishing to leverage such technologies. Radios, satellite, iridium, among other wireless transmission formats can provide the uplink required for IoT, but all of them face their own problem sets and challenges. Most of these formats are heavily bandwidth constrained, making 56K modem connections look fast. In addition, there is no standard type of network deployment for the United States. IoT sensors on the battlefield must be adaptable to all of these scenarios. Adapting into all of these scenarios with a product is very difficult to do. To do so successfully, you'd need an entire team dedicated to specific hardware implementations for whatever the flavor of the day is for the command you're deploying into. This very quickly becomes untenable. In many cases, every program within the Army has their own network protocol that will be leveraged by forces. While this is a difficult problem to solve, there is a path forward. As an IoT provider, adapting to demonstrate capabilities and deliver IoT data value without making a hardware change for every program is required. This can be successfully done by splitting the problem of network divergence into two issues – long and short ranges. Long range communications involve formats and paradigms of how data can be pushed rapidly and securely across large distances (miles) between a squad and command or command compute resources. The second issue is solving for short range communications. Short range specifically means intra-soldier – or communicating data wirelessly to other sensors, radios, or compute that is co-located on a single soldier. All Battlefield IoT companies should be primarily concerned with the latter issue. Solving for the long range networking challenges faced by the military is a challenging enough issue that it either requires the full attention of all the resources in a company, or it requires the deep pockets of a large prime with the backing of a program office to fund and approve the wireless format. Even then, it is unlikely that we will ever see a single type of format deployed across units and branches of the military. The networking requirements and radio frequency (RF) capability landscape is just too vast, too complex, and requires too many disparate configurations for any one format to meet every need. Certain units require ultra-long-range communications, others require wide band to make jamming more difficult, others require complicated frequency shifting paradigms to avoid or deter intercept attempts. By focusing on the short range communications challenges first, battlefield IoT systems are able to work within the confines of far more limited requirements. Short range wireless intra soldier communications are a given for sensor-to-sensor communications. However, even these short range approved communication mediums vary greatly between programs and military branches. While there are some up and coming formats becoming more established, namely the Army's new intra-soldier-wireless (ISW) format, there are many commercially available equivalents that work as demonstratable communications stand-ins. Leveraging Commercial Off The Shelf (COTS) solutions for short range communications like Bluetooth provides a fantastic jump-start opportunity for small companies to quickly build out IoT capabilities, allowing them to focus on providing data rather than transmitting that data. This substitution of proprietary, expensive, Government Off The Shelf (GOTS) hardware and software for cheap, well supported COTS solutions also means there are far greater variety in choices of suppliers, capability, and commercial or community support. The accessibility of these technologies allows teams to rapidly iterate on a particular product capability, bringing it to market five times faster and hundreds of times cheaper than what it would take to bring the same capability to market leveraging only GOTS solutions. As an added bonus, leveraging the commercially available short range networking solutions allows business development teams to take products and show them to interested buyers far more freely and with far greater ease than GOTS solutions that require additional specific hardware [6].

IoT sensors and weapons based IoT sensors in many cases are only as good as the data that they can provide. In many scenarios, especially in near-peer engagements, RF jamming is a very real and very prevalent possibility and threat. While GOTS solutions like an ultra-wide band ISW format is specifically designed to make jamming difficult, it isn't impossible. When an IoT sensor's capability to transmit data is impeded through jamming or interference, there is a need for capabilities that enable data redundancy and resiliency. Redundancy is the act of duplicating data in multiple locations, while resiliency refers to maintaining the accuracy of that data when compared to it's original source. In jamming scenarios, redundancy becomes difficult to achieve in ways that are not superficial. However, maintaining the resiliency of that data on the device is paramount to success. Storing data on the sensor in non-volatile memory is almost a no-brainer, but including methods to validate that data is still accurate is also important. Incorporating checksums and parity bit schemes in your data are always good ideas. When pulling large data sets from sensors that have been forced offline for extended periods of time, the extra comfort of knowing your data integrity remains secure and unaffected is very welcome. Especially so when that data tells a story around weapons usage and it's potential impact on the threat landscape.

The process of saving data to memory and incorporating error checking needs to be fully automatic. IoT sensor systems must be designed in ways that ensure this mode of operation any time the data they relay must be accurate and collected without gaps. Network issues, enemy jamming, or other RF interference can happen at any time. These same problems exist in the commercial IoT space, but in many cases the incidence is largely inconsequential apart from missing a light bulb toggle command. IoT systems must work in parallel with data flows to persistently push data off the device so that when a network connection does experience issues, the data is always saved somewhere. While this is an easier problem for IoT data that is relatively small in nature, it becomes very problematic for larger data sets such as video and audio data. Knowing how much data needs to be stored in a total network failure scenario is absolutely critical as a feedback loop to future hardware development. Board designs need to support the memory requirements of data storage in these scenarios. Data can't sit on the IoT device forever though. To meet redundancy requirements of IoT systems, that data needs to be pushed up to the end destination (like the cloud) as soon as possible. This presents yet another challenge in the Battlefield IoT space. While capturing and transmitting weapons based IoT data in real time, sensors must constantly write the same data to non-volatile memory in the event of network connectivity lapses. When a network connectivity lapse does occur, priority precedence of operations between historical data and real time data varies greatly between IoT data sets and operational concepts. In the case of real time weapons based IoT data, real time data typically carries a useful lifespan of around 3 s. Past that, and the event modeled happened sufficiently long enough ago that it's no longer relevant. For post-mission analysis though, the data will be extremely relevant. Managing both data streams will sometimes contest resources on embedded devices and lead to complex queue management in firmware. Every use case is different, but simplicity is generally the best approach in many of these cases. When timing is critical, pushing real time events is priority. Given the timeliness required for "real time" operations, a simple fire-and-forget approach to events avoids the cycle-intensive transmit retry loops, which quickly back up upon themselves causing other events to be reported late, and potentially backlogging data to be written to flash memory. The approach of fire-and-forget events works especially well for supporting adverse network conditions. Without a connection or a network present, fire-and-forget simply attempts once to transmit data, and without waiting for an acknowledgement, moves on to save that data to flash memory. If there is no network connection present, all the relevant data is saved in memory on the device, ready for offloading whenever mission requirements support it. All the data is pushed to required endpoints as quickly as possible, all the while never exposing the data or data pipelines to potential for backlogging.

2.6 Protection and Security for IoT Data Communication

Whenever considering any type of RF solution to transmit data, network security requirements work their way into the conversation quickly. There are a few types of security concerns when implementing IoT systems on the battlefield, many of which have equivalent counterparts in the commercial world. First, there is security on the IoT device itself. IoT device security consists of encryption of any IoT data stored on the device itself in memory to prevent unauthorized retrieval of sensitive data, authentication and authorization of services attempting to connect to the IoT device, and finally encryption of the firmware itself on the IoT device. In today's world, there is no excuse for not encrypting data. Encryption tools and keys have become the defacto standard for software implementations the world over, and for good reason. It is one of the most surefire ways to guard against any unauthorized access to data. Even in the event of an access breach, which is arguably far more difficult to protect against, the data itself remains safe and secure. It's also fairly simple to perform, as symmetric encryption with AES 128 (or AES 256 as approved by various DoD standards) is very well supported by many open source libraries. The difficulty in encrypting data on IoT devices is key management. Changing keys in the event of compromise is typically straightforward for backend software systems. Deployed IoT systems however are not so easily updated. Once shipped out the door to a customer, IoT systems and sensor configurations are locked in. Changing the firmware load or encryption keys is now a function of supporting over the air (OTA) updates and a security scheme. Even in cases where the keys can't easily be changed out for IoT devices, encryption of the data stored on the device is a crucial first step.

Encryption of IoT data in transit is the second form of IoT security to consider. Like encryption of data at rest on the IoT device in non-volatile memory, encryption of data in transit (or data that is being transmitted over a wireless connection) is equally critical and just as simple. Short range wireless formats such BlueTooth support encryption standards of at least AES 128 Encryption at rest and transit are commonplace and standard across industries as a best practice to guard against unauthorized access to sensitive information. Given the broad level of support from the community and the requirements for encryption from the DoD, it's common sense to provide this support in any fielded IoT device. Without it, data is passed over the network in the clear, or as-is. Anyone with a software defined radio or packet sniffer is able to intercept and parse information when it isn't encrypted utilizing one of these common methods. Worse yet, interception is usually completely invisible and undetectable. Given the severity of not encrypting data, it's easy to see why encryption at rest and in transit are commonplace requirements. Encrypted data in transit also ensures that any commands sent down to the IoT device from the network are also encrypted. In the case of human readable interfaces, or in some cases even memory overlay type interfaces, intercepted data could mean an intruder is able to identify exactly what commands are sent to and received from an IoT device, potentially compromising data as well as authorization/authentication to the device.

Authorization and authentication to IoT devices and weapons based sensors is the final security topic to discuss. Providing a means for an external system to connect to the IoT sensor is generally required by IoT sensors in both commercial and DoD spaces. Authentication is the security process by which a user's identity is verified, usually through a password login, key, or token. Authorization provides a particular user with certain data rights to interact with and manipulate data based on their security accesses. In some cases, IoT devices do not provide any authorization or authentication mechanisms, and do not allow direct connections to the device wirelessly. Without being able to connect with and communicate to the device, data is largely inaccessible rendering the device fairly useless. While secure, assuming physical security is not compromised, it is not very convenient. Generally speaking, the better way to provide authentication and authorization once again comes from established commercial markets. Using a well established authentication and authorization method such as OAuth 2.0 allows teams to take proven and very well supported authentication mechanisms and implement them in IoT spaces with ease. This process will typically require a user to be connected to the network to perform, which can be problematic considering the battlefield IoT space does not have guaranteed network access in most scenarios. For this reason, another layer of processing is required in the data flow so that users without consistent network access can continue to connect. Generally speaking, to use a login paradigm such as OAuth 2.0, at least a user's first time login must be connected to the network and backend resources that will provide the token. Once authenticated, the additional processing on top of the request should support maintaining that token for as long as customer requirements and mission can allow for. This keeps the user logged in, and allows users to log in again provided the token is still active. In addition to this, it's possible to design asynchronous systems that provide the same or similar functionality to, for example, provide support for a local download of encrypted accounts and information that allows for first time users to log in without requiring internet access.

Security is a big deal for Battlefield IoT devices. Unfortunately, the vast majority of IoT devices in the commercial market are woefully undersecured. Many devices lack any form of authentication or authorization, and many more lack any kind of encryption on the data in rest or transit. Part of the reason for this is that in commercial spaces, the data carried by IoT devices is usually inconsequential. These loose data practices are of significant issue and concern on the battlefield. Rather than a concern about a next door neighbor hacking your garage door, Battlefield IoT device security is life and death. The lackadaisical attitude displayed in commercial sectors for IoT is gravely concerning given how much commercial products contribute to the development of new products in the defense space. Bad habits and poor data security practices leave soldiers at risk. Known vulnerabilities in firmware communications protocol stacks also lead to significant security concerns, especially when coupled with poor security practices previously discussed. When there are vulnerabilities discovered on IoT firmware, it's problematic to say the least. Typically there are already devices downrange. To support application of these hotfixes, an OTA process to update IoT sensors is critical. Without it, vulnerabilities will continue to proliferate and exist, which over time increases the likelihood that a particular vulnerability will be exercised against you and the troops around you. This ranges from manipulation or deletion of data to silent intercept of data, or potentially installing unauthorized firmware on the device for other nefarious schemes. By properly securing devices with encryption at rest and in transit, coupled with proper authentication, many security risks are mitigated. Engaging with near peers however raises the bar. To ensure malicious firmware or code cannot be installed on a device, signing firmware provides a fantastic, easy, low effort solution. Signed firmware for IoT devices allows an organization to specify which firmware can be installed and which cannot. Without the proper signature on the firmware, the device cannot accept firmware modifications, foiling the attack. Other attacks are possible and vary greatly in vector and impact. Ideally, once an IoT sensor has been verified to be connected with it's short range receiving counterpart, RF communications can be shut down or made intermittent to communicate relevant data, without exposing the device to potential attackers. If the device can't be seen on the RF spectrum, and can't be connected to, it can't be hacked wirelessly.

In some cases, these security concerns are also coupled with the concern of being seen at all on the RF spectrum. Any identifiable RF signal from battlefield IoT sensors can lead to rather disastrous consequences ranging from unit detection to improvised explosive device (IED) triggers. These concerns are very real, and very valid during any type of engagement in the modern day world. Because these issues are inherent in the nature of RF systems – detecting the same signal used to transmit data – the only solution is to remove the signal entirely. Battlefield IoT devices must be able to completely shut off all RF emissions whenever needed by the soldier. Typically, this is performed with a switch or button on the device itself in a location that can't be accidentally turned back on. As discussed earlier with network solutions that are resilient to networks failures, this operational mode can effectively be considered an induced network failure mode. In this scenario, all of the data captured by weapons IoT sensors should still be captured for offload later. RF detection and identification methods performed by enemies are one of the greatest risks and threats to Battlefield IoT. This does not mean passive solutions such as radio frequency identification (RFID) are acceptable. On the contrary, RFID poses an even more significant risk to soldiers due to the fact that it is both detectable through magnetic field readers purchased off the shelf, and that it cannot be shut off because of the passive RF technology. RFID type solutions, while they do not actively emit any signal, actually complicate this problem far beyond a simple solution of turning a wireless RF signal off.

2.7 State of the Art

Today, there are a number of entities both commercial and defense focused that produce firearms sensors. Implementations vary greatly from picatinny and modular lock (MLOK) rail based sensors, holster sensors, and sensors embedded into weapons directly. These sensors support many different functions, but the most developed applications available today are ones that empirically determine physical or electrical state change in external variables. One example of this is discharge detection, or the detection of acceleration forces on the firearm that match the profile of a weapon discharge. These IoT sensors frequently operate on either commercial networks, or operate in isolation from one another, providing input for a single user or firearm without aggregation or mission context. These types of sensors are new, but available commercially to consumers. Within the DoD however, the concept of IoT firearm sensors is brand new. The first program to adopt a true IoT sensor requirement for detecting discharges is the Next Generation Squad Weapon Program (NGSW) [6]. With the first true DoD requirement for a shot counter, the entry point for other weapons based IoT sensing and reporting is open. NettWarrior and program manager (PM) Integrated Visual Augmentation System (IVAS) provide the full set of networking capabilities to soldier borne sensors and IoT for firearms that is required to provide the real time network communications for every dismounted unit. As the US Army moves forward in these programs and unveils the next generation of battlefield networks, the soldier borne sensors and IoT devices operating on these networks are provided a path forward to push data to the critical decision makers that rely on precise, up to date information. Firearms based IoT sensors that are detecting shots from every deployed unit equipped with the NGSW rifle will then be able to push real time discharge detection, weapon heading, and performance data directly to edge compute resources or to the cloud. There, data across echelons is aggregated and relayed in ways that minimize the sensor to shooter timeline.

2.8 Conclusion

For the United States military and its allies to continue progressing and remain ahead of the ever-advancing technological curve, it's imperative that our data sources improve. Without real data from the ground, the actual details of ground engagements are subject to communication gaps and delays. Any time there is a delay in real time intelligence, we pay the price in lives and opportunities. Back in Iraq 2010, Mike Canty would have had a very different experience if battlefield IoT was available. Instead of standing by for additional radio information on positioning and engagement data, a simple glance at an IoT enabled data display provides all the information and intelligence required to make command decisions. Immediately and automatically knowing friendly positions, maneuvering, and status, the threat would be efficiently located and neutralized with little to no risk of fratricide. Even the amount of ammunition remaining with the squad actively engaged would be a known detail, allowing timely engagement and support decisions to be made. Today, Armaments Research Company (ARC) is fielding state of the art weapons based IoT sensors for next-generation programs and technologies that enable these insights.

Sensor integrations have been built and designed for a variety of small arms and crew served weapons platforms fielded by the US Military today. Many of the solutions are currently undergoing active test and acceptance with the DoD so that tomorrow's warfighters can be enabled with these timely, critical insights. Until these technologies are fully deployed and fielded, we continue to leave a treasure trove of insights, intelligence, and usage data that could translate to billions of dollars saved whenever small arms are fielded without a feedback loop to provide the data on how they're used on the field. With weapons based battlefield IoT sensors, the whole game changes. In a world where every microsecond can count, this data and the means to effectively collect it, matters.

References

1 Canty, M. (2021). Interviewee, [Interview] 9 November 2021.

2 Wikipedia (2022). History of the Firearm. https://en.wikipedia.org/wiki/History_of_the_firearm#cite_note-21 (accessed 21 October 2022).

3 I. U. S. N. LCDR William Ayers (2022). Fratricide: Can it be stopped?. https://www.globalsecurity.org/military/library/report/1993/AWH.htm (accessed 21 October 2022).

4 Lopez, C.T. (2020). New Spectrum Strategy Reveals DoD's Plan to Master Airwaves. https://www.defense.gov/News/News-Stories/Article/Article/2404027/new-spectrum-strategy-reveals-dods-plan-to-master-airwaves/ (accessed 21 October 2022).

5 Program Executive Office Soldier. (2018). Next Generation Squad Weapons. *Industry Day*.

6 Program Executive Office Soldier. (2020). Adaptive Squad Architecture. *Industry Day #2*.

3

IoBT Resource Allocation via Mixed Discrete and Continuous Optimization

Jonathan Bunton and Paulo Tabuada

Department of Electrical and Computer Engineering, University of California, Los Angeles, CA, USA

Abstract

The fast-paced dynamics of the battlefield require constantly monitoring existing resources and reallocating them in response to adversarial actions. Many of these resource allocation problems consist of two coupled decisions: *which* resources to use and *how* to best use them. While the former is naturally formulated as a discrete optimization problem (e.g. should we use assets from class A or class B, should we deploy them to site C or site D), the latter is naturally formulated as a continuous optimization problem (e.g. which fraction of existing resources should we deploy to which site). An effective and optimal allocation of resources necessarily has to consider both aspects in its optimization: the discrete and the continuous. Although mixed discrete and continuous optimization problems are non-deterministic polynomial-time hard (NP-Hard) in general, in this chapter we show how to leverage submodularity and convexity to identify problem instances that can be solved exactly in polynomial time.

3.1 Introduction

When distributing available assets to a set of tasks, we naturally want to do so in the most effective and cost-efficient way possible. Resource allocation problems confront precisely this issue by describing an optimal way to divide a fixed budget of resources across a given set of tasks. These problems involve two fundamentally linked decisions: deciding which set of tasks we should allocate resources to, and then how much of our resources to assign to each of the selected tasks. To make the best use of our available budget, we must reason about both of these decisions – the choice of tasks and the best way to allocate resources to them – *simultaneously*.

More concretely, resource allocation is usually posed as an optimization problem, in which we want to minimize a function describing the cost of any particular allocation of the resources. If the amount of resources allocated to task i is represented by some continuous value $\mathbf{x}_i \in \mathbb{R}_{\geq 0}$ and there are n total tasks, then our decision variable is naturally a continuous vector $\mathbf{x} \in \mathbb{R}^n_{\geq 0}$. We then seek a solution to the optimization problem:

$$\begin{aligned} \underset{\mathbf{x}\in\mathcal{X}}{\text{minimize}} \quad & f(\mathbf{x}) + \lambda g(\mathbf{x}) \\ \text{subject to} \quad & \textstyle\sum_{i=1}^{n} W_i(\mathbf{x}_i) \leq B. \end{aligned} \tag{3.1}$$

In this problem, we measure the cost of an allocation $\mathbf{x} \in \mathbb{R}^n_{\geq 0}$ through the continuous function $f : \mathcal{X} \to \mathbb{R}$ that characterizes the cost of the allocation values (i.e. the continuous values of each $\mathbf{x}_i$),

IoT for Defense and National Security, First Edition. Edited by Robert Douglass, Keith Gremban, Ananthram Swami, and Stephan Gerali.

and a discrete[1] function $g : \mathcal{X} \to \mathbb{R}$ that assigns a cost to the combinatorial choice of which tasks to allocate resources to, with $\lambda \in \mathbb{R}_{\geq 0}$ as a tradeoff parameter. When resources are limited, the increasing functions $W_i : \mathbb{R} \to \mathbb{R}$ represent the cost of assigning $\mathbf{x}_i$ resources to task i, with some overall budget $B \in \mathbb{R}_{\geq 0}$.

If the cost function in (3.1) is purely continuous, then convex optimization is the leading technique for solving these resource allocation problems, due to the massive amounts of existing tools and modeling flexibility [1, 2]. Alternatively, in the context of purely discrete or combinatorial resource allocation, various specialized algorithms exist, many of which rely on either branch-and-bound methods or submodular function minimization as the backbone [3]. Our work bridges the gap between these two ends of the spectrum, leveraging the best existing algorithms from both approaches to solve the mixed optimization problem.

As an example, consider a simple internet of battlefield things (IoBT) architecture where a base station/local server would like to receive data from a set of n IoBT edge devices. Retrieving this data from the IoBT edge devices naturally requires choosing which devices should transmit and how much power (or at which data rate) should they transmit. Such a choice needs to be judicious so as to maximize some measure of the usefulness of the activated edge devices, and the data acquired by the base station/local server. We may also have natural upper bound constraints on the allocated power (data rate) assigned to each edge device of the form $\mathbf{x}_i \leq \overline{\mathbf{x}_i}$, since these lightweight IoBT devices are typically energy constrained.

A number of measures for the quality/utility of the activated IoBT edge devices have been proposed in the literature [4–6]. One simple choice is the *achievable data rate* at the base station/local server, which can be approximated with Shannon's formula [7]. In particular, the (negative, for consistency of notation) utility of powering IoBT edge device i could be modeled with the function $f_i : \mathbb{R}^n_{\geq 0} \to \mathbb{R}_{\geq 0}$ defined by:

$$f_i(\mathbf{x}) = -a_i \log_2\left(1 + h_i \mathbf{x}_i\right). \tag{3.2}$$

In this model, $a_i \in \mathbb{R}_{\geq 0}$ and $h_i \in \mathbb{R}_{\geq 0}$ represent the bandwidth and signal-to-noise ratio of IoBT device i, respectively. More generally, the utility functions f_i are often assumed to be increasing and concave functions of $\mathbf{x}$, a particular trademark of "elastic" IoBT networks and fairness-throughput tradeoffs [8].

In addition to the measured utility of the local server's incoming data, there is an inherent cost to powering any choice of IoBT devices. These start-up costs could, for example, include the logistical and power use required to send a transmission request. Similarly, some subsets of IoBT edge devices that we have chosen to power could be transmitting uninformative data, which we then receive and process at the base station unnecessarily. Moreover, since actively transmitting IoBT devices may alert an adversary to their presence, we may also want to restrict our choice of active devices to specific subsets to maintain covert operations. All of these costs depend exclusively on the discrete structure of the power allocation $\mathbf{x}$, in particular on the choice of which subset of IoBT edge devices in our arsenal we activate. If, for illustration, we set the "start-up costs" for each device to one, then (3.1) becomes:

$$\begin{aligned} &\underset{\mathbf{x}\in\mathbb{R}^n_{\geq 0}}{\text{minimize}} \ \textstyle\sum_{i=1}^n f_i(\mathbf{x}) + \lambda \|\mathbf{x}\|_0 \\ &\text{subject to} \ \textstyle\sum_{i=1}^n W_i(\mathbf{x}_i) \leq B, \end{aligned} \tag{3.3}$$

where $\|\cdot\|_0$ denotes the ℓ_0 pseudo-norm that counts the number of nonzero entries in the vector $\mathbf{x}$. The solution to this optimization problem is an allocation of power to the edge devices that maximizes the average utility of the deployed devices over their active period, while also considering the

1 In this paper, we use the term *discrete function* when referring to a function with finite image.

associated logistical and activation costs for each IoBT device. While we chose the ℓ_0 pseudo-norm here, we may naturally want to consider start-up costs with more complex forms of discrete structure. Moreover, this model easily extends to having m distinct device types by considering an appropriate allocation *matrix* $\mathbf{x} \in \mathbb{R}_{\geq 0}^{m \times n}$.

While we introduced an IoBT edge device allocation as a running example problem, more general resource allocations often fit the same optimization framework [3, 9, 10]. In Section 3.6.2, for instance, we consider a distributed computing scenario where we would like to optimally divide a large computational task between a set of n resources, considering both the computational speedup (or monetary costs of purchasing cloud-based resources) and the inherent power-draw costs for each resource [11]. Optimally solving this type of problem, then, implies the computational task is performed with both performance and start-up costs considered simultaneously.

Optimization problems with mixed structure in the cost function such as (3.3) are notoriously difficult and even NP-Hard in general. As a special example, if each function f_i were a least-squares cost, meaning $f_i(\mathbf{x}) = \|\mathbf{Ax} - \mathbf{b}\|_2^2$ for some matrix $\mathbf{A} \in \mathbb{R}^{m \times n}$ and vector $\mathbf{b} \in \mathbb{R}^n$, with $\| \cdot \| : \mathbb{R}^n \to \mathbb{R}_{\geq 0}$ the usual ℓ_2 norm, then problem (3.3) becomes the *basis pursuit*, or compressed sensing problem, which is well-known to be NP-Hard in general [12]. In our running example of allocating power to IoBT edge devices, the family of problems has been repeatedly established to be NP-Hard for various choices of utility functions f_i other than our simple example above [6, 13]. Even with relatively simple utility functions, adding discrete structure costs also quickly leads to intractable optimization problems [4, 14].

Traditionally, the difficulty of these problems is avoided by replacing the discontinuous function g with a convex relaxation of the discontinuous discrete penalty, such as the ℓ_0 to ℓ_1 norm relaxation [15]. While these convex relaxations are more amenable to optimization algorithms, they do not necessarily recover an optimal allocation for the originally posed problem (3.1) unless specific conditions are met [12, 16, 17]. Moreover, these conditions do not always extend to more generic discrete structure than the ℓ_0 pseudo-norm [18].

Other typical approaches to directly solve the problems (3.1) and (3.3) rely on suboptimal greedy algorithms [19], slow and suboptimal parameterizations [20], or require discretizing the problem solution [21, 22].

In contrast, our work identifies conditions under which we can achieve the best of both worlds: solving the mixed continuous and discrete optimization problem (3.1) both exactly and efficiently. To do so, we exploit submodularity, a property of functions that defines a boundary between efficiently solvable and NP-hard optimization problems in both continuous and discrete spaces [21, 23]. Our theory then guarantees that an agnostic pairing of convex optimization and submodular function minimization algorithms produces the exact global minimizer of (3.1) in polynomial time. In an IoBT context, for example, this implies that networks relying on optimization frameworks such as (3.1) can achieve the globally optimal solution in polynomial time, without resorting to approximation schemes. The optimal solution to the optimization problem fully specifying our desired network properties necessarily leads to an overall more efficient and effective IoBT.

Explicitly, in this chapter, we:

1. Identify sufficient conditions based on submodularity that guarantee that we can solve the *unconstrained* mixed continuous and discrete resource allocation problem (3.1) exactly and efficiently;
2. Use tools from submodular function theory to accommodate both discrete and continuous budget constraints on the resources;
3. Verify our theory on some simple proof-of-concept examples.

3.2 Lattices and Submodular Functions

The resource allocation problem we would like to solve inherently involves two domains – an infinite and continuous set such as $\mathbb{R}^n_{\geq 0}$, and a countable discrete set such as the set of possible nonzero entries of a vector. If the problem is defined purely with continuous functions and decision variables, convex optimization has proven to be a useful and generic tool for efficiently producing exact solutions [1, 2]. Alternatively, if the problem was defined with only discrete functions and decision variables (i.e. only using the structural costs in (3.1)), then the key property enabling us to efficiently compute exact solutions is submodularity [3, 23].

Submodularity, however, is a property of functions defined on both continuous and discrete domains. While it is usually defined for *set functions*, which are functions that map any subset of a finite set of elements V to a real number in $\mathbb{R}$, its definition more broadly applies to functions on a suitably structured space: a lattice.

To define a lattice, we first consider a set of elements $\mathcal{X}$ with a partial order defined on its elements, denoted by $\preceq$. Given this partial order and two elements $\mathbf{x}, \mathbf{x}' \in \mathcal{X}$, we define two fundamental operations: their least upper bound, or *join*:

$$\mathbf{x} \vee \mathbf{x}' = \inf\{\mathbf{y} \in \mathcal{X} \;:\; \mathbf{x} \preceq \mathbf{y},\ \mathbf{x}' \preceq \mathbf{y}\}, \tag{3.4}$$

and their greatest lower bound, or *meet*:

$$\mathbf{x} \wedge \mathbf{x}' = \sup\left\{\mathbf{y} \in \mathcal{X} \;:\; \mathbf{y} \preceq \mathbf{x},\ \mathbf{y} \preceq \mathbf{x}'\right\}. \tag{3.5}$$

A *lattice* is a set of elements with a partial order, written together as $(\mathcal{X}, \preceq)$, where for any two elements $\mathbf{x}, \mathbf{x}' \in \mathcal{X}$, their join, $\mathbf{x} \vee \mathbf{x}'$, and their meet $\mathbf{x} \wedge \mathbf{x}'$ exist and are in $\mathcal{X}$ [24]. When the partial order is clear from context, we omit the partial order and denote the lattice simply as $\mathcal{X}$.

A function $f : \mathcal{X} \to \mathbb{R}$ is then *submodular* on the lattice $\mathcal{X}$ if for all $\mathbf{x}, \mathbf{x}' \in \mathcal{X}$:

$$f(\mathbf{x}) + f(\mathbf{x}') \geq f(\mathbf{x} \vee \mathbf{x}') + f(\mathbf{x} \wedge \mathbf{x}'). \tag{3.6}$$

Moreover, f is *monotone* if it preserves the partial ordering of elements, meaning for all $\mathbf{x}, \mathbf{x}' \in \mathcal{X}$, we have:

$$\mathbf{x} \preceq \mathbf{x}' \quad \Longrightarrow \quad f(\mathbf{x}) \leq f(\mathbf{x}'). \tag{3.7}$$

The prototypical example of a lattice is the power set (the set of all subsets) of a finite "ground set" of elements V ordered by set inclusion, denoted $(2^V, \subseteq)$. The join and meet operations on this lattice are then set union, $\cup$, and set intersection, $\cap$, respectively. Following (3.6), submodular *set* functions $f : 2^V \to \mathbb{R}$ must satisfy, for all $A, B \in 2^V$:

$$f(A) + f(B) \geq f(A \cup B) + f(A \cap B). \tag{3.8}$$

Further, by (3.7), a set function is monotone if for all $A, B \in 2^V$:

$$A \subseteq B \quad \Longrightarrow \quad f(A) \leq f(B). \tag{3.9}$$

Arbitrary set function optimization is NP-Hard in general. Submodular set functions, however, can be approximately maximized with greedy algorithms [25, 26], and (as we will exploit later) exactly minimized with polynomial-time algorithms [27, 28]. In this way, submodularity defines a key threshold between computationally difficult and efficient optimization problems in discrete spaces.

While the subset lattice $(2^V, \subseteq)$ is the most commonly used example, the definition of submodular functions on lattices does not require its discrete nature. In particular, we can define submodular functions in continuous spaces, as long as we endow those spaces with an appropriate partial order. For example, consider the set $\mathbb{R}^n_{\geq 0}$, equipped the coordinate-wise partial order $\preceq$. More explicitly, any $\mathbf{x}, \mathbf{x}' \in \mathbb{R}^n_{\geq 0}$ are ordered according to the rule:

$$\mathbf{x} \preceq \mathbf{x}' \quad \Leftrightarrow \quad \mathbf{x}_i \leq \mathbf{x}'_i, \quad \text{for all} \;\; i = 1, 2, \ldots, n, \tag{3.10}$$

where here $\leq$ denotes the usual total order on $\mathbb{R}$.

For any $\mathbf{x}, \mathbf{x}' \in \mathbb{R}^n_{\geq 0}$, the join and meet operations resulting from this partial are the element-wise maximum and minimum, respectively:

$$(\mathbf{x} \vee \mathbf{x}')_i = \max\{\mathbf{x}_i, \mathbf{x}'_i\}, \quad \text{for all} \;\; i = 1, 2, \ldots, n, \tag{3.11}$$

$$(\mathbf{x} \wedge \mathbf{x}')_i = \min\{\mathbf{x}_i, \mathbf{x}'_i\}, \quad \text{for all} \;\; i = 1, 2, \ldots, n. \tag{3.12}$$

Further, submodular functions $f : \mathbb{R}^n_{\geq 0} \to \mathbb{R}$, must satisfy:

$$f(\mathbf{x}) + f(\mathbf{x}') \geq f(\max\{\mathbf{x}, \mathbf{x}'\}) + f(\min\{\mathbf{x}, \mathbf{x}'\}) \quad \text{for all } \mathbf{x}, \mathbf{x}' \in \mathbb{R}^n_{\geq 0}, \tag{3.13}$$

where the maximum and minimum operators are interpreted element-wise as in (3.11) and (3.12). If $f : \mathbb{R}^n_{\geq 0} \to \mathbb{R}$ is twice-differentiable, the inequality (3.13) is equivalent to the condition:

$$\frac{\partial^2 f}{\partial \mathbf{x}_i \mathbf{x}_j} \leq 0, \quad \text{for all} \;\; i \neq j. \tag{3.14}$$

As an example, recall that our IoBT setting considered the cost function:

$$f(\mathbf{x}) = \sum_{i=1}^{n} a_i \log_2 \left(1 + h_i \mathbf{x}_i\right). \tag{3.15}$$

This function is naturally submodular, since it is twice-differentiable and its cross-second derivatives are all zero. More succinctly, it is separable, and therefore a so-called "modular" function on $(\mathbb{R}^n_{\geq 0}, \preceq)$ [29].

As in the set function case, submodular functions on $\mathbb{R}^n_{\geq 0}$ also define a boundary between NP-Hard and efficiently solvable optimization problems. If $f : \mathbb{R}^n_{\geq 0} \to \mathbb{R}$ is submodular, then it can be approximately maximized within a constant-factor [30, 31], and efficiently minimized [21]. However, the existing algorithms for minimizing submodular functions on $\mathbb{R}^n_{\geq 0}$ rely on discretizing the space, which necessarily introduces error to the solution.

Now that we have identified submodularity as a key structure enabling efficient optimization in both continuous and discrete spaces, we return our attention to our resource allocation problem, defined over both of these spaces.

3.3 Problem Formulation

As established previously, the resource allocation problem in (3.1) involves intimately linked continuous and discrete decisions. Now that we have introduced submodularity and convexity as relevant theories in each domain, we use our IoBT allocation example to motivate the more general mathematical problem we tackle in this paper.

Recall that in our IoBT power allocation example, there were inherent costs for powering each IoBT edge device i that were incurred regardless of the continuous power allocation $\mathbf{x}_i$, and instead only depended on the binary decision to power device i. In this way, the start-up costs can be viewed as a function acting directly on the choice of nonzero entries of the vector $\mathbf{x} \in \mathbb{R}^n_{\geq 0}$.

More formally, for any $n \in \mathbb{Z}_{>0}$, we denote the set $\{1, 2, ..., n\}$ by $[n]$ and its power set by $2^{[n]}$. Then we define the support map supp : $\mathbb{R}^n_{\geq 0} \to 2^{[n]}$ as:

$$\text{supp}(\mathbf{x}) = \{i \in [n] \; : \; \mathbf{x}_i \neq 0\}, \tag{3.16}$$

which maps any $\mathbf{x} \in \mathbb{R}^n_{\geq 0}$ to the set of indices where it is nonzero. Then our function representing the startup costs is inherently a map $g : 2^{[n]} \to \mathbb{R}$, and we can re-write our disaster relief example (3.3) as:

$$\begin{aligned} \underset{\mathbf{x} \in \mathbb{R}^{m \times n}_{\geq 0}}{\text{minimize}} \quad & \textstyle\sum_{i=1}^{n} f_i(\mathbf{x}) + \lambda \, |(\text{supp}(\mathbf{x})| \\ \text{subject to} \quad & \textstyle\sum_{i=1}^{n} W_i(x_i) \leq B, \end{aligned} \tag{3.17}$$

where here $|\cdot| : 2^{[n]} \to \mathbb{Z}_{\geq 0}$ maps any set to its cardinality.

We can actually extend this problem setup to more general structures by replacing the sets $\mathbb{R}^n_{\geq 0}$ and $2^{[n]}$ with an arbitrary continuous space endowed with lattice structure $(\mathcal{X}, \preceq)$ and an associated discrete space also endowed with lattice structure $(\mathcal{Y}, \sqsubseteq)$, that are connected by a map $\eta : \mathcal{X} \to \mathcal{Y}$. Then, our more generic resource allocation problem with discrete allocation costs (3.1) is:

$$\begin{aligned} \underset{\mathbf{x} \in \mathcal{X}}{\text{minimize}} \quad & f(\mathbf{x}) + \lambda g(\eta(\mathbf{x})) \\ \text{subject to} \quad & \textstyle\sum_{i=1}^{n} W_i(\mathbf{x}_i) \leq B, \end{aligned} \tag{3.18}$$

where $W_i : \mathcal{X} \to \mathbb{R}$ is an increasing function defining the impact that allocating $\mathbf{x}_i$ amount of resources to task i has on our overall budget of $B \in \mathbb{R}_{\geq 0}$.

If the objective were a least-squares cost, i.e. $f(\mathbf{x}) = \| \mathbf{A}\mathbf{x} - \mathbf{b} \|_2^2$ for some $\mathbf{A} \in \mathbb{R}^{m \times n}$ and $\mathbf{b} \in \mathbb{R}^m$, and $g(\eta(\mathbf{x})) = |\text{supp}(\mathbf{x})| = \|\mathbf{x}\|_0$, the optimization problem in (3.18) becomes the NP-Hard *basis pursuit* (or compressed sensing) problem [12, 15, 16]. Therefore, without sufficient structure in the functions f, g, and the connecting function η, this problem is utterly hopeless. We imbue this structure on the problem by making several assumptions.

Assumptions Consider the lattices $(\mathcal{X}, \preceq)$ and $(\mathcal{Y}, \sqsubseteq)$ and the maps $\eta : \mathcal{X} \to \mathcal{Y}$, $f : \mathcal{X} \to \mathbb{R}$ and $g : \mathcal{Y} \to \mathbb{R}$. We assume:

1. The functions f and g are submodular on the lattices $\mathcal{X}$ and $\mathcal{Y}$, respectively,
2. The function g is monotone on $\mathcal{Y}$,
3. For all $\mathbf{x}, \mathbf{x}' \in \mathcal{X}$:

$$\eta(\mathbf{x} \vee \mathbf{x}') \sqsubseteq \eta(\mathbf{x}) \sqcup \eta(\mathbf{x}'), \quad \eta(\mathbf{x} \wedge \mathbf{x}') \sqsubseteq \eta(\mathbf{x}) \sqcap \eta(\mathbf{x}').$$

As we explain our theory in the proceeding sections, we will highlight the importance of each of these assumptions. We start by analyzing the unconstrained form of (3.18), and afterwards we show how to leverage submodular function theory to conveniently accommodate continuous or discrete budget constraints.

3.4 An Equivalent Parameterization

For simplicity, we start by examining the unconstrained problem, i.e.,:

$$\underset{\mathbf{x} \in \mathcal{X}}{\text{minimize}} \; f(\mathbf{x}) + \lambda g(\eta(\mathbf{x})). \tag{3.19}$$

The cost function in (3.19) suggests choosing a continuous allocation $\mathbf{x} \in \mathcal{X}$, then considering its incurred cost in the discrete space through $g(\eta(\mathbf{x}))$. One simple alternative is to instead choose a discrete value $\mathbf{y} \in \mathcal{Y}$ that this allocation will achieve, then select the associated continuous allocation $\mathbf{x} \in \mathcal{X}$ with $\eta(\mathbf{x}) = \mathbf{y}$. When considering all possible allocations that satisfy $\eta(\mathbf{x}) = \mathbf{y}$, we naturally ought to select the one that achieves the minimum cost as measured through $f(\mathbf{x})$. This intuition suggests re-expressing (3.19) as the following:

$$\underset{\mathbf{y}\in\mathcal{Y}}{\text{minimize}}\ g(\mathbf{y}) + \min_{\substack{\mathbf{x}\in\mathcal{X}\\ \eta(\mathbf{x})=\mathbf{y}}} f(\mathbf{x}), \tag{3.20}$$

where we have included the multiplicative factor $\lambda \in \mathbb{R}_{\geq 0}$ in the definition of g for brevity.

In our IoBT allocation example – neglecting power budget constraints – this corresponds to re-writing (3.3) as:

$$\underset{S\in 2^{[n]}}{\text{minimize}}\ g(S) + \min_{\substack{\mathbf{x}\in\mathbb{R}^n_{\geq 0}\\ \text{supp}(\mathbf{x})=S}} \sum_{i=1}^{n} f_i(\mathbf{x}). \tag{3.21}$$

Notice, however, that the second term (the inner-most minimization) in the cost function is ill-posed [2]. The inner-most minimization is over the set of vectors $\mathbf{x} \in \mathbb{R}^n_{\geq 0}$ such that $\text{supp}\,(\mathbf{x}) = S$, which requires some entries to be zero and *forces* others to be nonzero, and is not a compact subset of $\mathbb{R}^n_{\geq 0}$.

To remedy this issue, we propose the following similarly-inspired parameterization of the resource allocation problem:

$$\underset{\mathbf{y}\in\mathcal{Y}}{\text{minimize}}\ g(\mathbf{y}) + h(\mathbf{y}) \tag{3.22}$$

where we defined the function $h : \mathcal{Y} \to \mathbb{R}$ as:

$$h(\mathbf{y}) = \min_{\substack{\mathbf{x}\in\mathcal{X}\\ \eta(\mathbf{x})\sqsubseteq\mathbf{y}}} f(\mathbf{x}). \tag{3.23}$$

In our disaster relief problem, this corresponds to the optimization problem:

$$\underset{S\in 2^{[n]}}{\text{minimize}}\ g(S) + \min_{\substack{\mathbf{x}\in\mathbb{R}^n_{\geq 0}\\ \text{supp}(\mathbf{x})\subseteq S}} \sum_{i=1}^{n} f_i(\mathbf{x}). \tag{3.24}$$

The inner-most minimization defining h is now over a compact subset of $\mathbb{R}^n_{\geq 0}$, and therefore well defined.

This change in perspective on the decision variable in (3.24) corresponds to the following natural idea: if we select a subset of sites $S \in 2^{[n]}$ to send disaster relief to, we will necessarily allocate resources to those sites *optimally*. While a similar optimization problem, if we want to solve (3.22) instead of (3.19), we need to ensure that we recover the same optimal allocation. We show precisely this result in the following theorem.

Theorem 3.1 ***(Bunton and Tabuada [32])*** *Let the functions $g : \mathcal{Y} \to \mathbb{R}$ and $\eta : \mathcal{X} \to \mathcal{Y}$ satisfy Assumptions 2 and 3. Let $\mathbf{y}^* \in \mathcal{Y}$ be the minimizer of the parameterized problem (3.22), and let $\mathbf{x}^* \in \mathcal{X}$ be such that:*

$$h(\mathbf{y}) = \min_{\substack{\mathbf{x}\in\mathcal{X}\\ \eta(\mathbf{x})\sqsubseteq\mathbf{y}}} f(\mathbf{x}) = f\left(\mathbf{x}^*\right).$$

Then $\mathbf{x}^*$ is a minimizer of the original resource allocation problem (3.19).

Theorem 3.1 shows that indeed, under Assumptions 2 and 3, we can equivalently solve the parameterized problem (3.22) and still recover a minimizer to our original optimization problem (3.18).

The next concern is whether any of the structure enabling efficient solvability carries over to the new optimization problem. In particular, we care about submodularity, because the optimization problem (3.22) is over the discrete space $\mathcal{Y}$, and submodularity defines the critical boundary between a efficiently solvable and NP-Hard discrete optimization problems.

To this end, we would like to know when the function, $g + h : \mathcal{Y} \to \mathbb{R}$ is submodular, and thus the optimization problem (3.22) is efficiently solvable. In the following result, we show that under Assumptions 1 and 3, this structure does indeed carry over into the parameterized problem.

Theorem 3.2 ***(Topkis [29] and Bunton and Tabuada [32])*** *Consider the functions $f : \mathcal{X} \to \mathbb{R}$, $g : \mathcal{Y} \to \mathbb{R}$ and $\eta : \mathcal{X} \to \mathcal{Y}$. If Assumptions 1 and 3 are satisfied, then the function $g + h : \mathcal{Y} \to \mathbb{R}$, with h as defined in (3.23), is submodular on $\mathcal{Y}$.*

Theorems 3.1 and 3.2 together show that rather than trying to solve the mixed continuous and discrete resource allocation problem as first written, we can equivalently solve (3.22), which is a submodular function minimization problem over the discrete space $\mathcal{Y}$.

The submodularity of $g + h$ guaranteed by Theorem 3.2 implies that we can exactly minimize the function $g + h$ over $\mathcal{Y}$ in time that is polynomial in the size of the lattice, $|\mathcal{Y}|$, and the number of evaluations of the function $g + h$. However, by its definition (3.23), evaluating $h(\mathbf{y})$ for any $\mathbf{y} \in \mathcal{Y}$ requires solving an optimization problem over $\mathcal{X}$. In order to solve the problem (3.22) in fully polynomial time, we need some additional structure in the function $f : \mathcal{X} \to \mathbb{R}$ that makes evaluating h efficient.

In our IoBT power allocation example on the lattices $(\mathbb{R}^n_{\geq 0}, \preceq)$ and $(2^{[n]}, \subseteq)$ connected by the map $\text{supp} : \mathbb{R}^n_{\geq 0} \to 2^{[n]}$, evaluating the function $h : 2^{[n]} \to \mathbb{R}$ requires solving the minimization:

$$h(S) = \min_{\substack{\mathbf{x} \in \mathbb{R}^n_{\geq 0} \\ \text{supp}(\mathbf{x}) \subseteq S}} \sum_{i=1}^{n} f_i(\mathbf{x}). \tag{3.25}$$

If each of the functions f_i is a convex function over $\mathbb{R}^n_{\geq 0}$, as is the case in our disaster relief example, then this minimization is actually a convex optimization problem, for which we can apply any number of efficient minimization algorithms.

This observation guarantees we can exactly solve this specific parameterized problem (3.22) in polynomial time, as formally stated in the following corollary.

Corollary 3.1 ***(Bunton and Tabuada [32])*** *Let $f : \mathbb{R}^n_{\geq 0} \to \mathbb{R}$ be a convex and submodular function (satisfying Assumption 1), and $\mathcal{Y}$ be a finite lattice with $g : \mathcal{Y} \to \mathbb{R}$ and $\eta : \mathcal{X} \to \mathcal{Y}$ satisfying Assumptions 2 and 3. Further, assume that for any $\boldsymbol{y} \in \mathcal{Y}$, the set of vectors $\boldsymbol{x} \in \mathbb{R}^n_{\geq 0}$ such that $\eta(\boldsymbol{x}) \sqsubseteq \boldsymbol{y}$ is a convex subset of $\mathbb{R}^n_{\geq 0}$. Then evaluating the function $h : \mathcal{Y} \to \mathbb{R}$ defined in (3.23) is a polynomial time operation, and the problem:*

$$\underset{\boldsymbol{x} \in \mathbb{R}^n_{\geq 0}}{\text{minimize}}\, f(\boldsymbol{x}) + g(\eta(\boldsymbol{x}))$$

can be exactly solved in polynomial time.

Note that in our IoBT allocation example, we encountered an example of a fully separable cost, meaning each utility function $f_i : \mathbb{R}^n_{\geq 0} \to \mathbb{R}_{\geq 0}$ was only a function of the power allocated to device i. While this phenomenon is not uncommon, more complex utility functions exist in literature that may satisfy the assumptions of Corollary 3.1 [4, 5, 7].

3.5 Returning to Constraints

The previous section considered general lattices and, for simplicity, disregarded the constraints in the resource allocation problem. We now focus on the case of the lattices $(\mathbb{R}^n_{\geq 0}, \preceq)$ and $(2^{[n]}, \subseteq)$ connected by the map supp : $\mathbb{R}^n_{\geq 0} \to 2^{[n]}$, and reconsider the budget constraints:

$$\begin{aligned} &\underset{\mathbf{x}\in\mathbb{R}^n_{\geq 0}}{\text{minimize}}\ f(\mathbf{x}) + \lambda g\,(\text{supp}\,(\mathbf{x})) \\ &\text{subject to}\quad \textstyle\sum_{i=1}^{n} W_i(\mathbf{x}_i) \leq B, \end{aligned} \tag{3.26}$$

where $W_i : \mathbb{R}_{\geq 0} \to \mathbb{R}_{\geq 0}$ is an nondecreasing function, and $B \in \mathbb{R}_{\geq 0}$ represents some finite budget.

Unfortunately, general constrained submodular function optimization is NP-Hard, so we cannot hope to perfectly solve (3.26), even with our parameterization method [33]. We can, however, take the usual approach and augment the cost with a Lagrange multiplier $\mu \in \mathbb{R}_{\geq 0}$ associated with the budget constraint, creating the unconstrained optimization problem:

$$\underset{\mathbf{x}\in\mathbb{R}^n_{\geq 0}}{\text{minimize}}\ f(\mathbf{x}) + \lambda g\,(\text{supp}\,(\mathbf{x})) + \mu \textstyle\sum_{i=1}^{n} W_i(\mathbf{x}_i). \tag{3.27}$$

Despite the non-convexity of the functions, there exists a choice of $\mu \in \mathbb{R}$ that renders the unconstrained and constrained optimization problems equivalent [22]. Determining this μ, however, is difficult in practice.

Rather than searching for this particular μ, we instead leverage a result from submodular function theory that connects the solution of one convex optimization problem to the solution to (3.27) for all possible regularization values $\mu \in \mathbb{R}_{\geq 0}$.

The result we apply uses the Lovász extension of a submodular set function, a critical concept from submodular function theory. Given a submodular set function $\phi : 2^{[n]} \to \mathbb{R}$, its Lovász extension is a continuous and *convex* function $\phi_L : \mathbb{R}^n_{\geq 0} \to \mathbb{R}$ [34]. In particular, it is defined as the *tightest convex envelope* of the set function ϕ in the following manner: for any set $A \in 2^{[n]}$, define the indicator vector $\mathbf{1}_A \in \mathbb{R}^n_{\geq 0}$ as:

$$(\mathbf{1}_A)_j = \begin{cases} 1, & j \in A \\ 0 & j \notin A, \end{cases} \tag{3.28}$$

and the Lovász extension evaluates to exactly the set function on the same set, $\phi_L(\mathbf{1}_A) = \phi(A)$.

Theorem 3.3 ***(Proposition 8.4 in [34])*** *Let $\phi : 2^{[n]} \to \mathbb{R}$ be a submodular set function, and $\phi_L : \mathbb{R}^n \to \mathbb{R}$ be its Lovász extension. If, for some $\epsilon \geq 0$, $\psi_i : \mathbb{R}_{\geq\epsilon} \to \mathbb{R}$ is nondecreasing everywhere but strictly increasing on an interval $[\epsilon, c_i]$ for all $i = 1, 2, \ldots, n$, then the minimizer $\mathbf{u}^* \in \mathbb{R}^n$ of the convex optimization problem:*

$$\underset{\mathbf{u}\in\mathbb{R}^n_{\geq 0}}{\text{minimize}}\ \phi_L(\mathbf{u}) + \sum_{i=1}^{n} \int_{\epsilon}^{\epsilon+\mathbf{u}_i} \psi_i(\mu) d\mu, \tag{3.29}$$

is such that the set $S^\mu = \{i \in [n] : \mathbf{u}^*_i > \mu\}$ is the minimizer with smallest cardinality for the submodular set function minimization problem:

$$\underset{S\in 2^{[n]}}{\text{minimize}}\ \phi(S) + \sum_{i\in S} \psi_i(\mu), \tag{3.30}$$

for any $\mu \in \mathbb{R}_{\geq\epsilon}$.

Theorem 3.3 states that discrete optimization problems consisting of a submodular set function plus an additive function with a single parameter as in (3.30) are fundamentally linked with the convex optimization problem (3.29). In particular, we can threshold the solution of (3.29) by any $\mu \in \mathbb{R}_{\geq\epsilon}$ to recover the minimal (in the sense of set cardinality) minimizer for the set function optimization problem (3.30) with parameter value μ.

Note here that evaluating the Lovász extension and computing its subgradient are possible in linear time for submodular functions. Moreover, this process only requires evaluations of the function on various subsets in $2^{[n]}$. We can therefore apply any convex minimization algorithm that uses subgradients to solve the optimization problem (3.30).

This same result was previously used to recover some solutions to cardinality-constrained set function minimization [33], then extended and applied to functions that are submodular on the continuous lattice $(\mathbb{R}^n_{\geq 0}, \preceq)$ [21, 22]. Our work, in contrast, straddles both of these cases. To further leverage this result, we show how two classes of budget constraints and cost functions f allow us to write the mixed continuous and discrete problem with Lagrange multiplier μ in the required form of (3.30).

3.5.1 Knapsack Constraints

The first form of constraint we consider is a separable knapsack constraint, which occur when the functions $W_i : \mathbb{R}_{\geq 0} \rightarrow \mathbb{R}$ only depend on the zero pattern of their arguments. In particular, knapsack constraints are of the form:

$$\sum_{i=1}^{n} W_i(\mathbf{x}_i) = \sum_{j \in \text{supp}(\mathbf{x})} \mathbf{w}_j, \tag{3.31}$$

for some vector $\mathbf{w} \in \mathbb{R}^n_{\geq 0}$. As a special case for our IoBT allocation example, when $\mathbf{w}_i = 1$ for all i, the budget constraint corresponds a limit $B \in \mathbb{R}_{\geq 0}$ on the *number* of edge devices we can power. This particular example is often written as an ℓ_0 pseudo-norm constraint.

The resulting Lagrange multiplier problem (3.27) is:

$$\underset{x \in \mathbb{R}^n_{\geq 0}}{\text{minimize}}\ f(\mathbf{x}) + g(\text{supp}\,(\mathbf{x})) + \mu \sum_{j \in \text{supp}(\mathbf{x})} \mathbf{w}_j. \tag{3.32}$$

If we follow the steps outlined in the previous sections and write the parameterized problem (3.22) for the Lagrange multiplier problem (3.32), we have:

$$\underset{S \in 2^{[n]}}{\text{minimize}}\ g(S) + h(S) + \sum_{j \in A} \psi_j(\mu), \tag{3.33}$$

where we defined the functions $\psi_j : \mathbb{R} \rightarrow \mathbb{R}$ as $\psi_j(\mu) = \mu \mathbf{w}_j$ for all $j = 1, 2, ..., n$. Because the vector $\mathbf{w} \in \mathbb{R}^n_{\geq 0}$ is nonnegative, the optimization problem (3.33) is in precisely the form (3.30) with $\phi = g + h$, $\epsilon = 0$, so $c_i = \infty$, thus we can apply Theorem 3.3.

Corollary 3.2 *The solution $\boldsymbol{u}^* \in \mathbb{R}^n_{\geq 0}$ of the convex optimization problem:*

$$\underset{\boldsymbol{u} \in \mathbb{R}^n_{\geq 0}}{\textit{minimize}}\ g_L(\boldsymbol{u}) + h_L(\boldsymbol{u}) + \frac{1}{2} \sum_{i=1}^{n} \mathbf{w}_i \boldsymbol{u}_i^2, \tag{3.34}$$

is such that the set $S^\mu = \{i \in [n] \ :\ \boldsymbol{u}_i^* > \mu\}$ is the minimizer of smallest cardinality for the problem (3.32) for all possible values of $\mu \in \mathbb{R}_{\geq 0}$.

Algorithmically, we solve the convex optimization (3.34) to recover the solutions S^μ of (3.32) for all $\mu \in \mathbb{R}_{\geq 0}$, then select the smallest $\mu \in \mathbb{R}_{\geq 0}$ such that the budget constraint, $\sum_{j \in S^\mu} \mathbf{w}_j \leq B$, is satisfied.

3.5.2 Continuous Budget Constraints

Another family of naturally occurring budget constraints arise when the functions $W_i : \mathbb{R}_{\geq 0} \to \mathbb{R}_{\geq 0}$ are *continuous*, convex, and strictly increasing. Following the same Lagrange multiplier approach as above, we construct the unconstrained problem:

$$\underset{\mathbf{x}\in\mathbb{R}^n_{\geq 0}}{\text{minimize}} \; f(\mathbf{x}) + g(\text{supp}\,(\mathbf{x})) + \mu \sum_{i=1}^{n} W_i(\mathbf{x}_i). \tag{3.35}$$

Proceeding again to the parameterized problem (3.22) leads to:

$$\underset{S\in 2^{[n]}}{\text{minimize}} \; g(S) + h(S, \mu), \tag{3.36}$$

where we have modified the definition of the function $h : 2^{[n]} \times \mathbb{R}_{\geq 0} \to \mathbb{R}$ to accommodate the multiplier μ as:

$$h(S, \mu) = \min_{\substack{\mathbf{x}\in\mathbb{R}^n_{\geq 0} \\ \text{supp}(\mathbf{x})\subseteq S}} f(\mathbf{x}) + \mu \sum_{i=1}^{n} W_i(\mathbf{x}_i). \tag{3.37}$$

If we would like to match the structure required to apply Theorem 3.3, we need h to be separable. This is only possible when the function $f : \mathbb{R}^n_{\geq 0} \to \mathbb{R}$ in (3.37) is also separable.

If we make this required assumption, in particular:

$$f(\mathbf{x}) = \sum_{i=1}^{n} f_i(\mathbf{x}_i), \tag{3.38}$$

then we can consider the separable minimizations in (3.37) component-wise as:

$$h_i(\mu) = \min_{v\in\mathbb{R}_{\geq 0}} f_i(v) + \mu W_i(v), \tag{3.39}$$

and match the structure required by Theorem 3.3, as (3.36) becomes:

$$\underset{S\in 2^{[n]}}{\text{minimize}} \; g(S) + \sum_{j\in S} h_j(\mu). \tag{3.40}$$

Finally, we require each function $h_i : \mathbb{R}_{\geq 0} \to \mathbb{R}$ to be nondecreasing everywhere and strictly increasing on some interval $[\epsilon, c_i]$.

Lemma 3.1 *Let $W_i : \mathbb{R}_{\geq 0} \to \mathbb{R}_{\geq 0}$ be strictly increasing and convex for each $i = 1, 2, \ldots, n$. Then the functions $h_i : \mathbb{R}_{\geq 0} \to \mathbb{R}$ defined in (3.39) are nondecreasing everywhere, and strictly increasing on the interval $[0, c_i]$, where $c_i \in \mathbb{R}_{\geq 0}$ is the smallest constant such that $h_i(c_i) = h_i(0)$. In addition, each h_i is constant and equal to $h_i(0)$ on the interval $[c_i, \infty]$.*

Therefore, by Lemma 3.1, we can apply the results and suggested algorithmic approach of Theorem 3.3.

Corollary 3.3 *Let $W_i : \mathbb{R}_{\geq 0} \to \mathbb{R}_{\geq 0}$ be strictly increasing and convex for each $i = 1, 2, \ldots, n$. Then the solution of the convex optimization problem:*

$$\underset{\boldsymbol{u}\in\mathbb{R}^n_{\geq 0}}{\textit{minimize}} \; g_L(\boldsymbol{u}) + \sum_{i=1}^{n} \int_{\epsilon}^{\epsilon+\boldsymbol{u}_i} h_i(\mu)d\mu, \tag{3.41}$$

is such that the set $S^{\mu} = \{i \in [n] \; : \; \boldsymbol{u}_i^* > \mu\}$ is the minimizer of smallest cardinality in the problem (3.40) for all $\mu \in \mathbb{R}_{\geq \epsilon}$.

It may be alarming to see the *integral* of the function h_i appear in the cost function of (3.41). However, to compute a subgradient of this integral, we only need to evaluate h_i at a given value of $\mathbf{u}_i$, which is tractable by the convexity of f_i and W_i.

As before, we can use any desired convex optimization routine to produce the entire family of minimizers S^μ. We then select the smallest value of μ whose minimizer produces (via Theorem 3.1) an allocation $\mathbf{x}^* \in \mathbb{R}^n_{\geq 0}$ such that the budget constraint is satisfied.

3.6 Computational Examples

In this section, we illustrate our theory and evaluate the proposed approaches with some proof-of-concept examples on the lattices $(\mathbb{R}^n_{\geq 0}, \preceq)$ and $(2^{[n]}, \subseteq)$, connected by the map $\text{supp} : \mathbb{R}^n_{\geq 0} \to 2^{[n]}$.

3.6.1 Unconstrained Optimization

We begin with a simple unconstrained quadratic programming problem with a monotone and submodular set function regularizer. Because we are considering the unconstrained problem, we can compare against other current state-of-the art algorithms.

Existing algorithms for minimizing continuous submodular functions discretize each dimension of the domain $\mathbb{R}^n_{\geq 0}$ into k discrete points. To apply this algorithm, we consider the bounded subset of $\mathbb{R}^n_{\geq 0}$ and use the fastest algorithm, which is a Pairwise Frank–Wolfe approach to solving the discretized optimization problem [21]. We then plot all the corresponding results in using the middle shade of gray and label them *Cont Submodular*.

We also compare against the approach outlined in [20], which amounts to a particular choice of convex and submodular minimization algorithms in our theory. To implement this approach, we use IBM's CPLEX 12.8 constrained quadratic program solver in MATLAB® to evaluate the function h (as defined in (3.23)) and use Polyak's step-size rule for updating the step size [2]. We label the corresponding results with *Projected (Sub)Gradient* and plot using the lightest gray in figures.

Unlike [20], our proposed approach is agnostic to the choice of convex and submodular minimization algorithms. Therefore, we also use CPLEX to evaluate h, but instead use faster specialized algorithms, in particular the minimum-norm point algorithm [27] as implemented in MATLAB® by Krause [35], and a semi-gradient lattice pruning strategy [36]. Our results are plotted in black, and labeled *Min Norm + CPLEX* in figures.

We consider a simple denoising example, where we consider a signal $\mathbf{x} \in \mathbb{R}^n$ corrupted by some additive normally distributed noise $\mathbf{w} \sim \mathcal{N}(0, 0.1\mathbf{I})$. We would like to recover the smooth signal $\mathbf{x}$ from the noisy measurements $\mathbf{y} = \mathbf{x} + \mathbf{w}$, where smoothness here means variations between adjacent entries of $\mathbf{x}$ should be small. We would also like to endow the optimization problem with our knowledge that the true signal $\mathbf{x}$ consisted of a small number of adjacent nonzero values, separated by portions of signal that were zero.

The fit and smoothness we seek can be promoted with the convex and submodular cost function $f : \mathbb{R}^n \to \mathbb{R}$:

$$f(\mathbf{x}) = \frac{1}{2} \|\mathbf{x} - \mathbf{y}\|_2^2 + \sum_{i=1}^{n-1} (\mathbf{x}_i - \mathbf{x}_{i+1})^2. \tag{3.42}$$

We also consider searching for a vector $\mathbf{x}$ that consists of a small number of contiguous nonzero entries, which we can promote with the monotone and submodular set function regularizer $g : 2^{[n]} \to \mathbb{R}$ defined as:

$$g(A) = |A| + \#\text{int}(A), \tag{3.43}$$

where the function $\#\text{int}(A)$ counts the number of *sets* of sequential indices in the set A.

For experiments, we use the signal $\mathbf{x} \in \mathbb{R}^n$ shown in the top plot of Figure 3.1, with the noise-corrupted measurements $\mathbf{x} + \mathbf{w} = \mathbf{y} \in \mathbb{R}^n$ with an example shown in dotted orange. In this case, for the continuous submodular algorithm we discretize the compact set $[-1, 1]^n \subseteq \mathbb{R}^n$ into $k = 101$ distinct values per index.

We show the signals reconstructed by each algorithm from the noisy measurement in the second, third, and fourth plots in Figure 3.1, with the running time comparison over a small window of problem dimensions in the bottom right. As expected, the discretization that the continuous submodular minimization algorithm employs creates artifacts in the reconstructed signal. Moreover, as we expected, our approach serves as a compromise between the high speed of the continuous submodular minimization algorithm and the optimality of the projected subgradient descent method.

We also plot the cost function over iterations of each of the algorithm for an example instance in the bottom left plot of Figure 3.1 with $n = 100$. Again, the minimum-norm point algorithm converges rapidly to the minimum alongside the projected subgradient method, while the continuous submodular function minimization approach's discretization error prevents it from ever achieving global optimality.

3.6.2 Knapsack-Constrained Allocations

In many IoBT scenarios, we are confronted with large computation tasks that would take an excessive amount of time to run on a single, resource-constrained edge device. To combat this issue, we are asked to *split* the computation among a possible set of n computational resources that are available on an IoBT [37, 38].

If we let the value $\mathbf{x}_i \in \mathbb{R}_{\geq 0}$ represent the amount of computation we send to each available resource, then the computation required from all of the available resources can be thought of as a nonnegative vector $\mathbf{x} \in \mathbb{R}^n_{\geq 0}$. Our task then, is to determine how much compute we will send to each of our n available resources.

Several past works proposed a model based on modern portfolio theory for this problem [11, 37]. In this model, the cost function is:

$$f(\mathbf{x}) = \mathbf{x}^T \mathbf{Q} \mathbf{x} + \mathbf{p}^T \mathbf{x}, \tag{3.44}$$

where $\mathbf{p}$ is the cost of sending a unit of computation to resource i and $\mathbf{Q} \in \mathbb{R}^{n \times n}$ is the covariance of these costs. In the previously cited papers, these costs describe the actual monetary cost of purchasing cloud computation resources from large vendors such as Amazon or Google. More generally, however, we can view $\mathbf{p}_i \in \mathbb{R}_{\leq 0}$ as the expected reduction in computation time achieved by $\mathbf{x}_i$ amount of computation on resource i, with $\mathbf{Q} \in \mathbb{R}^{n \times n}$ a positive semidefinite matrix describing the interactions between these computational resources' performance.

It is also natural to assume that the available computational devices have some shared start-up costs. For example, if we access one server in a particular location, it costs us effectively no extra effort to access an additional server in the same location. Consider m groups of resources with shared start-up costs, i.e. where each group is represented by a set $G_j \in 2^{[n]}$ with start-up costs

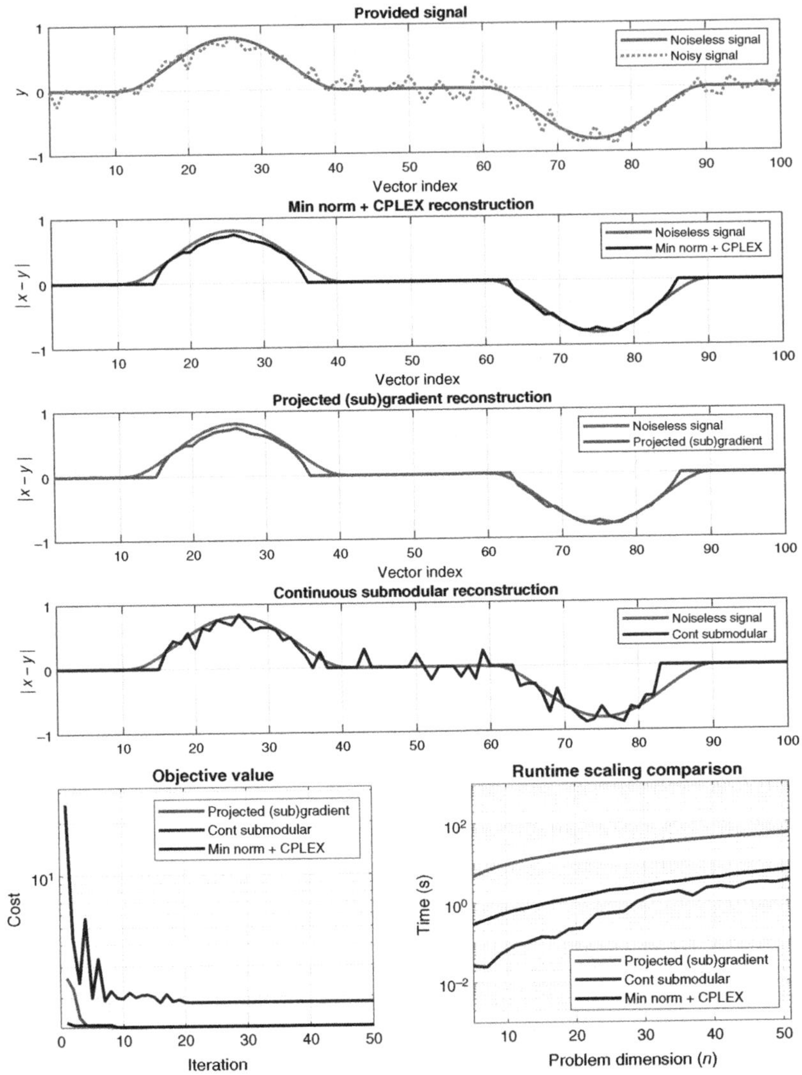

Figure 3.1 Unconstrained optimization problem results. The true signal and its noisy counterpart are shown in the top plot. The second, third, and fourth plots show the signals produced by each of the algorithms, with the lines in black corresponding to our proposed algorithm. Note that the results from the minimum-norm point algorithm and the projected subgradient descent method are globally optimal and thus identical. The bottom left plot shows the cost over iterations of each algorithm for an instance with $n = 100$, and bottom right shows the running times of each algorithm across some sample problem dimensions.

$\mathbf{v}_j \in \mathbb{R}_{\geq 0}$ for $j = 1, 2, \ldots, m$. Then the function measuring the shared start-up cost that a specific choice of computational resources incurs is:

$$g_{groups}(S) = \sum_{\substack{j=1 \\ G_j \cap S \neq \emptyset}}^{m} \mathbf{v}_j. \tag{3.45}$$

Each selected computational device will also require some initial effort to divide the task, access the resource, and power the device throughout the entire computation process. We represent the total cost of these efforts with the value $\mathbf{w}_i \in \mathbb{R}_{\geq 0}$ for each available resource $i = 1, 2, \ldots n$. Because we inherently have a limited amount of powering and logistical capabilities in our resource-constrained IoBT, say some $B \in \mathbb{R}_{\geq 0}$, we naturally face the constraint:

$$\sum_{j \in \text{supp}(\mathbf{x})} \mathbf{w}_i \leq B. \tag{3.46}$$

Combining the portfolio-inspired cost (3.44) with our shared start-up costs (3.45) and the support knapsack constraint, we have the optimization problem:

$$\begin{aligned} &\underset{\mathbf{x} \in \mathbb{R}^n_{\geq 0}}{\text{minimize}} \ \mathbf{x}^T \mathbf{Q} \mathbf{x} + \mathbf{p}^T \mathbf{x} + g_{groups}(\text{supp}\,(\mathbf{x})) \\ &\text{subject to } \textstyle\sum_{j \in \text{supp}(\mathbf{x})} \mathbf{w}_j \leq B. \end{aligned} \tag{3.47}$$

By Corollary 3.2, we can add a Lagrange multiplier $\mu \in \mathbb{R}_{\geq 0}$ associated with our knapsack constraint and solve a single associated convex optimization problem to recover the solution of (3.47) for all values of $\mu \in \mathbb{R}_{\geq 0}$. We take precisely this approach, starting with an initial $\mathbf{u}^{(0)} \in \mathbb{R}^n_{\geq 0}$ with each $\mathbf{u}^{(0)}_i \sim \text{unif}(0, 1)$ and applying generic projected subgradient descent:

$$\mathbf{u}^{(i+1)} = \max\left(0, \mathbf{u}^{(i)} - \alpha^{(i)} \xi(\mathbf{u}^{(i)})\right), \tag{3.48}$$

with step sizes $\alpha^{(i)} = 1/\sqrt{i}$, where $\xi : \mathbb{R}^n_{\geq 0} \to \mathbb{R}^n$ is a subgradient of the convex cost function. Again, we use CPLEX to evaluate the parameterized h function (3.23).

To generate a problem instance, we consider a moderately-size problem with $n = 50$ and construct $\mathbf{Q}$ according to:

$$\mathbf{Q} = \mathbf{C} + \mathbf{C}^T + n\mathbf{I}, \quad \mathbf{C}_{ij} \sim \text{unif}(-1, 0), \ \text{for all } i, j = 1, 2, \ldots, n. \tag{3.49}$$

Similarly, we generate the expected reduction in computation time as $\mathbf{p}_i \sim \text{unif}(-1, 0)$ for each resource $i = 1, 2, \ldots, n$.

This construction guarantees the quadratic form in (3.47) is both convex and submodular. This pattern of primarily zero and negative covariance matrix entries is also present in empirical data for cloud compute pricing [37].

We then create a random instance of the group-based start-up cost function g_{groups} by randomly selecting $m = 10$ subsets $G_j \in 2^{[n]}$ and weighting them with values $\mathbf{v}_j \sim \text{unif}(0.1, 0.15)$ for all $i = 1, 2, \ldots, m$.

In the left-hand plot of Figure 3.2, we show the cost of the convex optimization problem across iterations of projected gradient descent. At each iteration, we perform the thresholding step suggested by Corollary 3.2 on the current iterate $\mathbf{u}^{(i)}$ to find the value of μ such that the set S^μ satisfies the budget constraint. We then record the cost achieved by these corresponding discrete solutions over iterations in the right-hand plot of Figure 3.2.

As expected, we see a continued decrease of the convex function's cost, but since the solutions recovered by Theorem 3.3 only depend on the *ordering* of the vectors $\mathbf{u} \in \mathbb{R}^n_{\geq 0}$, the minimizing discrete set recovered with our thresholding technique does not often change. Note that the optimal value is not necessarily zero, since we are measuring the cost function directly and not the suboptimality (since computing the true optimizer requires exponential time). In fact, in some instances, the *ordering* of the minimizer $\mathbf{u}^*$, and therefore the minimizing discrete solution, was discovered after only a few iterations.

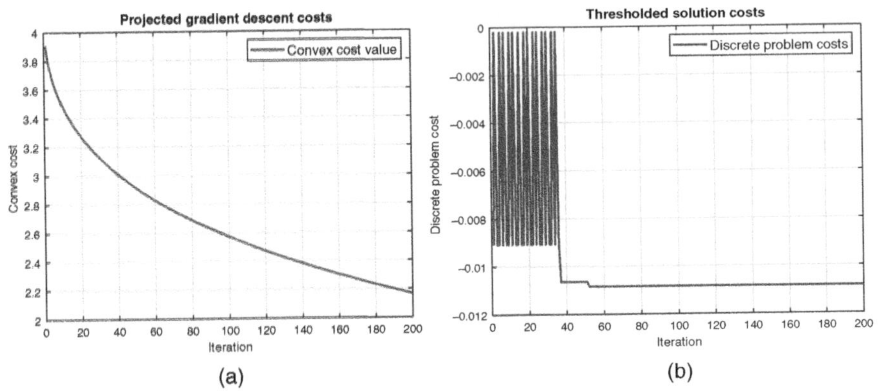

Figure 3.2 Results of the portfolio optimization problem. (a) The values of the convex optimization problem's cost function during the iterations of the projected subgradient descent. (b) The behavior of original constrained optimization problem's cost for the sets S^μ selected according to Theorem 3.3 with the current iterate, where μ is the smallest value of regularizer such that the knapsack constraint is satisfied.

3.6.3 Continuous Budget-constrained Allocations

We now return to the running example of a disaster relief scenario. Rather than simply considering the ℓ_0 norm of the allocation, we consider group-structured start-up costs g_{group} for the set of disaster sites of the form (3.45). In this context, shared start-up costs express that sending resources to site i may cost the same amount initially as sending resources to both sites i and i' if they are in a group $i, i' \in G_j$. This phenomena could happen naturally if the sites require the same physical equipment, or if they are nearby in location and therefore take no additional logistical effort to organize.

Together, this creates the optimization problem:

$$\begin{aligned} \underset{\mathbf{x}\in\mathbb{R}^n_{\geq 0}}{\text{minimize}} \ & \textstyle\sum_{i=1}^n a_i e^{-r_i \mathbf{x}_i} + g_{groups}(\text{supp}\,(\mathbf{x})) \\ \text{subject to} \ & \textstyle\sum_{i=1}^n \mathbf{w}_i \mathbf{x}_i \leq B. \end{aligned} \tag{3.50}$$

For proof-of-concept, we consider a moderately sized problem with $n = 100$. To generate a random instance of (3.50), we randomly select $m = 25$ subsets $G_j \in 2^{[n]}$ and weight them each with a value $\mathbf{v}_j \sim \text{unif}(0.05, .15)$ for each $j = 1, 2, \ldots, m$. We then randomly generate the expected number of initial issues according to $a_i \sim \text{unif}(1, 1.25)$, and the expected rate of issue resolution as $r_i \sim \text{unif}(0.75, 1.25)$ for each site. Finally, we construct the continuous budget constraint coefficients as $\mathbf{w}_i \sim \text{unif}(0, 1)$ and allow a maximum budget of $B = 1$.

Because the h_i functions as defined in (3.39) are scalar convex optimization problems, we explicitly derive expressions and for both h_i and its integral for simplicity and clearer visualization of the algorithm's progress. In practice, however, any arbitrary optimization engine could compute each h_i.

As before, we leverage Corollary 3.3 to solve a single convex optimization problem, thus recovering the solutions to the Lagrange multiplier problem for all values of $\mu \in \mathbb{R}_{\geq 0}$. We use the same projected gradient descent algorithm outline in (3.48) with identical step size selection. On the left-hand plot in Figure 3.3, we show the cost of the convex optimization problem over iterations after starting at a random initial $\mathbf{u}^{(0)} \in \mathbb{R}^n_{\geq 0}$ with each $\mathbf{u}_i^{(0)} \sim \text{unif}(0, 1)$.

At each iteration we perform the thresholding process of Corollary 3.3 on the current iterate $\mathbf{u}^{(i)}$ and construct the discrete set S^μ corresponding to an allocation satisfying the budget constraint. We then plot the achieved costs by these allocations in the right-hand plot of Figure 3.3. Note again that because the thresholded sets only depend on the *order* of the values in the optimal $\mathbf{u}^*$, we may discover an optimal discrete set before converging to the minimizer of the convex problem $\mathbf{u}^*$.

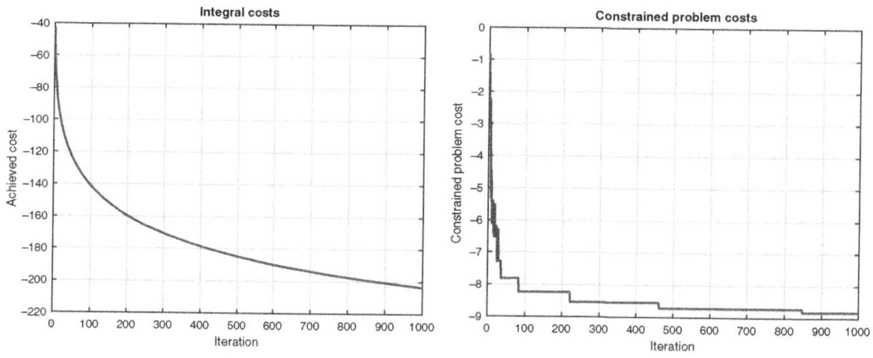

(a) (b)

Figure 3.3 Results of the portfolio optimization problems. (a) The values of the convex optimization problem's cost function during iterations of the projected subgradient descent algorithm. (b) Show cost achieved by the set S^{μ} selected according to the thresholding approach in Theorem 3.3 applied at each iteration, with μ the smallest value such that the continuous budget constraint is satisfied.

3.7 Conclusions

In this work, we expressed resource allocation problems with mixed continuous and discrete costs as optimization problems defined over two fundamentally connected lattices. We then derived sufficient conditions under which submodular and convex minimization algorithms can be easily married to produce a guaranteed optimal solution in polynomial time. In the context of internet of things (IoT), we showed that this class of optimization problems naturally arises when determining the optimal allocation of power/data rates on edge devices for an IoBT. Efficiently solving these problems means we can rapidly and optimally design, deploy, and operate IoBTs to meet mission requirements.

To accommodate realistic resource allocation scenarios, we also addressed two types of constraints that could be handled by our framework: knapsack and separable continuous budgets. We used a typical Lagrange multiplier technique to recover unconstrained optimization problems, but used a result from submodular function theory to recover the solution for all possible values of the Lagrange multiplier from a single convex minimization problem.

One of the most interesting directions of future research is relaxing the strong assumption of both convexity and submodularity in the cost function for resource allocation. While notions of approximate submodularity exist for discrete functions, they are less well-defined for functions over continuous lattices. Because the conditions in this work are sufficient but not necessary, the same algorithm-agnostic approach may be applicable up to some tolerable deviation from the assumed submodularity on $\mathbb{R}^n_{\geq 0}$.

References

1 Xiao, L., Johansson, M., and Boyd, S.P. (2004). Simultaneous routing and resource allocation via dual decomposition. *IEEE Transactions on Communications* 52 (7): 1136–1144. https://doi.org/10.1109/TCOMM.2004.831346.

2 Boyd, S. and Vandenberghe, L. (2004). *Convex Optimization*. Cambridge University Press.

3 Katoh, N., Shioura, A., and Ibaraki, T. (2013). *Resource Allocation Problems*, 2897–2988. New York: Springer. https://doi.org/10.1007/978-1-4419-7997-1_44.

4 Zhuang, B., Guo, D., Wei, E., and Honig, M.L. (2018). Large-scale spectrum allocation for cellular networks via sparse optimization. *IEEE Transactions on Signal Processing* 66 (20): 5470–5483. https://doi.org/10.1109/TSP.2018.2868046.

5 Kelly, F. (1997). Charging and rate control for elastic traffic. *European Transactions on Telecommunications* 8 (1): 33–37. https://doi.org/https://doi.org/10.1002/ett.4460080106.

6 Eriksson, K., Shi, S., Vucic, N. et al. (2010). Globally optimal resource allocation for achieving maximum weighted sum rate. *2010 IEEE Global Telecommunications Conference GLOBECOM 2010*, pp. 1–6. https://doi.org/10.1109/GLOCOM.2010.5683826.

7 He, Y., Zhang, S., Tang, L., and Ren, Y. (2020). Large scale resource allocation for the Internet of Things network based on ADMM. *IEEE Access* 8: 57192–57203. https://doi.org/10.1109/ACCESS.2020.2982293.

8 Mo, J. and Walrand, J. (2000). Fair end-to-end window-based congestion control. *IEEE/ACM Transactions on Networking* 8 (5): 556–567. https://doi.org/10.1109/90.879343.

9 Kondaveti, R. and Ganz, A. (2009). Decision support system for resource allocation in disaster management. *2009 Annual International Conference of the IEEE Engineering in Medicine and Biology Society*, pp. 3425–3428. https://doi.org/10.1109/IEMBS.2009.5332498.

10 Altay, N. (2013). Capability-based resource allocation for effective disaster response. *IMA Journal of Management Mathematics* 24 (2): 253–266. https://doi.org/10.1093/imaman/dps001.

11 Ali-Eldin, A., Westin, J., Wang, B. et al. (2019). SpotWeb: Running latency-sensitive distributed web services on transient cloud servers. *HPDC '19*, pp. 1–12. New York, NY, USA: Association for Computing Machinery. ISBN 9781450366700. https://doi.org/10.1145/3307681.3325397.

12 Rauhut, H. (2010). Compressive sensing and structured random matrices. *Theoretical Foundations and Numerical Methods for Sparse Recovery* 9: 1–92.

13 Hayashi, S. and Luo, Z. (2009). Spectrum management for interference-limited multiuser communication systems. *IEEE Transactions on Information Theory* 55 (3): 1153–1175. https://doi.org/10.1109/TIT.2008.2011433.

14 Shi, Y., Zhang, J., and Letaief, K.B. (2013). Group sparse beamforming for green cloud radio access networks. *2013 IEEE Global Communications Conference (GLOBECOM)*, pp. 4662–4667. https://doi.org/10.1109/GLOCOMW.2013.6855687.

15 Bach, F., Jenatton, R., Mairal, J., and Obozinski, G. (2012). Structured sparsity through convex optimization. *Statistical Science* 27 (4): 450–468.

16 Candes, E.J. and Tao, T. (2005). Decoding by linear programming. *IEEE Transactions on Information Theory* 51 (12): 4203–4215. https://doi.org/10.1109/TIT.2005.858979.

17 Zhao, P. and Yu, B. (2006). On model selection consistency of lasso. *Journal of Machine Learning Research* 7: 2541–2563.

18 El Halabi, M., Bach, F., and Cevher, V. (2018). Combinatorial penalties: which structures are preserved by convex relaxations? In: *Proceedings of the Twenty-First International Conference on Artificial Intelligence and Statistics, volume 84 of Proceedings of Machine Learning Research (PMLR)* (ed. A. Storkey and F. Perez-Cruz), pp. 1551–1560. PMLR (09–11 April 2018).

19 Elenberg, E.R., Khanna, R., Dimakis, A.G., and Negahban, S. (2018). Restricted strong convexity implies weak submodularity. *The Annals of Statistics* 46 (6B): 3539–3568.

20 El Halabi, M. and Jegelka, S. (2020). Optimal approximation for unconstrained non-submodular minimization. *International Conference on Machine Learning (ICML)*.

21 Bach, F. (2019). Submodular functions: from discrete to continuous domains. *Mathematical Programming* 175 (1–2): 419–459.

22 Staib, M. and Jegelka, S. (2019). Robust budget allocation via continuous submodular functions. *Applied Mathematics & Optimization*. https://doi.org/10.1007/s00245-019-09567-0.

23 Schrijver, A. (2003). *Combinatorial optimization: polyhedra and efficiency*, vol. 24. Springer Science & Business Media.

24 Davey, B. and Priestley, H. (2002). *Introduction to Lattices and Order*. Cambridge University Press.

25 Nemhauser, G.L., Wolsey, L.A., and Fisher, M.L. (1978). An analysis of approximations for maximizing submodular set functions—I. *Mathematical Programming* 14 (1): 265–294.

26 Fisher, M.L., Nemhauser, G.L., and Wolsey, L.A. (1978). An analysis of approximations for maximizing submodular set functions—II. In: *Polyhedral Combinatorics* (ed. M.L. Balinski and A.J. Hoffman), 73–87. Springer.

27 Fujishige, S. and Isotani, S. (2011). A submodular function minimization algorithm based on the minimum-norm base. *Pacific Journal of Optimization* 7 (1): 3–17.

28 Schrijver, A. (2000). A combinatorial algorithm minimizing submodular functions in strongly polynomial time. *Journal of Combinatorial Theory, Series B* 80 (2): 346–355. https://doi.org/https://doi.org/10.1006/jctb.2000.1989.

29 Topkis, D.M. (1978). Minimizing a submodular function on a lattice. *Operations Research* 26 (2): 305–321.

30 Bian, A.A., Mirzasoleiman, B., Buhmann, J., and Krause, A. (2017). Guaranteed non-convex optimization: submodular maximization over continuous domains. Proceedings of the 20th International Conference on Artificial Intelligence and Statistics, volume 54 of *Proceedings of Machine Learning Research*, pp. 111–120. PMLR (20–22 April 2017).

31 Bian, A., Levy, K., Krause, A., and Buhmann, J.M. (2017). Continuous dr-submodular maximization: Structure and algorithms. *Advances in Neural Information Processing Systems (NeurIPS)*, volume 30. Curran Associates, Inc.

32 Bunton, J. and Tabuada, P. (2020). Joint continuous and discrete model selection via submodularity. Volume 23: 1–42. http://jmlr.org/papers/v23/21-0166.html.

33 Nagano, K., Kawahara, Y., and Aihara, K. (2011). Size-constrained submodular minimization through minimum norm base. *International Conference on Machine Learning (ICML)*, pp. 977–984. Madison, WI, USA: Omnipress.

34 Bach, F. (2013). Learning with submodular functions: a convex optimization perspective. *Foundations and Trends® in Machine Learning* 6 (2–3): 145–373.

35 Krause, A. (2010). SFO: A toolbox for submodular function optimization. *Journal of Machine Learning Research* 11: 1141–1144.

36 Iyer, R., Jegelka, S., and Bilmes, J. (2013). Fast semidifferential-based submodular function optimization: extended version. *ICML*.

37 Sharma, P., Irwin, D., and Shenoy, P. (2017). Portfolio-driven resource management for transient cloud servers. *Proceedings of the ACM on Measurement and Analysis of Computing Systems* 1 (1). https://doi.org/10.1145/3084442.

38 Harlap, A., Chung, A., Tumanov, A. et al. (2018). Tributary: spot-dancing for elastic services with latency SLOs. *Proceedings of the 2018 USENIX Conference on Usenix Annual Technical Conference*, USENIX ATC '18, pp. 1–13. USA: USENIX Association. ISBN 9781931971447.

4

Operationalizing IoT Data for Defense and National Security

Steve Morgan[1] *and Jaime Wightman*[2]

[1] *Chief Technology Office, Raft LLC, Herndon, VA, USA*
[2] *Chief Data and Analytics Office, Lockheed Martin Corporation, Bethesda, MD, USA*

Abstract

Operationalizing IoT Data for Defense and National Security is the common language among all processes and digital systems, yet our military and systems were never designed to fully communicate end-to-end. Due to information protection policies enforced by government agencies and their defense contractors, even leveraging technology effectively across domains has become excessively complex. Foundational data architectures which can enable digital threads from the product or service concept or need, through the supply chain, to the military exercise, bringing parts to the point of need, have not been effectively implemented. There are many obstacles that are currently preventing Defense and National Security from using sensor data from their own operations to reduce the cycle time of the Observe–Orient–Decide–Act loop. This chapter will outline what many of those obstacles are, and what will need to be done to deliver scalable infrastructure to enable IoT data as the common language for the Defense and National Security enterprise.

4.1 Introduction

Data has been generated by sensors in military operations for decades. Attempts to capture data from sensors on a military asset, join it with other sensors, and use it to create a competitive advantage is how Internet of Things (IoT) data for defense and national security is being defined. This chapter will outline the problems being faced by the U.S. defense complex. The following section will enumerate the challenges faced with solving those problems. Special consideration will be given to the security constraints considered in defense, particularly to the threats that are present and the resources of our adversaries. A suggested strategy will be discussed, including recommendations on how to prioritize strategy based upon concepts that are already being developed. A discussion of the desired end state describing the advantages gained from effectively operationalizing IoT data for defense will lead into a conclusion.

IoT for Defense and National Security, First Edition. Edited by Robert Douglass, Keith Gremban, Ananthram Swami, and Stephan Gerali.

4.2 Problem Statement

The United States and our allies are facing significant challenges in leveraging data in a way to create an information advantage over our adversaries. The military systems that were designed 30–50 years ago were never thought to be able to communicate amongst each other seamlessly. There are dozens of tactical data links using microwave, Satellite Communications (SATCOM), fiber optic, and radio network transport establishing point to point and broadcast communications [1]. Continued investment into 5G has enormous promise to remediate the issues that exist with existing network protocols, but that does not address the challenges faced in joining two datasets, a fundamental aspect of system design which has been greatly overlooked when considering our military systems of systems, and the need for it to act as one system.

The problem begins with in-theatre communications. Battle management systems (BMS) have been designed at best to suit a single branch of the military, even those systems are really comprised of multiple lower-level battle management environments which do not actually communicate with each other. The "system integration" is handled by an operations analyst turning to a different monitor. Exercise debriefs are not data driven, and simple metrics like threat deterrence rate are not captured consistently and used to improve operations. In other words, IoT data is captured by the BMS, but that data is never shared amongst other BMS. Further adding to the problem, any IoT data that was captured in a given BMS, may not be useful in another BMS because it lacks additional contextualizing datasets.

The U.S. Military has claimed to be the world's greatest logistics organization. Moving the massive amounts of resources around the globe to respond to a threat or a disaster is an immense undertaking. Supporting operations once the military has established operations requires continued logistics support to supply basic needs as well as to repair any platforms deployed to the field. IoT data allows for insights into operations at near-real time speeds, hypothetically. There are signals being produced from assets in field. Planes, helicopters, drones, and any ground or sea equipment are expensive to operate. Any downtime is increasingly expensive. A drone being down for maintenance may mean that a special operations team could be operating in a contested space without aerial surveillance, making a mission too dangerous to operate. With the use of artificial intelligence (AI), it is becoming increasingly possible to predict when preventative maintenance can be performed. That predication can be made by gathering IoT signals from the device. Rather than waiting for a part to fail to begin the procurement process, it can be ordered before failure by predicting when a part will need to be replaced and shipped to the location of the aircraft. Thereby, drastically reducing the downtime of the aircraft and reducing costs.

Many of the processes of moving data today involve shipping portable hard disk drives. Some military bases do have server environments which can connect to broader infrastructure, but the reality is that moving data from the field to a location where it can be analyzed is not a priority when soldiers have been on-duty for the better part of a day. The means for automating the transfer of data to a long-term storage location are not broadly established. Data is moved from an asset by someone logging onto the technical system and initiating a transfer to disk is an assumed standard. That disk is then shipped at leisure to a place where data can be leveraged for improved operations. Military furnished 5G antennas available at commodity cost can be retrofitted to a wide variety of applications. The transfer of data will no longer be a manual process, but a process that is automated based upon the completion of a mission [11].

To improve logistics in the field requires incorporation of the military's suppliers into the data feed provided from the field. For the Original Equipment Manufacturers (OEM) that support the development and sustainment of the military to be able to deliver parts proactively and to reduce the time to develop new capabilities, they must be able to integrate their systems into the military

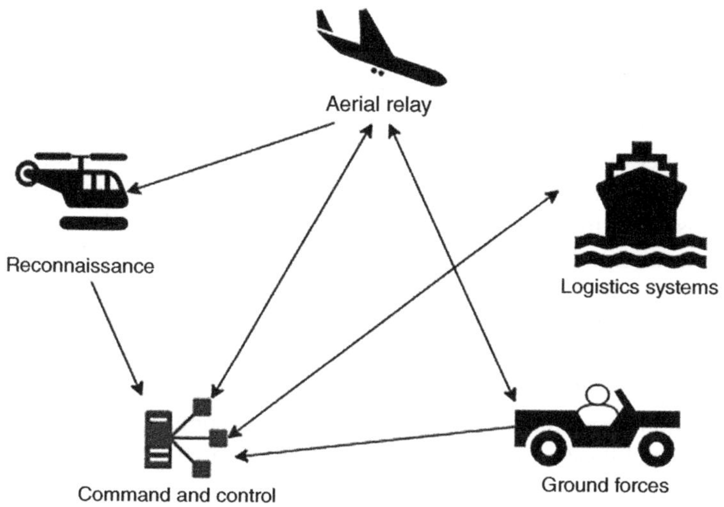

Figure 4.1 Current state: supply chain and military campaign. Source: Adapted from [2].

operations, and downstream into their own suppliers. Today, these integrations are limited. Even within military suppliers, there can be many unique part numbers across the engineering, product lifecycle management, and enterprise resource planning systems. These systems were designed as point solutions without consideration of the needs of the enterprise and today's extended and integrated missions. The result is that there are not common threads throughout the operational systems of our suppliers because they lacked the foresight to create data models and master data management systems that could support the level of integration required. In addition, the supplier model is continually driven by motivation leading to unique products that cannot communicate amongst the various platforms utilized for full mission effectiveness.

The process for ensuring the right parts and services are at the right place at the right time is enormously labor intensive, given the massively complex supply chain (from OEM, second, and third tiers suppliers). Gaining visibility into key digital threads that cross the extended supply chain is a huge challenge. Understanding the risk that flows into those suppliers would allow for OEMs to mitigate challenges proactively. Again, the reality is that using data as a common language is saddled with enormous technical debt and divergent priorities amongst suppliers. However, there needs to be direction and realization on how supply chain partners can collaborate in areas with standards to improve interoperability. Figure 4.1 depicts a high-level illustration of the problem, fundamental disconnects, and complexity. Each of the systems can provide real-time information which can be used to drive more precise and timely decisions. To operationalize IoT data, the defense system itself needs to be approached as the sum of all of its parts allowing for common data threads to be created.

The focus of the solutions has been largely centered around the technology, but that has nearly exclusively led to unique point-to-point connectivity which has essentially exacerbated the problem. The technology evolution of today will improve that situation, but there is more needed. The development of technology within siloed domains has led to IoT defense systems that are unable to communicate with one another. The problem this paper addresses is that our military systems now operate with limited visibility due to the lack of a holistic approach which has created large inefficiencies and caused for a truly connected IoT defense system to struggle to develop. The chapter will describe strategies including a holistic data fabric, improvements in the information security backbone, technology advancements, and treatment of data as a common language.

4.3 Challenges

Now that we have expanded the problem statement, we next review the challenges that need to be considered before discussing a solution strategy [3]. First, we should emphasize what addressing these challenges means. Will Roper explained the crushing costs of sustainment in his paper "There is No Spoon: The New Digital Acquisition Reality". He cited the American aerospace businessman Norman Augustine for his quote in his famous 1984 Laws: "In the year 2054, the entire defense budget will purchase just one aircraft." The paper stated the costs were from addressing the sustainment of an aging aircraft fleet. Addressing the challenges below provide the opportunity of addressing those costs through digital enablement to improve the decision making and use of data.

The Observe–Orient–Decide–Act (OODA) Loop, defined by Air Force Colonel John Boyd, has impacted the Department of Defense (DoD) [4]. The OODA Loop is essentially a four-step approach to decision-making that focuses on filtering available information, putting it in context, and quickly making the most appropriate decision while also understanding that changes can be made as more data becomes available. This is often used at the operational level during military campaigns but has been applied in the business domains to improve decision making.

Figure 4.2 illustrates how the OODA loop can be used to inform military decisions within the battlespace and the supply chain being disconnected from the process. Defense IoT and supply chain data can be joined to understand the precise timing of materials arriving for in-theatre operations. This realization with the operational tempo of today's military, and the complexity of the OEMs and the extended supply chain, is that the OODA Loop described in Figure 4.2 needs to extend to include the entire landscape. Later in the chapter the strategy for this extension is described. Figure 4.2 also illustrates the communication breaks across and within the entire supply chain.

It has taken a significant amount of time to evolve into the current situation. That is both from the standpoint of the landscape of technology, longevity of the various platforms, and the extensive landscape of technical debt that needs to be considered. The pace of technology change continues and is increasing in release and capability, and the state of industry has incorporated changes to start to bend the curve. IoT has only become a household phrase in the past decade. Many of the systems used by the military have been in service for dozens of years. The existing assets in the field that are producing data for the military IoT will need to be updated to conform to a new set

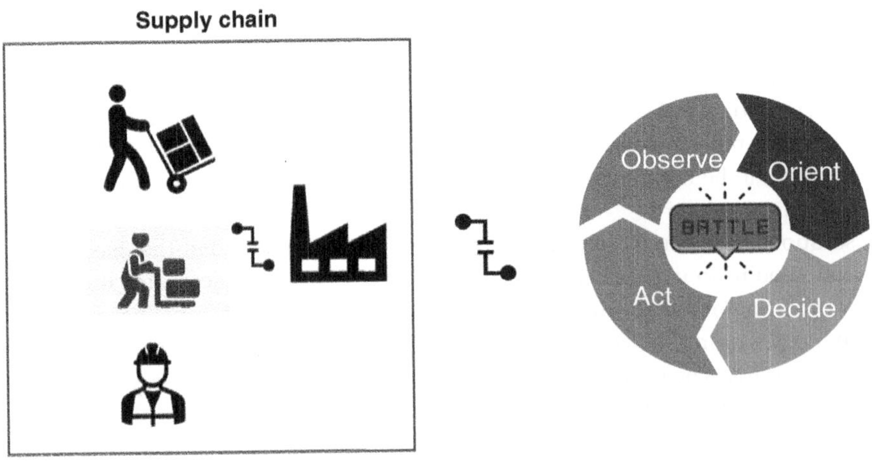

Figure 4.2 Current state: supply chain and military campaign (OODA included).

of standards. Technology alone will not address the needs or get to the desired end state, especially considering the lack of direction, availability, or consistency of open standards.

The area of standards is of particular importance going forward. It should be noted this is not just technology standards, but data and process standards need to be incorporated. Data standards can be described as documented agreements on representation, format, definition, structuring, tagging, transmission, use, and management of data. Process or business process standards can be described as documented agreements on representation, input, output, and outcomes from the executing business processes (e.g. purchase requisition, product delivery…). The ideal is not to expose supply chain intellectual property, but standardize the expectation for the business process execution to improve interoperability from well thought out data standards.

To make the need for data standards relatable, consider the efforts that were undertaken to create standard weights and measures. Great Britain has created 24 weights and measures acts since 1803, with the most recent taking place in 1985. Laws describing the need for standard measures were included in the Magna Carta, 588 years before the first weights and measures act. The importance of standard measures comes down to commerce, mostly [5]. Before there was a standard measurement for a yard, how could anyone compare prices between two markets? It would be likely that the yards from each market are two different lengths. The same can be said for anything that is purchased in measurable quantities. Something as simple as following a recipe is impractical if there are a dozen different definitions of an ounce.

The United States did not officially adopt standard time until 1918, but it was the railroad industry that first developed time zones for North America in 1883 [6]. The railroad companies had no choice. Scheduling was an absolute nightmare. Different towns and states were allowed to develop their own time systems. That generated massive inefficiencies and confusion. Can you imagine landing from a flight within your home time zone, and every clock reads differently than your watch? The need for standards translates into our data systems. Standards do not exist because solutions were developed for a specific need. Once those issues had been addressed, higher level problems are then exposed. Creating interoperable systems based on data standards has similarities to developing commerce, communication, and travel systems. Time is now spent on deciding how to translate data between different sources, and what the most efficient means of doing that will be.

The pace of standards changes, and development can be quite lengthy. History can help reinforce that idea with the amount of time it has taken to develop standard weights and measures. This would indicate the need to remove the approach of standards for all and focus on defining and prioritizing digital threads where standards can be of high impact (again, as history has shown). Typically, digital threads refer to the technology or communication framework that permits a connected data flow and integrated view of the asset's data throughout its lifecycle across usually siloed functional perspectives or organizations. Digital threads aim is to digitize and provide traceability throughout the entire lifecycle.

As illustrated in Figure 4.3, standardization of key digital threads adds to the overall supply chain effectivity. Digital threads can within the supply chain crossing one to many supplier tiers and extending to the OEMs. Digital threads need to first be identified to be able to standardize their definitions for implementation. Ideally, it would be of more initial value to focus on digital threads that cross more organizations within the supply chain.

The identification and standard definition of digital threads needs to focus on process standards foremost. This process standards first approach allows greater understanding and reference to the data that crosses the digital threads. Also, you can more effectively implement metrics and measures to determine process and data effectivity. Some key resources include the American Productivity & Quality Center (APQC) end-to-end process map and performance

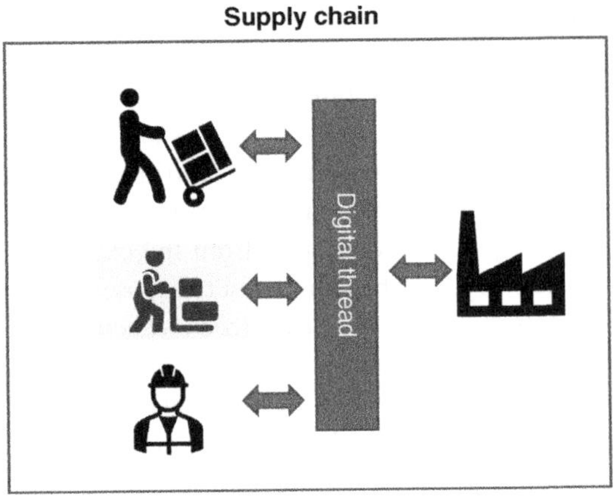

Figure 4.3 Standards based digital threads.

metrics. Particularly, resources like APQC can assist with process maps and aligning key performance indicators (KPIs) to measure effectivity improvements [7]. Secondly, define the key data/information elements that are required for your identified and defined digital threads and align those information elements to data standards.

Probably the hardest challenge is changing mindset. While all participants have the best intent, clarity needs to be brought about on the driving need for overall, end-to-end interoperability across platforms through the supply chain. The reward for giving up short term autonomy needs to be made clear and obvious. The same difficulty existed in creating standard weights and measures. People did not want to change the way they functioned. It would cost them in the short term, and the benefits were unclear. There was increased fear of taxation and dependency upon the government. This paper is not advocating for government intervention for standardized data but acknowledging that some level of governance is required for standards to be followed. When the appropriate level of standards is established and followed, then higher order problems can become the focus of our efforts.

This could include the need to support additional operating models which focus on integrated performance and continued delivery. These changes include contracting models, commercial-like contract models, and sustainment type contracting models like Performance Based Logistics (PBL). These types of contracting models are more "built" for how platforms and services perform and require continual integration or macro-level System of Systems type environment requirements.

4.4 Security Considerations

To be able to exploit digital threads the next challenge to be considered is the management of risks and Information Security requirements. The need for the utmost security requirements cannot be over-emphasized. However, if we define a security model and capabilities that encompass the end-to-end prioritized digital threads, security is not only protecting the warfighter, but it is an enabler to get to the desired end state.

This point alone is not without enormous challenges, but properly implemented security models can be an overall performance enabler. It is also noted that to be successful the need for well-designed architecture and cyber models requires focus on digital threads and their interactions. The current state perimeter-based security models are causing inefficiency in this very decentralized situation. Simply put, security evaluations must start from the holistic system and end-to-end processes and what it will achieve. A digital thread cannot be enabled by focusing on securing specific components – blind to the overall system. This brief section was not meant to answer what the security model should be but rather the need to jointly improve operational effectivity and security by considering holistic interoperability. Security is discussed in many sections because it must be integrated and not separated.

4.5 Developing a Strategy for Operationalizing Data

Mathematics is said to be the universal language. It can be understood throughout the world because common symbols are used regardless of how they are spoken. Data can become a common language amongst our systems and the users of those military systems. By developing and following industry standards, best practices, and protocols, differing organizations can begin to develop systems and data products which can be leveraged in a common way. Technology can provide the backbone for enabling systems to communicate in a ubiquitous manner. We are no longer constrained by the ability to scale data across multiple data centers and access points, what is ultimately missing is the ability to staff the appropriate engineering talent to fully leverage technology at scale. The tech industry and other peers in industry have made incredible advances in short periods of time. The defense community will need to continue developing partnerships outside of their own industry to accelerate capability deployment.

An analogy of data as a common language is Aviation English, which is the international language for civil aviation. Aviation English is a type of English for the specific purpose of communicating within civil aviation and is analogous to a data taxonomy (common language) on communications within civil aviation. The 300 phrase taxonomy allows for safe communication among pilots and air traffic control [8].

Creating the best practices, standards, and protocols to enable data as a common language will lead to data sets which are already following common formats, using common primary and secondary keys, enforce change data capture, provenance, version control, define meaningful quality standards, have consistent structures and processes for including metadata. Each of these areas are currently pain points which are generally dealt with in isolation. Not all of these are industry specific either. Creating common data structures, for example, could be done for supply chain data regardless of industry.

On the contrary, metadata definitions may be specific to the application and the program itself. Special Access Programs, which are generally thought of as more secure than Top Secret, will have their own metadata definitions depending on what need an application fills. For example, applied computer vision needs to capture information about what is in an image, and that computer vision application may be operating on photography from drones or manned aircraft. The same program may build special purpose weaponry and capture data from the factory floor regarding the build process. Sensor data from the factory floor would not be aware of the part it is working on. In the case of metadata management, industry may only be able to provide best practices. In the case of common formats for data types or critical data assets, industry standards could be followed.

The strategic implementation of data as a common language manifests itself as the implementation of a data fabric. A data fabric is fundamentally a modern data architecture and set of data services that provides consistent functionality across a variety of environments. From an overall perspective a data fabric delivers the following:

- Ability to adapt to new requirements.
- Capability to connect data at a data layer.

- Support cross functional data connections.
- Work within the constraints of much of the technical debt.

Fundamentally, a data fabric is simplification, in terms of adding meaning to the data and not just data. This means the data fabric is the data layer, more specifically it represents a semantic layer. Semantic layers map complex enterprise data into familiar business terms to provide a unified, consolidated view of data across systems and environments.

Organizations across the supply chain are struggling to manage and access all the data they control. Nearly 75% of data never gets used or analyzed [9]. This means, organizations are creating enormous amounts of data and never benefiting from it. This has led to a substantial gap between data sources and business users. Organizations undertaking digital transformation are using semantic layers to help improve this gap. There are more benefits to a semantic layer.

Semantic layers reduce complexity and prevent business users from having to spend time trying to become a data engineer when they need access to data. This means semantic layers can enhance productivity and collaboration while helping teams gain faster and more valuable insights from their data.

An enhanced semantic layer means businesses and the DoD can work together more effectively. By enabling ease of tagging, searching, and sharing data, collaboration is improved across the participating organizations, and teams can make better decisions in less time. At a basic level, semantic layers remove complexity from the process of finding, sharing, and analyzing data.

One of the largest problems with data is sharing, accessibility, and security. That can occur within an organization or be compounded across the supply chain. If the security model is too complex, users won't be able to do their jobs and will try to bypass security rules and make multiple copies of data. This can mean loss of integrity of the data and uncertainly of the system or record for the data. It is without question that accepting more risk is not an option, the needed desired state is a security model that is both efficient and effective. A semantic layer does foster improvements to the security model. Users can make changes to the data at a virtual level, which leads to less corrupting of data sources and systems of record.

An effective semantic layer, and the enablement of frictionless security controls, depend upon creating seamless means to capture metadata. Metadata can quickly become a loaded term, but the simplest definition is data about data. This does not mean that there is only one way to describe data sets. Metadata gives context. The context has a dependency on the object which it is describing. Describing an image may require time, location, source (machine or person), and lens, but also information describing what is in the image, in addition to any business or mission context. For example, suppose a photo was taken from an automated feed on a Blackhawk helicopter during a training mission at Hanscom Air Force Base on 12/5/2021 at 0900 Romeo as part of Operation Data Storm. The mechanisms to consistently and efficiently capture metadata are not well established. Data like that given in the example is often carried on as tribal knowledge, or knowledge that only lives within the people who were involved in the initial data capture process, unless they put that information into a data store when the data is captured. Labelling data is not a process that AI can take over until ground truth is established. That is a labor-intensive process when applied to creating bounding boxes of imagery and verifying the content of an image (for example, what's in the bounding box?). Once ground truth is established, it will need to be continuously validated by people to ensure that any model still is performing well and is not identifying false positives as true positives.

The concept of data sheets has been used to collect data during the COVID-19 pandemic. The spirit of the data sheet is to have a consistent series of questions for understanding the source of datasets and how they were expected to be used [10]. Understanding how a dataset is expected to be used plays a key role in consistent use of data. Certain aspects of metadata will need to have direct input, but the semantic layer cannot fully depend upon manual input. Use of simple automation when files are uploaded could give location and time information if desired. If a submitter is tied to certain programs or organizations, that information is perceivably maintained in other systems as well. AI can be used to augment missing information. Predicting missing values in a fixed domain is something that machine learning applications can excel at. The application of algorithms to do that is still in its infancy. Given the many gaps still existing in metadata capture, a robust and meaningful semantic layer is challenge that will take many years to address.

To enable consistent handling of metadata to create an effective semantic layer that is built on a common and standard data architecture, metadata creation protocols will need to be developed. For every data creation process, our ability to add context to the data relies upon bespoke practices that were tailored to a specific process in a specific organization. Most metadata may not be retained at all. The technology landscape to support a semantic layer and data catalogs, is relatively immature. To aid in the creation of effective semantic layers, protocols need to be developed to simplify the capture of metadata. When protocols are consistent, the underlying data structures will not require customization unless it is truly warranted.

This position on a semantic layer does not detract from the need for continued, vast improvements and adoptions in technology. The utmost is the promise of 5G. Availability of data is not necessarily the biggest problem that the DoD community faces. The pipelines to move the data to the right place at the right time are being solved by 5G. We now have the opportunity to operate with immensely improved bandwidth, more dynamically than we have in the past. To that end, any recommendations for 5G would be consistent with the DoD Strategy Implementation Plan:

1. Promote technology development,
2. Assess, mitigate, and operate through 5G vulnerabilities,
3. Influence 5G standards and policies, and
4. Engage partners.

The 5G Network Data Layer (NDL) presents the ability to dynamically add or drop nodes to a network without loss of data or unnecessary replication. That combined with the promise of bandwidth speed and concurrency of modern broadband Internet, is what makes 5G such a compelling opportunity. Bandwidth constraints are significantly reduced. Not just that, localized networks can be created rapidly. 5G antennas could be retrofitted to assets to allow them to share data amongst assets in a targeted theatre. For example, a drone swarm could come online at a desired point in time, perform coordinated actions alongside other assets, and return only the relevant data back to command and control [12]. The current process may not be able to communicate back to a battle management system within days. The information could remain stagnant for days. Reducing the cycle time of key information will allow for military leaders to expedite operations.

Revisiting how this impacts supply chain, when an asset is damaged or requires maintenance, that information can be relayed back to a supplier or base immediately. Assuming that suppliers can accept that information from the field or logistics management systems owned by the DoD, cycle time can be reduced dramatically. Order placement may only require approvals at certain points instead of making multiple phone calls between various organizations. The entire process can become automated.

The application of AI at the edge becomes entirely more practical in a 5G operating environment. The current process is bottle necked by network connectivity. It is common practice for data to be taken from an asset in field and loaded to hard disks. This may be a viable path with the volumes greater than 2 terabytes, as a rule of thumb. Most datasets that will be pulled from the field are unlikely to exceed that volume, particularly if data is sent back to a cloud environment on regular intervals (e.g. daily).

For AI to be reliable in the field, rapidly deploying updated and improved models is critical. The current process relies upon a cycle time which can be days in the best case, but likely weeks to months given dependencies on getting disks shipped to locations which they could be uploaded to centralized storage for training. Once data is uploaded to a centralized location, the work has really just begun. Data is coming in condensed formats which need to be extracted to pull any available metadata. If there is an absence of metadata, it may need to be reintroduced to the dataset which is a labor-intensive process. If there are issues with formatting or using standard formats, that work will need to be done as well. Once those steps complete (and many others), machine learning engineers can begin to train new models on datasets. Getting newly trained models back to the field faces the same challenges as returned data to a centralized environment in the first place. It could take weeks at minimum if the best path is through the movement of physical storage media. In an environment which is connected with 5G, cycle time can be reduced by weeks, effort will then be spent on higher order problems like ensuring data quality is enforced and models are fully reliable. The process of automatically updating models in a connected environment is not a conjecture, Tesla is putting it into practice with their self-driving systems today. The only primary difference is a dependence on Wi-Fi when a Tesla is parked at your home [13].

In a highly connected yet interrupted environment with national security implications, both in field and within the supply chain, threat vectors increase exponentially. The existing military systems have been designed to act with a high level of trust amongst differing assets. The communication methods are relatively fixed. That does not mean that those existing system are void of cyber threats, but that the new paradigm with 5G is much more dynamic with more risk. As data becomes more interoperable, the need for protecting the channels of communications increase as well. The rate at which technology changes will continue to increase. The DoD has accepted that need to match the rate of change experienced in commercial markets. Increasing our rate of change must not be hindered by security evaluation processes. Nicholas Chaillan, the outgoing Chief Software Officer of the Air Force and Space Force, recognized the need and championed Platform One. Platform One is a recognition of the need for the DoD to adopt DevOps practices. These automated scanning and deployment methods are imperative to the success of operationalizing IoT data for defense [14].

Our major strategic recommendations include a robust data strategy that promotes data as a common language across all organization trading partners (e.g. Supply Chain), the implementation would be promoted through a modern data architecture that is based on a data fabric that propagates a data layer that is based on a semantic model and continues expanded development of leading technology namely exploiting the use of 5G. In addition, there will likely need to be protocols for the extension and connectivity of the data architecture from the proposed data fabric through a 5G network.

> "The simple requirement will be from this day forward, all data produced by the Department of Defense, all data produced by every weapon system in the Department of Defense will be accessible, period... And the reason it has to be that way is because without that data, and without that data accessibility, we will not achieve the speed that we need to deal with the future we face."
>
> Vice Chairman of the Joint Chiefs of Staff, General John Hyten

4.6 Precedence

What has been presented in this chapter is a complex problem with a proposed approach to a strategic set of solutions. Next, we need to review the precedence or suggested priority to that approach.

What has been recommended are very strategic solutions and require a strong and active industry consortium. Developing an industry consortium will enable open discussions and provide the ability for industry partners to work together. Potentially, scope could be added to an existing industry consortium, if a new consortium is developed it might be a consortium or industry consortiums. The analogy for a consortium of consortiums is like the System of Systems model. This would ideally exploit more collaboration and use of best practices from multiple existing bodies. It is if of utmost importance to include diverse suppliers (i.e. multiple tiers) from this very diverse and complex supply chain. Without active participation from multiple tiers of suppliers, any outcome from the industry consortium might not be very inclusive.

There are a great many definitions to ontology. In summary, ontology, as a branch of philosophy, is the science of what is the types and structures of objects. In simple terms, ontology seeks the classification and explanation of entities. In computer science, ontology is a technical term denoting an artifact that is designed for a purpose, which is to enable the modeling of knowledge about some domain, real or conceptual. One of the initial tasks is the ontological representation of digital threads. This allows the depiction of core digital threads, and the eventual development of the proposed semantic layer. This would be good first action for the proposed industry consortium. There are several ontologies that could be used depending on the specific industry, below are a few examples:

- Interoperability Framework (IOF) Website (industrialontologies.org)
- Basic Formal Ontology (BFO)|Home (basic-formal-ontology.org)

It should also be noted that an enterprise canonical model standard (ISO-5054) could be used for a semantic model [e.g. Home (oagi.org)]. An enterprise canonical model is a design pattern that allows the authoritative communication of data between two different data formats.

Collaboration with commercial technology providers will enable the continued evolution of leading technology. Including, 5G networks and the extension and connectivity into the discussed data fabric, and the continued evolution of development processes (e.g. DevOps, DevSecOps…) which have accelerated many projects and technology initiatives.

Given the state of this solution, especially the decentralized nature of 5G networks, the traditional perimeter-based security model cannot be depended upon to manage risks. This requires the evolution of a Zero Trust model where participants are not specifically trusted based on their position of the network. Zero Trust is a security architecture based on validation of all users and devices no matter their position on the network. Zero Trust architecture is not only needed because of the decentralized state of 5G but to incorporate the supply chain and the decentralized state of the supply chain, which are key participants.

Zero Trust end-user capabilities improve visibility, control, and the risk posture of application and data usage [15]. These capabilities provide a secure environment for mission execution. To assess how these Zero Trust capabilities benefit the data interoperability that has been described in this section let's look at the key goals from the DoD Zero Trust Reference Architecture:

- *Modernize Information Enterprise to Address Gaps and Seams:* There is a need to modernize the extended information enterprise to deal with the situation of today but the dynamic new reality. This situation has evolved with decentralization and building infrastructure and solutions along organizational and operational lines with multiple security and support tiers and decentralized networks. These capabilities developed in silos, have resulted in disconnects and gaps in the processes and responsiveness.

- *Simplify Security Architecture:* Consider the fragmented approach to information technology and cybersecurity, and the technical complexity, of the situation depicted in Figure 4.1. Not only does this create cyber vulnerabilities, but there is a long delay of responsiveness to the warfighter and general unresponsiveness to the end user.
- *Produce Consistent Policy:* If we are going to improve interoperability across organizations, cyber policies must be consistently applied across environments (organizations) for secure and effective interoperability. The continued application of security policies based on perimeter defense systems that apply implicit trust based on network location will change the security posture or ability to interoperate across the supply chain.
- *Optimize Data Management Operations:* The success of missions, and supply chain participation, is overwhelmingly dependent on data availability and data that is understood. While data management practices and data and process standards exist, they are not consistent and inconsistently implemented. This results in the continual frustration and business delays across the entire landscape, and hampered abilities to fully leverage the benefits of cloud computing, data analytics, AI, and machine learning.

We described a strategic approach that includes technology, architecture, and processes changes. There is more needed. To that end we previously described a mindset that must change. This manifests into changes in contracting models, including commercial like contract models and sustainment type contracting models like PBL. These types of contracting model are more "built" for how platforms and services perform for this type of required continual integration (i.e. an overall System of Systems type environment requirements).

4.7 End State

Overall, to achieve the desired end state, as depicted by Figure 4.4, the need is a robust data strategy adoption by all the cognizant organizations. Figure 4.4 illustrates how OODA Loop information process should encapsulate the entire scenario of the supply chain and the active military campaigns. The perspective is inclusive in the decisions and information process as opposed to the

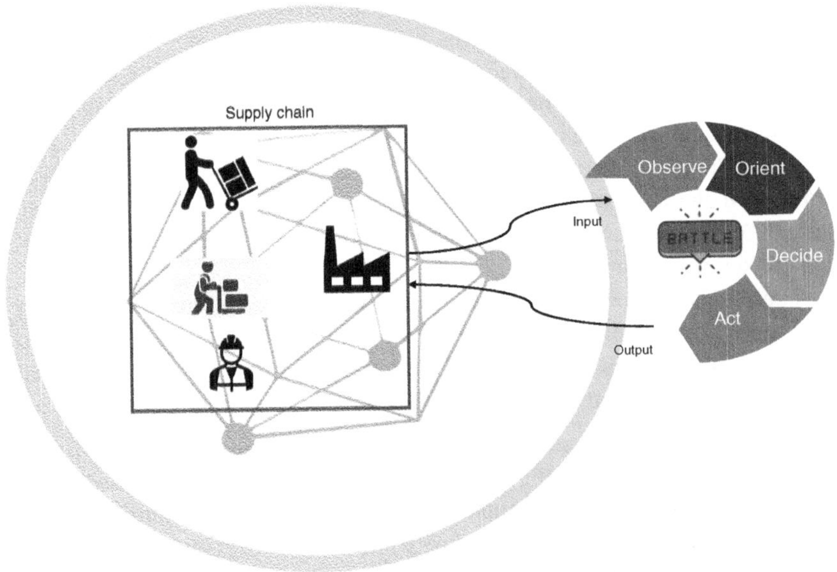

Figure 4.4 Future state: supply chain and military campaign (OODA included).

current state separation. This means the consumption of input into the OODA Loop and providing output to supply chain for further action.

Inside this OODA Loop you will see the deployment of a recommended modern data architecture based on a data fabric. This data fabric provides the application of a wide-ranging data layer instead of the current data and communication breakages and separations. The data fabric will require a security model based on a Zero Trust Architecture where permissions are earned rather than today's state security, based perimeters with minimal integration across perimeters.

4.8 Conclusion

This chapter was meant to define the problem statement and characteristics around operationalizing data for defense IoT. The current state and characteristics of the current state were summarized. The emphasis of this chapter was to describe a strategy and approach around operationalizing data for defense IoT. In addition, security considerations and ties to efficiency were described, but security was a thread that ran throughout this chapter. While the use cases focused on supply chain integration, the strategy described here is still usable and applicable to many other use cases.

This is a complex problem which developed over many decades. However, given the evolution of technology and data management practices, the problem can begin to be addressed by initiating the recommended strategic approach to improve data interoperability and decision making.

References

1 White, B.E. (1999). Tactical Data Links, Air Traffic Management, and Software Programmable Radios. https://www.mitre.org/sites/default/files/publications/white_tactical_data_links.pdf (accessed 29 July 2022).

2 Schantz, R.E. (2002). Characteristics of Military System of Systems. https://www.researchgate.net/figure/Characteristics-of-Military-System-of-Systems_fig8_2565962 (accessed 29 July 2022).

3 Roper, W. (2020). There is No Spoon: The New Digital Acquisition Reality. https://software.af.mil/wp-content/uploads/2020/10/There-Is-No-Spoon-Digital-Acquisition-7-Oct-2020-digital-version.pdf (accessed 29 July 2022).

4 Richards, C. (2004). *Certain to Win: The Strategy of John Boyd, Applied to Business*. Xlibris, Corp.

5 Wikopedia (2022). Weights and Measures Acts (UK), 30 June 2022. https://en.wikipedia.org/wiki/Weights_and_Measures_Acts_(UK) (accessed 29 July 2022).

6 History.com Editors (2009). Railroads Create the First Time Zones. https://www.history.com/this-day-in-history/railroads-create-the-first-time-zones (accessed 29 July 2022).

7 APQC (2021). End-to-End Process Maps and Measures: Procure-to-Pay, 10 March 2021. https://www.apqc.org/resource-library/resource-listing/end-end-process-maps-and-measures-procure-pay (accessed 29 July 2022).

8 International Civil Aviation Organization (2013). Language Proficiency Requirements (LPR). https://www.icao.int/safety/lpr/Pages/Language-Proficiency-Requirements.aspx. (accessed 29 July 2022).

9 Barrett, J. (2018). Up to 73 Percent of Company Data Goes Unused for Analytics. Here's How to Put It to Work| Inc.com, 12 April 2018. https://www.inc.com/jeff-barrett/misusing-data-could-be-costing-your-business-heres-how.html (accessed 29 July 2022).

10 Gebru, T., Morgenstern, J., Vecchione, B. et al. (2018). Datasheets for Datasets. https://arxiv.org/pdf/1803.09010.pdf.

11 DoD (2020). Department of Defense 5G Strategy Implementation Plan, 15 December 2020. https://www.cto.mil/wp-content/uploads/2020/12/DOD-5G-Strategy-Implementation-Plan.pdf (accessed 29 July 2022).

12 Lopez, D., Sun, T., Park, J.-G., and Henry, M. (2018). A Network Data Layer Concept for the Telco Industry. https://www.ngmn.org/wp-content/uploads/Publications/2018/180831_NDL_White_Paper_v1.0.pdf (accessed 29 July 2022).

13 Tesla (2022). *Artificial Intelligence & Autopilot*. Tesla. https://www.tesla.com/AI (accessed 29 July 2022).

14 Department of the Air Force (2022). Platform One. https://software.af.mil/team/platformone/ (accessed 29 July 2022).

15 DoD (2021). Department of Defense (DOD) Zero Trust Reference Architecture, February 2021. https://dodcio.defense.gov/Portals/0/Documents/Library/(U)ZT_RA_v1.1(U)_Mar21.pdf (accessed 29 July 2022).

5

Real Time Monitoring of Industrial Machines using AWS IoT

Stephan Gerali

Enterprise Operations, Lockheed Martin Corporation, Bethesda, MD, USA

Abstract

How can companies and the national defense implement an Industry 4.0 smart manufacturing solution for IoT data capture, transformation, analytics and visualization for real-time monitoring of manufacturing machines? This chapter covers multiple technologies in the IoT solution deployment to include: KepServerEX for edge connectivity to industrial protocols, AWS IoT Core for IoT data processing, Amazon S3 for scalable storage of IoT Data, MT Connect for a common data model for industrial machine telemetry data, Amazon EMR for interactive analytics of IoT data, Tableau for visualization of IoT data and unsupervised machine learning algorithms for automated anomaly detection for industrial telemetry data.

5.1 Problem Statement

The Internet of Things (IoT) has successfully disrupted our daily lives with the convenience, comfort and the valuable insights that it offers from smart thermostats to manage our homes, to remote door locks that lock our homes when we leave and to application-controlled appliances to help with cooking and cleaning for our everyday life experiences.

The same transformation that we are seeing in our daily lives is now making its way into industrial use cases to help businesses and the national defense become more efficient with the deployment of their capital in the way that they design, manufacture and sustain their operations going forward. The problem for most businesses and the national defense is how to get started connecting their equipment, identifying valuable data with the equipment to save, sharing the data across the business, visualizing the data to change business behaviors, identifying issues during the manufacturing process and to support further automation of their business processes.

This chapter will walkthrough an example use case to show how companies and the national defense can begin their IoT transformation process for their business to support connectivity to equipment, enable conditioned based monitoring of that equipment, developing a federated data storage mechanism for sharing data across the business, development of data analytics for automated anomaly detection, development of dashboards for management of equipment and support automation capabilities of the industrial equipment going forward.

It is important to note that any cloud based IoT solution must be interoperable in both public cloud deployments for commercial activities and in classified clouds for the national defense.

IoT for Defense and National Security, First Edition. Edited by Robert Douglass, Keith Gremban, Ananthram Swami, and Stephan Gerali.

The entire solution demonstrated in this paper supports both use cases for public cloud and classified cloud deployments using Amazon Web Services (AWS) [1].

5.2 Solution Statement – Overview

In order for business and the national defense to start the industrial transformation using IoT devices, they must begin to connect these devices to the network. The biggest issue faced with doing so is that many of these devices may be fairly old, use outdated versions of operating systems that cannot be patched, are no longer supported by the manufacturer or require add-ons in order to enable connectivity to the equipment itself.

The first step that most businesses and national defense must perform is to inventory all of their existing equipment assets and to categorize the equipment by the type of machine that it is and the tasks it performs. Once inventorying is complete, businesses and national defense should determine which devices that they need to gain more valuable insight from. These could be devices that require lots of manual labor today to expensive devices that need a full accounting of their utilization.

Once you have mapped which devices are needed, we must begin the process of determining how to connect those devices. Some may already have ethernet connections in which case simply bringing network connectivity is making sure you run the cables directly to the device. For some devices, you may need to purchase additional hardware in order to enable connectivity to that device. Further, some devices might even connect completely wirelessly. No matter what, having a plan of attack is critical so that you focus on the highest value targets and determining how to connect those devices to your network.

As was said before, many of these devices are using outdated operating systems and do not have existing connectivity to the Internet to support regular patching/maintenance of the device. Often patching requires downloading software to a Universal Serial Bus (USB) storage device and manually running the software on the intended machine to improve its security posture.

You will need an extra layer of protection and this is done by using firewalls or by creating Virtual Local Area Networks (VLANs) [2] where you can isolate the industrial equipment from all other devices and only enable whitelisted devices to connect to the industrial equipment to ensure the risk tolerances for outdated equipment is minimized.

5.3 Solution Statement – Edge Computing

In Figure 5.1, it visually depicts the devices that we want to connect (i.e. Computer Numerical Control devices [3], 3D printers [4], lathes [5], etc.). For each of those device types, we provide network connectivity to those devices and enable a firewall rule to enable connectivity from the industrial devices to a connectivity platform (i.e. KEPServerEX) that knows how to communicate with the device to manage and monitor its data (as shown in step 1).

KEPServerEX is an industry leading connectivity platform that provides a single mechanism to connect devices, manage, monitor and control diverse automation devices using automation standards like Open Platform Communications (OPC) [6]. OPC is an interoperability standard for the secure and reliable exchange of data between multiple vendor devices and controls applications without any proprietary restrictions [7]. An OPC server like KEPServerEX can communicate data continuously among multiple equipment devices. We use KEPServerEX to connect to the industrial equipment using the OPC standard and then rely on an edge computing platform that can take that data and store it within the cloud.

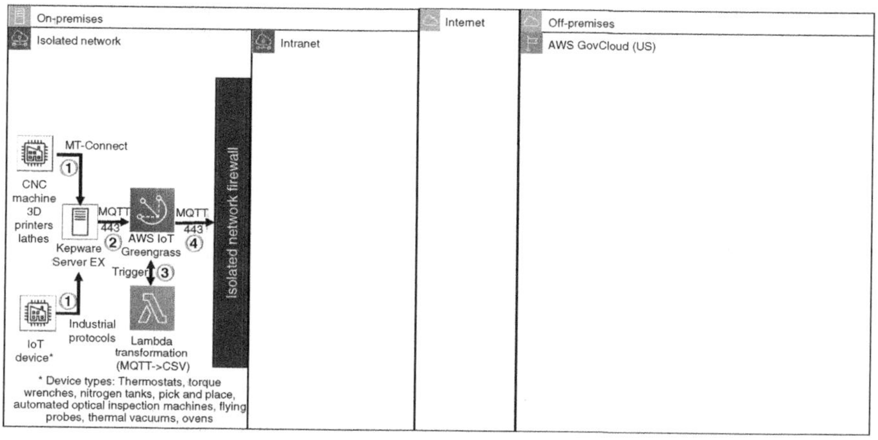

Figure 5.1 Edge computing is a distributed computing paradigm that brings computation and data storage closer to the sources of data to support better real-time processing [14]. Source: Stephan Gerali.

AWS has an open-source edge runtime service for connecting industrial data to the greater AWS ecosystem of services called AWS IoT Greengrass. AWS IoT Greengrass runs at the edge and provides connectivity to KEPServerEX to allow the industrial equipment data to come into its runtime service (as shown in step 2) [8].

AWS IoT Greengrass provides a serverless event-driven compute service called AWS Lambda that lets you run code for any type of backend service without provisioning or managing services [9]. We leverage AWS Lambda to transform data from AWS IoT Greengrass using the Message Queue (MQ) Telemetry Transport (MQTT) [10] message format which is an Open Applications Group Integration Specification (OAGIS) [11] standard messaging protocol for IoT into a common data format for all devices to a MTConnect [12] data standard which is an American National Standards Institute (ANSI) [13] standard for manufacturing equipment to provide structured, contextualized data with no proprietary format (as shown in step 3).

Identifying how you want to store your IoT data, leveraging a common data format and then transforming the data as it comes into a common data format allows you to propagate that data downstream to any other systems that need to process it in a common way.

Once the data is formatted into a common data format, AWS IoT Greengrass can propagate that data from the edge network outside into the cloud (as is shown in step 4) using MQTT message format as the intended wrapper for the message.

5.4 Solution Statement – Cloud Connectivity

In Figure 5.2, it visually depicts the movement of data from the edge network directly into your cloud-based services. Since AWS IoT Greengrass is running at the edge network, it will need to leverage a proxy server to get outside the firewall (as shown in step 5). The traffic is then pushed across the Internet (as shown in step 6) and finally reaches the intended cloud service you intend to use to store and manage the associated data (as shown in step 7).

For the purposes of this example, we are leveraging AWS IoT Core which allows you to connect millions of IoT devices and route messages to AWS services without having to manage the underlying infrastructure [15]. AWS IoT Core provides secure device connections using mutual authentication and end-to-end encryption to ensure only allowed devices can connect with the service and the data is protected as it goes across the Internet. AWS IoT Core will allow us to

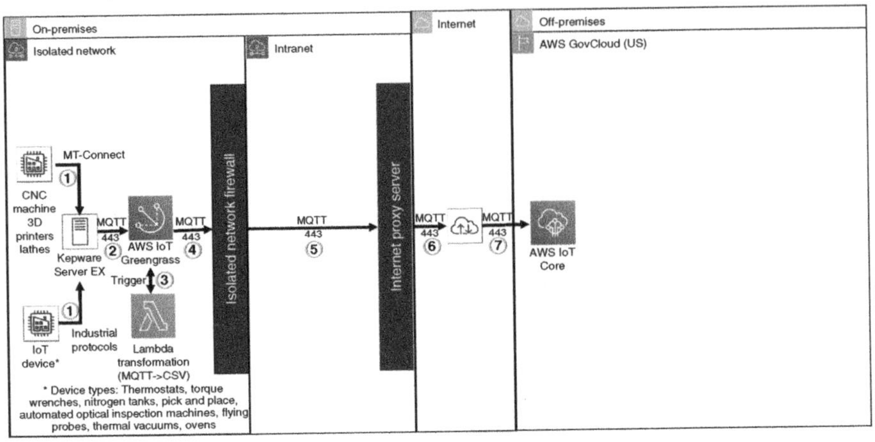

Figure 5.2 Cloud connectivity from edge computing to cloud services for on-demand availability of compute and storage services for data processing [16]. Source: Stephan Gerali.

filter, transform and act upon device data as it streams in based on business rules setup within the system.

In the next section, we will cover how we can define the business rules to support streaming analytics for anomaly detection, support near real time storage of the data and cold data storage options to make it easier for teams to consume and leverage the data to meet their particular needs in building dashboards/reports.

5.5 Solution Statement – Streaming Analytics and Data Storage

In Figure 5.3, it visually depicts how rules can be created to enable particular processes. Within AWS IoT Core, you can create IoT rules to access any of the available AWS services to interoperate with that cloud service. For our purposes, we want to be able to have data streaming in for doing analytics, data that can store real time data in a performant way and a storage mechanism for archiving data to slower yet cheaper methods of managing large data sets.

To support streaming, you can setup an IoT rule to send any incoming data from industrial equipment to an Amazon Kinesis Data Stream (as shown in step 8) which is a serverless streaming

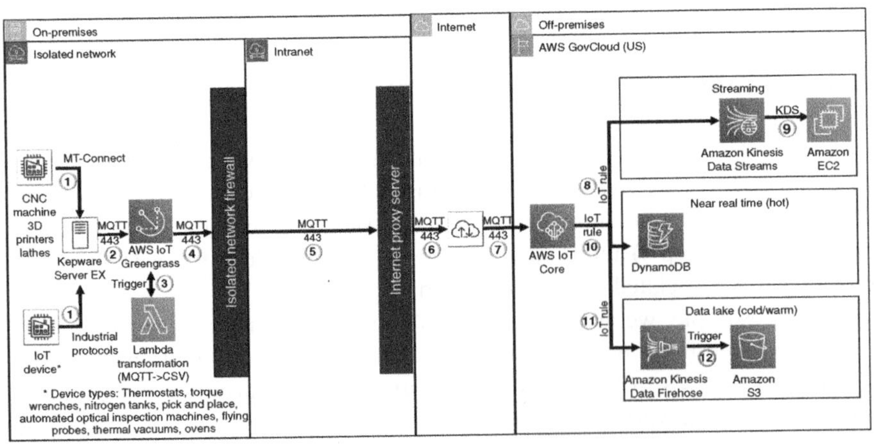

Figure 5.3 Processing of IoT data for streaming event-based analytics and storage of IoT data using hot, warm and cold storage methodologies [26]. Source: Stephan Gerali.

data service to capture, process and store data streams at any scale [17]. Today, many of the Artificial Intelligence/Machine Learning (AI/ML) algorithms make use of a Graphics Processing Unit (GPU) [18] based environment in order to perform their function. Often, teams will need to leverage Amazon Elastic Compute Cloud (EC2) [19] instances with GPU enablement to process the streaming data and do automated anomaly detection using AI/ML [20] methodologies (as shown in step 9).

To support near real time storage of the IoT time series data, AWS IoT Core provides the ability to store such data using Amazon DynamoDB service which is a NoSQL database service that supports key-value and document data structures to store the time series data (as shown in step 10) [21].

To support archiving of data, AWS IoT Core supports the ability to capture, transform and load streaming data into batches and save that data in different data formats to include Apache Parquet [22], Comma-Separated Values (CSV) [23], etc. (as shown in step 11). When the batch size is hit, Amazon Kinesis Data Firehose [24] can save the data into the data format of your choosing without having to build your own processing pipelines and save it into scalable storage services like Amazon S3 (Simple Cloud Storage) [25] to grow on-demand as needed (as shown in step 12).

5.6 Solution Statement – Data Visualization

In Figure 5.4, it visually depicts the visualization capabilities offered by the system for reporting of anomalous conditions along with dashboards/reports on the utilization and maintenance of the industrial equipment. For notifications of anomalous conditions with the industrial equipment, the AI/ML algorithms process the streaming data and upon identification of issues send notifications out to the shop floor personnel to notify them of problems and what work needs to be performed to remedy the issue (as shown in step 13).

Just as important for identification of anomalous conditions is the need to identify the utilization of the industrial equipment to see if there are opportunities to share such assets across all the work that needs to be performed by the business and national defense. Many of the dashboards/reports built today can leverage standard business intelligence reporting tools such as Tableau [27] to visualize the time series data in its raw format and aggregate the data to help with follow on decision making on behalf of the business and national defense (as shown in step 14).

Typically, you will need to integrate data from your Enterprise Data Warehouse [28] with the time series data for the industrial equipment. Since most of our Enterprise Data Warehouse is within

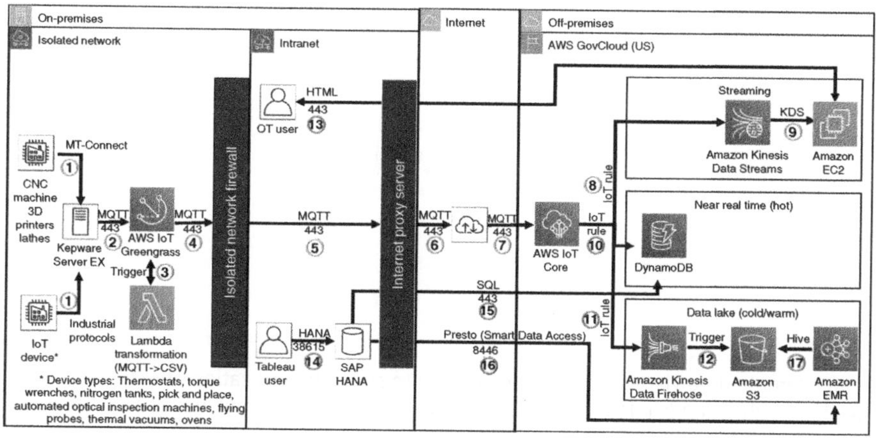

Figure 5.4 Data visualization that graphically represents the data (usually in a time series format) as charts and tables using Tableau to identify utilization of industrial equipment [34]. Source: Stephan Gerali.

SAP HANA [29], we can leverage that database to virtualize data from both our near real time data storage in Amazon DynamoDB (as shown in step 15) and cold data storage using Amazon Elastic Map Reduce (EMR) [30] (as shown in step 16) which includes a service called Presto for making data available in your data lake via structured queries (as shown in step 17) [31]. Hot storage is data that is accessed frequently that needs fast storage to be accessed quickly and cold storage is data that is rarely accessed and can be stored on cheaper storage but its query performance will be slower [32]. Having the ability to store data based on its needed query performance allows us to save money by leveraging cold storage for data that is not used relatively often or does not have high performance requirements allowing us to save money on deployed infrastructure.

Integrating your Enterprise Data Warehouse with your Operational Data Store for IoT will enable you to do more complex reporting on the utilization of your industrial equipment [33]. For instance, you can use this data to identify which industrial equipment is being used or not used enough for business and national defense purposes. You can begin to quantify how much it costs to use the industrial equipment when building out parts and materials. It can even let you know how much you use the industrial equipment to help with changing out serviceable parts to help improve the maintenance of your equipment based on utilization. All of these use cases and more become available as you integrate your IoT data with the existing data you already have within your business.

5.7 Solution Statement – Example Data Visualizations

In Figure 5.5, it visually demonstrates how we can take the data from different work centers and visualize how the machine state transitions are used throughout the day for using machines, how much of the availability capacity is being used on a regular basis and ensuring optimal utilization of such capital resources.

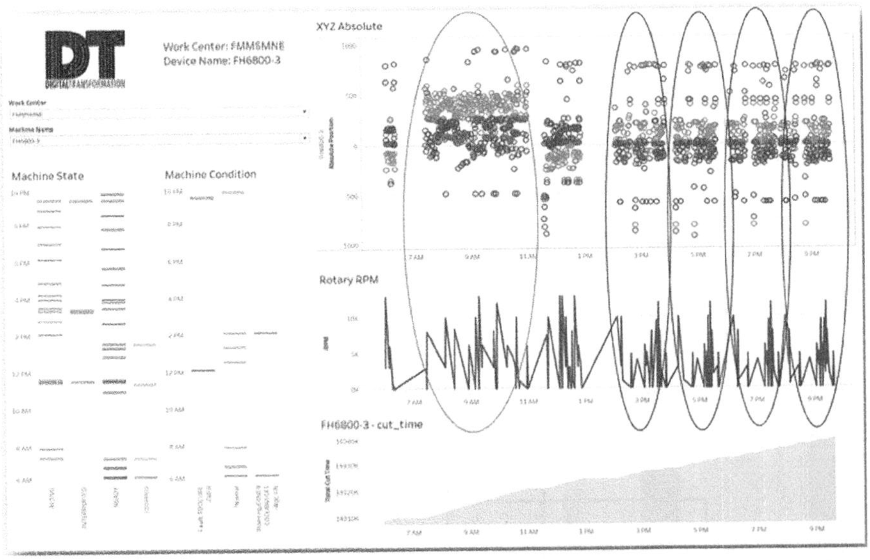

Figure 5.5 Provides an example data visualization using Tableau to show the utilization of industrial assets using graphs to show utilization (circled in picture) for data-driven decision making. Source: Stephan Gerali (Author).

5.8 Results

Leveraging a common platform like AWS IoT services for connecting industrial equipment at the edge and supporting different outcome-based options like stream processing for real time analytics and a hybrid hot/cold data storage mechanism for storing of IoT allows businesses and the national defense to cost effectively process the data as needed but also store data in a cost-efficient way. This chapter gave examples on how you can utilize AWS services to enable such use cases in the processing of IoT data.

Ensuring security of the data in transit and in storage enables businesses and the national defense to safely acquire new operational data sets to help improve the manufacturing business processes. Having a common method to isolate industrial equipment using firewalls combined with certificates to allow edge-based processes to save data into the cloud-based services makes it easier for businesses to grow the utilization of IoT data services. It also ensures adequate security policies for both users and devices that can be enabled throughout the platform to ensure data in being collected in a safe and efficient manner.

Utilizing cloud-based services allow businesses to "pay as they go pricing" meaning it is easy to quickly get an initial service deployed with a few industrial machines and grow it over time as the maturity of the business and organization grows while only paying for what you need. Since all cloud services are inherently scalable, this means the entire environment can scale to meet as many IoT devices as are required with responsive and reliable communication methods.

Leveraging AWS IoT services provides seamless integration with other AWS services so businesses and the national defense can quickly integrate new capabilities into their overall approach for IoT to take advantage of new cloud-based services as needed. In addition, storing of such operational data can allow businesses and the national defense to quickly identify where assets are being utilized and allow businesses and the national defense to integrate their own data analytics into the process to enable more insight to the overall manufacturing process.

5.9 Next Steps

Developing a common platform for quickly connecting devices, managing devices, monitoring devices and controlling devices will enable businesses and the national defense to add further automation to their already existing manufacturing business process. Having a common edge computing environment makes a lot of sense to support upgrading and maintenance of those environments going forward so working towards having a common physical architecture for edge-based deployments to help with the maintenance of those environments becomes critical.

Further, having common data modeling for all equipment types and ensuring all the relevant data from that equipment is being stored for both analytical and reporting purposes ensures that businesses and the national defense have a common methodology to the management of such assets and will enable them to gain the most from their IoT investments.

Next, developing reporting across business areas and segments on utilization of capital equipment and developing standard dashboards to enable more of an Enterprise perspective on how these resources are being leveraged can enable businesses and the national defense to find opportunities to improve the use of their industrial assets. In addition, having a continuous improvement mindset to finding opportunities to improve existing processes and baselining those improvements will enable the business and the national defense to become more efficient over time.

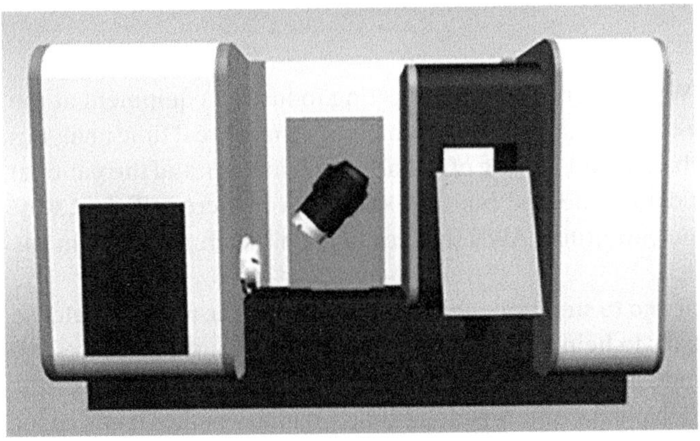

Figure 5.6 Developing a digital twin or virtual representation of industrial machines as used during the production process to overlay current conditions of the equipment [35]. Source: Stephan Gerali (Author).

Finally, looking at new visualization methods of showing the industrial equipment in action using Augmented Reality (AR) and Virtual Reality (VR) as show in Figure 5.6 to be able to show the asset in use and allowing remote operations of those devices to support on-demand needs of the business and the national defense allows businesses to be flexible with their business operations.

As with many endeavors within technology, being versatile in an approach that will enable the business and the national defense to grow its use of IoT devices and gain insight from those devices will have downstream impact to the business to continually improve its manufacturing and sustainment operations. Even though the technology is growing in its maturity, it will have a direct benefit to businesses and the national defense to streamline their operations and improve their way of doing business for many years to come.

References

1 AWS (2022). Start Building on AWS Today. https://aws.amazon.com/ (accessed 8 July 2022).
2 Wikipedia (2022). VLAN. https://en.wikipedia.org/wiki/VLAN (accessed 8 July 2022).
3 Wikipedia (2022). Computer Numerical Control. https://en.wikipedia.org/wiki/Numerical_control (accessed 8 July 2022).
4 Wikipedia (2022). 3D Printing. https://en.wikipedia.org/wiki/3D_printing (accessed 8 July 2022).
5 Wikipedia (2022). Lathe. https://en.wikipedia.org/wiki/Lathe (accessed 8 July 2022).
6 Kepware (2022). KEPServerEX. https://www.kepware.com/en-us/products/kepserverex/ (accessed 8 July 2022).
7 OPC Foundation (2022). What is OPC?. https://opcfoundation.org/about/what-is-opc/ (accessed 8 July 2022).
8 AWS (2022). AWS IoT Greengrass. https://aws.amazon.com/greengrass/ (accessed 8 July 2022).
9 AWS (2022). AWS Lambda. https://aws.amazon.com/lambda/ (accessed 8 July 2022).
10 Wikipedia (2022). MQTT. https://en.wikipedia.org/wiki/MQTT (accessed 8 July 2022).
11 OAGi - Open Applications Group (2022). OAGIS. https://oagi.org/ (accessed 8 July 2022).
12 MTConnect (2022). MTConnect Standardizes Factory Device Data. https://www.mtconnect.org/ (accessed 8 July 2022).

13 American National Standards Institute (ANSI) (2022). The American National Standards Institute Oversees Standards and Conformity Assessment Activities in the United States. https://www.ansi.org/ (accessed 8 July 2022).
14 Wikipedia (2022). Edge Computing. https://en.wikipedia.org/wiki/Edge_computing (accessed 8 July 2022).
15 AWS (2022). AWS IoT Core. https://aws.amazon.com/iot-core/ (accessed 8 July 2022).
16 Wikipedia (2022). Cloud Computing. https://en.wikipedia.org/wiki/Cloud_computing (accessed 8 July 2022).
17 AWS (2022). Amazon Kinesis Data Streams. https://aws.amazon.com/kinesis/data-streams/ (accessed 8 July 2022).
18 Wikipedia (2022). Graphics Processing Unit. https://en.wikipedia.org/wiki/Graphics_processing_unit (accessed 8 July 2022).
19 AWS (2022). Amazon EC2. https://aws.amazon.com/ec2/ (accessed 8 July 2022).
20 Wikipedia (2022). Machine Learning. https://en.wikipedia.org/wiki/Machine_learning (accessed 8 July 2022).
21 AWS (2022). Amazon DynamoDB. https://aws.amazon.com/dynamodb/ (accessed 8 July 2022).
22 Apache (2022). Apache Parquet. https://parquet.apache.org/ (accessed 8 July 2022).
23 Wikipedia (2022). Comma-Separated Values. https://en.wikipedia.org/wiki/Comma-separated_values (accessed 8 July 2022).
24 AWS (2022). Amazon Kinesis Data Firehose. https://aws.amazon.com/kinesis/data-firehose/ (accessed 8 July 2022).
25 AWS (2022). Amazon S3. https://aws.amazon.com/s3/ (accessed 8 July 2022).
26 Wikipedia (2022). Analytics. https://en.wikipedia.org/wiki/Analytics (accessed 8 July 2022).
27 Tableau (2022). The World's Leading ANalytics Platform. https://www.tableau.com/ (accessed 8 July 2022).
28 Wikipedia (2022). Data Warehouse. https://en.wikipedia.org/wiki/Data_warehouse (accessed 8 July 2022).
29 SAP (2022). SAP HANA. https://www.sap.com/products/hana.html (accessed 8 July 2022).
30 AWS (2022). Amazon EMR. https://aws.amazon.com/emr/ (accessed 8 July 2022).
31 Presto (2022). Distributed SQL Query Engine for Big Data. https://prestodb.io/ (accessed 8 July 2022).
32 Ctera (2019). The Differences Between Cold, Warm and Hot Storage. https://www.ctera.com/company/blog/differences-hot-warm-cold-file-storage/ (accessed 8 July 2022).
33 Wikipedia (2022). Operational Data Store. https://en.wikipedia.org/wiki/Operational_data_store (accessed 8 July 2022).
34 Wikipedia (2022). Data and Information Visualization. https://en.wikipedia.org/wiki/Data_and_information_visualization (accessed 8 July 2022).
35 Wikipedia (2022). Digital Twin. https://en.wikipedia.org/wiki/Digital_twin (accessed 8 July 2022).

6

Challenges and Opportunities of IoT for Defense and National Security Logistics

Gisele Bennett[1], William Crowder[2], and Christina Baxter[3]

[1]*MEPSS LLC, Indian Harbour Beach, FL, USA*
[2]*Logistics Management Institute, Tyson, VA, USA*
[3]*Emergency Response TIPS, LLC, Melbourne Beach, FL, USA*

Abstract

The Internet of Things (IoT) for use in logistics has gained traction with the introduction of RFID for logistics traceability. At the same time that commercial industries were struggling with deploying RFID, the US Department of Defense (DoD) was also working towards utilizing track and trace for various applications. This chapter covers the many opportunities and challenges that IoT has in supporting DoD applications and logistics to include improved visibility for predicting state of a system, improved sustainment of equipment, and efficient utilization of resources.

6.1 Introduction

The Internet of Things (IoT) has the potential to allow senior leaders in the military to increase readiness through understanding of the condition of their things (people, equipment, facilities, and anything relevant for normal operations and battlefield uses). The ability to know the past, current and future performance is a powerful tool for tailoring the force to meet the mission. The past and current performance can be measured; however, the future performance can be calculated based on existing data and models to some degree of acceptable confidence interval. There are a couple of critical points that must be made clear throughout this chapter. The first is that organizational leaders will still make the decisions of how and where things are used; however, what we are proposing is that the decision makers will have greater information backed with data on the current and future condition, or state, of the organization (people and equipment). The second premise is that the definition of IoT, as used in this chapter, like most technologies, is not new and there is no universal definition. In fact, IoT is an outgrowth from the rise of radio frequency identification (RFID) and gained momentum in the early 2000s. IoT in this chapter is not just a sensor, data, embedded systems, cloud-based system, database, connectivity between devices or anything else but it should be viewed as an integrated system of systems providing data needed for information generation for decision makers. Sometimes IoT is interchanged with M2M or IoE or

IoT for Defense and National Security, First Edition. Edited by Robert Douglass, Keith Gremban, Ananthram Swami, and Stephan Gerali.

Figure 6.1 IoT in the battlefield configuration to demonstrate the diversity of IoT at the system implementation level. Source: [8]/from Collaborative Research Alliance (CRA).

RFID. A good depiction of how the functionality changes is presented in an article by Lueth [1]. He presents definitions that are used in IoT and include:

M2M (Machine to machine) communication; Web of Things
Industry 4.0; Industrial internet (of Things); Smart systems; Pervasive computing; and
Intelligent systems. In a diagram, Lueth [1] presents a mapping of who/what is impacted by the concepts, and what is being altered by the concept. Some of the areas for opportunities for warfighter effectiveness include situational awareness, better mission control and autonomy, ability to fight more effectively, and improved soldier performance.

Numerous studies have been conducted by the Defense Science Board (DSB) and the Army Science Board (ASB) on the uses of IoT by the Department of Defense (DoD) [2–5]. The Defense Business Board (DBB) conducted a study on Logistics as a Competitive War Fighting Advantage [6]. Although the DBB does not call out IoT, it will become apparent with the applications outlined in this chapter that IoT can be an enabler for logistics. All the studies concluded with opportunities for cost savings and utility for use of IoT systems by the DoD. Although not specifically mentioned, IoT should be considered a strategic advantage for both strategic and expeditionary maneuver by providing situational awareness on the condition of things (people and equipment) [7].

The uses of IoT for the military are diverse and seen in applications that include robots/autonomous vehicles, wearable devices, weapons, vehicles, sensors, and munitions as depicted in Figure 6.1 [8]. The application layer is at the system level that is supported by functions that include sense, attach, collect and process, sustain, fix, defend, and collaborate between people and systems as shown in Figure 6.1. Other areas of use for military IoT systems include: training, autonomous vehicles, smart inventory, and business systems [9–11]. Other countries are implementing IoT for military logistics, readiness, and cost savings [12–17].

6.2 Linking Industry and DoD Uses of IoT

Industry has widely deployed IoT for various applications with return on investment in visibility of system operations. It is time for DoD to harvest what industry has been able to use for commercial applications for military applications. In some applications plug and play options are available and in other cases a transition for more secure and rugged systems will be required. The crawl, walk, run concept can easily be applied to deployment of IoT. In the early 1900s Henry Ford and others created the mass-produced automobile and truck. At that time the Army was using horses and wagons. The Army did not need to invent the automobile nor the truck, however, they needed to figure out how to use them and what modifications they required. IoT is in a similar relative position today. DoD needs to adapt potential industry solutions and define any required modifications or security features that are required. Some of those solution benefits provide cost savings and others provide a tactical advantage.

There are numerous examples of industry use of IoT that are easily translated to DoD applications. The translation between industry and DoD applications should be straightforward with a return on investment that provides operational and strategic efficiencies, cost savings, and situational awareness for decision making. Some of the challenges in translating commercial applications to DoD use are the DoD requirements for device usage have environmental and operational requirements that necessitate the redesign or modification for use in the field. Additional considerations include testing requirements which are more stringent and can increase cost for deployment, security, standardization, and data processing [18, 19]. However, most of the examples incorporated within this chapter provide opportunities to leverage maintainability and sustainment operations.

GE Digital uses IoT for asset performance management, service optimization, and supply chain management. The key concepts in these areas allow for the sales of usage by power vs. just a non-recurring sale of assets, e.g. engines for planes and trains [20]. This concept is not new, it started in the late 90s. With more data, the ability to create models that provide a digital twin for parallel intelligence enables predictive and condition-based management of individual assets as shown by the numerous submissions to the first Digital Twin Parallel Intelligence Conference [21].

ThyssenKrupp and Microsoft developed a solution to connect thousands of sensors and systems in its elevators to improve maintenance and up time. The outcome is a framework to sell services. The resulting benefit to the customer is reliability and cost reductions in maintaining skills for sustainment of equipment. The concept of power by the hour was introduced and changed the business model for maintainability and reliability. Cost savings were achieved with increased sensor data and modeling of performance for predictive maintenance.

6.3 Situational Awareness

Situational awareness is prevalent in all the examples provided in this chapter. The outcomes are implied in the examples provided in the following sections. Important aspects, that can be chapters in themselves, are related to the policies, legal, and challenges associated with sensors, monitoring, and processing of data when people are part of the system. The DoD Data strategy calls out the DoD as a "data-centric organization that uses data at the speed and scale for operation advantage and increased efficiency [22]. Data ethics and collective data stewardship are called out as 2 of the 8 guiding principles. There are more extensive resources available for reviewing policies and use cases [23].

6.3.1 Policy and Legal Implications

To achieve good situational awareness, sharing of information from personal smart devices, smart platforms, and smart infrastructure must be available and protected. This in turn means that personal information must be protected which will require adjustments in both policy and law. For example, collecting information on US persons is controlled by Title 50. The law was written to protect citizen rights to privacy. If DoD wants to integrate data from smart devices of their soldiers, civilians, or individuals that train around installations, policies and laws must be adjusted to protect the collector as well as the individual.

6.3.2 Challenges and Considerations

In many ways industry leaders like Meta (Facebook) and Google have had a free ride in the collection and use of data spewing from IoT devices, smart devices, and internet usage. DoD will not get the same free legal ride on collection and protection of data. Some levels of the data can be atomized and non-person attributable, but other data is going to be person and platform specific. Once collected, this data will have to be protected. Equally important will be to prevent others from collecting against DoD.

6.4 Applications for DoD

The applications outlined in this section are not inclusive but provide a range of opportunities for the DoD. The proliferation of sensors and cloud computing has made the utilization of IoT systems possible. Edge computing offers a level of capability at the point of need that was not available decades ago and thus has evolved decision making options to near real-time. Continuous sensing and data gathering that informs operational processes and enables decision making is a game changer for operations, situational awareness, and sustainability to include logistics supportability. Numerous programs have been initiated by the DoD starting with the Internet of Battlefield Things (IoBT) solicitation released by the Army Research Lab (ARL) in 2017 [8]. An interesting survey of the resiliency needed for the IoBT lists a series of assurance needs such as scalability, responsiveness, resilience, and heterogeneity that must operate in an environment that is adversarial, obstructed, and contested, e.g. megacity [25]. Deploying IoT is not without technical and regulatory challenges.

6.4.1 Situational Awareness of People and Equipment for Maintainability and Sustainability

Industry has exploited data for visibility and decision making in the areas of energy management, predictive maintenance, improved security, and traffic management. There are extensive examples of the benefits associated with the investment in sensors that provide the necessary data to lead to information and intelligent dashboards. The focus of this section is to provide insights into maintainability of equipment and people for uninterrupted service. There is a need to create models of the equipment performance, collect data on system usage, and service lifecycle. The collection of the data can take advantage of the diverse IoT feeds built into modern systems. Organizations can migrate to providing equipment with minimum downtime, increased reliability, and advanced planning for replenishment of parts required.

There are three examples that show the art of the possible, demand-based toll road pricing, driver performance monitored based insurance pricing, and dealer maintenance monitoring of vehicles. All three examples draw performance, position, and location data from onboard sensors, smart devices, and fixed readers. Add to that mix real-time monitoring of pacemakers and other medical devices over smartphones, fitness devices, and smart watches. A situational status of performance of our people and equipment can be assembled to provide the current condition and predict a future performance as a function of external conditions.

Sensors are essential for data collection. Traditionally sensors have been costly, require connectivity, and were not always easily retrofitted to legacy equipment or buildings, much less people. However, the ubiquitous use of smartphones and their sophisticated built-in sensors have opened the aperture for data collection that is efficient, timely, and cost effective. For the case of driver or vehicle monitoring, today's systems include capabilities to monitor performance of the platform and the driver without retrofitting the system with additional sensors. Platforms with non-real-time sensor feeds could be married to an operator's real-time feed to infer causal relationships and performance of both.

Crowdsourcing is not a new concept; however, it can be used as a method for integrating near real-time data for sensing and condition monitoring. Smart devices/phones can be outfitted with additional sensors or realign sensors already embedded, such as cameras, to allow for monitoring of gasses or other emissions to obtain a more complete picture of potential hazards. The applications for dual use capabilities are extensive.

6.4.2 Data Collection for Real-time and Predictive CBM

Condition based maintenance (CBM) is an emerging concept and significant departure from traditional maintenance practice. CBM seeks to use real-time data derived from sensors on equipment to define its useful life remaining and the maintenance actions necessary to maximize the equipment's productive operating time. Because it uses actual equipment conditions, and seeks to prevent system failure, the resulting maintenance and support processes for this approach are also different from traditional practice. The key components of a CBM platform are:

Prognostics and Health Monitoring: The analysis and design of sensors and data collection systems based on the equipment failure modes and effects analysis, operating environment and economic analysis.

Data Fusion and Analysis: The process of developing the means to turn the raw sensor data into real-time processed data sets, archiving the data sets and passing it to the 'analysis engine' in usable formats. Allowing the analysis process to develop the algorithms and means to take the data sets and turn them into meaningful information to be used as the basis for maintenance and operational decisions.

Distribution of the Information: Taking the results of the analysis and distributing it to the various elements of support: engineering (for Repair Cycle Maintenance (RCM), design use and user notification such as service bulletins), material support (the 'Supply Chain' that ensures the right material is available to maximize operating time), and training (for operators of the equipment).

Applying CBM for infrastructure protection and monitoring is not directly related to battlefield operations but critical for monitoring and maintaining DoD infrastructure. There have been recent events that offer a view of what might be possible if IoT monitoring could be applied to infrastructure, such as fuel and water storage and distribution. Both systems suffer losses of product due to leakage. In some cases that leakage can be catastrophic, causing a problem in the surrounding community. For example, a leakage with toxic components can cause cross contamination with local water systems that first surface when residents downstream of the leak and the cross contamination become ill. Two examples of this type of event are in Fairfax, Virginia and Honolulu, Hawaii. In the first case a commercial pipeline terminus storage facility leaked into an aquifer where the community was serviced by well water. In the second case, a DoD fuel reserve facility leaked into the island aquifer. In both cases, residents started complaining of bad water and resultant illness. In effect, the detection was made by a human sensor. We can postulate that if both the water and the fuel facilities had state of the art IoT based sensor systems to detect loss and cross contamination, the leaks and the contamination could have been detected and minimized before the population was adversely affected.

Alternatively, applying CBM to the chemical sensors utilized by deployed forces could greatly enhance battlefield operations by minimizing sensor down time. In this case, collection of calibration data across the suite of chemical sensors (e.g. ion mobility spectrometers, photoionization detectors, electrochemical cells) would allow the maintainers to recognize when sensor drift, sensitivity losses, or humidity affects are plaguing a detector and replace components prior to the systems becoming non-functional. This is especially important when performing dismounted operations or security sweeps as the data is time critical and technical support is not readily available.

IoT processing at the edge: industry is developing approaches to IoT data processing at the edge to minimize the need to pipe raw data back to a hub (cloud or computing facility.) This is an area that we have drawn some inferences on how it might be accomplished using soldier oriented smart devices. This is an area that DoD requirements for security and masking are going to be more rigorous than industry requires. Therefore, it is a logical investment area that DoD must provide incentives to industry to find solutions. Available bandwidth on the battlefield is limited and normally oversubscribed, For IoT to be seen as value added, it must minimize its demand for bandwidth and be robust for battlefield operations without disclosing true situational awareness to adversaries.

6.4.3 Prepositioning and Planning for People and Supplies (Prepo-in-motion)

Prepositioning of people and material provides a potential tactical advantage. This advantage can be fully realized with data, sensing, and analytics that leverage and utilize past data and outcomes, current conditions, system models, and analytics that factor or predict the requirements for executing a mission [7]. IoT enables tracking of people and conveyances carrying equipment and supplies, allowing military decision makers leverage of dynamic prepositioning vs. static repositioning.

Today, most smart devices are capable of some level of biometric monitoring of the operator of that device. Simple examples are physical activity, heart related indicators, and stress related feedback. Newer platforms have enabled low bandwidth paths to off load platform data that can be captured by linked smart devices. Therefore, by linking the two feeds in the smart device, personalized fusion at the smart device level incorporates human and machine interaction. This in turn can then be transmitted to the appropriate organization level to achieve a fused picture of the organization's performance. The DoD will need to investigate the cyber challenges and implications of soldier performance monitoring. Soldier readiness and resilience is primarily a function of sleep deprivation and lack of physical fitness. One third of the active-duty soldiers get less than five hours of sleep per night which increases the risk of mental and physical health [2]. Real or near real-time soldier health monitoring gives the opportunity for early intervention for health and mental mission readiness and planning. The collective situational awareness at all levels provides a system view at the appropriate organizational level to have personnel who are ready to execute the mission. Much research, development, and operational testing programs are underway utilizing these types of data for human performance optimization (HPO). HPO programs cover the continuum from health sustainment to performance sustainment resulting in performance enhancement and optimization at the individual and unit level. Uniformed Services University's Human Performance Resources, which is a Consortium for Health and Human Performance, link performance to Unit Mission essential tasks that link

Human Health to Health Sustainment (limited ability to be healthy) to a warfighter core task which corresponds to performance sustainment, and they link unit mission essential task lists (METL) to performance enhancement Consortium for Health and Military Performance (CHAMP)[1] [27]. This ranking of health aids in the readiness of the troops.

In addition to a customizable troop configuration for executing a mission, biometric monitoring across the battlefield or in emergency response situations allows for near real-time triage assessment and provides a potential to minimize casualties in the field from extraction efforts. In a triage situation, IoT sensors and smart data fusion platforms would allow for routing of injured personnel to the appropriate medical facility.

Cloud based reporting from biometric IoT devices will allow for early intervention, leadership engagement, and near real-time visibility into unit readiness. Security concerns have already shown up and will have to be addressed to minimize exposure by adversaries of the moment through interception of personally worn devices as, for example, the Strava Fitness exploitation [28].

IoT can enable DoD to shift from reliance on fixed large inventory approaches to a dynamic inventory approach that offers the potential to minimize physical aggregation of supplies (iron mountains) in zones of danger where operational conditions allow. Amazon is the ideal case study that the DoD can benefit from duplication of services and methodology in approaching distribution of goods. As an example, the best distribution of goods is a multivariate problem that requires considerations of demand, availability, timing, and optimization of the supply system. When we enable processing at the edge and fusing data at the point of need, we will shift to anticipatory repairs instead of reacting to unplanned failures as outlined in the CBM section of this chapter. Further evolving to a smart pre-positioning model will require DoD to rethink their traditional maintainability and supply transaction-based approach to a sustainment environment that leverages the industry advances to achieve smart supply chain management focused on virtual end user demand (mission support). Internationally, military operations rely on logistical support for tactical and sustainment advantages. The list of published papers in this area is extensive. [13, 14, 16, 24], IoT is an enabler to improve visibility and readiness in the supply chain. Spares are an important element in readiness. An example of lack of situational awareness of spare parts is the improper storage of engine containers. High value assets such as aircraft engines, transmissions, and rotor blades assets are frequently misplaced while in transit and storage. In addition, knowledge of the asset's condition, while in-storage or in-transit, enables intervention to correct undesirable conditions before additional damage is sustained. Preparation for the right repair resources to be in place when the asset is inducted for overhaul and total asset visibility into location, status, and condition, will enable better asset distribution decisions [26]. Some of the problems include fundamental issues of location of the containers since the markings on the containers are not complete or incorrect. Assuming you can find the container with the engine, which is a high value asset, that engine might not be ready for issue since it was improperly stored. The improper storage results in the engine returned to the depot for repair and a replacement engine sent, if available. This not only costs the agency funding but compromises the readiness of the fleet. However, IoT offers the opportunity to monitor the location and environmental conditions of the engine stored in the container to provide early warning of a compromise of the asset and thus early intervention.

---

1 Human Performance Resources by CHAMP (HPRC) is the human performance optimization (HPO) educational arm of CHAMP, a DoD Center of Excellence located at the Uniformed Services University. HPRC provides holistic, performance optimization resources that help members of the military community stay physically and mentally fit, fuel and hydrate properly, maintain social ties, and stay resilient—all pieces of the puzzle that make up Total Force Fitness.

6.4.4 IoT at DoD Installations

Smart cities are taking over by providing connectivity through networks, sensors, and data analytics to provide situational awareness [29]. Smart cities are growing through investments by the local governments and industries. The benefits of smart cities include public safety, smart surveillance, and transportation providing crowd control and surveillance capabilities thereby improving public safety. In addition, controlling perimeter access can provide efficient monitoring of restricted areas and detecting non authorized personnel in those areas, detecting a person entering a restricted area while carrying an unallowable object, and rerouting traffic away from a restricted or controlled area. Smart surveillance provides an ability to track people and events in real-time which provides safety personnel with the ability to look at trends such as correlations between crime and events, tracking persons of interest, detecting unusual activity such as drones or other vehicles that would normally not appear in the city. From an efficiency standpoint the ability to monitor traffic and change traffic flows, both pedestrian and automotive, can provide for efficient movement in a region. This is easily achieved through control of traffic lights in a city for intelligent intersections and with the use of predictive algorithms through monitoring of traffic the ability to synchronize lights sufficiently and to alter traffic flow to accommodate emergency vehicles.

6.4.4.1 Energy Management

The Army is positioned to implement smart installations if resources are allocated for sensors and data analytics tools to provide real-time information and predictive data analytics for energy savings, installation utilization, and improved utilization of resources with the surrounding municipalities. The implications are cost savings and increased collaborations for efficient use of resources provided by the adjacent city. The Assistant Secretary of the Army for Installations, Energy and Environment [ASA (IE&E)] in conjunction with Army Corps of Engineers Engineer Research and Development Center (ERDC) built the smart and resilient installations (SaRI), an architecture that provides a virtual testbed for installation management effectiveness (V-TIME) [30]. ASA IE&E created a roadmap for using IoT at installations [31].

IoT has benefits for maintainability and automation for fixed infrastructure. An example is Heating, Ventilation, and Air Conditioning (HVAC) equipment monitoring and usage. Imagine a scenario, which is currently possible in homes, of controlling HVAC settings based on the day, expected occupants, and external factors that change normal patterns such as traffic delays, holidays, or travel out of office which would reduce when people were expected in the building. Furthermore, as installations are integrated with the surrounding community, capturing data to formulate requirements for an installation can maximize the usage at and around the installations. The utilization of chemical sensors integrated with HVAC systems allows for the rapid shut-off of systems when warranted due to chemical releases, whether accidental or intentional.

In addition to HVAC control, reliability of infrastructure equipment is critical at installations in and out of the US. Air quality, water quality, pumps, air volume control, broilers, and other equipment needed to support an infrastructure are important for daily activities and in combat situations. Other automation opportunities utilizing IoT include infrastructure security (buildings, installations, intermediate support bases, or any other temporary basing, e.g. fire bases). Security includes access, monitoring, intruder detection, and overall situational awareness. Key enablers for small and medium infrastructures include cheap wireless sensors; local controls at the device level; true plug and play integration (i.e. common standards); wireless communication to the Internet; cloud-based data storage, analytics, and remote control.

6.4.4.2 Installations as Training Platforms

The significance of smart cities and the warfighter challenge is understanding how we operate and defend an attack [32]. The challenges of megacities are obvious in most cases. They are larger and more complex thus increasing the challenges of navigation, maneuverability, and the ability to distinguish between noncombatants and combatants. Many of the features that smart cities provide will help reduce the complexity and infrastructure for warfighting engagements in megacities. The characteristics of megacities include density, scale, context, threats, connectedness, and flow as outlined in the report to the Chief of Staff of the Army [32]. These characteristics intersect to define the level of complexity of engagement in cities.

How to exploit a city's sensor infrastructure Is one element to winning a battle in a complex environment such as a city or megacity. We can exploit existing sensors by creating arrays that utilize the current infrastructure, augment the existing sensors, leverage edge computing for distributed decision making, and develop an environment for machine-to-machine (M2M) trust.

Installations are equivalent to cities in operations, requirements, infrastructure, utilization, and challenges. DoD units spend much of their non-deployed time on installations, to include training areas. As IoT proliferates on and around the installation, it will affect how soldiers train and can exploit the infrastructure to be better prepared to fight in a city. A simple example is provided by the unit recall roster. This starts as a document published by the unit to identify who calls who when a commander initiates a unit level recall. Soldiers have taken that document and turned it into groups on their personal smartphones to speed the process. Military medical services use smart devices to control patient scheduling. Soldiers will continue to find ways to ease their workload by leveraging applications on their personal devices. The challenge is to expand the efforts quickly to find the early adaptors and encourage them[2]. Further the tools and techniques must be transferable and transportable to allow soldiers to take the capabilities used at an installation to a training area, and subsequently into the deployed area of operations so that soldiers fight with the same tools as they train with. It is this requirement that will require unique DoD attributes that can be integrated into the industry IoT devices.

6.4.5 IoT and Emergency Response

Data accessibility to first responders has increased rapidly over the past ten years. The first area with the most dramatic changes is hazardous materials response. In this case, the number of real-time and near-real-time sensors has increased exponentially. With this, data availability has also increased. Unfortunately, there has not been a parallel increase in sensor fusion and data analytics. Currently, most data are evaluated as individual data points vs. utilizing the data for enhanced decision making. FEMA has developed two free programs for emergency response personnel to allow for data sharing for radiological and chemical sensing systems that are wirelessly enabled, but the tools, RadResponder and CBRNResponder respectively, allow the operator to visualize data on maps but do not provide decision support capabilities. Vlahi System's Chemical Emergency Response E-Service (CERES) provides a similar capability but also allows the user to use the real-time data to develop plume dispersion models estimating future material movement based upon sensor readings and real-time weather station data. In addition, the data can be used in reverse to identify the likely/possible emission sources. CERES also allows for optimized sensor deployment and enhanced planning. The first tool to provide advanced Decision Support to the hazardous materials community is the Emergency Response Decision Support System

2 https://www.army.mil/article/35148/g_6_launches_apps_for_the_army_challenge/

(ERDSS) developed by MEPSS, LLC. This system takes information from the disparate sensors and recommends operational best practices from protection to detection, decontamination, and remediation. While ERDSS currently requires data to be input manually, past field trials have demonstrated the capability to employ real-time sensor feeds into the decision support matrix thus rounding out an IoT system. Integration of all these datasets into a common operating picture with the decision support tools evaluating the data and providing recommended best practices remains elusive at this time.

Considerable federal funds have been utilized to develop tools to track responder locations to reduce total line of duty deaths. The most recent success in this area has been with NASA Jet Propulsion Laboratory's (JPL) program which tracks responders in three dimensions with an accuracy of a few inches inside buildings. The DHS funded program utilizes magnetoquasistatic (MQS) fields to determine positions of firefighters wearing receivers. Transmitters are located on apparatus and a computer is used for data visualization at a command center. Tracking systems using a variety of technologies including LIDAR, GPS, and others have been under development over the past 15 years to provide personnel accountability in firefighting situations and to identify officers in active shooter events. While complete integration is still in development, these systems can be a source of personnel accountability but will also represent a source of data overload if not managed properly.

Wearable sensors focusing on the physiological health of responders have been under development for many years with funding predominantly coming from DoD, DHS, and others. While the wearable sensing systems are often used for athletes, they are not widely deployed in emergency response scenarios. Current systems capture data using environmental and physiological sensors (e.g. oxygen, carbon monoxide, respiratory rates, heart rate, and skin temperature) and then use advanced algorithms to estimate other parameters such as core body temperature or use artificial intelligence to understand the responders needs. The DHS Wearable Alert Monitoring System (WAMS) working with NASA JPL's Assist for Understanding Data through Reasoning, Extraction, and Synthesis (AUDREY) is an example of an AI-assisted wearable sensor system. The U.S. Army Research Institute of Environmental Medicine (USARIEM) has performed a considerable amount of work using non-invasive measures to optimize military performance.

The major hurdles that remain in the development and deployment of IoT systems in emergency response applications include the integration of disparate sensor data into one usable format, the integration of disparate video feeds (e.g. bodycams, dash cams, helmet cams, traffic cameras, security cameras, etc.) into the same temporal and spatial dimensions, the development of decision support systems to make these large amounts of data operationally relevant (vs. data overlaid), and the collection of historical data sets upon which machine learning algorithms can be trained. The first step towards the future is the adoption of the DHS Next Generation First Responder (NGFR) ecosystem to discover, connect, fuse, and understand different IoT domains.

6.4.6 IoT and Disaster Response

All phases (preparedness, response, recovery, and mitigation) of disaster management can benefit from IoT technologies. In the preparedness and prevention stage, early implementation of asset tracking and sensor systems can be used to manage the locations of large chemical storage containers. Following events like hurricanes, considerable assets are used to track and monitor these hazards to ensure that they are still providing the structural integrity required for storage of hazardous chemicals. With the implementation of asset tracking along with sensing systems, this entire process can be automated thereby freeing up emergency assets for other critical life safety initiatives. Unfortunately, applying these assets post incident will not provide the same level of benefit. During the response phase, IoT technologies can help with early notifications of an incident (e.g. wildfire sensing), asset tracking at the operator and apparatus level, overarching situational awareness, and supply chain management. The FEMA Urban Search and Rescue programs have slowly implemented a variety of IoT technologies from RFID tagging of assets through supply chain management to ensure operational readiness at the time of need. During the recovery phase of disaster management, supply management is key to success. In this case, the supplies can range from food, water, shelter, clothing, etc. Without these necessities of life, the recovery process cannot begin.

Throughout history, there have been many major events that could be used as defining examples of how IoT technologies could be used to mitigate future disasters. This is critical for the prevention and planning phases of natural and man-made disasters preparedness. The events in Fukamushi in 2011 (e.g. earthquake, tsunami, hydrogen explosions, nuclear reactor core meltdowns, and radiation dispersal to air, land, and water assets) are another great example of where IoT technologies could have provided great benefits. For example, buses used for evacuation of personnel hung Radiation Dosimetry Cards in the door to demonstrate that the radiation level in the vehicle was below concern. Once the radiation cards reached levels of concern, the buses were removed from service. Unfortunately, colorimetric responses tend to have high (up to 25%) error partially due to color interpretation and due to the chemistries themselves. In addition, the radiation dosimetry cards also were plagued with UV-radiation effects. Had electronic dosimeters been available and installed in the evacuation vehicles, real time assessments of radiation exposure and contamination spread could have been made saving considerable time and resources. The ongoing cleanup program to decontaminate the affected areas currently has real-time monitors in place. While those monitors are integrated and can provide situational awareness as to the current state of contamination, the integration of the sensors along with the delta in radiation in comparison to the decontamination measures would provide a great source of data that could be used in future events to ensure quick and efficient mitigation efforts. Investigations following this event noted "poor communication and delays in releasing data on dangerous radiation leaks at the facility" as significant failures in the response efforts - each of which could have been mitigated by the incorporation of IoT technologies before the incident. While radiation was released during this event through deliberate venting to relieve pressure, discharge of contaminated coolant water, and uncontrolled events, monitoring, and mapping of it in real time would have served as an excellent early warning system for others in the path of the radiation and as a mitigation monitoring tool for those at and close to ground zero.

6.5 Observations on the Future

As Vint Cerf said in an interview with the Army Science Board on the future of IoT, "you will not be able to be unconnected by 2025."[4] Many would hold that is already true in more ways than most comprehend. Today and tomorrow the challenge is going to be to understand how best to use the data that is being generated and how realistic the picture is for the user of the data. In many ways, today is much like when the phones first came into use with party lines. Then we moved to private lines and legislation was created to protect the privacy of individuals. One thing is clear: the adage that the more people that know a secret, the more likely that what everyone thought was secret, is not.

Patterns of life in a community of interest will be discoverable. How we make use of the information becomes the challenge. First responders may use it to determine the initial scope of an event. That initial assessment will then have to be fleshed out with additional data from the community and the responders to determine a course of action. Disaster relief may use it in the same manner for an initial assessment of the event and of the potential resources available to respond. There are those that will say, in both cases, that this is an intrusion into citizens' lives. Most people would agree and would offer that the need for this type of information must be balanced against the common good of the community.

When the data use shifts to law enforcement, the challenge on usage and admissibility becomes more difficult. Then protection of rights to privacy and non-self-incrimination come into play. This is also true when the military wants to use the data. The laws that govern both situations will need to be refined and rules established. In both cases, because data knows few boundaries and is stateless in the cloud, all sides of a situation will try to use the data to shape the story or response. This is not abstract and has been visible in our lives for the last 4 years on an increasing scale. Cell phones and WiFi are radios with all the features that entails. The key is that anyone can capture the signal, but can they understand it and the purpose of the signal?

That leads to what needs to happen in the IoT world. End-to-end encryption, explicit permissions on usage, and disclosure of usage are some areas that must be defined and enforced. One rule that we will have to follow is, assume the data is out on the internet and that the next use of the data will be different than the last. A second rule is that once the data is out in the internet, it cannot be erased or forgotten. Finally, IoT comes in many flavors so DoD must take a system of systems approach and integrate data sources with well-defined APIs for interoperability and not a stovepipe platform.

We have laid out the challenges and opportunities that IoT presents to the nation in Defense and National Security. IoT systems for military applications can provide platforms for operational and strategic advantage at many levels and across agencies. The ability to sense and analyze our things, people, and the environment, combined with analytics and performance modeling provides the ability for commanders or other leadership chains to tailor their responses to meet the mission with the right equipment and personnel at the right time. The mission can be for combat, emergency response, or humanitarian and disaster relief. In addition to readiness, there is a potential for cost savings in preventive maintenance of our people and equipment through sensing and analytics. However, in this era where data is readily available, data sources are growing exponentially, and information overload is a consistent threat to operations, we need to focus on the development of situational awareness aggregators and decision support tools that minimize the risk of decision paralysis when it most matters - during military operations or emergency response.

Acknowledgement

The authors would like to thank members of the Army Science Board for their dedication and the numerous lively discussions over the years on this topic area which lead to the publication and recommendations in the reports referenced in this chapter. In addition, the authors thank Mike Snow at LMI for his feedback.

References

1 Lueth, K.L. (2014). Why the Internet of Things is called the Internet of Things: Definition, history, disambiguation. https://iot-analytics.com/internet-of-things-definition/#:~:text=The%20term%20Internet%20of%20Things%20is%2016%20years%20old.&text=But%20the%20actual%20term%20%E2%80%9CInternet,new%20exciting%20technology%20called%20RFID (accessed 26 October 2022).

2 Army Science Board (ASB) Study Report (2018). The Internet of Things: Creating Smart Installations.
3 Army Science Board (ASB) Study Report (2017). Capabilities to Operate in Megacities and Dense Urban Areas.
4 Army Science Board (ASB) Study Report (2016). The Military Benefits and Risks of the Internet of Things.
5 Defense Science Board (DSB) Study Report (2016). Autonomy.
6 Defense Business Board (DBB) Study Report (2017). Logistics as a Competitive War Fighting Advantage.
7 Army Science Board (ASB) Study Report (2014). Decisive Army Strategic and Expeditionary Maneuver.
8 Army (2017). Internet of Battlefield Things (IoBT) Collaborative Research Alliance (CRA) Opportunity Day, 27 March 2017. https://www.arl.army.mil/business/collaborative-alliances/current-cras/iobt-cra/ (accessed 26 October 2022).
9 Kott, A., Swami, A., and West, B. (2016). The internet of battle things. *Computer* 49: 70–75. https://doi.org/10.1109/MC.2016.355.
10 Pradhan, M., Gökgöz, F., Bau, N., and Ota, D. (2016). Approach towards application of commercial off-the-shelf Internet of Things devices in the military domain. *2016 IEEE 3rd World Forum on Internet of Things (WF-IoT)*, pp. 245–250.
11 Suri, N., Tortonesi, M., Michaelis, J. et al. (2016). Analyzing the applicability of Internet of Things to the battlefield environment. *2016 International Conference on Military Communications and Information Systems (ICMCIS)*, 2016, pp. 1–8.
12 Raja, P. and Bagwari, S. (2018). IoT based military assistance and surveillance. *2018 International Conference on Intelligent Circuits and Systems (ICICS)*, pp. 340–344. https://doi.org/10.1109/ICICS.2018.00076.
13 Jalaian, B., Gregory, T., Suri, N. et al. (2018). Evaluating LoRaWAN-based IoT devices for the tactical military environment. *2018 IEEE 4th World Forum on Internet of Things (WF-IoT)*, pp. 124–128.
14 Cao, L., Zheng, G., and Shen, Y. (2016). Research on design of military ammunition container monitoring system based on IoT. *2016 Prognostics and System Health Management Conference (PHM-Chengdu)*, pp. 1–4.
15 Tian, G. (2013). The research of Internet of Things in military logistic management. *Applied Mechanics and Materials.* 303–306: 2170–2176.
16 Yushi, L., Fei, J., and Hui, Y. (2012). Study on application modes of military Internet of Things (MIOT). *2012 IEEE International Conference on Computer Science and Automation Engineering (CSAE)*, pp. 630–634.
17 Zheng, D.E. and Carter, W.A. (2015). Leveraging the Internet of Things for a more efficient and effective military. Center for strategic and International Studies Report.
18 Bennett, G. and Herkert, R. (2008). Deployment considerations for active RFID systems. In: *RFID Technology and Applications* (ed. S. Miles, S. Sarma and J. Williams). Cambridge University Press, pp. 101–112.
19 Lixianli, Wei, P., Jianyong, A., and Ping, W. (2020). The application research on military Internet of Things. *2020 17th International Computer Conference on Wavelet Active Media Technology and Information Processing (ICCWAMTIP)*, pp. 187–191.
20 Ruh, B. and Speaker, P. (2018). GE implementation of IoT for Condition Monitoring. World Forum on IoT.
21 *2021 IEEE 1st International Conference on Digital Twins and Parallel Intelligence (DTPI)*. DOI:10.1109/DTPI52967.2021.

22 DoD Data Strategy Report, DTIC Accession Number AD1112684, https://apps.dtic.mil/sti/citations/AD1112684, (2020).

23 Chatfield, A.T. and Reddick, C.G. (2019). A framework for Internet of Things-enabled smart government: a case of IoT cybersecurity policies and use cases in US federal government. *Government Information Quarterly* 36 (2): 346–357.

24 Wang, J., Cao, L., Shen, Y., and Zheng, G. (2018). Research on design of military logistics support system based on IoT. *2018 Prognostics and System Health Management Conference (PHM-Chongqing)*, pp. 829–832.

25 Abdelzaher, T., Ayanian, N., Basar, T. et al. (2018). Toward an internet of battlefield things: a resilience perspective. *Computer* 51 (11): 24–36.

26 Bennett, G. (2006). *Active RFID: Development and Deployment Lessons Learned*, Active RFID Summit, Track. Locate, Sense, IDTechEx, Atlanta, GA.

27 CHAMP (2020). Human performance optimization: moving left of bang. https://www.hprc-online.org/total-force-fitness/tff-strategies/human-performance-optimization-moving-left-bang (accessed 26 October 2022).

28 Hsu, J. (2018). The strava heat map and the end of secrets. *Wired Magazine*.

29 Mohammadi, N., Francisco, A., Taylor, J.E. et al. (2019). IoT integration of infrastructure systems in smart cities: the impact of interdependencies in building energy systems. *International Conference on Sustainable Infrastructure 2019. Leading Resilient Communities through the 21st Century*.

30 Marrano, L.R., Koster, A.P., Wolters, S.R. et al. (2019). Summary Report. Army installations of the Future Industry Day, https://erdc-library.erdc.dren.mil/jspui/handle/11681/38582.

31 Report: Army Installations Strategy- supporting the Army in Multiple Domains, December 2020. chrome-extension://efaidnbmnnnibpcajpcglclefindmkaj/viewer.html?pdfurl=https%3A%2F%2Fwww.asaie.army.mil%2FPublic%2FSI%2Fdoc%2FArmy_Installations_Strategy_(AIS)_FINAL_Signed.pdf&clen=1876239&chunk=true.

32 Strategic Studies Group Report (2014). Megacities and the United States Army, Preparing for a Complex and Uncertain Future. Chief of Staff of the Army, Strategic Studies Group (June 2014).

7

Digital Twins for Warship Systems: Technologies, Applications and Challenges

Sara Ferreno-Gonzalez, Alicia Munin-Doce, Marcos Míguez González, Lucía Santiago Caamaño, and Vicente Diaz-Casas

Grupo Integrado de Ingeniería, CITENI, Campus Industrial de Ferrol, Universidade da Coruña, Ferrol, Spain

Abstract

Thanks to the development of new technologies such as the Internet of things, it is possible to consider an in-depth knowledge of the operation of a ship, through its replication in the digital world, both hull, systems and equipment on board. It is a technological challenge due to the complexity of warships. It is necessary to define architectures (storage in the cloud), type of deployment (up to what extend the ship should be sensorized), use or not of open-source platforms, scalability, etc. The physical part connected to the virtual part, with the first sending real data to the second part, allows reliable predictions of the behavior of equipment and systems on board. The digital twin seeks to virtually reproduce the functions of the ship systems, as well as all the conditions involved in the ship's mission (route to follow, maintenance times, modes of navigation and sea conditions).

7.1 Introduction

The development of IoT technologies, and with it the ability to have connected products and devices have made possible the birth of digital twin concept. The combination of IoT, simulation and data analytics will allow the development of fully functional digital twins.

The connection between the physical system and its virtual replica, the digital twin, is possible thanks to the IoT, that allows to cover the gap between physical and virtual worlds. The digital twin is a compendium of technologies that starts from the connection with the real world and reaches up to the joint analysis of real and simulated data, to faithfully replicate the behavior of any system. This is a very interesting concept in the naval world as it ensures the proper functioning of the ship's equipment and systems. On a ship, especially a navy ship, it is a capital issue to ensure that a severe failure will not occur. It is important to anticipate possible anomalies in the operation of the equipment to a certain extent, to minimize the probability of the ship becoming inoperable. It is for this reason that research efforts are more focused on the concept of the digital twin, rather than on the direct application of IoT technology. This application is understood to achieve a more ambitious goal such as the digital twin of a ship. The relation between IoT and Digital twin is explained in detail in Section 7.2.

The digital twin concept is relatively new, dating back to 2002, when it was firstly streamlined by Grieves and Vickers [1] in a presentation regarding product lifecycle management (PLM) systems.

IoT for Defense and National Security, First Edition. Edited by Robert Douglass, Keith Gremban, Ananthram Swami, and Stephan Gerali.

According to his definition, as stated in [1], "a digital twin is a set of virtual information constructs that fully describes a potential or actual physical manufactured product from the micro atomic level to the macro geometrical level. At its optimum, any information that could be obtained from inspecting a physical manufactured product can be obtained from its Digital Twin". So, basically, when we refer to a digital twin, it is necessary to define a physical system to work with, a complete virtual environment that reproduce the behavior of the physical system, an architecture for the transfer of data from the physical system to its virtual representation and for the transfer of information in the opposite direction, and some additional subsystems providing different functionalities.

The initial applications of the digital twin concept originated with NASA, and were aimed at obtaining newly designed space vehicles which, at the same time, should be lighter than their predecessors and able to withstand harsh environments and longer operational times [2]. After these initial applications, and within the framework of the 4th industrial revolution, the use of this concept has been spread and is considered, nowadays, as one of the Industry 4.0 main drivers, amalgamating many of the so-called Industry 4.0 enabling technologies, which include the Internet of Things, big data, cloud, robotics, artificial intelligence, and additive manufacturing [3].

As described in [4] and [5], five main characteristics and capabilities of the digital twins can be defined, including the following: first of all, the fact that a digital twin represents one single physical system, in its as-built configuration (unicity); the capability for CAD representation of its corresponding physical system; the digital twin must have the capability to obtain, using a set of sensors, real time information from the physical system and to describe the context it is in; and last, but not least, the capability of the digital twin to reproduce the behavior of the physical system.

The latter is considered one of the most important functionalities of the digital twin, which exploits models to simulate the behavior of the physical system and to analyze the deviations of its performance, models which could be trained or that could learn from real operational data, and that could also be used for making future predictions of the behavior or maintenance needs of the physical system. Currently, digital twins make use of real time data and machine learning/artificial intelligence tools to fulfill these needs [4].

Although the application of the digital twin concept to the civil maritime sector is very limited, there exist, however, some practical approximations and mainly, a large interest of the sector in the development of these type of systems. One of these cases is that of the Open Simulation Platform [6], created as a collaboration between different Northern Europe companies and institutions, to develop a cloud simulation platform which, in its current version, was applied to 3 use cases: the design of a hybrid ferry propulsion system, a virtual commissioning of a coastal service vessel and an operational planning for crane operation.

On this matter, the involvement of the Classification Society DNV GL must be highlighted. On one hand, they consider digital twins as a basic future tool for the assessment of regulatory and standard compliance, during the process of verification and certification, and also for the safe and reliable operation of autonomous vehicles [7]. And on the other hand, they coach and support the policies of some local authorities which empower the use of digital twins within the whole set of involved stakeholders [8].

As can be observed from the above, and some other references herein [9], the applications of digital twins to the civil maritime sector are scarce. In the case of the naval vessels sector, the situation is similar, with the difference that there is a general interest of all the involved stakeholders in digital twins. Some examples of this could be the global approach of the U.S. Navy, or the Spanish shipbuilding company Navantia.

In the case of the U.S. Navy, there is a general governmental strategy to empower the use of "model-based techniques, digital practices, and computing infrastructure" to change "the way we conceive, build, test, field, and sustain our national defense systems". This strategy, initiated in 2018, is being adopted by suppliers as well. The strategy was initially applied in the shipbuilding process as in Newport News Shipyard, [10]) and, after that, in the product itself (the naval vessel [11].

In the case of Navantia Shipyard in Spain [11], the approach is similar to the previous one, aiming to generate both process and product digital twins which, in the near future, could be developed in parallel with their physical counterparts and delivered to the final client. The digital twins would then be maintained and updated during the whole life cycle of the ship.

These examples show that a digital twin is something more than a digital copy of a given ship or an individual system of that ship. That is, a digital twin is an architecture which assembles tools and information, creating a system with higher capabilities than each of its constituent parts by themselves.

In the case of the marine sector, and is described in [9], a digital twin has to provide decision support capabilities from large scale gathering of real operational data from sensors, be able to integrate complex systems and different sources of information, possibly to assess future designs on the basis of performance and to be able to remotely monitor the physical counterpart.

In this work, a description of the tools, methodologies and technologies needed to set up a digital twin of a naval vessel, which fulfils all the requirements above, is presented. The description includes data acquisition and transfer from the physical system to the digital twin (IoT), simulation models, data processing tools and techniques, and information flow from the digital twin to the operator or the physical asset.

7.2 A Digital Twin Architecture for Implementation

7.2.1 Physical Level

As the simulation models are designed and built, the input variables of these models need to be defined. Once the input data has been determined, it is possible to start evaluating the origin of raw data. This starting point is the sensors and hardware devices that measure some physical property of the monitored system's state as temperature, pressure, humidity or fluid flow. These devices are IoT devices and are capable of measure and gathered all needed data with a low-level data processing (Figure 7.1).

Traditionally, automated systems used hardwired sensor systems. Recent advances in electronic and wireless communications allow the implementation of wireless sensor networks (WSN). These networks have become essential in different domains such as industrial automation, environment, infrastructure, or military domain.

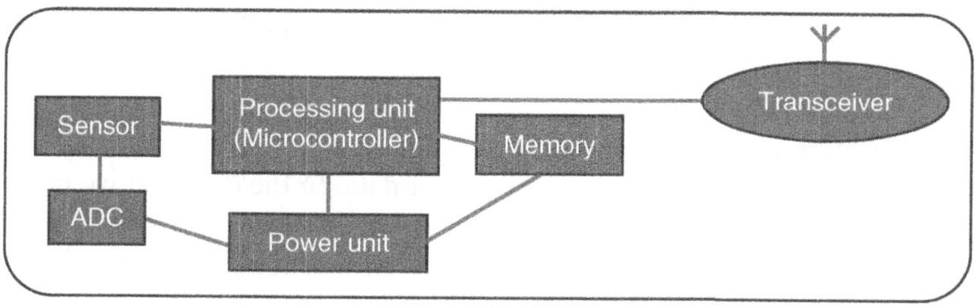

Figure 7.1 General schema of a sensor node main components.

Table 7.1 Resume of main communication technologies capacities at physical layer.

Technology	Sensitivity (dBm)	Data rate	Spectrum strategy
WiFi	−95	1–54 Mb/s	Wide band
Bluetooth	−97	1–2 Mb/s	Wide band
BLE	−95	1 Mb/s	Wide band
ZigBee	−100	250 kb/s	100 m
SigFox	−126	100 b/s	Ultra narrow band
LoRa	−149	18 b/s–37.5 kb/s	Wide band
Cellular data (2G, 3G)	−104	100 kb/s–3 Mb/s	Narrow band
Cellular data (4G, 5G)	−120	8 Mb/s–100 Mb/s	Wide band

There are three ways of approaching system sensorisation [12]: from the machine point of view; from the internet connection point of view; and from the data analysis point of view.

The machine approach focusses on the idea that every object/machine can have several sensors installed to provide real time information. This can be accomplished by embedded electronic devices applying RFID (Radio Frequency IDentification), NFC (Near Field Communication) or other technologies (see Table 7.1). Here the role of the microcontroller is fundamental because of their low power consumption, programmability, reusability, and cost-effectiveness. Probably the microcontroller most effective is Arduino [13].

The basis of the internet approach is that all devices can be connected through the network and so become "smart machines/objects" (SO). This implies that every machine has a unique IP that allows data integration for monitoring them in real time.

The data analysis approach refers to the possibility of obtaining meaningful information through data analysis. For this, it is important to filter raw data. Sometimes readings from two or more sensors show inconsistencies (data acquisition [DAQ] noise data), in this case it is necessary to remove some of the data from a data set [14]. The goal is to obtain enough accurate data that can be used in simulation models. Therefore, data requirements are quality, noise-free, and uninterrupted data [15]. Due to the affordability and availability of sensors and actuators nowadays, data acquisition has become relatively easier.

To obtain a good set of data, we need to bear in mind the volume of data needed and the frequency of data acquisition. On the one hand, data volume depends on acquisition frequency - data recorded at 5 Hz implies smaller datasets than data recorded at 20 Hz and smaller storage capacity required. The acquired data must be converted, managed and stored into a format that can be used for algorithms or simulations [16]. On the other hand, the sampling rate should be higher than the expected frequency to accurately record the signal.

Other important issue is the distance between transmitter and receiver, the chosen protocol depends on the range of the technology as can be seen in Figure 7.2.

There are different communication protocols to transmit sensor data. In the case that we have low data rates and need low power consumption a good option could be Zigbee or LoRa. Zigbee operates on an IEEE standard, IEEE 802.15.4. In contrast, Bluetooth technology is useful when high-rate transfer is needed. Right now, there is an improved version (Bluetooth Low Energy, BLE), which requires less energy (see Table 7.1.).

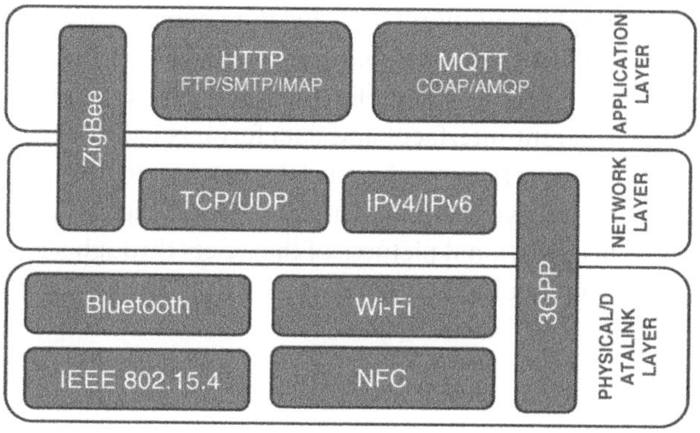

Figure 7.2 Schema of communication protocols for digital twin.

7.2.2 Physical World/Virtual World Interface

The present and future perspectives of digital twin technology go hand in hand with evolution of the Industrial Internet of Things (IIoT) because it creates the possibility of connecting computer automated control systems for remote monitoring and rapid response to events requiring real-time handling [17]. This implies that cyber-physical systems (CPS) are autonomously exchanging information facilitating the convergence of the physical and virtual worlds (Figure 7.3).

To fulfil the continuous transmission of the data, an infrastructure is required that enables a real time interaction between virtual image and physical object. The basic architecture of digital twin comprises a combination of three layers. One is the Edge layer consisting of a physical layer and a data link layer, this means sensors and actuators which are connected to an IoT gateway through communication protocols under an IEEE standard. The next one is the network layer which facilitates the transmission of data gateway to the cloud middleware for data ingestion and encapsulation [18]. There is a family of control system gateways, and the choice should be based on performance, reliability, security, and scalability requirements for a particular application. The protocols used are ethernet, Wi-Fi or cellular. Middleware solutions are used for integration communication architectures and message routing in CPS [19]. The last layer is the application layer which is responsible for data analysis and presentation and is usually based on HTTP or MQTT protocols.

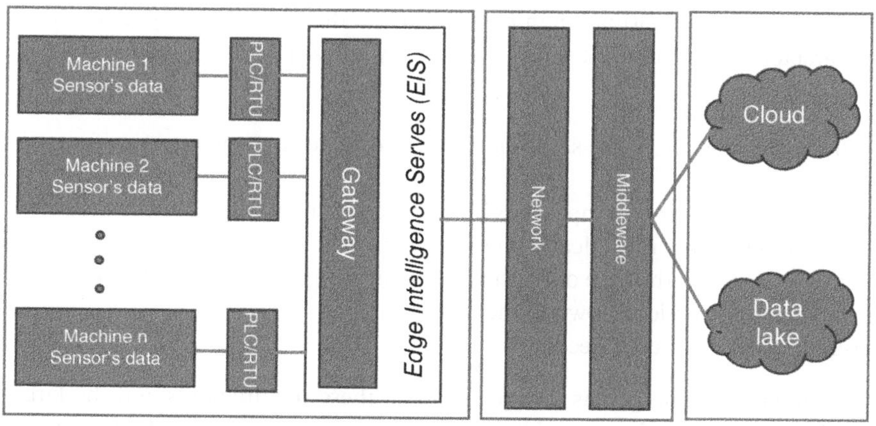

Figure 7.3 Description of the main components in a digital twin architecture.

Information flow is a key issue in digital twin implementation [20]. This flow comes from the fusion of raw sensor data and information from different sources and goes to high-level understanding (machine learning algorithms, big data analysis, etc.). Therefore, it is still a big challenge to combine the data from different sources with different interfaces and data formats in real-time applications [16]. An improved IoT system is crucial to carry out real-time data acquisition through its smart gateway and edge computing devices. To achieve the main functionality of digital twin implementation, the system must provide accurate operational pictures of the assets through simulation models (physics-based model or data-driven models) [21].

7.2.3 Digital Twin

7.2.3.1 Integration of Functionalities: User Interfaces

In the industry 4.0 and for the digital twin, the interaction between human and cyber-physical systems is crucial to enhance machine efficiency and operation safety [22, 23]. This interaction can be by direct manipulation or with the aid of a user interface; the second option is the most common in digital twins.

This subsection is focused on user interfaces, also called Human-Machine Interfaces (HMIs). An HMI can be understood as the means by which the user can communicate with a machine (or a virtual model and the sensors in the case of a digital twin) and includes all the points of contact between the user and the system [22, 24].

HMIs have significantly changed along the years. In particular, four stages of development can be recognized [23, 25]:

1. HMI 1.0 (until 1990): includes buttons and lights.
2. HMI 2.0 (1990–2010): represented by desktop visualization and panels.
3. HMI 3.0 (2010-future): characterized by web applications and portable devices.
4. HMI 4.0 (future): dominated by virtual and augmented reality.

For the digital twin the current era is HMI 3.0, the fourth stage is rare as there are still some problems in the implementation. Portable devices, such as mobile phones or tablets, add new ways of interaction between human and CPSs that are very advantageous such as touchscreens, voice recognition, gesture recognition, image processing and object recognition. These capabilities improve the usability, making the communication more intuitive.

In general, an HMI must provide information in real-time about the status of the cyber-physical system and its key characteristics are usability and transparency [22, 24].

Other requirements for an HMI are [22]:

- Accessibility to different systems from the same user interface, instead of one interface for each of them.
- User-friendly design, as the number of components increases continuously, the complexity of the interface also rises, and the worker must deal with this rising.
- As everything is decentralized (in the case of physical systems), the position of each component must be tracked, and this information shown to the user.
- Tracking of the user's position as well if needed.

Context-sensitive user interfaces emerged as a response to all these requirements and the large amount of data available. This kind of interface filters the information actively in order to provide to the user only the relevant information for his current situation [22, 24].

The last issue is that the interfaces for portable devices are developed using apps. The main reason is the multiple operating systems that exist make it difficult to create a manufacturer-independent and multi-platform interface. There are three types of apps [22]:

1. Web-based apps: mobile website that is run using a client and a server.
2. Native apps: it must be programmed in the same language as the operating system. By contrast, it gives full access to the components of the device and allows more ways to interact with the user.
3. Hybrid apps: it is a mix of the other two.

7.2.3.2 Simulation Models

Digital twins generate an increase in a product's value throughout its lifecycle. They make it possible by taking advantage of the data generated by the IoT devices (real data) and by simulation models (simulated data) to report about the operation of a system and at the same time better predict its behavior.

In many cases, digital twins that offer predictive maintenance and analytics are based on historical data and statistical models [26, 27]. However, there is a problem: it cannot be applied in the early stages of product development or system start-up, due to the time required to obtain enough data to allow this type of analysis. In addition, this data must be processed and put into a usable format. The main challenge in the use of digital twins based on historical data lies in obtaining enough data to cover all the possible scenarios that the product may face in real life.

Alternatively, digital twins that incorporate physics-based simulation models have important advantages. First, they may complement or even replace historical data with data obtained in simulations. This will reduce the number of physical tests required to collect sufficient data, as well as the processing time for that data. Additionally, the ability of a physics-based digital twin to simulate system conditions can even reduce the number of sensors required in the IIoT device. Now, instead of having to wait for enough field data to build and implement a digital twin, only a small amount of data is needed to validate it. This speeds development and time to market.

In addition, it is often the case that the products in the real world have a limited number of sensors, and some components may not even have suitable sensors due to cost or other limitations. In these cases, a real-time virtual replica provides information and value by combining known data with simulation to "fill in the gaps." Then we will have an unlimited number of "virtual sensors" provided by the simulation. Components can be monitored, diagnosed and also future predictions can be made [28, 29].

The possibility of combining both types of models in the development of the digital twin should be considered. Depending on the nature of the asset to be simulated, using hybrid models that are based on physics but that also incorporate models based on statistical data, will allow simulations of the real behavior of the system considered with a high degree of reliability [30].

In general, physics-based simulation is based on computer programs that replicate the behavior of a system, or a set of systems, so that we can check how it behaves. Robert Shannon and James D. Johannes [31] defined simulation in the 70s as: "the process of designing a model of a real system and carrying out experiences with it, in order to understand the behavior of the system or evaluate new strategies - within the limits imposed by a certain criterion or a set of them - for the functioning of the system".

The world of simulation is very large, since the simulations can have the degree of detail, the nature, and the objective that we want. We can simulate a physical phenomenon, the operation of a machine, a manufacturing process, etc. We can refer to simulation when it means a small program in which we represent the movement produced by a force on a pendulum. But we can also talk about simulation when we are in front of a sophisticated software program that allows us to emulate the piloting of an aerospace vehicle.

But focusing on the simulation models that are applied in the development of digital twins in ships and in ship's systems, these will be Multiphysics simulation models, that is, models that simulate products or systems characterizing the physical behavior of their different components and interactions between them.

In general, there are usually two types of simulation models:

Behavior Prediction Models: are those in which a simulation can be run in an acceptable amount of time, and that simulates the behavior of the system over a longer time interval. These models allow an operator to:

- Acquire experience and knowledge about the operation of the system that it is simulating.
- Identify problems in a complex process, in which using only experience or expert knowledge is not enough.
- Propose alternative scenarios for a system to explore the behavior of complex systems, for which making even a prototype would not be affordable economically and technically.
- Apply simulation in any phase of the life cycle of the system: design, construction, and operation.

Models for Real-time Simulation: are those which allow comparison in real time of the sensors signals with their simulated equivalents. These models also allow identification of variables for which we do not have sensors or, for which cost represents an important limitation when it comes to being installed in a system.

Furthermore, we may be interested in simulating the entire system, or some critical part of it. In some cases, it may be relevant to use complex simulation models, such as computational fluid dynamics, or in other cases, the use of simple numerical models may be sufficient. Depending on the objective, the following tools may be applicable:

- **Tools for Generating a Simulation Model for a System:** This type of tool supports the development of multidomain 1D models, which seek to simulate the behavior of a system, and which can cover different disciplines (multiphysics models) and their integration. They can also incorporate the simulation of the control system. These types of tools include Matlab® Simulink, and Amesim or, for example, those tools that are based on the Modelica language (OpenModelica, Dymola, Maplesim). These tools will allow us, in the development environment of a digital twin, to simulate systems in a simplified way from functional diagrams and/or specifications of the equipment that make up the real system.
- **Tools for Complex Systems Simulation**: This category includes simulation tools that are used for a specific discipline, for example: finite element modeling (FEM), computational fluid dynamics (CFD) or Simulation of multi-body systems (MBS). On one hand, this type of software allows us to acquire much greater precision, but on the other hand, they require time to carry out the simulation and, in some cases, computers with high processing and storage performance. This makes this type of model not directly usable in real-time operation/monitoring simulations or in predictive maintenance. A possible solution to these problems is the use of Reduced Order Models, or ROM [32]. A ROM is a simplified version of a high-fidelity computational model that preserves essential behavior and dominant effects, while at the same time can be used to reduce solution time and required computational storage capacity. These models are simplifications of complex models, which still maintain the reliability of their behavior. ROMs can enable the use of complex systems simulations in a digital twin environment, where simulation times must be very short.

7.2.3.2.1 Integration of Simulation Models in the Digital Twin Environment

Once the simulation models have been developed, they must be run on the platform in which the digital twin is integrated. In addition, it is reasonable to think that at the time when the simulation models must be implemented, it will be necessary to integrate component simulation models that may have been developed on different software tools, such as those mentioned in the previous sections. In recent years, work has been done on a standard that allows this integration to be carried out, while maintaining the quality of the models. A consequence of the above is the development of the Standard FMI: Functional Mockup Interface [33]. The FMI standard allows the exchange of simulation models between different software's and allows the co-simulation of different models (regardless of the tool with which they have been developed). This standard was developed by ITEA (Information Technology for European Advancement) in the MODELISAR project [34], that it was focused on the study of exchanges of simulation models of multiphysics systems in the automotive sector. Once the MODELISAR project was completed, the development of the standard continued, creating its own entity related to the FMI. Today, the main simulation tools allow the import and/or export of FMI-based models, which are called FMUs (Functional Mockup Units). Some of the simulation platforms that allow export of FMU models are [35]: 20-sim, Adams, AMESim, ANSYS, CATIA, Dymola, EcosimPro, MapleSim, OpenModelica, Scilab, SimulationX, and Simulink.

A FMU file is actually a compressed file that contains [36] these elements:

- A Xml file in which variables are stored.
- C-code corresponding to equations.
- Other data, such as tables or comments.

The implementation of these FMUs can be done by developing code, or by using any of the tools and libraries already developed for this purpose. There are libraries for C/C++ or Python languages that have been widely tested and that can be implemented in a platform for the digital twin (such as an IoT platform).

There are also specific tools and software for this type of model, such as Livetwin from Siemens, Rexcore from Rexygen, ETAS from Labcar, and dSPACE, which are generally designed for the execution of real-time models to support digital twins.

The simulation of FMU models in the environment of the digital twin can be carried out, mainly from 2 perspectives:

- Edge computing.
- Cloud computing.

The advantage of edge computing is the reduction of latency, since the real signals only have to travel from the physical layer to the Edge layer, and the simulation models are run in that same layer.

The advantage of running the models, on the IoT platform in the cloud, is that greater processing capacity is available, since in the case of complex models, an Edge platform may be somewhat challenged. Although to solve this limitation, we have ROM-type models available. In addition, in this case the challenge of installing software in the edge layer is avoided, with everything centralized in the cloud.

In the case of having several FMU models for an entire system, co-simulation is necessary. The output of the system depends on running each of the models together. The FMI standard considers two types of models: Model-Exchange or Co-Simulation.

In Model-Exchange models, the numerical solver is not included in the model, but the tool that imports the model must provide it. In the Co-simulation type models, the solver is integrated.

According to the latest version of the FMI standard (version 2.0) there are models that can be both types at the same time.

A master algorithm is required for Co-Simulation type models and to communicate with each other. This algorithm is not included in the standard. The master algorithm contains the running sequence of the models and the communications between them.

The disadvantages of Co-Simulation models include:

- The simulation time is increased when sequencing the executions of each FMU, increasing proportionally according to the number of models to be sequenced. If the number of models is remarkably high, it is possible that time limits are exceeded, and that would not meet temporary requirements for running real time models.
- Some FMU models exported by simulation software may need a license to run it (especially if they need a particular type of "numerical solver").
- It is necessary to program the master algorithm, the running sequence, and the communication between models, with their input and output variables between the models. This algorithm must figure out data problem compatibilities between models to perform the correct orchestration of all data.

7.2.3.3 Data Storage and Data Lakes

A data lake is a repository that stores raw, unprocessed data of different types:

- Structured data is stored in a relational database, such as MySQL, MariaDB, PostgreSQL, Microsoft SQL Server.
- Unstructured data such as images or video, text, handbooks, or technical documents are stored as PDFs.
- Semi structured data is data without a fixed schema, but with labels, for instance, data in JSON, XML format.

Data lakes can be made up of different databases and optimized to store a specific type of data.

In the case of a digital twin, it is interesting to store simulation results and aggregated data from sensors. In both cases data is structured. For simulation models, data is defined as input and outputs variables. Simulation results also have a time stamp associated to each variable; therefore, they can be considered as a time series. There are databases optimized to be efficient both in terms of storage space and in reading and writing speedily by storing time series such as InfluxDB (Figure 7.4).

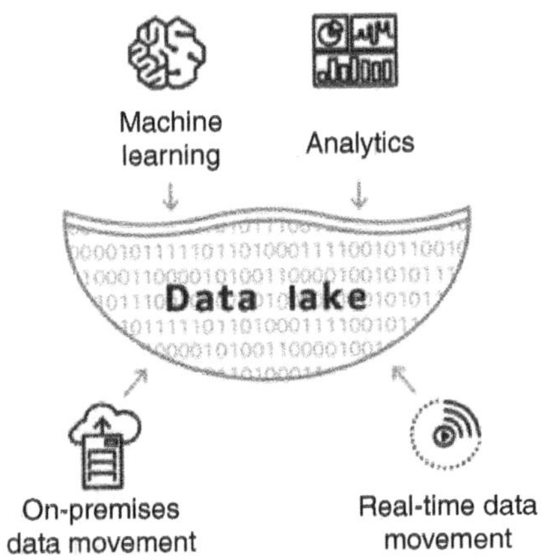

Figure 7.4 Scheme of information sources and data flow to the data lake.

Figure 7.5 Diagram with the phases of the ETL process.

A concept related to the data lake, but with a different purpose is the concept of data warehouse. In a data warehouse, the data is stored organized by subject and is prepared for data analytics and company decision-making. In data warehouse environments ETL (Extract Transform Load) processes are usually used. These processes are necessary because data is often served from different heterogeneous sources. An example of an ETL tool is Apache Nifi (Figure 7.5).

Data lake consumers are the applications that will query the data that is stored. For example, a machine learning model can extract a large volume of data to use for a training set by querying the data lake or querying the latest data in real time that can serve as input to simulation models. The process of extracting the data to create a training set is not straightforward, but once the data is in the data lake it could be automated.

Another example of a consumer of data can be an application with dashboards to visualize the simulation data. These dashboards can make the queries to the stored data and the show graphs of this data.

Relational databases provide APIs for writing or reading data by other external tools.

7.2.3.4 Data Analysis, Machine Learning, and Predictive Algorithms

Technologies related to data analytics/big data, or machine learning have received considerable attention in recent years, and they are an important part of the current landscape of the digital twin. The significant increase in the number of sensors on board and the growing implementation of the IIoT, makes it possible to use AI algorithms to analyze the information collected from the ship [4].

The integration of AI in the digital twin implies an important qualitative leap. Authors such as [37] or [38] place at a higher level those digital twins that incorporate learning algorithms or artificial intelligence. In general, an intelligent digital twin must have two main capabilities with respect to data processing: on the one hand, it must be able to apply appropriate algorithms to perform data analysis and extract new knowledge from it. On the other hand, it must be able to improve the behavior and performance of the real system.

The use of AI in the digital twin opens a wide range of possibilities of use, with special consideration for algorithms oriented to:

- Process & operation optimization. By using real variables read from the real system, models can be run hundreds or thousands of simulations aimed at optimizing or controlling system operations and thus mitigating risks, reducing costs, or improving efficiencies.
- Predictive maintenance. In Industry 4.0 applications, models can determine the remaining useful life and report the best time to perform maintenance or replace equipment.

- Detection of anomalies. The model runs in parallel to real system and immediately points to any operational behavior that deviates from the simulated behavior.
- Diagnosis and correction of anomalies: AI-based models will enable identification of the cause of a possible failure or anomaly detected, as well as to determine the best option to try to mitigate that failure.

The application of data-based methods should not be seen as a contrast to those that use physics-based simulation models, but rather as something complementary. Taking advantage of both will be very important in the development of an optimized digital twin. The use of physics-based simulation models is justified above all, in the case of those processes in which the nature of their operation is faithfully known. In addition, in the design stages of the ship, before data are available, they will be the models that can be used. However, once it is possible to have data, and as it has been seen, data-based models have important advantages, among which it is worth highlighting that it is possible to model complex systems that are not easily modellable using physics-based models.

In addition, the use of physics-based models in combination with data-based models, such as what is proposed in [58, 59], will allow the readjustment of the behavior models based on the results/analysis performed by AI algorithms based on real data and those of the models that are being analyzed.

In addition, the strength that AI-based algorithms have compared to other analysis methods is the fact that it is not necessary to carry out a complex and sometimes difficult task of defining rules (which usually go through an exhaustive almost manual process). Instead, data can be labelled in bulk, or even use unsupervised artificial intelligence algorithms, which are capable of learning without the need for supervised training.

7.3 Ship Digital Twin Implementation

7.3.1 Physical Level

Nowadays, the level of automation of ships allows not starting from scratch in the sensorisation of the systems. There are control systems on the ship with sensors, data acquisition systems and actuators, that are connected through the on board communications network [39]. Data sources such as AIS (Automatic Identification System), GNDSS or ECDIS (Electronic Chart Display System) enhance monitoring both on board and from shore.

Ships already have engine monitoring control systems (unattended engine room), their sensors provide an important data set including basic measures such as main engine and auxiliary fuel consumption, main engine and auxiliary power, shaft speed, etc [40]. The same thing happens with route and weather data sets. Adding to this the voyage data recorder (VDR) as a source of the ship speed, relative wind speed and direction [41], etc., we have a large-scale data sets that could be analyzed to evaluate ship performance under different weather conditions and its speed loss [27]. For instance, it is possible to use this data set as input data for a propulsion system model for condition-based maintenance [42]. All these sources of information allow to know ship performance and navigation parameters.

Usually there is an on-board server to collect acquired data. This server is an autonomous industrial embedded computer which is connected to the sensors and machines of the ships and stores the collected operational data [43]. The critical ship components are connected through wired networks (Ethernet), but the lowest level devices are connected with a switch or gateway through a variety of protocols [39]. In the process of building a modern warship, a significant volume of cabling is used (450,000-1,000,000 meters of cable). This comes with a cost and weight increase and the problem will become even more pressing in the new ships that are to be developed with specifications of the "Ship 4.0" type, associated with the concepts of monitoring, hyper-sensing, and information redundancy. These new specifications translate into a notable increase in the number of devices (sensors and actuators) that must be wired. Likewise, an exponential increase in the volume of data/information whose traffic must be managed and distributed through the ship's communication networks can also be expected and which, consequently, will have to increase their capacity in the same proportion. In this context, both in current and in future ships, any line of action aimed at reducing wiring and increasing the capacities of data networks is of great interest. Therefore, the tendency is to reduce wired sensor on board, changing the traditional architecture to the wireless sensor networks (WSN's) [44]. This is a challenging issue because a ship environment is quite different from an industrial one.

Therefore, the first thing is to assess what data flow is already available on the ship and what new data flow needs to be incorporated and how. For instance, to measure ship motions, we can implement motion reference units (MRUs), this kind of device is capable of collecting the six degrees of freedom (heave, surge, sway, roll, yaw and pitch) [14]. Or wireless sensors can be applied such as triaxial accelerometers and gyroscopes to measure ship responses (motions, accelerations and global loads) [45]. However, other kinds of sensor like structural health sensors for hull monitoring can be very expensive and difficult to maintain [46]. Nevertheless, a literature review shows that most models (data-driven and physical- based models) use data sources already installed in the ship.

7.3.2 Physical World/Virtual World Interface

The ship digital twin architecture does not differ so much from a digital twin for a production factory, both are complex environments with challenging conditions. So, a war ship, as any other CPS, will need the sensor/actuator level, the edge level, and an application level. The difference lies in consequences of interchanging data through communication networks outside the ship and the metallic environment where hardware devices must work. Therefore, there are important issues that must be considering in the selection of a specific architecture to implement a digital twin:

- Performance: ensures that the system is efficient and can carry out its intended tasks timely and correctly.
- Availability: ensures that the systems and information contained within them are available to authorized users. This is especially important for a war ship because access to the data is paramount in critical systems.
- Scalability: due to the continuous increase of data sources that can be included as a part of a digital twin architecture, it is necessary to implement a scalable digital framework to meet this growth [47].
- Fault-tolerance: ensures that the systems are robust and can continue operating at a reasonable level in the event of a failure [48].
- Security: the interconnectivity required for implementing the desired CPS functionalities introduces new hazards, as cyber-attacks that can exploit vulnerabilities in the communication links and directly affect the integrity or availability of the data and control systems [49]. This concept includes database security that involves extensive use of information security controls to protect against a compromising confidentiality, data integrity and availability [44].

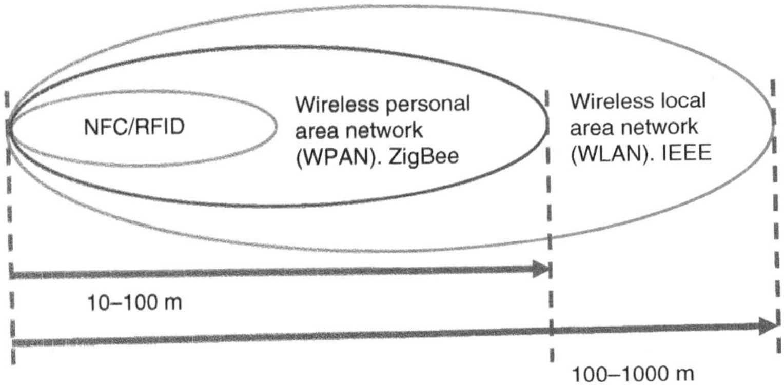

Figure 7.6 Use of communication protocols for sensors based on distance ranges.

The inside the ship wireless sensor network must deal with metal shielding. Therefore, communications range will be affected and have not the same functionality as in industrial networks (Figure 7.6).

7.3.3 Integration of Functionalities and the User Interface

The main objective of the HMI in digital twins in ships is to provide a clear vision of the situation to the user in order to allow him/her to make decisions. This is due to the fact that most of digital twins in the maritime sector are oriented to decision support systems or for training purposes.

The most common type of HMI in digital twins in ships is based on web services. For the implementation, the recommended characteristics are [50]:

- Visualization: humans have a better understanding if data is visually represented. For this reason, data should be presented using graphics.
- Modularization: following the context-sensitive approach and considering the increasing functionalities, the interface should be modular. This characteristic allows integration of a new component without disturbing the original system.
- Object-oriented approach: this is a consequence of modularization. In this type of programming, the ship could be represented by an object/class and the subsystems by the methods inside the object or subclasses.

An example of HMI in digital twins in ships is the Open Simulation Platform developed by a partner agreement between DNV GL, Kongsberg Maritime, SINTEF and NTNU [51].

7.3.4 Simulation Models

Focusing on the simulation models that are applied in the development of digital twins in ships and in ship's systems, there will be a high proportion of multiphysics models, that is, models that allow simulation of products or systems characterizing the physical behavior of their different components and interactions between them. Moreover, data-driven models also can be used. In this way and in general, on the ship we will find models of different origins, developed on heterogeneous platforms. For this reason, the implantation of simulation models in the digital twin ship environment, together with the co-simulation will play a very important role.

One of the main investigations being carried out in this field is the work of the OSP platform. The Open Simulation Platform (OSP) has been the result of the collaboration of Rolls-Royce Marine,

NTNU (Norwegian University of Technology Science), DNV GL, and Hyundai Heavy Industries through a joint industrial project (OSP-JIP) agreement to allow simulations of digital twins and solve the challenges of design, in terms of operation and assurance of complex and integrated systems. Within this project, a set of libraries has been developed for FMU co-simulation, through a graphical interface or through xml code.

7.3.5 Data Analysis, Machine Learning, and Predictive Algorithms

Since the use of data-driven techniques depends primarily on the availability of a significant number of sensors and on the possibility of collecting that information in near real time, data-driven models are likely not applicable in all ship systems, although as expected, the ship's critical systems will generally have the necessary infrastructure for this.

Some developments based on AI applications in shipbuilding are described below.

The SOPRENE Project (UDC, Indra & Spanish Navy) aims to investigate the application of artificial intelligence techniques to improve ship maintenance. The data available at the Center for Monitoring and Analysis of Monitored Data of the Spanish Navy (CESADAR) will be used. These data come from the sensorised equipment of the vessels and are recorded while the vessels are navigating. The goal is to study the advantages that their analysis can provide in order to reinforce the predictive maintenance of the vessels, avoiding unforeseen breakdowns, increasing their availability and saving costs [52].

Fujitsu & Maritime and Port Authority of Singapore have developed a solution to detect ship collision risks and predict areas where risks are concentrated as "critical points of dynamic risk." This technology has the potential to be implemented in a Vessel Traffic Service system to help maritime controllers proactively with traffic management, with the aim of improving navigation safety [53].

Stena Line, in collaboration with the technological company, Hitachi, has developed an artificial intelligence assistance system that will help determine what is the most efficient way in terms of fuel to operate a ship on a specific route [54]. It considers the currents, weather conditions, water depth and speed, combining data in a way that would be impossible in a manual system. Although it is an advance that is still under development, a first pilot is already being carried out on Stena Scandinavica on its route between the Swedish port of Gothenburg and the German port of Kiel.

The company that has advanced the most in the implementation of AI in ships is Rolls Royce. At this moment it has already carried out the first tests with an autonomous ship, Falco, a ferry of about 54 meters in length, which has the Rolls-Royce Ship Intelligence system, [55–57] and includes a series of cameras and sensors that are placed along the ship, and are responsible for scanning the waters looking for other ships. The Falco can navigate the waters safely, considering the presence of other vessels, and make the usual trip between Parainen and Nauvo. In addition, it can perform all the necessary procedures for docking autonomously.

References

1 Grieves, M. and Vickers, J. (2017). Digital twin: mitigating unpredictable, undesirable emergent behavior in complex systems. In: Kahlen, J., Flumerfelt, S., and Alves, A. (eds). *Transdisciplinary Perspectives on Complex Systems*. Springer: Cham.

2 Glaessgen, E.H. and Stargel, D.S. (2012). The digital twin paradigm for future NASA and U.S. Air force vehicles. *Collection of Technical Papers - AIAA/ASME/ASCE/AHS/ASC Structures, Structural Dynamics and Materials Conference*, p. 432938.

3 Martinelli, A., Mina, A., and Moggi, M. (2021). The enabling technologies of industry 4.0: examining the seeds of the fourth industrial revolution. *Industrial and Corporate Change* 30 (1): 161–188.

4 Erikstad, S.O. and Ove, S. (2017). Merging physics, big data analytics and simulation for the next-generation digital twins. High-Performance Marine Vehicles, October 2017, pp. 11–13. https://www.predix.io (accessed 27 Ocotber 2022).

5 Cabos, C. and Rostock, C. (2018). Digital model or digital twin? *Proceedings of the 2018 Conference on Computer Applications and Information Technology in the Maritime Industries (COMPIT).*

6 Perabo, F., Park, D., Zadeh, M.K. et al. (2020). Digital twin modelling of ship power and propulsion systems: application of the Open Simulation Platform (OSP). *IEEE 29th International Symposium on Industrial Electronics (ISIE)*, pp. 1265–1270.

7 Smogeli, Ø. (2017). The Internet of Big Things. Digital Twins at work in maritime and energy. DNV - GL.

8 DNV-GL (2018). Digital Twin Report for DMA: Digital Twins for Blue Denmark.

9 Taylor, N., Human, C., Kruger, K. et al. (2020). Comparison of digital twin development in manufacturing and maritime domains. In: *Service Oriented, Holonic and Multi-agent Manufacturing Systems for Industry of the Future SOHOMA 2019 Studies in Computational Intelligence*, vol. 853 (ed. T. Borangiu, D. Trentesaux, P. Leitão, et al.). Cham: Springer.

10 Debbink, N. and Coleman, C. (2020). Strategy for an intelligent digital twin (IDT). Newport News Shipbuilding.

11 Drazen, D. (2018). *Cyber-Physical Systems: Navy Digital Twin.* Naval Surface Warfare Center, Carderock Division. US Navy.

12 Mohammad, G.B. and Shitharth, S. (2021). Wireless sensor network and IoT based systems for healthcare application. *Materials Today: Proceedings.*

13 Kondaveeti, H.K., Kumaravelu, N.K., Vanambathina, S.D. et al. (2021). A systematic literature review on prototyping with Arduino: applications, challenges, advantages, and limitations. *Computer Science Review* 40: 100364.

14 Brandsæter, A. and Vanem, E. (2018). Ship speed prediction based on full scale sensor measurements of shaft thrust and environmental conditions. *Ocean Engineering* 162: 316–330.

15 Mathupriya, S., Banu, S.S., Sridhar, S., and Arthi, B. (2021). Materials Today: Proceedings Digital twin technology on IoT, industries & other smart environments: a survey. *Materials Today: Proceedings.*

16 Adamenko, D., Kunnen, S., Pluhnau, R. et al. (2020). Review of methods of designing the Digital Twin and comparison of the methods designing. *Procedia CIRP* 91: 27–32.

17 Sosa-Reyna, C.M., Tello-Leal, E., and Lara-Alabazares, D. (2018). Methodology for the model-driven development of service oriented IoT applications. *Journal of Systems Architecture* 90: 15–22.

18 Kiesel, R., Weiss, A., and Schmitt, R.H. (2021). Filling the semantics gap in industrial communication: middleware+ for an internet of production. *IFAC-PapersOnLine* 54 (1): 432–437.

19 Stock, D., Schel, D., and Bauernhansl, T. (2020). Middleware-based cyber-physical production system modeling for operators. *Procedia Manuf.* 42: 111–118.

20 Liu, Z., Meyendorf, N., and Mrad, N. (2018). The role of data fusion in predictive maintenance using digital twin. *AIP Conference Proceedings*, Volume 1949 (April 2018).

21 Farsi, M., Daneshkhah, A., Hosseinian-Far, A., and Jahankhani, H. (2020). *Internet of Things Digital Twin Technologies and Smart Cities*, 217. Springer Ebook.

22 Gorecky, D., Schmitt, M., Loskyll, M., and Zühlke, D. (2014). Human-computer-interaction in the industry 4.0 era. *12th IEEE International Conference on Industrial Informatics*, pp. 289–294.
23 Ma, X., Tao, F., Zhang, M. et al. (2019). Digital twin enhanced human-machine interaction in product lifecycle. *Procedia CIRP* 83: 789–793. https://doi.org/10.1016/j.procir.2019.04.330.
24 Krupitzer, C., Müller, S., Lesch, V. et al. (2020). A survey on human machine interaction in industry 4.0, pp. 0–45.
25 Papcun, P., Kajáti, E., and Koziorek, J.Ĝ. (2018). Human machine interface in concept. *2018 World Symposium on Digital Intelligence for Systems and Machines (DISA)*, pp. 289–296.
26 He, R., Chen, G., Dong, C. et al. (2019). Data-driven digital twin technology for optimized control in process systems. *ISA Transactions* 95: 221–234. https://doi.org/10.1016/j.isatra.2019.05.011.
27 Coraddu, A., Oneto, L., Baldi, F. et al. (2019). Data-driven ship digital twin for estimating the speed loss caused by the marine fouling. *Ocean Engineering* 186: 106063.
28 Tiwana, M. (2019). How to quickly create a simulation-based digital twin of an IIoT-connected product.
29 Tiainen, T., Miettinen, J., Viitala, R. et al. (2019). Digital twin and virtual sensor for a rotor system. *Annals of DAAAM and Proceedings of the Internatioal DAAAM Symposium* 30 (1): 1115–1121.
30 Wang, P. and Luo, M. (2021). A digital twin-based big data virtual and real fusion learning reference framework supported by industrial internet towards smart manufacturing. *Journal of Manufacturing Systems* 58 (PA): 16–32. https://doi.org/10.1016/j.jmsy.2020.11.012.
31 Shannon, R. and Johannes, J.D. (1976). Systems simulation: the art and science. *IEEE Transactions on Systems, Man, and Cybernetics* 6 (10): 723–724.
32 Hartmann, D., Herz, M., and Wever, U. (2017). Model order reduction a key technology for digital twins. *Reduced-Order Modeling (ROM) for Simulation and Optimization*, pp. 1–179.
33 Bertsch, C., Ahle, E., and Schulmeister, U. (2014). The Functional Mockup Interface - seen from an industrial perspective. Proceedings of the 10th International Models Conference, Volume 96, Lund, Sweden (10–12 March 2014), pp. 27–33.
34 Chombart, P. (2012). Multidisciplinary modelling and simulation speeds development of automotive systems and software. *ITEA2 Innovation Report* 33: 52–54. https://itea3.org/project/result/download/6202/MODELISARInnovationReport.pdf.
35 Junghanns, A. and Blochwitz, T. (2018). 10 Years of FMI Where are we now? Where do we go? *Japanese Modelica Conference 2018*.
36 Negri, E., Fumagalli, L., Cimino, C., and MacChi, M. (2019). FMU-supported simulation for CPS digital twin. *Procedia Manufacturing* 28: 201–206. https://doi.org/10.1016/j.promfg.2018.12.033.
37 Ashtari Talkhestani, B., Jung, T., Lindemann, B. et al. (2019). An architecture of an intelligent digital twin in a cyber-physical production system. *At-Automatisierungstechnik* 67 (9): 762–782.
38 Madni, A., Madni, C., and Lucero, S. (2019). Leveraging digital twin technology in model-based systems engineering. *Systems* 7 (1): 7.
39 Sahay, R., Meng, W., Estay, D.A.S. et al. (2019). CyberShip-IoT: A dynamic and adaptive SDN-based security policy enforcement framework for ships. *Future Generation Computer Systems* 100: 736–750.
40 Bui, K.Q. and Perera, L.P. (2021). Advanced data analytics for ship performance monitoring under localized operational conditions. *Ocean Engineering* 235: 109392.
41 Perera, L.P. and Mo, B. (2018). Ship speed power performance under relative wind profiles in relation to sensor fault detection. *Journal of Ocean Engineering and Science* 3: 355–366.

42 Cipollini, F., Oneto, L., Coraddu, A. et al. (2018). Condition-based maintenance of naval propulsion systems with supervised data analysis. *Ocean Engineering* 149: 268–278.

43 Shaw, H.J. and Lin, C.K. (2021). Marine big data analysis of ships for the energy efficiency changes of the hull and maintenance evaluation based on the ISO 19030 standard. *Ocean Engineering* 232 (3): 108953.

44 Wang, D. (2021). Ship machinery detection and diagnosis technology based on wireless sensors. *Microprocessors and Microsystems* 80: 103599.

45 Bennett, S.S., Brooks, C.J., Winden, B. et al. (2014). Measurement of ship hydroelastic response using multiple wireless sensor nodes. *Ocean Engineering* 79: 67–80.

46 Vanderhorn, E., Wang, Z., and Mahadevan, S. (2021). Towards a digital twin approach for vessel-specific fatigue damage monitoring and prognosis. *Reliability Engineering and System Safety* 219: 108222.

47 Scime, L., Singh, A., and Paquit, V. (2022). A scalable digital platform for the use of digital twins in additive manufacturing. *Manufacturing Letters* 31: 28–32.

48 Qiao, L. and Yang, Y. (2018). Fault-tolerant control for T – S fuzzy systems with sensor faults: application to a ship propulsion. *Journal of the Franklin Institute* 355 (12): 4854–4872.

49 Bolbot, V., Theotokatos, G., Boulougouris, E., and Vassalos, D. (2020). A novel cyber-risk assessment method for ship systems. *Safety Science* 131: 104908.

50 De Oliveira, F.F. (2021). *An Open Web Platform Aimed at Ship Design, Simulation and Digital Twin*. Faculty of Engineering, Norwegian University of Science and Technology.

51 DNV GL AS. Open Simulation Platform, (2020).

52 Novoa, D., Eiras, C., Fontenla, Ó., and Lamas, F. (2020). Mantenimiento predictivo de motores de buques mediante aprendizaje automático. *Congreso Nacional de I+D en Defensa y Seguridad*, Zaragoza.

53 Fujitsu (2020). Reducing risk on the high seas by reimagining everyday operations.

54 Hitachi (2021). Value Creation Story " AI Captain " to Curtail Fuel Costs and Optimize Travel Routes.

55 Levander, O. (2016). Ship intelligence - a new era. *Smart Ship Technology*.

56 Rolls-Royce (2016). Ship Intelligence.

57 Rolls-Royce (2018). Benefits of Ship Intelligence Solutions. https://www.rolls-royce.com/~/media/Files/R/Rolls-Royce/documents/customers/marine/RR-Ship-Intel-Broch-Oct2018.pdf (accessed 27 October 2022).

58 Sun, C., Shi, V.G. (2021). PhysiNet: A combination of physics-based model and neural network model for digital twins. *Int J Intell Syst* 37: 5443–5456.

59 Mas, P., and Sobie, C., et al (2018). Connecting physics based and data driven models: The best of two worlds. Siemens AG.

Section 2

Introduction: Artificial Intelligence and IoT for Defense and National Security

Robert Douglass

Alta Montes, Inc., Sandy, Utah, USA

Internet of Things (IoT) uses Internet protocols and networking infrastructure to join three elements: sensing, processing, and action. The explosion of smartphones drives the performance of sensors up while driving down their cost, size, weight, and power. Smartphones also introduce many additional rich sources of information that can be extracted from phones and their apps that feed IoT networks. The expanding smartphone market also drives the expanding bandwidth and coverage of digital cellular communications that in turn support larger and more powerful IoT networks. But the most spectacular IoT gains in the past decade have come from increasing processing power supporting increasingly intelligent algorithms. In terms of increased computational power, measured in floating point operations per second or the number of bytes of memory, processing gains are merely a continuation of a Moore's Law expansion of computing chip densities. However, the expanding compute power and increasing amounts of available data have allowed intelligent algorithms to cross an inflection point. Algorithms, such as deep-learning neural networks, exhibit increasingly intelligent behavior in many new domains. These algorithms, commonly referred to as artificial intelligence (AI), solve many tasks previously thought to require human intelligence. The expanding network of sensors and information extracted from applications and data bases now provides the prodigious amounts of digital data that AI machine-learning algorithms need for training to achieve high performance.

Cloud and edge technology shares enhanced processing power with IoT devices without having to incur the cost, size, weight, and power of embedding high-performance processors in each device. IoT can use intelligent processing to close the loop between sensing and action to accomplish missions currently requiring people intimately involved in the loop. Closing this loop creates new and greater possibilities for automating defense systems and creating autonomous and semi-autonomous battlefield IoT networks. Such networks have the potential of not only amplifying humans on the battlefield but also moving them back from some of the most dangerous missions and environments to positions of greater safety. Intelligent processing also increases the efficiency and lowers the cost of tasks outside of combat, such as maintenance and resupply. This section includes chapters on the use of AI in IoT applications, specifically for controlling robots and processing acoustic data in new ways. This section also addresses some of the difficulties of special concern for defense and national security that arise when inserting AI algorithms into IoT networks.

IoT for Defense and National Security, First Edition. Edited by Robert Douglass, Keith Gremban, Ananthram Swami, and Stephan Gerali.

AI holds great promise for amplifying IoT's power. However, the nature of deep-learning networks, the most common form of AI in IoT, can make AI algorithms brittle, untrustworthy, and vulnerable as Chapter 8 explains. Deep-learning algorithms are trained on large data sets of exemplars and can recognize people, objects, and events using video, sounds, and radar signals. They can recognize and generate spoken and printed language. They can string together complex sets of actions to defeat the most skilled humans in tasks such as chess, poker, predicting protein folding, and simulated air combat. On the negative side, brittleness and vulnerabilities often arise because of biases or limitations in an algorithm's training set. Challenges also arrive because of the way deep-learning networks store their knowledge. A deep-learning network uses a distributed set of nodes analogous to artificial neurons that embody its knowledge in the weights of connections between thousands of nodes arranged in layers. This strategy is particularly powerful for finding and storing patterns in large, complex data sets. Its disadvantage arises from the difficulty a human has in understanding what knowledge such a network contains and what knowledge it lacks. Because a deep-learning network is not organized into discrete rules, understandable by people, it is difficult to understand why such an AI neural network makes the decisions it does and why it takes the actions it does. A deep-learning network can rarely explain its decisions and actions to people. Because its knowledge is impenetrable and its actions are unexplainable, it is difficult to gain confidence by testing an AI-driven IoT network and therefore difficult to trust it in essential roles in combat and national security. Rob Brooks, one of the principal developers of both military and commercial robots, as quoted in the 2021 September issue of *IEEE Spectrum*, points out that today the successful applications of AI occur in domains where the consequences of failure are small, such as game playing, vacuuming households, and speech recognition to command home appliances. Fulfilling the promise of AI-empowered IoT for military operations requires that their AI algorithms can be trusted, and their limitations anticipated. In Chapter 8, a combined U.S. Army, University of California at Los Angeles, University of Illinois, and SRI International team describe principles and techniques that make machine learning models robust, resilient to adversarial attacks, and more interpretable for human-on-the-loop decision-making. The chapter also identifies the key challenges in developing trustworthy machine learning for the Internet of Battle Things (IoBT).

Chapters 9 and 10 review research that uses AI to drive applications of robotics and sound perception, both key emerging defense applications of IoT. Chapter 9 describes how distributed AI technology at the edge of IoT networks can bypass disruptions and limitations of battlefield networking and communications, specifically for robots. Robotic devices play an increasingly important part in air warfare in the form of unmanned aerial vehicles, commonly called drones. A variety of defense programs in the United States, China, and elsewhere are exploring the use of semi-autonomous and autonomous intelligent aircraft to provide comprehensive surveillance, coordinate ground attack, and serve as combat associates for manned fighter aircraft. Many countries now incorporate semi-autonomous, stand-alone robotic aircraft into their standard military strategy and tactics, for example the Predator and Reaper aircraft in the U.S. Air Force and the Turkish Bayraktar TB2 used by several nations' militaries. The Defense Advanced Projects Agency (DARPA) through the Air Combat Evolution (ACE) program is developing autonomous jet fighters that can assist human pilots in dog fights. These aircraft are essentially IoT networks in themselves that couple sensors, processing, and actuators including weapons. They are also linked to larger IoT networks that include external IoT nodes and that retain humans in top-level control. Limited bandwidth and unreliable communications on the battlefield prevent low-level human control of robots. The increasing autonomy and behavioral complexity of these robotic aircraft underlies their widespread adoption. This autonomy results principally from advances in AI algorithms. Strong advantages of robotic aircraft for air warfare push the rapid adoption of more intelligent unmanned air vehicles. They can be built more cheaply than manned aircraft, fly longer missions, and they clearly save pilots' lives in contested airspace. Maritime operations have been slower to adopt robotic technology than air operations, but autonomous surface and subsurface vessels are now in experimental use in many navies. Some navies may already be using unmanned vessels in routine operations.

Military ground robots trail years behind their aerial brethren. The U.S. Defense Department sponsored the first driverless ground vehicle as long ago as 1985 and subsequently matured the core technology underlying the driverless cars of today. Despite that early and leading role of the military, nations are just beginning to widely incorporate robots for military operations on the ground. Most of today's ground robots are used for explosive ordinance removal, a job so hazardous to humans that even fully teleoperated robots with limited autonomy have great value in preserving soldiers' lives and limbs. Emerging use of robots for ground operations include robotic convoying of supplies, small-unit surveillance, logistic support vehicles, mine clearance, and robotic sentries for perimeter patrol. Ground robots lag robotic air vehicles largely because negotiating the ground environment is more complex than maneuvering in the air. Communications and networking on the ground are also more challenging, especially in urban terrain. However, limited communications bandwidth and uncertain connectivity constrain the utility of all types of robotic IoT systems in all combat environments (a key challenge to all IoT in military operations, as discussed in Section 4). As AI algorithms grow in intelligence, they overcome connectivity problems by making the IoT systems increasingly autonomous, which requires less bandwidth for control. However, AI algorithms require significant processing power. For commercial IoT applications, processing power is easily available from resources in the cloud, remote from the IoT edge devices. For defense systems in the face of uncertain communications, IoT devices must find ways to power intelligent processing at the edge of the network. An international team from IBM outlines these challenges in Chapter 9 and presents a distributed AI approach that can overcome them.

Ambient sounds provide a wealth of information which can be useful in many IoT solutions for both commercial and defense missions. The application of AI-based techniques to traditional acoustic signal processing provides many interesting use-cases for defense and national security. Uses include detecting and locating small arms fire, efficient process management of naval facilities and equipment, and detecting possible intruders at borders. AI provides an augmented capability for a data-driven understanding of the environment through acoustics, but also comes with several challenges. For example, AI models need to operate in environments that may be different from the environment within which they are trained, and these models need to be able to retrain themselves or adapt themselves dynamically in deployed situations. In Chapter 10, a U.S.–Japanese team of IBM researchers describe a system for deploying AI based acoustics in real-world environments and reviews lessons learned from using their system.

The European Union, UK, China, and the U.S. are making substantial investments in AI for both commercial and national security. AI is most useful for national security and most powerful when coupled to sensors and actuators in an IoT system where it can perceive, control, and alter its environment. Basic research results at universities are widely available to the public. Some commercial research is publicly available, but much of it is protected as proprietary intellectual property (IP). In the defense and national security arena, little visibility exists for the public into the advances and application of AI to IoT, for example in naval applications such as the DARPA Ocean of Things program. The three chapters in Section 2 provide a glimpse into developments for defense using AI for IoT. Resources on AI for defense IoT, beyond Section 2 of this book, include what little information is publicly accessible on specific defense programs, such as the ACE program, and reports and papers on basic research from the U.S. Army's IoBT participants beyond the ones included in this book.

8

Principles of Robust Learning and Inference for IoBTs

Nathaniel D. Bastian[1], Susmit Jha[2], Paulo Tabuada[3], Venugopal Veeravalli[4], and Gunjan Verma[5]

[1]*Army Cyber Institute, United States Military Academy, West Point, NY, USA*
[2]*Neuro-symbolic Computing and Intelligence, CSL, SRI International, Menlo Park, CA, USA*
[3]*ECE Department, University of California at Los Angeles, Los Angeles, CA, USA*
[4]*ECE Department, University of Illinois at Urbana-Champaign, Champaign, IL, USA*
[5]*U.S. Army DEVCOM Army Research Laboratory, U.S. Army Futures Command, Austin, TX, USA*

Abstract

The Internet of Battlefield Things (IoBTs) operate in an adversarial rapidly-evolving environment, necessitating fast, robust and resilient decision-making. The success of machine learning, in particular deep learning methods, can improve the performance and effectiveness of IoBTs, but these models are known to be brittle, untrustworthy, and vulnerable. In this chapter, we discuss the principles and methodologies to make machine learning models robust, resilient to adversarial attacks, and more interpretable for human-on-the-loop decision-making. We also identify the key challenges in developing trustworthy machine learning for IoBTs.

8.1 Internet of Battlefield Things and Intelligence

The Internet of Battlefield Things (IoBTs) [1, 2] aims at providing a pervasive, heterogeneous sensing and actuation capability to enhance command and control system autonomy and agility, information analytic capabilities against adversarial influence and control of the information battle-space; delivering intelligent, agile, and resilient decisional overmatch at significant standoff and optempo. While the traditional approaches have focused on either centralized or decentralized decision-making, with the decision structure either fixed vertical stovepipes or dynamic task organized, and the information dissemination either limited (need to know) or broadcasted (need to share), IoBTs aim at providing options across these extremes of the spectrum to provide an adaptive mission-oriented network of sensors and actuators. Thus, the discovery, composition and adaptation of available network nodes for sensing, secure information sharing, and actuation is a critical capability for IoBTs.

This has motivated enabling intelligent services as core components of IoBTs to make them autonomous and to enable services necessary for effective command and control. The examples of such artificial intelligence (AI) services that need to be supported by the complex autonomic IoBTs include intelligent analytics, anomaly detection in broadly heterogeneous and varied data that may be unknown combinations of sparse and voluminous, and centralized and distributed

IoT for Defense and National Security, First Edition. Edited by Robert Douglass, Keith Gremban, Ananthram Swami, and Stephan Gerali.

decision-making on whether received data is trustworthy or suspect. Further, the adversarial nature of the contested environment in which IoBTs operate requires enriching the resiliency of the IoBT, such that it can be hardened against tampering and adversarial compromise, continue operating under attacks, and provide bounded guarantees of performance.

The tremendous success of machine learning, in particular deep learning methods, make them a promising paradigm to develop and deploy the intelligent services in an IoBT. But these machine learning models are known to be brittle, untrustworthy, and vulnerable to adversarial attacks. These limitations have fueled research into principles and methodologies to make machine learning models robust, resilient to adversarial attacks, uncertainty-aware, and more interpretable for human-on-the-loop decision-making. In particular, there has been significant progress towards learning new representations and techniques motivated by dynamical systems, information theory, and formal methods to train machine learning models with guarantees on their robustness and performance not just in their training environment but also in new environments. We describe a holistic view of the problem, which addresses the challenge of high-assurance machine learning in IoBTs not in isolation but in the context of the overall system in which the learning models are integrated.

The challenge of trustworthy, robust and interpretable AI and machine learning (ML) models is not limited to IoBTs but is also important in other high assurance applications such as autonomous vehicles, health care, cybersecurity, and medical devices. In some of these applications, it is possible to decompose the overall system into a safe base system whose performance is enhanced using AI/ML models. Such systems can, then, safely deploy AI/ML models using architectures such as Simplex [3] which rely on detecting conditions under which the system must fall back to the safe base system. But IoBTs are substantially different from other high-assurance systems. The scale and speed of acquisition, assessment, aggregation, state estimation, and decision-making in IoBTs operating in a rapidly-evolving high-tempo environment necessitates the use of AI/ML models as the core components. Thus, the development of trustworthy, robust and interpretable AI/ML models is central to the successful development and deployment of IoBTs.

8.2 Dimensions of Responsible AI

AI/ML models, particularly deep learning, have demonstrated near human-level performance in many domains of relevance to IoBTs such as computer vision, sparse multimodal sensing, signal compression, tracking, and adaptive decision-making under uncertainty. However, deep learning models are known to be brittle to even a small change in input distribution and vulnerable to adversarial attacks. Unlike the commercial applications of machine learning, IoBTs operate in inherently adversarial and contested environments. Further, AI's responsible and ethical use within the Multi-Domain Operations (MDO) effects loop [4] in IoBTs requires effective partnership with humans and AI, and the need to certify AI for safe behavior consistent with the operational rules and regulations. These create a unique set of fundamental research challenges for robust learning and inference in the context of IoBTs.

Consider an example scenario where an IoBT network is responsible for detecting and tracking a platoon of enemy vehicles, troops, and other assets, using a mixture of trusted blue and third-party gray nodes. How can an IoBT robustly detect the vehicles even when faced with physically-realizable attacks on the perception system or significant change in the environment due to perturbations such as low visibility due to smoke, increased background vibrations, and adversary-introduced radio interference? How can a vehicle-class detector in an IoBT recognize that the current input does not belong to any of the training classes and is an out-of-distribution input on which its decision cannot be trusted? What are the learning architectures, training methods, and runtime monitoring algorithms that can provide guarantees on their performance, robustness, and resilience? How can we exploit the distributed nature of sensing in IoBTs to improve robustness and resilience? We need a combination of inference-time and training-time approaches to address such challenges and ensure safe autonomous IoBT improvisation to meet commander intent for fast-paced missions.

8.2.1 Research Challenges in IoBTs

The unique research challenges in developing responsible AI for IoBTs can be decomposed into two fundamental research questions.

- *First, how do we make machine learning models self-aware of their limitations and cognizant of the IoBT system-context in which they are deployed so that the overall AI-stack in IoBTs is robust and resilient to new environments, out-of-distribution data, and adversarial inputs?*
  While most data-driven statistical ML algorithms exhibit poor generalization, the remarkable performance of deep neural networks (DNNs) on in-distribution inputs similar to training data makes their lack of generalization very noticeable and likely to induce misleading confidence on AI/ML models if they are only tested on in-distribution data. This problem is further exacerbated by extremely high confidence (softmax values) exhibited by DNNs while making incorrect predictions on novel inputs outside their training distribution [5, 6]. The responsible deployment of DNNs in IoBTs necessitates the detection of out-of-distribution (OOD) data so that DNNs can abstain from making decisions on those. Beyond this lack of robustness, DNNs are also susceptible to adversarial attacks [7–9] that can change the prediction of a DNN via small imperceptible perturbations. Physically-realizable attacks [10, 11] can exploit this vulnerability without cyberattacks. Attacks on ML models can also target cyber components in an IoBT [12, 13] or the network intrusion detection systems using AI/ML models [14–16]. An adversary will readily exploit any vulnerability in a battlefield. The environment in a battlefield is also constantly evolving. Thus, the robustness and resilience of ML models are critical in IoBTs. The distributed nature of IoBTs provides an opportunity to add system-level robustness to ML models. Traditionally, ML is used in either low-assurance recommendation systems or extremely high-assurance systems with well-defined non-ML fallback. In contrast, we need to smoothly trade off robustness and performance in IoBTs, depending on the mission's acceptable risk-level. These make the ML robustness and resilience challenges unique for IoBTs.
- *What learning representations and architectures for deep learning are more data-efficient, capable of deployment in resource-constrained platforms, provably robust, and amenable to scalable formal and mathematical analysis?*
  Several new architectures of deep learning have been proposed that make learning efficient or allow encoding domain-specific properties such as translational and scale invariance in convolutional neural networks [17], symmetric relational knowledge in graph convolution neural networks [18], attention over input parts in transformers [19], and gate-regulated memory cells in long short-term memory networks [20]. But how do we design new architectures that make the analysis of DNNs easier and achieve provable robustness and learnability? How do we create new architectures that can integrate background knowledge in the form of physics models or symbolic rules to minimize the amount of data needed for machine learning? How do we exploit unsupervised embedding and latent space learning in our new architecture to

decrease the amount of supervision required for training DNNs for IoBTs? In contrast to many applications, IoBTs cannot rely on purely supervised learning, since continuously providing labels in a rapidly-evolving environment is infeasible. Further, many applications of learning in IoBTs are in domains with rich scientific knowledge gathered over hundreds of years, such as fluid dynamics and radiometry. This necessitates developing new DNN architectures that enable the integration of scientific knowledge, dynamics models, and logical rules. The architecture should also enable operation on resource-constrained platforms by trading off the ML model's computational complexity with its performance and robustness.

8.2.2 Trust, Resilience and Interpretability

While well-trained models provide correct answers on the examples from the training distribution, the accuracy of the model drops to almost zero on the adversarial examples. Until recently, machine learning was difficult, even in a non-adversarial setting and consequently, adversarial examples were not interesting to most researchers because mistakes were the rule, not the exception. But the human-level performance of deep learning models has fueled significant investigation into adversarial attack and defense methods. The attempts to integrate AI/ML models into IoBT and similar safety-critical military and civilian applications further highlight the need for resilient AI/ML models. Further, machine learning models exhibit confounding failure modes with difficult to explain failures on some inputs. This has created challenges in trusting machine learning models to make fair and reliable decisions. Weather perturbations or other environmental disturbances such as glare can cause ML based vision pipelines to fail due to distribution-shift, making the inputs out of distribution with respect to the training set used to train the vision models. Another failure mode of these models is when they are provided novel inputs comprising new classes or new contexts. In addition to the fragility of deep learning models to adversarial attacks, the good performance of the model on the test data from the training distribution does not transfer to out of distribution data. Further, rapid and adaptive decision-making and information gather provided by AI/ML models in IoBTs also needs the ability to explain the choices and decisions that are made. This is particularly important when the results returned by these algorithms are counter-intuitive, and require human-in-the-loop or human-in-the-loop collaboration. This leads to the third element, interpretability, of the trinity of challenges for AI/ML models. Several recent studies have demonstrated that these challenges are not isolated but strongly interconnected, and improvement in trust and robustness leads to improvement in interpretability while improvement in interpretability leads to improvement in resilience. Further, improvement in resilience itself makes the model robust and more interpretable.

As mentioned earlier, this trinity of the challenges of increasing the trustworthiness, robustness and interpretability of machine learning and inference in IoBTs has additional facets. The models need to be trained and verified after training to ensure they can generalize to new environments and are not susceptible to adversarial attacks. Such a robust training must not employ techniques such as ensembles that face calibration challenges in low data regime [21] but rather exploit background knowledge available from operation experience and scientific knowledge. Ensembles relying on subsampling datasets need more data, and a more detailed study of the limitations and strengths of deep ensembles is presented in [22]. Monitoring techniques need to be developed to detect any shift in distribution of the environment or presence of surprising novel inputs that make the learned models no longer trustworthy. The inference nodes themselves may be subject to adversarial attacks and so, compromised nodes need to be detected to enable resilient distributed inference. Many applications of IoBTs such as identification and tracking need the integration of physics models into inference to make it robust and resilient.

We briefly describe the work done by the authors in an ongoing U.S. Army Collaborative Research Alliance on the Internet of Battlefield Things[1] on the topic of building trusted and responsible AI for IoBTs. This effort has made significant advances in addressing some of the challenges discussed in this chapter, with focus on deployment in IoBTs. We developed methods for verifying DNNs for robustness as well as DNNs composed with physical plant models for safety [23, 24]. We also developed a novel approach that exploits attribution of decisions made by DNNs to compute confidence on its prediction without the need for additional data for calibration or training of ensembles. This confidence metric was used to detect adversarial inputs and out of distribution data [25]. We developed methods to use latent space learning and manifold projection [8] to defend against adversarial attacks, and theoretically established the limitations of manifold-based defenses [26]. We have also developed techniques to detect surprise from out-of-distribution and novel inputs [27, 28].

We developed statistical methods for efficiently adapting to slowly changing machine learning problems [29, 30]. We formulated efficient methods for detecting distribution-shift and model change [31] and secure state estimation [32]. We developed new information-theoretic bounds on the generalization error of machine learning algorithms [33]. We demonstrated how physical models can be used to aid detection of compromised nodes for resilient distributed inference [34]. We have also investigated generalizability and safety in the context of reinforcement learning and inverse reinforcement learning [35–37].

Based on our prior and ongoing work on integrating AI in IOBTs, we identify and describe key research problems solving which would be critical to the deployment of responsible AI in IoBTs in the rest of this book chapter.

8.3 Detecting Surprise: Adversarial Defense and Outlier Detection

DNNs exhibit overfitting in negative log-likelihood space [5, 6], which makes them overconfident on wrong predictions. Coupled with their limited generalization, this makes the models vulnerable to adversarial attacks and brittle to novel inputs. Further, robust statistical approaches to learning such as inductive conformal prediction [38, 39] needs calibration set in addition to training set, and the ensemble methods such as bootstrap bagging [40] that rely on data sub-sampling to make learning more robust use only a part of the dataset for training. Hence, their use in domains such as IoBT is challenging, where we expect scarcity of supervisory data. While most of the research literature on adversarial defense and outlier detection has investigated these problems separately, the root cause for both of these is the failure of ML models to self-monitor their inference and detect when they are surprised. Motivated by predictive processing models of human cognition [41], we posit that the path to making ML models robust and resilient is to enable it to identify surprises by building models of its internal inference and to abstain from decision-making when an alarming deviation is detected. Such runtime monitors can improve the trustworthiness and resilience of AI/ML models.

Unlike uncertainty quantification techniques such as Bayesian NNs and calibration methods, the aim of runtime monitoring is not to approximate the true posterior distribution of the decision of a machine learning models but instead perform hypothesis testing to detect unusual inference using internal features of the model and the attribution of the model's decision over these features. Runtime monitoring focuses on detecting surprise for atypical inferences to enable the model to abstain when the input is suspicious. In [25], we had adopted a top-down introspection approach. We used the model's decision on an input to assign importance/attribution to the features for

1 https://iobt.illinois.edu/

this decision, and then constructed a neighborhood using this attribution. Conformance in this neighborhood yielded confidence of the model. We have also investigated non-neighborhood approaches [27, 42] based on Mahalanobis distance of internal features from the features associated with a particular decision. While neighborhood methods are good at detecting aleatoric uncertainty on confusing inputs and adversarial examples, Mahalanobis distance does well on epistemic uncertainty when inputs are completely novel.

We recently proposed such a hypothesis testing and conformal prediction based approach for surprise detection called iDECODe [43]. The idea is to use conformal prediction with transformation equivariance learned on in-distribution (iD) data. Equivariance of outputs to certain geometric data transforms is a general desired property of ML systems. For example, it is desirable for a classifier trained on images of upright cats to also correctly classify rotated images of cats. In other words, classifiers should learn a representation that is invariant to the orientation of the training data. Sharing of kernels in convolutional neural networks (CNNs), and more generally group CNNs, leads to learning features equivariant to translations, and more generally to group transforms. Another common approach to encode these transformations is data augmentation [17, 44–46]. This is not guaranteed to lead to equivariance for all inputs, and is more likely to work for *in-distribution data* used for training than for *out-of-distribution data* dissimilar to that used for training. This is the crucial insight used in iDECODe. To get formal guarantees on the false detection rate, iDECODe leverages conformal prediction [47, 48], which is a general methodology to test if an input conforms to the training data. It uses a non-conformity measure (NCM) to quantitatively estimate how different an input is from the training distribution. Commonly used NCMs are based on the properties of the input's k-nearest neighbors from the training data [48, 49] and kernel density estimation methods [50]. Inductive conformal anomaly detection (ICAD) [51] uses an NCM to assign a non-conformity score to the input for computing its p-value indicating anomalous behavior. The performance of ICAD can depend strongly on the choice of the NCM [48]. iDECODE uses the deviation (or error) in the predictable behavior of a model equivariant in-distribution (iD) with respect to a set G of transformations as the NCM for OOD detection. It also deploys a novel approach to increase surprise detection performance by aggregating n scores computed from the proposed base NCM on n IID transformations sampled from a distribution over G, leading to an aggregated NCM.

We envision the need for three techniques to detecting surprise of machine learning models: first, hypothesis testing to detect deviations in features across layers of a DNN; second, data-efficient methods to approximate tail distributions corresponding to surprise using unlabeled data based on deep generative models and extreme-value theory; and third, hierarchical monitoring approach that enables trading-off robustness with performance.

8.4 Novel Deep Learning Representation: Dynamical System

Many applications of machine learning in IoBTs can benefit from scientific models both to reduce their need for supervisory data, and to increase their robustness. For example, detection and targeting of long-range entities must model effects such as atmospheric refraction, and mission planning might need to model chaotic dynamics such as fire propagation. These problems are well-studied in dynamics literature, but neither purely ML models nor purely analytical models are sufficient to learn such models. In the dynamical system view of DNNs, the inference through layers of a DNN can be seen as the evolution of the corresponding dynamical system. This view enables integration of physics models more easily as prior background knowledge. It also enables learning with much lower memory footprint, as intermediate gradients don't need to be stored. Further, it enables the use of decades of research in mathematical tools from control theory and dynamical systems to analyze DNN models. Despite some recent progress in this area [52, 53], this approach is in its infancy. We need to develop a novel representation that allows integration of these to reduce data needed for learning and make models more interpretable, and hence, amenable to human-on-the-loop decision-making in IoBTs.

While we have earlier investigated compositions of DNNs and differential equation models for formal verification of safety requirements [23, 24], the dynamical systems view of DNNs provides a new uniform representation of the environment and ML model and opens more scalable analysis techniques. The goal is to not just analyze ML models in isolation but to establish system-level safety guarantees. Recently, we have shown how controllability results can be used for establishing theoretical bounds on learnability [52]. We have also recently shown that the dynamical systems view of DNNs produces more interpretable internal representations and more clearly separable conditional distributions of internal features given the model output [54, 55].

We identify the following research directions in the study of new deep learning representations based on dynamical systems. The first direction is using control theory and results from dynamical systems to theoretically characterize learnability, robustness, generalizability and resilience of machine learning models. Second, we can use unsupervised latent space learning via encoders and decoders that are represented as differential equations and hence, enable a more accurate characterization of probability transformations. Third, integration of other symbolic knowledge in the form of logical rules over the states and physics models of the evolutions of the dynamics can be more easily integrated with deep learning models. Finally, these can be leveraged for the development of abstraction methods that allow inference on resource-constrained platforms with performance guarantees, such as the use of early-exit strategies during inference.

8.5 Robust Secure State Estimation

We have made advances on the problem of detecting sensor attacks by using mathematical models of the underlying physical phenomenon being observed. In particular, we showed that this problem, although NP-hard in general, can be solved in polynomial time for a wide class of linear models for the physical dynamics [32]. Moreover, we extended these results to the context of networks where adversaries can spoof sensors, can change messages sent in the network, and can even make network nodes behave arbitrarily (i.e. in a Byzantine manner) [34]. Even though this work focused on networks, its solution relied on collecting all the data in a central location for processing. The problem of detecting sensor attacks in a decentralized manner has not been sufficiently investigated (see [56, 57] for some recent work). Another important problem to be investigated is how to handle the effect of measurement noise. Most work on secure-state estimation has focused on either noise free settings or on bounded noise models. However, there are many instances where stochastic noise models are preferable. This brings the question of how to optimally use a stochastic noise model combined with the model of the underlying physical phenomenon being sensed to detect sensor attacks.

A primary purpose of AI/ML models in IoBTs is to sense and estimate the state of the world. Most algorithms for secure state estimation are centralized, i.e. they require all the sensor data to be collected in a central node for processing. This represents a single point of failure that can be exploited by an adversary. To further improve resiliency, we have developed decentralized algorithms for sensor attack detection. Our approach is based on first reaching consensus on all the nodes about a minimal set of information, extracted from sensor measurements (including attacked ones), that is sufficient for detecting which sensors have been attacked. To do so, we leverage recent advances in the design of consensus algorithms for tracking time-varying signals, such as [58]. In parallel with the consensus process, we also need a light-weight attack detection algorithm that takes as input the minimal information over which consensus is sought. As all the nodes reach consensus on the minimal information, they will also reach consensus on which sensors are under attack. This algorithm is predicated on an attack model where all the nodes are honest, and only the sensors are under attack [59].

There is also a need for extensions of this problem to the more challenging case where nodes can be malicious and actively attempt to prevent consensus to be reached. This will be especially important when using gray nodes such as third-party infrastructure repurposed by IoBT to collect information. In order to further enhance the resiliency of secure state estimation, we need to extend the existing algorithms to the case of stochastically modeled sensor noise by leveraging recent results on robust mean estimation [60, 61]. To see why robust mean estimation is relevant, assume that we have k identical sensors, some of which are subject to attacks. If we could compute the mean of the sensed values in a way that is resilient to attacks, we would both address the presence of the attacks as well as the presence of sensor noise by using the mean as the estimate of the true values being sensed. However, in the secure state estimation problem, the quantity to be estimated are not the sensed values, but rather the state of a dynamical system. One can extend the ideas in robust mean estimation to the secure state estimation problem by using our previous work [32] that allows simplifying the relationship between sensor measurements and the state. One aspect of the secure state estimation problem that makes it more challenging is the fact that the effective measurement vectors whose mean is being robustly estimated may have missing entries due to the inherent sparsity in the sensor measurements. This requires development of a new theory for robust mean estimation with missing entries to address this challenge.

8.6 Distributionally Robust Learning

Given training data from an unknown distribution P, the goal in supervised learning is to choose the parameters of the learning algorithm (a.k.a. hypothesis) to minimize the population risk, which is the expected loss over P. However, since the distribution of the data is unknown, the empirical risk, which is the average loss over the training data, is usually minimized instead; this procedure is referred to as empirical risk minimization (ERM). An alternative to ERM for approximating the minimization of population risk was first proposed in [62], based on a minimax optimization framework. In this framework, an ambiguity set $A(P_n)$ is defined around the empirical distribution P_n of the training data, which is assumed to contain the distribution P with high probability for large sampleset size n. The worst case expected loss over $A(P_n$ is minimized, thereby also guaranteeing good performance at P. Different forms for the ambiguity set have been proposed in the literature, and optimization techniques for solving the corresponding minimax optimization problem have been given [63, 64]. The problem of especial relevance in IoBT settings is one where the distribution of the data at the time of inference (say Q) may be different from that at the time of training (P) in some bounded manner. How should we modify/replace ERM to provide robustness to such distributional uncertainty at inference time?

The goal of robust learning is to provide robustness to distributional uncertainty at inference time, i.e. we want our learning algorithm to work well when the distribution at inference time Q is within some "ball" around the distribution P that governs the training data. A natural approach to this problem is to simply extend the minimax alternative to ERM. In particular, we can enlarge the ambiguity set $A(P_n)$ around the empirical distribution P_n to include the distribution Q, and then explore efficient solutions to the corresponding minmax optimization problem to determine the parameters of the learning algorithm. Among the minmax alternatives to ERM, a promising approach given in [65] is based on exploiting the equivalence (which was first shown in [66]) between the problem of minimizing the worst case expected loss and maximizing entropy. In particular, this equivalence is exploited in [65] to establish the form of the worst-case distribution for moment constrained ambiguity sets. Can such results be derived for other ambiguity sets? More importantly, can the maximum entropy approach provide learning algorithms that are robust to distributional uncertainty at inference time? These research questions are critical to building robust learning.

Another approach to such distributionally robust learning is through a direct modification of the ERM algorithm that tries to take into account distributional uncertainty. There have been a few empirical studies that show that such modifications can result in robustness [67, 68]. Can we develop a systematic methodology for designing robust modifications of ERM that are provably good? This is where recent work on information-theoretic bounds on the performance of machine learning algorithms [33, 69, 70] can play an important role. We need to extend existing information-theoretic bounds to allow for distributional uncertainty at inference time. A result in this direction was given in [71]. However, the bound just involves the addition of a term involving the Kullback-Leibler (KL) divergence between P and Q, which is not useful for characterizing and comparing the robustness of different modification to ERM. An important research question is whether we can derive new information-theoretic bounds on generalization that will allow us to design provably robust modifications to ERM.

8.7 Future Directions

The lack of generalization in ML models and their lack of robustness and resilience is a natural consequence of their overfitting to training data. This overfitting is severe in the state-of-the-art larger models with millions and billions of parameters. We hypothesize that information-theoretic measures such as mutual information and conditional generative entropy can be used to regularize training of large models to avoid overfitting while retaining good performance and provide mathematical bounds on generalizability. Beyond model training, we can develop statistical and formal monitors that introspectively examine the inner state of the DNN to measure its level of surprise on new inputs and report its quantitative confidence in addition to its output. This can detect adversarial attacks and novel inputs on which DNNs are likely to fail, and can provide certified defenses against adversarial resilience. A combination of information-theoretic and formal approaches can enable training and runtime monitoring that make ML models robust and resilient to adversarial attacks, novel inputs and new environments.

A promising new representation for learning is to regard deep neural networks as discretizing an ordinary differential equation (ODE). This idealization enables mature mathematical tools from dynamical systems, optimal control, and formal methods to analyze the robustness, resilience, performance, and interpretability of DNNs. Learning in this framework can be viewed as a mean-field optimal control problem. Controlled addition of noise into these models enables making the model more robust and also improves their interpretability. Further, we can use other symbolic models beyond ODEs, such as logical rules, to make ML models robust and to lower the model complexity (number of parameters). This will facilitate direct integration of scientific knowledge and reduce

ML models' complexity, enabling their execution on resource-constrained platforms. Thus, the dynamical systems view of DNNs can be exploited to develop new neurosymbolic representation for learning that is more amenable to mathematical and formal analysis.

8.8 Conclusion

This chapter discusses the principles and methodologies to make machine learning models trustworthy, robust, resilient to adversarial attacks, and more interpretable for human-on-the-loop decision-making. We have also identified the key challenges in developing reliable machine learning for IoBTs that can enable AI-human teams to operate in an adversarial rapidly-evolving environment of a battlefield effectively. While the need for fast, robust, and resilient decision-making in IoBTs makes it critical to incorporate AI/ML models in sensing, state estimation, decision-making, and actuation, the high-assurance and distributed nature of this application of AI/ML models raise unique challenges. The techniques described in this book chapter present a promising first step towards creating responsible AI for IoBTs. Further, this need for trustworthy, robust, and interpretable AI and ML models is not limited to IoBTs but is also crucial in other high assurance applications such as autonomous vehicles, health care, cybersecurity, and medical devices. The techniques and methods for responsible AI described in this book chapter will also be helpful in these other high-assurance applications.

References

1 Abdelzaher, T., Ayanian, N., Basar, T. et al. (2018). Toward an internet of battlefield things: a resilience perspective. *Computer* 51 (11): 24–36.

2 Russell, S. and Abdelzaher, T. (2018). The internet of battlefield things: the next generation of command, control, communications and intelligence (C3I) decision-making. *MILCOM 2018-2018 IEEE Military Communications Conference (MILCOM)*, pp. 737–742. IEEE.

3 Sha, L., Goodenough, J.B., and Pollak, B. (1998). Simplex architecture: meeting the challenges of using cots in high-reliability systems. *Crosstalk* 7–10.

4 Russell, S., Abdelzaher, T., and Suri, N. (2019). Multi-domain effects and the internet of battlefield things. *MILCOM 2019-2019 IEEE Military Communications Conference (MILCOM)*, pp. 724–730. IEEE.

5 Guo, C., Pleiss, G., Sun, Y., and Weinberger, K.Q. (2017). On calibration of modern neural networks. *arXiv preprint arXiv:1706.04599*.

6 Hendrycks, D. and Gimpel, K. (2016). A baseline for detecting misclassified and out-of-distribution examples in neural networks. *arXiv preprint arXiv:1610.02136*.

7 Goodfellow, I.J., Shlens, J., and Szegedy, C. (2014). Explaining and harnessing adversarial examples. *arXiv preprint arXiv:1412.6572*.

8 Jha, S., Jang, U., Jha, S., and Jalaian, B. (2018). Detecting adversarial examples using data manifolds. *MILCOM 2018-2018 IEEE Military Communications Conference (MILCOM)*, pp. 547–552. IEEE.

9 Moosavi-Dezfooli, S.-M., Fawzi, A., and Frossard, P. (2016). Deepfool: a simple and accurate method to fool deep neural networks. *Proceedings of the IEEE Conference on Computer Vision and Pattern Recognition*, pp. 2574–2582.

10 Tu, J., Ren, M., Manivasagam, S. et al. (2020). Physically realizable adversarial examples for LIDAR object detection. *Proceedings of the IEEE/CVF Conference on Computer Vision and Pattern Recognition*, pp. 13716–13725.

11 Evtimov, I., Eykholt, K., Fernandes, E. et al. (2017). Robust physical-world attacks on machine learning models. *arXiv preprint arXiv:1707.08945*, 2 (3): 4.

12 Devine, S. and Bastian, N. (2021). An adversarial training based machine learning approach to malware classification under adversarial conditions. *Proceedings of the 54th Hawaii International Conference on System Sciences.*

13 Bierbrauer, D.A., Chang, A., Kritzer, W., and Bastian, N.D. (2021). Cybersecurity anomaly detection in adversarial environments. *Proceedings of the AAAI Fall 2021 Symposium on AI in Government and Public Sector.*

14 Schneider, M., Aspinall, D., and Bastian, N. (2021). Evaluating model robustness to adversarial samples in network intrusion detection. *Proceedings of the 2021 IEEE International Conference on Big Data.*

15 Talty, K., Stockdale, J., and Bastian, N.D. (2021). A sensitivity analysis of poisoning and evasion attacks in network intrusion detection system machine learning models. *Proceedings of the 2021 IEEE Military Communications Conference.*

16 Alhajjar, E., Maxwell, P., and Bastian, N. (2021). Adversarial machine learning in network intrusion detection systems. *Expert Systems with Applications* 186: 115782.

17 Krizhevsky, A., Sutskever, I., and Hinton, G.E. (2012). Imagenet classification with deep convolutional neural networks. In: *Advances in Neural Information Processing Systems*, vol. 25, 1097–1105.

18 Henaff, M., Bruna, J., and LeCun, Y. (2015). Deep convolutional networks on graph-structured data. *arXiv preprint arXiv:1506.05163.*

19 Jaderberg, M., Simonyan, K., and Zisserman, A. (2015). Spatial transformer networks. In: *Advances in Neural Information Processing Systems*, vol. 28, 2017–2025.

20 Cheng, J., Dong, L., and Lapata, M. (2016). Long short-term memory-networks for machine reading. *arXiv preprint arXiv:1601.06733.*

21 Rahaman, R. and Thiery, A.H. (2021). Uncertainty quantification and deep ensembles. In: *Advances in Neural Information Processing Systems*, vol. 34, 20063–20075.

22 Abe, T., Buchanan, E.K., Pleiss, G. et al. (2022). Deep ensembles work, but are they necessary? *arXiv preprint arXiv:2202.06985.*

23 Dutta, S., Jha, S., Sankaranarayanan, S., and Tiwari, A. (2018). Output range analysis for deep feedforward neural networks. *NASA Formal Methods Symposium*, pp. 121–138. Springer.

24 Dutta, S., Chen, X., Jha, S. et al. (2019). Sherlock-a tool for verification of neural network feedback systems: demo abstract. *Proceedings of the 22nd ACM International Conference on Hybrid Systems: Computation and Control*, pp. 262–263.

25 Jha, S., Raj, S., Fernandes, S. et al. (2019). Attribution-based confidence metric for deep neural networks. In: *Advances in Neural Information Processing Systems*, 11826–11837.

26 Jang, U., Jha, S., and Jha, S. (2019). On the need for topology-aware generative models for manifold-based defenses. *International Conference on Learning Representations.*

27 Kaur, R., Jha, S., Roy, A. et al. (2021). Imagenet classification with deep convolutional neural networks. *Uncertainty and Robustness in Deep Learning Workshop at ICML.*

28 Kaur, R., Jha, S., Roy, A. et al. (2022). iDECODe: In-distribution equivariance for conformal out-of-distribution detection. *36th AAAI Conference on Artificial Intelligence.*

29 Wilson, C., Bu, Y., and Veeravalli, V.V. (2019). Adaptive sequential machine learning. *Sequential Analysis* 38 (4): 545–568.

30 Bu, Y., Lu, J., and Veeravalli, V.V. (2019). Active and adaptive sequential learning with per time-step excess risk guarantees. *2019 53rd Asilomar Conference on Signals, Systems, and Computers*, pp. 1606–1610. IEEE.

31 Bu, Y., Lu, J., and Veeravalli, V.V. (2019). Model change detection with application to machine learning. *ICASSP 2019-2019 IEEE International Conference on Acoustics, Speech and Signal Processing (ICASSP)*, pp. 5341–5346. IEEE.

32 Mao, Y., Mitra, A., Sundaram, S., and Tabuada, P. (2019). When is the secure state-reconstruction problem hard? *2019 IEEE 58th Conference on Decision and Control (CDC)*, pp. 5368–5373. IEEE.

33 Bu, Y., Zou, S., and Veeravalli, V.V. (2020). Tightening mutual information-based bounds on generalization error. *IEEE Journal on Selected Areas in Information Theory* 1 (1): 121–130.

34 Mao, Y., Diggavi, S., Fragouli, C., and Tabuada, P. (2020). Secure state-reconstruction over networks subject to attacks. *IEEE Control Systems Letters* 5 (1): 157–162.

35 Jha, S. and Lincoln, P. (2018). Data efficient learning of robust control policies. *2018 56th Annual Allerton Conference on Communication, Control, and Computing (Allerton)*, pp. 856–861. IEEE.

36 Vazquez-Chanlatte, M., Jha, S., Tiwari, A. et al. (2017). Learning task specifications from demonstrations. *arXiv preprint arXiv:1710.03875.*

37 Jha, S. and Rushby, J. (2019). Inferring and conveying intentionality: beyond numerical rewards to logical intentions. *AAAI Symposium on Towards Conscious AI Systems.*

38 Papadopoulos, H. (2008). *Inductive Conformal Prediction: Theory and Application to Neural Networks.* INTECH Open Access Publisher Rijeka.

39 Vovk, V. (2012). Conditional validity of inductive conformal predictors. *Asian Conference on Machine Learning*, pp. 475–490. PMLR.

40 Bühlmann, P. (2012). Bagging, boosting and ensemble methods. In: *Handbook of Computational Statistics*, 985–1022. Springer: Berlin, Heidelberg.

41 Friston, K. (2010). The free-energy principle: a unified brain theory? *Nature Reviews Neuroscience* 11 (2): 127.

42 Kaur, R., Jha, S., Roy, A. et al. (2021). Are all outliers alike? On understanding the diversity of outliers for detecting OODs. *arXiv preprint arXiv:2103.12628.*

43 Kaur, R., Jha, S., Roy, A. et al. (2022). iDECODe: In-distribution equivariance for conformal out-of-distribution detection. *AAAI.*

44 Baird, H.S., Bunke, H., and Yamamoto, K. (1992). Document image defect models. In: *Structured Document Image Analysis*, 546–556. Springer, Berlin, Heidelberg, 1992.

45 Cireşan, D.C., Meier, U., Gambardella, L.M., and Schmidhuber, J. (2010). Deep, big, simple neural nets for handwritten digit recognition. *Neural Computation* 22 (12): 3207–3220.

46 Chen, S., Dobriban, E., and Lee, J. (2020). A group-theoretic framework for data augmentation. In: *Advances in Neural Information Processing Systems, JMLR*, vol. 33.

47 Vovk, V., Gammerman, A., and Shafer, G. (2005). *Algorithmic Learning in a Random World.* Springer Science & Business Media.

48 Balasubramanian, V., Ho, S.-S., and Vovk, V. (2014). *Conformal Prediction for Reliable Machine Learning: Theory, Adaptations and Applications*, 1e. San Francisco, CA, USA: Morgan Kaufmann Publishers Inc. ISBN 0123985374.

49 Papernot, N. and McDaniel, P. (2018). Deep k-nearest neighbors: towards confident, interpretable and robust deep learning. *arXiv preprint arXiv:1803.04765.*

50 Smith, J., Nouretdinov, I., Craddock, R. et al. (2014). Anomaly detection of trajectories with kernel density estimation by conformal prediction. *IFIP International Conference on Artificial Intelligence Applications and Innovations*, pp. 271–280. Springer.

51 Laxhammar, R. and Falkman, G. (2015). Inductive conformal anomaly detection for sequential detection of anomalous sub-trajectories. *Annals of Mathematics and Artificial Intelligence* 74 (1): 67–94.

52 Tabuada, P. and Gharesifard, B. (2020). Universal approximation power of deep neural networks via nonlinear control theory. *arXiv e-prints*, pp. arXiv–2007.

53 Chen, R.T.Q., Rubanova, Y., Bettencourt, J., and Duvenaud, D. (2018). Neural ordinary differential equations. *arXiv preprint arXiv:1806.07366*.

54 Jha, S., Ewetz, R., Velasquez, A., and Jha, S. (2022). On smoother attributions using neural stochastic differential equations. *30th International Joint Conference on Artificial Intelligence*.

55 Jha, S., Ewetz, R., Velasquez, A. et al. (2022). Shaping noise for robust attributions in neural stochastic differential equations. *36th AAAI Conference on Artificial Intelligence*.

56 An, L. and Yang, G.-H. (2019). Distributed secure state estimation for cyber–physical systems under sensor attacks. *Automatica* 107: 526–538.

57 Lee, J.G., Kim, J., and Shim, H. (2020). Fully distributed resilient state estimation based on distributed median solver. *IEEE Transactions on Automatic Control* 65 (9): 3935–3942.

58 Kia, S.S., Van Scoy, B., Cortes, J. et al. (2019). Tutorial on dynamic average consensus: The problem, its applications, and the algorithms. *IEEE Control Systems Magazine* 39 (3): 40–72.

59 Mao, Y. and Tabuada, P. (2021). Decentralized secure state-tracking in multi-agent systems. *CDC*.

60 Liu, J., Deshmukh, A., and Veeravalli, V.V. (2020). Robust mean estimation in high dimensions via L_0 minimization. *arXiv preprint arXiv:2008.09239*.

61 Diakonikolas, I. and Kane, D.M. (2019). Recent advances in algorithmic high-dimensional robust statistics. *arXiv preprint arXiv:1911.05911*.

62 Ben-Tal, A., Hertog, D.D., De Waegenaere, A.D. et al. (2013). Robust solutions of optimization problems affected by uncertain probabilities. *Management Science* 59 (2): 341–357.

63 Namkoong, H. and Duchi, J.C. (2016). Stochastic gradient methods for distributionally robust optimization with f-divergences. In: *Advances in Neural Information Processing Systems*, vol. 29, 2208–2216.

64 Lee, J. and Raginsky, M. (2017). Minimax statistical learning with Wasserstein distances. *arXiv preprint arXiv:1705.07815*.

65 Farnia, F. and Tse, D. (2016). A minimax approach to supervised learning. In: *Advances in Neural Information Processing Systems* vol. 29, 4240–4248.

66 Grünwald, P.D. and Dawid, A.P. (2004). Game theory, maximum entropy, minimum discrepancy and robust bayesian decision theory. *the Annals of Statistics* 32 (4): 1367–1433.

67 Volpi, R., Namkoong, H., Sener, O. et al. (2018). Generalizing to unseen domains via adversarial data augmentation. *arXiv preprint arXiv:1805.12018*.

68 Li, T., Beirami, A., Sanjabi, M., and Smith, V. (2020). Tilted empirical risk minimization. *arXiv preprint arXiv:2007.01162*.

69 Xu, A. and Raginsky, M. (2017). Information-theoretic analysis of generalization capability of learning algorithms. *arXiv preprint arXiv:1705.07809*.

70 Negrea, J., Haghifam, M., Dziugaite, G.K. et al. (2019). Information-theoretic generalization bounds for SGLD via data-dependent estimates. *arXiv preprint arXiv:1911.02151*.

71 Wu, X., Manton, J.H., Aickelin, U., and Zhu, J. (2020). Information-theoretic analysis for transfer learning. *2020 IEEE International Symposium on Information Theory (ISIT)*, pp. 2819–2824. IEEE.

9

AI at the Edge: Challenges, Applications, and Directions

Dhiraj Joshi[1], Nirmit Desai[1], Shyama Prosad Chowdhury[3], Wei-Han Lee[1], Luis Bathen[2], Shiqiang Wang[1], and Dinesh Verma[1]

[1] *IBM Thomas J. Watson Research Center, Yorktown Heights, New York, USA*
[2] *IBM Research – Almaden, IBM, San Jose, CA, USA*
[3] *IBM GBS, IBM India, Kolkata, WB, India*

Abstract

Robotic devices have several applications in commercial and defense IoT establishments. However, robots may not always have the capacity to run complex AI based applications, and high speed connectivity to exploit applications in cloud or data centers may not always be present. This situation arises in both defense and commercial contexts, with defense environments lacking sufficient network stability, and commercial environments concerned about data privacy and communications costs. The exploitation of AI capabilities at the edge can enable many use-cases by bypassing issues with network connectivity. At IBM Research, we have been developing Distributed AI technology and working on several IoT use cases using a robotic dog to enable applications such as visual and thermal inspections to identify anomalous conditions in industrial assets. In this chapter, we will lay out those use-cases, and discuss key Distributed AI components that enable such IoT applications, and how a robotic environment allows for new capabilities such as re-positioning a robotic sensor for optimal sensing. We will also discuss relevance of the use-cases to a military environment.

9.1 Introduction

Operations across many commercial and defense industries are getting ever more instrumented with IoT (Internet-of-things) devices and sensors embedded in various equipment and processes. With analytics and machine learning techniques, data collected from such connected operations can be analyzed to assist decision making, automation, and efficiency in businesses.

However, in many environments, the equipment is mission-critical, having legacy technology, or too costly to instrument, e.g. nuclear power plants, large chemical furnaces, or large power grid stations. In such environments, having a roaming IoT device (RID) with versatile multi-modal sensing capability, e.g. Boston Dynamics Spot robot, may be a cost-effective solution. In other words, instead of deploying a large number of IoT sensors in legacy equipment or replacing legacy equipment, a small number of RIDs can walk around the environment autonomously to sense the operational situations across multiple modalities, e.g. visual, acoustic, thermal, and depth-sensing and collect important operational data that can be analyzed.

IoT for Defense and National Security, First Edition. Edited by Robert Douglass, Keith Gremban, Ananthram Swami, and Stephan Gerali.

Figure 9.1 Examples of roaming IoT devices (RIDs). Source: Innox / Adobe Stock.

A RID (Figure 9.1) is a generalization of a robotic device with the ability to sense it's environment (possibly across multiple modalities), physically move (autonomously or guided by a human), and optionally execute computational workloads on-board. RIDs are becoming ubiquitous in many industries today and together with fixed-IOT sensors, they are capable of assisting operations at manufacturing and defense facilities. However, at present, they are largely limited to performing pre-defined tasks while navigating well-defined trajectories in predictable environments. RID operation in completely unknown environments additionally poses mobility related challenges.

With the explosion of a variety of IoT devices that constantly inter-operate within multi-vendor cloud ecosystems, a new computing paradigm is taking shape. This paradigm is being referred to as Distributed Cloud Ecosystems [1] which includes fixed and s RIDs, edge clouds, data centers, and multi-cloud environments (e.g. AWS and IBM Cloud). Within such ecosystems, there is a critical need to develop adaptive AI methodologies that can optimally utilize available computational resources, operate across heterogeneous IoT devices, incorporate human interaction effectively, and perform complex tasks in semi-structured and dynamic settings, such as factories, fulfillment centers, and defense establishments. Moreover, the AI developed for such complex ecosystems needs to be adaptable to new domains and environments with little human effort. Achieving all this still remains the holy grail of IoT and edge-computing at large.

In this chapter, we will focus our discussion on certain key challenges, solutions, and insights we have gathered by working on some real IoT use-cases involving a robot dog. We will then describe novel technologies that we have created for efficient AI at the edge. We will also discuss some interesting research directions in this space which we have started to work on.

9.2 IoT Applications

We begin by describing certain relevant industrial inspection use-cases using robotic IoT devices. Our discussion will draw inspiration from some real projects which we have worked on that require IoT-robotics as the key element for inspection. While these use-cases are described from a commercial/industrial standpoint, they are clearly very relevant to defense establishments as well.

Asset inspection is a recurring theme across many different industries. Regular inspection of industrial equipment is essential for safe operations. Historically, human experts equipped with domain knowledge have performed asset inspection. In the era of the fourth industrial revolution, electronic substitution of human visual monitoring of industrial assets is a must especially when inspection involves going to dangerous areas (such as nuclear power-plants and industrial furnaces).

In this section, we will focus our discussion on inspection of industrial assets by a robotic dog (e.g. Spot robot). While Spot robot has been used in the industrial use-cases that we worked on, readers should note that our discussion is not limited to Spot and is generalizable across different classes of RIDs that can navigate in industrial settings (at least semi-automatically) and are capable of multi-modal sensing. Spot robot has been designed to navigate efficiently in complex environments using predefined missions. Spot is also equipped with a variety of sensors (visual, thermal cameras, LiDAR etc.) which can be used based on need. For visual inspection, PTZ (pan-tilt-zoom) cameras will be used. Similarly for thermal inspection, IR/thermal cameras will be needed. Where the requirement is to map the surrounding region in a 3D plane, LiDAR sensors can be used. Examples of assets that need to be monitored include fire extinguishers, analog gauges, switches, and transformers. As discussed, inspection can incorporate visual, thermal, or a combination of different modalities.

9.2.1 Visual Inspection of Assets

Visual inspection is perhaps the most common inspection use-case encountered across different domains. Using a robot for industrial visual inspections is especially important where instrumenting assets (with IoT camera sensors) is cost prohibitive. Moreover, it is a scalable solution that can potentially be used across different environments with certain adaptation. In our experience, the most important industrial use-case for visual inspection is identifying visual anomalies (e.g. when an asset such as a fire extinguisher is not in proper working condition).

9.2.1.1 Visual Recognition

Typically, in such scenarios, a state-of-the-art deep learning model (e.g. Inception V3 [7] or Yolo V3 [8] as described in Section 9.4.3.4) can be used to model visual scenes or objects. Visual recognition could involve a classification task (such as classifying good vs bad scenarios visually) or object recognition task (recognizing instances of an object or anomalous condition) or a combination of both. The visual inference pipeline can be manually instrumented (using domain knowledge of experts) or automatically built using AI techniques. Visual classification typically involves global analysis of images vs. object recognition which involves detecting instances of objects (so analyzing local regions).

9.2.1.2 AI Optimization

Model training typically ranges from being fully supervised to being completely unsupervised. Few-shot learning approaches can be especially helpful for creating robust models. The need for model optimization (pruning, quantization etc.) is imperative given that robotic IoT devices such as Spot robot have a short battery life which needs to be conserved. Model optimization approaches (with experiments) which are directly applicable here will be described in detail in Section 9.4.3.

9.2.1.3 Fixed IoT Sensors vs. RIDs

One key difference between visual inspection using fixed cameras vs. RIDs such as Spot is that the latter can maneuver and position themselves at optimal locations for taking photos of assets which is not always possible using fixed cameras. This higher degree of freedom can provide clearer views of assets often resulting in more accurate inference. Secondly, visual inference using RIDs can be combined with an OOD detector (e.g. described in Section 9.4.2.1) to get the best views of assets to be passed through the visual model (while ignoring views captured erroneously or under conditions which are unknown to the visual model). Some interesting research ideas in this space will be discussed in Section 9.5.

9.2.2 Thermal Inspection of Assets

Inspecting surface temperature of assets (or specific regions in assets) is an important industrial process which is largely manually performed today. This inspection process comprises reading of temporal changes of temperature at a point, surface, or region. In industrial processes, use of thermal cameras (for thermal inspection) is quite prevalent as it can generate a 2D thermal map which can then be mapped to the 3D object (in question) to get accurate temperature readings at each point, surface, or region. Readers should note that thermal cameras will only work if temperature of the intended object is available at its surface (i.e. it will not work for shielded objects). Thermal inspection can be used to monitor a variety of assets (indoor/outdoor, planar/non-planar) such as switchboards, transformers, and assets at electric substations.

9.2.2.1 Inspection at Electric Substations

Power distribution substations have multiple connectors to make connections between different electrical lines among the power distribution network on demand. This connection by the switches stays in binary form which is either in ON or OFF form. This requirement is maintained through two connector junctions which are hinges and clips. Hinges and clips in a switch are responsible for establishing connections between two electrical line segments. Excess heat at these power junctions is dangerous as a heated wire increases resistance and reduces the functional life of the connector wire. This can also cause severe accidents, frequent outages, and other operational challenges.

The current inspection process is to manually monitor these junctions with handheld remote thermal sensors. In this manual process, a human operator must come periodically to each location of interest at a power substation. Facing towards a particular junction, the person must target a handheld thermal camera towards it. The handheld remote monitoring device can identify the maximum temperature in the targeted zone and display it which then needs to be manually validated as safe or risky. This brings in possibility of human error as well as safety risks associated with such areas. A robotic device like Spot (or another RID) equipped with appropriate thermal sensors can substitute a human operator efficiently and perform thermal inspections tirelessly. We next discuss a possible approach to achieve this automation.

9.2.2.2 Proposed Automation

As mentioned before, Spot is capable of performing pre-recorded missions (going to specific points, pointing to specific assets/regions, taking visual and thermal images). For thermal inspection, accurate mapping of visual and thermal image pairs is necessary. In other words, it's important to register every pair of visual and thermal images capturing a specific view of an asset for accurate attribution of temperature to the visual content. Visual and thermal registration requires calculation of a transformation matrix (which is a hardware property as it directly depends on the relative positioning of the visual and thermal cameras in a robot). However, the key AI challenge in this use-case occurs from the fact that even though Spot follows a pre-defined mission to photograph assets of interest, there is some variance in the positioning of the robot at each mission instance, which results in somewhat shifted asset views every time a mission is executed (in both thermal and visual domains). Therefore, the key challenge is to be able to register every pair of visual and thermal capture to a predefined template view of an asset without human intervention. One way to achieve this is through visual key-point detection using a scale invariant feature transform (SIFT)-type descriptor [9] and feature extraction followed by feature matching across image pairs. Key-point detectors have demonstrated their capability to identity and characterize (though features) important landmarks in an image.

9.2.3 Inspection of Analog Meters and Gauges

Analog gauges are widely used across a wide variety of industrial environments to monitor processes. There is provision for using digital gauges too. However, analog gauges are preferable in most scenarios; one of the main reasons being that analog gauges don't require any external energy sources to activate the reading mechanism. Another reason is the prohibitive cost of re-instrumenting legacy equipment with digital gauges. These analog gauges display the reading in the center mounted moving scale and have no inbuilt mechanism to transmit the result automatically to an edge node for further analysis. Current inspection style involves taking manual readings and analyzing them as per requirement. This can be automated using an RID (e.g. Spot) with visual camera. Spot can go to a particular location periodically, focus on the desired analog gauge, and take an image which can be analyzed in-situ, or transmitted to a remote site for further analysis. As a specific use-case, we focus on reading analog pressure gauge images. Examples of real industrial pressure gauge images from the web are shown in Figure 9.2. In practical scenarios, when an RID photographs pressure gauges in real settings, the images can appear in several different orientations with the possibility of mild to severe perspective distortion. This is because in several situations gauges are located in awkward, far-off, or tough to navigate places (e.g. too high, behind big machinery, lying down on the ground etc.).

A number of AI challenges need to be addressed in order to create a functional automatic analog gauge reading system. Even though Spot may be programmed to take a zoomed photo of a particular gauge, there may be other objects present in the background creating distractions for the AI algorithms. Illumination of the scene may also vary in different scenarios. For outdoor gauges, changes in daylight intensity and artificial illumination (surrounding illumination as well as reflection of light from the gauge protective glass) can make the gauge detection process even more difficult. Furthermore, the protecting glass of the gauge often becomes hazy and dirty over time creating difficulties for gauge dial and pointer detection.

9.2.3.1 Gauge Detection

The first step in automatic gauge reading is accurate gauge detection and localization from a scene. As explained before, variability of gauges and their complex background makes a template based approach unusable. However, deep learning algorithms for object detection (such as Faster R-CNN [10] or Yolo v3 [8]) can be used to reliably detect gauges. A good design choice here would be highest intersection over union (IOU) score over a controlled validation set of gauges. Images with multiple gauges can sometimes pose challenge for object detection due to factors such as occlusion.

Figure 9.2 Examples of analog pressure gauges. Source: (a) LittleVisuals / Pixabay; (c) ImageParty / Pixabay; (e) klausdie / Pixabay.

9.2.3.2 Perspective Correction

Because of the differences in the orientation of a gauge's optical axis and Spot's camera optical axis, there could be significant perspective distortion introduced in a captured image. Perspective distortion can lead to erroneous readings and needs to be identified and corrected. This can be done using multiple approaches such as, correction based on circularity assumption of the dial and deformation correction. A circular dial which has been distorted to appear elliptical can be thus corrected by perspective transformation.

9.2.3.3 Pointer Detection and Text Recognition

A pressure gauge comprises of several key points and identification of these key points can be an important cue for gauge reading. These key points are the minimum and maximum angular positions of the gauge scale, center of the pointer, and position of the pointer tip on the scale. In addition, for obtaining an accurate numerical reading, the numbers printed on the scale need to be recognized using number recognition techniques [11]. Once the pressure values at different angular positions is determined, the current gauge reading can be determined using the angular position of the gauge pointer.

9.2.4 Other Defense and Commercial Use Cases

Visual and thermal inspections are very generic and not limited to the specific assets and use-cases described in Sections 9.2.1 and 9.2.2. We want to emphasize that by providing a few inspection examples, we do not want to narrow down the broader scope and enormous potential of robotic multimodal inspections. In particular, any current manual inspection (e.g. of unwanted objects within a given territory) can be replaced by a robotic visual inspection method. In addition, we have described how a combination of multiple sensors can additionally facilitate deeper inspections (such as thermal inspections).

One important type of inspection especially important in defense is identification of potentially dangerous objects at the tactical edge (without prior knowledge of what dangerous could mean). A tactical edge is a dynamically changing environment. Any anomalous event can potentially have massive financial and operational repercussions for military. In addition, human life can be at risk. An RID such as Spot can be immensely useful for such an application as it can detect anomalies in places where it's risky to send people and can spot things obscured from aerial view (e.g. under shrubbery, exploded buildings etc.). Subtle anomalies which would otherwise elude the human eye can also be detected using advanced sensing (e.g. IR imaging). Anomalies can appear in many forms such as (i) presence of suspicious devices, (ii) soldiers who need assistance (during a conflict), or (iii) overheating military equipment. Hence analysis of multi-modal data is essential.

A potential solution for unsupervised anomaly detection is to build temporal multi-modal models of *normal* operating conditions at locations of interest within a facility. An RID then roams around the tactical edge facility and captures multi-modal information (visual, audio, thermal etc.) at those specific places. This is sent to an edge node which incrementally updates an AI model of what is *normal* at these locations. If at a future time point, some anomalous event occurs, the multimodal anomaly detection module flags it and raises an exception in the defense enterprise asset management system so that next steps are taken.

9.3 Distributed AI Architecture

Based on the high-level challenges identified in Section 9.1 and motivated by the exemplar use case scenarios described in Section 9.2, we are ready to present the architecture of Distributed AI and key components needed to address the challenges. First, we introduce the key background and

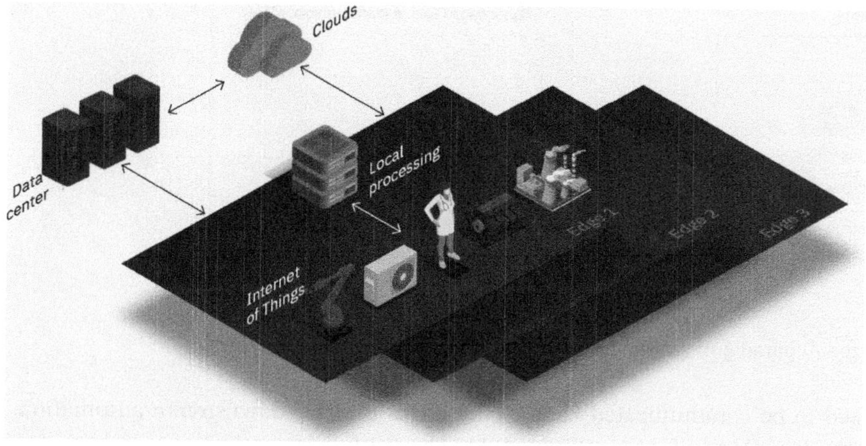

Figure 9.3 Internet of things at the edge.

evolution of current AI deployment patterns and challenges they face. Next, we motivate the novel pattern of Distributed AI and describe how such a pattern overcomes the challenges. Finally, we present the key components and a high-level description of the functionality they offer. Section 9.4 provides a deeper dive into the key components.

9.3.1 Background: Centralized AI and Edge AI

Reference to the term *AI* in this context is interpreted broadly to encompass machine learning, data processing and analysis, or rule-based decision-making applications. Over the last two decades, the meteoric growth in data coupled with advancements in algorithms and availability of computing resources such as GPUs has led to a growing infusion of AI into business processes and decision-making. Much like the industrial revolution, greater automation is one of the key drivers for the increasing penetration of AI in the real-world. However, the paradigm of how the AI capabilities are created, deployed, and managed across the computing environments has greatly evolved. In particular, with the contemporary growth of the Cloud computing paradigm, a majority of AI-infused applications are consumed via the *Centralized AI* paradigm. However, since a vast amounts of data originates outside of the central locations such as Clouds or data centers, having to consume the data via a Centralized AI pattern leads to friction. As a result, the *Edge AI* paradigm has emerged (Figure 9.3).

Let us examine these patterns closely to understand this evolution. In such an exercise, focusing on how data is created, stored, or migrated and how it is used for downstream processing or analysis allows us to identify the essential elements of the paradigms. For example, mobile connected vehicles start producing rich telemetry data as soon as they are turned on. Such data may be streamed to the nearest regional data center. The data center may then analyze the data and produce actionable results, e.g. turn-by-turn navigation instructions. These results need to be relayed back to the vehicle where they may be presented to an end user. In this case, the AI application is the turn-by-turn navigation, the edge is the vehicle itself, and the core location is the data center.

9.3.1.1 Centralized AI

In Centralized AI, all the data, regardless of where it originates, is collected in a central core location. The exact nature or size of the location does not matter, i.e. it could be a data center, a public Cloud, or even an on-premise system. Data is stored and eventually processed and analyzed to train analytical models. AI applications leverage one or more trained models to produce useful insights. When deployed, the models make inferences on input operational data from the edges.

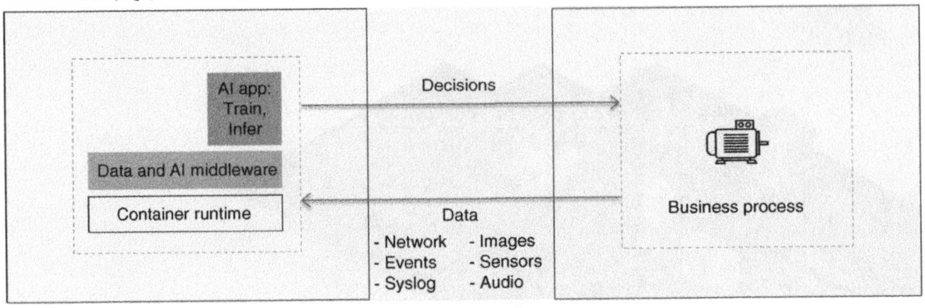

Figure 9.4 Centralized AI paradigm, centralized application and centralized data-plane.

Such inferences need to be communicated back to the edges to drive downstream automation. Figure 9.4 depicts the key elements of Centralized AI. The container runtime, e.g. Kubernetes, offers lifecycle management of containerized applications. The data and AI middleware offers common data and AI platform services, e.g. data storage and retrieval, analytics runtime such as Spark, model training, and model serving runtime such as Tensorflow.

Since the AI application is deployed only at the core location, we say that the application deployment is centralized. Also, since all data is being collected centrally, we say that the data-plane is centralized. Since the edges are dependent on the core for operational automation, any intermittent connectivity issues are a challenge. Further, since all of the operational data is continuously being pushed to core, scaling to a large number of edges is another challenge. There may also be regulatory or sensitivity issues in moving the data out of the edges. Due to these limitations, Centralized AI paradigm falls short.

9.3.1.2 Edge AI

With the advent of advances in control-plane, e.g. multi-cloud management in Kubernetes runtime [12], it is possible to have truly distributed application lifecycle managed from a "single-pane-of-glass" centrally. Hence, the data and AI middleware as well as the AI application including the analytical models can be deployed to the edge locations and managed from the core location. As shown in Figure 9.5, the application deployed at the edge can make inferences based on the operational data to localize decision-making at the edge.

We refer to this paradigm as Edge AI wherein the application is distributed but the data is still centralized. Edge location has greater autonomy in Edge AI as it is no longer dependent on the core

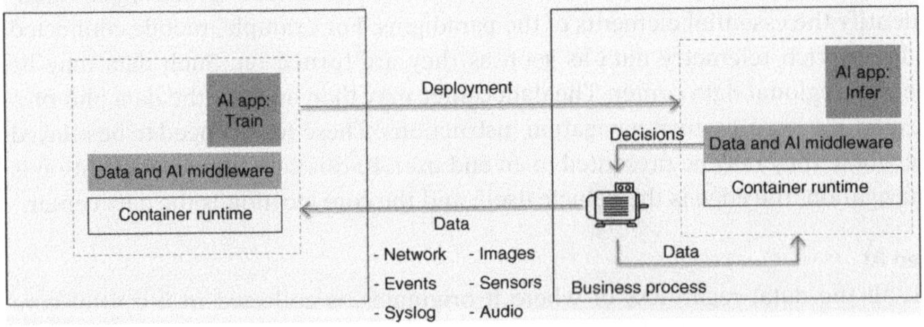

Figure 9.5 Edge AI paradigm, distributed application and centralized data.

location for operational decisions and can withstand intermittent loss of connectivity to the core location. Further, since the decision loop is localized, decision-making latency is lower compared to Centralized AI. Despite these improvements, Edge AI still requires the operational data to be collected in the core location as the training of the analytical models requires all training data to be available centrally. Additionally, Edge AI runs into several challenges when the number of edge locations increases or when multiple applications leveraging a variety of data modalities need to be deployed to the edge. In the following, we describe such challenges and introduce the novel paradigm of Distributed AI.

9.3.2 Open Challenges in Edge AI

Across industries, there is a growing need to enable analytics and AI applications at the edge to drive greater automation and accelerate decision-making. However in practice, the number and nature of edge locations varies greatly across industries. A manufacturing enterprise can have tens of plants spread globally, each focused on a unique product mix catering to their target markets. Such plants typically have an on-premise or a regional data centers acting as the edge. A retail enterprise, on the other hand, operates hundreds of retail stores each having a rack of servers acting as the edge. In the emerging space of software-defined vehicles, each vehicle is the edge with a limited compute resource combined with a rich array of sensing capabilities. Applying the paradigm of Edge AI to develop, deploy, and manage enterprise-wide analytics and AI applications across industries does not scale due to the following key challenges.

Data Gravity: Edges generate vast amounts of raw data, possibly with different modalities. The raw data is often repetitive, noisy, sensitive. As a result, collecting all of the raw data in the core location for training the analytics and AI pipelines runs into the challenges of costs, connectivity, and regulatory constraints. These challenges are exacerbated as the number of edges and the number of applications deployed to the edge grows.

Heterogeneity: Operational environment varies across each edge location, leading to material changes in data and feature distributions, e.g. differences in product mix across plants. As a result, an analytics algorithm trained in the core location may not be suitable for all edge locations. Also, the operational environment evolves dynamically, e.g. changes in the assembly lines to accommodate new products being manufactured. Thus, the accuracy of decisions being made by an AI pipeline may decline over time.

Resource Constraints: Computational resources at the edge locations may be limited. Also, due to heterogeneity, the available resources vary across edge locations. Hence, an analytics algorithm created in the core location may not function at the edge locations due to unmet resource dependencies.

Scale: As the number of edge locations grow, the above challenges become increasingly acute. Orthogonal to the number of edge locations, the challenge of scale is manifested in the need to support a variety of applications, possibly involving different data modalities. Either way, a hand-crafted manual approach to address the challenges of heterogeneity, data gravity, and resource constraints does not scale.

The challenges identified above are non-trivial and correspond to an area of active research in the AI research community. Thus, addressing these challenges calls for the development of new algorithms packaged as reusable components, especially in the data and AI middleware layer. As described above, one of the key challenges is in developing the algorithms such that they are agnostic to the specific application or data modality.

9.3.3 New Paradigm: Distributed AI

Since the emergence of edge computing, terminology such as Network edge, Near edge, Cloudlets, Fog computing, and extreme edge (among others) has been used to describe the various flavors of edge computing [13]. These flavors characterize the physical location and capacity of the computing resource available at a given location. Although the physical location and compute capacity are important considerations, we argue that the location of data and the role played by a location in either producing the data or orchestrating its management and analysis are core to defining a computing paradigm. The fundamental technical challenges encountered in managing analytics and AI applications across multiple public Cloud environments are no different from those faced in doing the same across a public Cloud and several on-premise environments, regardless of the capacity or physical location of the on-premise and Cloud environments.

We propose to characterize the Distributed AI paradigm in terms of a single *Hub* location and one or more *Spoke* locations. Spokes are closer to data sources and business operations and offer a variety of data-plane functions to localize operational decision-making. Hub plays the role of a central location and offers control-plane functions for managing the lifecycle of applications across the Hub and Spokes. The roles of Hub and Spokes are orthogonal to the physical location or the capacity of the compute resources. Hence, each Hub and Spokes could be public Cloud, on-premise, Network edge, or other environments with a variety of compute resources.

Figure 9.6 depicts the novel paradigm of Distributed AI and identifies the key technology components across Hub and Spokes that address the above-described challenges in scaling Edge AI. Distributed AI preserves and builds on the idea of localizing operational decision-making and a "single-pane-of-glass" management of application lifecycle. Specifically, four novel technology components are introduced to address the challenges: Data Ops, Model Ops, Federated Learning, and AI Optimization and adaptation. In the following, we briefly introduce the main ideas behind each of the components. Section 9.4 dives deeper into the specific algorithms and techniques. An initial implementation of these capabilities is made available on a trial basis at the IBM API Hub [14].

Data Ops: The key idea in Data Ops is to automatically identify "interesting" data at the Spokes and summarize it before exporting to the Hub. Also Data Ops ensures the enterprise policies for data localization, replication, and retention are enforced across the Hub and Spokes. These capabilities address the challenges of data gravity and scale.

Model Ops: The key idea in Model Ops is to automate the lifecycle of analytics and AI models. Monitoring the accuracy of models deployed at the Spokes is essential, even when

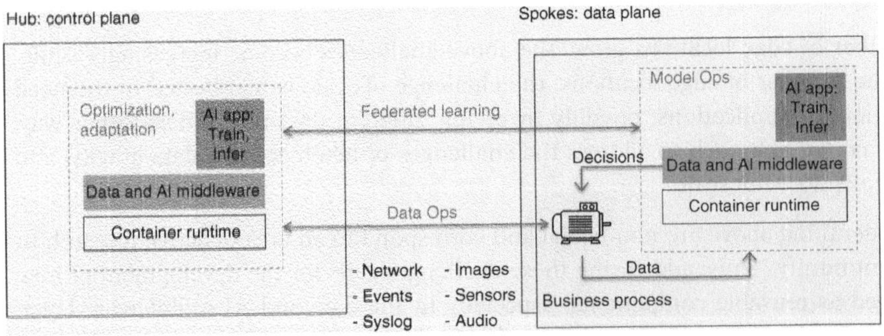

Figure 9.6 Distributed AI paradigm, distributed application and distributed data-plane.

"ground truth" is not available. Further, automatic detection of drift in performance and remediation via re-training the model with "relevant" training data is essential. These capabilities address the challenges of heterogeneity and scale.

Federated Learning: The key idea in Federated Learning is to train AI models from data residing across Spokes without needing the share or export the raw data with another Spoke or the Hub. Instead of sharing the raw data, model parameters and other metadata may be shared with the Hub while minimizing the overhead associated with doing so. Federated learning helps address data gravity challenge as well as the challenge of resource constraints.

Optimization and Adaptation: The key idea in Optimization is to minimize the compute resources required during the training, inference, and data processing at the Hub and Spokes. Since resources are always limited at the Hub and especially at the Spokes, this is an essential requirement. The key idea in Adaptation is to adapt the pre-trained analytics and AI models to the target Spoke environments, accounting for differences in the operating environment and data distributions. These capabilities help address the challenge of resource constraints as well as heterogeneity.

9.4 Technology

In this section, we dive deeper into the Distributed AI components introduced in Section 9.3. Specifically, in Section 9.4.1, we describe the data summarization techniques within the Data Ops component. Section 9.4.2 then describes in detail the model fingerprinting technique in the context of the Model Ops component. Model optimization is presented in Section 9.4.3 as part of the Optimization and Adaptation component. Finally, Section 9.4.4 describes the Federated Learning component of Distributed AI.

9.4.1 Data Ops

As introduced in Section 9.3.3, Data Ops consists of data summarization techniques as well as policy-based automation to enforce data management policies. Here, we focus on the data summarization techniques. Data summarization is useful for two purposes: (i) to provide high-level summaries of the data that are interpretable by human users, and (ii) to train models on the reduced space/size of summaries which are more efficient than training on the raw data samples. Broadly speaking, data summarization can be categorized into three categories: statistical summaries, dimensionality reduction, and sampling from original space [15], as shown in Figure 9.7.

9.4.1.1 Statistical Summaries

Statistical summaries capture useful statistics from the data. Its most basic form is summary statistics, which can include basic information such as average, variance, number of samples, etc.

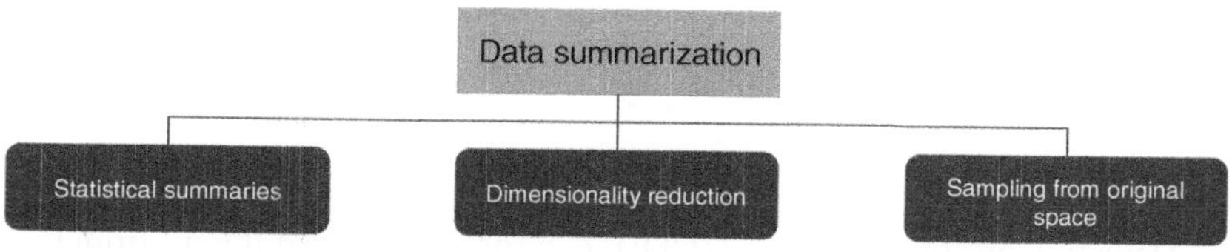

Figure 9.7 Data summarization techniques.

The statistics can be also computed on meta-information, e.g. the time and duration of data collection. This information allows human experts to understand basic characteristics of the data, so that they can query specific parts of the collected data for further investigation, to train a model for a specific purpose, etc. More advanced forms of statistical summaries include histograms, which can capture the distribution and similarity between data samples. Some applications use these summaries (e.g. bag-of-words can be considered as a histogram) as model inputs for classification tasks, in a similar way as features maps.

9.4.1.2 Dimensionality Reduction

Dimensionality reduction projects the input data onto a lower-dimensional space, to provide a more condensed representation that usually captures the essential information better than the raw data sample. Different from statistical summaries, the output of dimensionality reduction may not be easily interpretable by humans, but they can carry useful knowledge for machines to understand the data. A basic way of dimensionality reduction is linear projection, which includes projection by random matrices as well as principle component analysis (PCA). Advanced operations such as feature extraction using the convolutional layers of a (possibly pre-trained) convolutional neural network (CNN) can usually reduce the dimensionality too. The embeddings obtained from dimensionality reduction can be fed into a model for further training and inference. In the context of edge computing, this can be used for several purposes. For example, it may be difficult to train a good model using raw inputs from a small amount of data, due to the high degree of freedom caused by the high input dimensionality. Training on the embeddings (with reduced dimensionality compared to the raw data) may produce a model with higher accuracy. In addition, it consumes much less storage to store the embeddings instead of the original data, and also much less bandwidth when sending the embeddings to other locations via the communication network.

9.4.1.3 Sampling from Original Space

A third method of data summarization is to sample from the original data space. In essence, this method selects a few representative data samples from a large dataset. The selected samples can be used by human experts to annotate labels, which are then propagated to similar data samples that are not selected as representative ones, for instance. These representative samples can be also used for training machine learning models, which can be useful in cases where it is not possible to train on the full dataset directly, such as due to constraints on storage and/or data sharing policies. Approaches of data sampling include: (i) random sampling, (ii) model-based sampling, and (iii) coreset-based sampling. While random sampling is the easiest to implement, its selected data samples may not represent the whole population well enough, because small populations can be easily missed by this approach. Model-based sampling selects data samples that are representative for a specific machine learning model, whereas coreset-based sampling selects those data points that are important for a class of models that conform to certain common assumptions [16].

The above summarization techniques can be combined to achieve their advantages altogether. For example, a combination of dimensionality reduction and coreset-based sampling was studied by Lu et al. [17].

9.4.2 Model Ops

As motivated in Section 9.3.3, a key requirement in Model Ops is to monitor the accuracy of models deployed across Spokes. Detecting out-of-distribution (OOD) records, which are different from the data involved in the training process, is essential to assess how well the models are performing at

Spokes, without having access to ground truth. However, existing OOD detection methods either assume the availability of OOD records during training or require to retrain the model itself, thus limiting their application in practice.

In this section, we introduce a novel OOD detection method, NeurlFP [18], without requiring any access to OOD records for training. It constructs non-linear fingerprints of neural network models to memorize the information of training data (assumed to be normal data). The key idea of NeuralFP is to exploit the difference in how the neural network model responds to data records in its training set vs. data records that are anomalous. Specifically, NeuralFP builds autoencoders for each layer of the neural network model and then carefully analyzes the error distribution of the autocoders in reconstructing the training set to identify OOD records. To validate the effectiveness of NeuralFP, we conduct experiments on multiple real-world datasets.

9.4.2.1 OOD Detection Algorithm

In this section, we will first define the problem, then show a motivating example and describe the key steps of NeuralFP.

9.4.2.1.1 *Problem Statement*

NeuralFP aims to predict whether data records at the Spokes are anomalous or not by using a model trained at the Hub. Here, we assume the training data at the Hub as $\mathcal{D} = \{(\boldsymbol{x}_i, y_i)\}_{i=1}^N$, based on which an L-layer neural network model is learnt. The model parameter vector for the l-th layer is denoted as $\boldsymbol{\theta}_l$ and the overall model parameter vector is thus $\boldsymbol{\theta} = \left[\boldsymbol{\theta}_1^T, \cdots, \boldsymbol{\theta}_L^T\right]^T$ where $.^T$ is the transpose.

NeuralFP aims to construct fingerprints of the model applied to the training set, denoted by $\mathcal{FP}(\boldsymbol{\theta}, \mathcal{D})$. For a given data record $\boldsymbol{x}^*$, the Spokes will communicate with the Hub and utilize $\boldsymbol{\theta}, \mathcal{FP}(\boldsymbol{\theta}, \mathcal{D})$ to determine whether it is OOD data or not.

9.4.2.1.2 *Motivating Example*

The key intuition of NeuralFP is the difference that the neural network model responds to the in-distribution data and the OOD data. We use a 7-layer neural network model trained from the MNIST training data as an example to validate the intuition. We first pass the MNIST training data through the model and get the activations at each layer. Then we construct an autoencoder for each layer by using the corresponding activations. Then, we feed the MNIST testing data (in-distribution data) and the Fashion-MNIST testing data (OOD records) to the autoencoders and show the distribution of reconstruction errors in Figure 9.8.

From Figure 9.8, we have the following observations to motivate the design of NeuralFP: (i) the reconstruction errors are useful for distinguishing in-distribution data (MNIST) from OOD records (Fashion-MNIST); (ii) the reconstruction error of in-distribution data may not always be smaller than that of the OOD data (see Layer-6 and Layer-7); (iii) different layers have different capability in distinguishing OOD data (see Layer-1 and Layer-3).

9.4.2.1.3 *Design Details*

The architecture of NeuralFP is shown in Figure 9.9). NeuralFP is composed of two modes: fingerprinting on the Hub and OOD detection at the Spokes using the fingerprints.

9.4.2.1.4 *Fingerprinting on the Hub*

The fingerprints of NeuralFP on the Hub consist of two components: (i) deep generative models; and (ii) the reconstruction error distributions of the training set. NeuralFP first passes all the training data through each layer of a neural network model to obtain the corresponding activations,

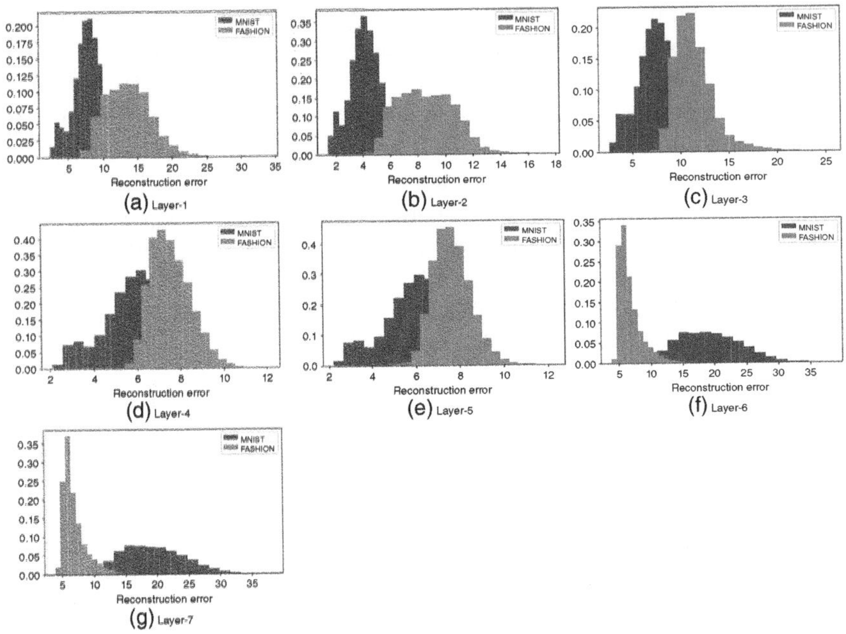

Figure 9.8 The distribution of reconstruction errors for MNIST testing data (blue) and Fashion-MNIST testing data (orange) applied to the model fingerprints of MNIST training data.

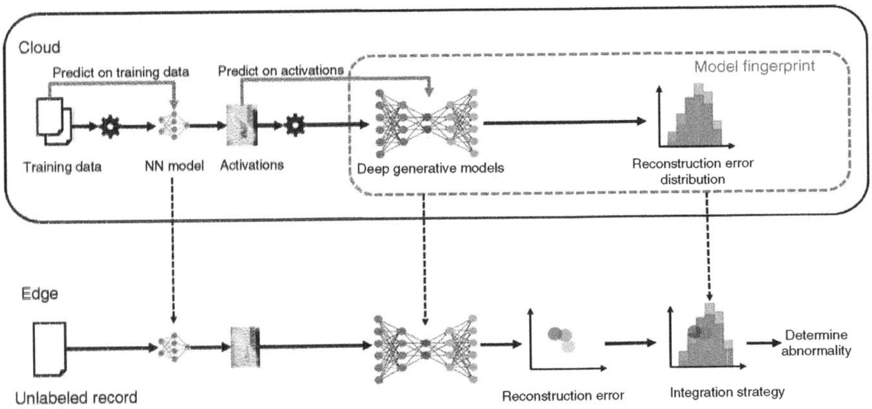

Figure 9.9 The framework of NeuralFP.

based on which deep generative models will be constructed. The deep generative models will then be used to obtain latent information (such as the distribution of reconstruction errors) of the training data, which will serve as comparison benchmarks when new unlabeled data is coming for prediction. Therefore, we view NeuralFP as a *model fingerprinting* method where the fingerprints consist of the deep generative models and the reconstruction error distribution of the training set.

9.4.2.1.5 OOD Detection on the Spokes

For a given data record $\boldsymbol{x}^*$, it will pass through the neural network model to compare with the stored model fingerprints for determining abnormality. According to the second observation in Section 9.4.2.1, the reconstruction error of an in-distribution record may not always be smaller than the OOD data. Therefore, we consider data records with reconstruction errors locating outside of the error distribution of the training data as outliers. Specifically, we define threshold parameters $\{\mu_l, \tau_l\}$ (where $\tau_l > \mu_l$) to bound the reconstruction error distribution of the training set at the l-th layer, and a data record $\boldsymbol{x}^*$ would be classified as outlier if its reconstruction error $e_l^* = \mathcal{L}(\boldsymbol{a}_l(\boldsymbol{x}^*), \hat{\boldsymbol{a}}_l(\boldsymbol{x}^*))$ satisfies

$$e_l^* > \tau_l \quad \text{or} \quad e_l^* < \mu_l. \tag{9.1}$$

Next, we investigate how to integrate the detection results across multiple layers to enhance the accuracy of OOD detection. According to the third observation in Section 9.4.2.1, some layer may incorrectly recognize the OOD records as normal data. Therefore, we propose the *one-out* integration strategy where a data record would be labeled as *outlier* if it is recognized as outlier by at least one layer, i.e. we consider $\boldsymbol{x}^*$ as OOD record if

$$\exists\, l, \quad e_l^* > \tau_l \quad \text{or} \quad e_l^* < \mu_l. \tag{9.2}$$

9.4.2.2 Experiments

To validate the effectiveness of NeuralFP, we use four real-world datasets including MNIST dataset [19] (handwritten digits), SVHN dataset [11] (street-view house numbers), Fashion-MNIST dataset [20] (fashion items), and CIFAR-10 dataset [21] (various types of vehicles, animals, etc).

Training data of each dataset is used to learn a neural network model and construct the model fingerprints according to Section 9.4.2.1. Then its testing data together with the testing data of other datasets (considering as OOD records) is used to evaluate detection performance. Note that we have transformed the images of SVHN and CIFAR-10 datasets from 32 × 32 color images to 28 × 28 gray scale images for fair comparison with MNIST and Fashion-MNIST datasets.

We evaluate the neural network model trained on MNIST and Fashion-MNIST with various OOD data (SVHN, CIFAR-10, Fashion-MNIST, MNIST). The AUC (area under curve) values are presented in Figure 9.10. From the expeimental results, we have the following import observations: (i) NeuralFP achieves good performance in detecting outliers. (ii) NeuralFP is robust under various settings of parameters since the high AUC scores (over 0.94 for all cases). (iii) MNIST data is more representative than Fashion-MNIST where the corresponding fingerprints created by NeuralFP are relatively easier to be distinguished from outliers.

9.4.3 Optimization and Adaptation

Model optimization and adaptation plays a particularly important role in Distributed AI due to the following factors: (i) resource constrained devices, (ii) limited battery power, (iii) need for near real-time performance, (iv) potential deployments in mission critical environments, and (v) potential limited connectivity, thus, requiring local inference. This section will dive deeper into various model optimizations that aim to shrink the size of the model and reduce the memory footprint, the compute requirements, and the overall inference execution time, while at the same time minimizing impact on model accuracy.

Training \ OOD	MNIST	FASHION	CIFAR10	SVHN
MNIST	N.A.	0.9826	0.9921	0.9921
FASHION	0.9498	N.A.	0.9442	0.9809

Figure 9.10 AUCs of detecting various OOD records.

9.4.3.1 Model Pruning

Reed [22] states that a rule of thumb for training ML models in order to achieve good generalization is to train small-enough models that will fit the data just right, with the caveat that no one knows what the "right size" is. Thus, models have traditionally been over-designed and over-trained, leading to over-fitting. Pruning aims to reduce the size of these models by removing unnecessary neurons and connections, a concept with roots in neuroscience [23]. As deep learning architectures continue to improve and their number of layers and trainable parameters increase, techniques such as pruning become ever so important. This is particularly true for CNNs, whose popularity has skyrocketed in the past decade [24–37].

There are two types of pruning; structured and unstructured pruning. Unstructured pruning relies on zeroing out weights in order to mimic deletion behavior, leading to sparse matrices. In the case of convolutional layers, sparse matrix operations require specialized hardware support in order to take full advantage of such pruning [38], else the number of operations remains the same as non-sparse matrices. Moreover, sparse matrices are still represented as multi-dimensional arrays, thus, the memory and storage space remains the same as non-sparse matrices. Structured pruning on the other hand, relies on the modification of the network architecture as removing neurons and weights will lead to changes downstream [25–27, 29–31, 34, 35, 39]. Channel and filter pruning are similar structured pruning techniques, where filters/channels are often ranked using various metrics (e.g. L1/L2-norm [24], Frobenius norm [25], Average Percentage of Zeros (APoZ) [30], Geometric Median [34], energy impact [37, 40], etc.). The benefits of pruning a channel or a filter for the current layer is that the outbound features are pruned and so are the inbound features in the next layer. A downside is very much the same in the sense that when skip connections, residual layers [41], or inception layers [42] are introduced, removing channels or filters would impact not just the immediate layer, but all the layers downstream that are immediately connected. Thus, structured pruning requires surgical precision.

9.4.3.2 Model Quantization

Quantization has its roots in digital signal processing where signals have to be reduced in magnitude in order to process them or transfer them over a network [43]. The premise behind quantization in ML is that models may not need the full 64-bit representation, as the human brain is believed to store information in quantized form [44]. [45] were among the first to use quantization in neural networks. Though their intent was to be able to use optical computers, they showed that their simple neural networks could indeed converge despite the reduced weight value representation. One-shot quantization, or post-training quantization will often dynamically adjust the rate as values may have different ranges and may be uniform or non-uniform.

A well-known scheme for quantization targeting $k - bit$ hardware is to look at which bits in the $k - bit$ word are more active. The process starts by converting the $k - bit$ float into its binary representation. A histogram is then creating for the $k - bit$ word, and a window with the highest hit rate is chosen. Of course, like any quantization scheme, there will be underflow/overflow instances, however, the hope is that those outliers are kept at a minimum. Several pieces of work have found that the ranges across layers vary [46, 47], thus requiring tuning ranges on a per-layer, per-weight, per-activation, even per-gradient basis; [46] represent different data granularities using different precision levels.

Much like pruning, quantization can be an iterative approach, where models are first trained, then quantized using post-training quantization, following with a re-training round using the quantized weights [48]. Others schemes train with quantized values [46, 47, 49], Jacob et al. [49] for example, introduce quantization nodes into the input and output layers of the network, while keeping the weights and biases with full precision. The idea is to train using full precision, while fine-tuning the network to assume integer-based inference. Once the network is to be deployed, inference is done using integer-only arithmetic. More recent work has looked at going beyond 8-bits. For example, Banner et al. [50] propose post-training 4-bit quantization that individually selects range clippings for each activation/weight on a per-channel basis. Mixed precision was also shown to outperform fixed 4-bit and 8-bit quantization schemes [51]. In recent years, binarized neural networks (where the weights and activations are constrained to be binary) have also shown a lot of promise for model optimization [52–54].

9.4.3.3 Other Schemes

Neural architecture search (NAS) attempts to generate optimal models given some target architecture. The challenge being that searching architectures for large datasets such as ImageNet can take thousands of GPU hours [55]. Often NAS-based schemes try to perform search using a subset of the dataset, or some type of proxy task. ProxylessNAS [55] over-parameterizes large neural networks, and searches for redundant paths. The goal is to start big, trim the network slowly, and finish with a slimmer network. Once-for-all [56] starts with a large network architecture as well, and slowly goes through the process of selecting specialized subnets that are optimized for different hardware architectures.

Layer fusion [57] attempts to optimize data reuse during convolutions by fusing layers. Given the fact that most convolutions are matrix-based operations over a stream of data, it makes sense to do as many computations on the data as possible before going off-chip. SkyNet [58] groups layers into bundles, which are implemented as hardware logic. Bundles then make up back-bones, which are then responsible for operating over the data fed after re-ordering. Pruning, quantization, and Huffman coding were combined in [40] to produce compressed models. Low-rank approximation is also used to compress the weights of fully connected layers by keeping only the most prominent components of the decomposed matrices [59–61].

9.4.3.4 Experiments: Model Optimization for Asset Inspection

Model optimization was an important component of asset inspection use-cases as discussed in Section 9.2. In this section, we discuss the approach and results. From an AI standpoint, the asset inspection use-cases involved visual recognition using Boston Dynamics Spot robot cameras: (i) a classification task where Spot had to classify fire extinguishers according to various safety categories, (ii) a recognition task where Spot had to detect transformers as it walked through a building. In order to detect objects (here transformers), a Yolo V3 [8] architecture was adopted. For classification, Inception V3 [7], another well-known CNN classifier, was used. For each model we started with the default weights obtained by training the network on the COCO dataset [62] for Yolo V3 and ImageNet [63] for Inception V3. In both scenarios we applied transfer learning in order to train and establish a baseline model. Optimization was then performed for each model using channel pruning using L-1 norm as the importance score, on a layer basis. For this example, we used one-shot pruning on the entire model, followed by a re-training phase.

Figure 9.11 shows our model pruning results. Inference was done on the CPU (x86). In both cases, we can observe significant reduction in (i) number of parameters, (ii) inference time, while maintaining accuracy almost at par with the baseline model.

9.4.4 Federated Learning

In an agile sensing system, data can be collected by various sensors over time. For example, such sensors can be cameras and microphones located on IoT devices or robots. As the robot moves

Inception V3	Parameters	Size (MB)	Inf. Time (s)	Accuracy (%)
Baseline	21,791,751	312.70	0.82	99.05
Pruned	11,918,511	191.37	0.54	97.62
Yolo V3	**Parameters**	**Size (MB)**	**Inf. Time (s)**	**Accuracy (%)**
Baseline	61,581,727	243	0.39	100
Pruned	33,781,103	184	0.27	100

Figure 9.11 Pruning Inception V3 and Yolo V3 for Spot; 8th Gen. Intel Core i5; PyTorch 1.8 and TensorFlow 2.3. Readers can note that pruning results in reduction of model size as well as inference time without compromising accuracy.

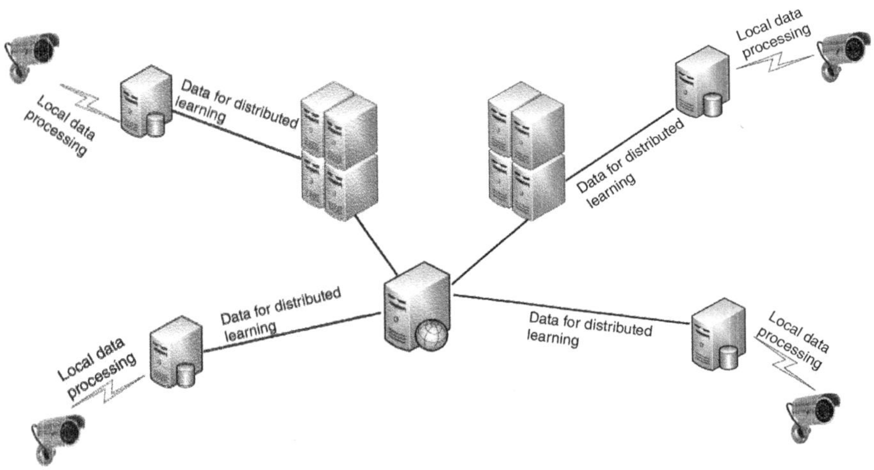

Figure 9.12 Example of a federated learning system. Source: pixelrobot/Adobe Stock.

around, new data get collected. Federated learning (FL) is a way of training machine learning models from such dispersed datasets collected at local devices [64], as illustrated in Figure 9.12. Compared to the centralized learning setting, the benefit of FL is that the data remains local and is not shared with the server or other devices, thereby preserving data privacy and saving communication bandwidth.

The main idea of FL is to interleave local training on each device's local dataset with model parameter synchronization between devices and the server. At a high level, the canonical federated averaging (FedAvg) algorithm includes the following steps:

1. The server sends the current model parameters (usually represented as a vector or multiple matrices) to the devices.
2. Each device trains its own model for a certain number of iterations, starting from the global parameter vector received from the server.
3. Each device transmits its current model parameters, or the difference between the current parameter vector and the previous global parameter vector received at the beginning of the round, to the server.
4. The server aggregates the parameter updates, to obtain the new global parameter vector, and resume from Step 1.

The above Steps 1–4 are often referred to as a *round* of FL. It usually takes many rounds before reaching convergence. This procedure is an extension to the stochastic gradient descent (SGD) method that is widely used for deep learning in the centralized setting. If the number of local iterations per round is equal to one, then the progression of model parameter vectors is the same as in centralized SGD (strictly speaking, in expectation). Theoretical analysis and empirical evaluations have shown that FedAvg also converges when the number of per-round local iterations is larger than one, which has the benefit of saving communication overhead, because modern models usually have a large number of parameters and transmitting them frequently would consume a lot of communication.

9.4.4.1 Resource Efficiency of FL

Even with infrequent communication, transmitting the full set of model parameters between all devices and the server can be still prohibitive in many practical scenarios, especially in edge computing environments where both the computational capability and the network bandwidth can be limited. To tackle these problems, several approaches to further reduce the resource consumption of FL at the Spoke can be applied. We discuss some of these techniques as follows.

Partial Device Participation: When multiple IoT devices are located in the same physical space, it is likely that they collect similar data samples. In this case, having only one (randomly chosen) device participating in each FL round can be sufficient, which gives a similar model accuracy compared to the case of all devices participating in all rounds. In general, especially when the data distribution across different devices are unknown, the FL system can randomly choose a small percentage of devices to participate in each round in a controlled manner. As long as the aggregated parameter vector at the server remains unbiased, the model training still converges. Partial participation can be also involuntary, such as when a device disconnects from the network for a period of time. Model training can still succeed in these cases, by properly offsetting the bias caused by unavailable devices, such as increasing the weight of those devices' updates when they join back.

Compressed Model Updates: Another method for improving the communication efficiency is to transmit compressed updates of model parameters. While standard data compression methods can be used, their complexities are often high and requires both compression and decompression operations. An interesting observation in FL is that simple compression and sparsification techniques, e.g. transmitting randomly selected k components in the parameter vector (also known as random-k), are often sufficient and guarantees training convergence. Efficiency may be further improved by using the class of biased compressors, such as top-k that transmits k components of the update vector with the largest magnitudes, together with an error feedback mechanism that keeps components that are not transmitted locally and they may be transmitted in a future round. The intuition behind this top-k method is that the important updates get transmitted first so that the different devices remain synchronized to a large extent. In addition to compressing the parameter updates for communication efficiency, the model itself can be compressed by pruning unimportant weights in the neural network (Section 9.4.3.1), which improves both computation and communication efficiency, and also generates a small model for efficient inference.

9.4.4.2 Privacy Considerations

Apart from training efficiency, privacy is a major aspect and a key motivation for FL. Although gradients and parameter updates are less sensitive than the raw data, they can still leak a substantial amount of information about the local training data. Therefore, FL is often implemented with a secure aggregation algorithm, which allows the server to only obtain the aggregated parameter vector, but not the individual parameters from devices. This prevents the server from knowing from which device a specific piece of information comes. Other types of privacy preserving techniques, such as differential privacy, can also be used together with FL.

9.5 Research Directions

The confluence of IoT, Distributed AI, and robotics is still a very new research area. The most closely related works are referred to as fog-robotics which uses a spectrum of computing capabilities from Spoke to Hub for robotic applications [2–6]. Most of these mainly focus on creating fog-robotics solutions using standard deep learning models e.g. object recognition etc. to be deployed in robots. Models are stored somewhere in the continuum between Hub and Spoke. There is not much research on AI at the Spoke in the context of robotics as well as on how advances in areas such as self-supervised learning and few-shot learning can be leveraged in these scenarios. In this section, we will discuss some research directions in this space.

9.5.1 Learning with Resource Optimization

Designing a robotic IoT ecosystem capable of perception, planning, and learning holistically at the Spoke to perform tasks in real and dynamic industrial/defense environments involves many moving parts. In such cases, assuming that the robot possesses sufficient computing capabilities may not always be realistic. It's also critical to understand that computational resources (and energy requirements) may vary across robots and other IoT devices (at the Spoke). The key challenge is to allocate the resources appropriately and robustly across the Spoke and the Hub. A good design choice would be to prioritize tasks based on their criticality for safety, latency tolerance, and their overall resource requirements for successful completion. Distributed AI techniques that leverage robots and available SPOKE devices in their current vicinity or Hub resources as needed will be critical.

For robotic IoT devices applications to be useful across diverse domains, it is imperative that the robotic learning is capable of adaptation to the constantly changing operating environment. In this context, self-supervised learning methodologies which leverage multimodal sensor data to build semantic representations automatically without the need for expensive human annotation can be extremely helpful [65, 66]. In order to adapt AI models across different tasks and domains, few-shot learning methodologies which have proven to be very powerful in several fields of AI can be leveraged [67].

One of the key differentiating features of RIDs is the ability for adaptive sensing (dynamic adjustment of sensing parameters via situational awareness and current task context). The AI solutions thus need to be especially designed in light of adaptive sensing capabilities within such environments.

9.5.2 Collaboration Among Humans and Robots

Designing learning solutions for collaborative tasks that require close coordination between multiple IoT robotic devices is very challenging especially when resources are distributed across the Hub and Spoke. To address these challenges, multi-agent planning methods [68–70] and coordination strategies [71, 72] will need to be extended for the Spoke/Hub environment. Potentially conflicting robot trajectories can be detected by evaluation of robot plans using a combination of traditional methods and perception-based motion prediction [73–76]. An important AI design strategy here is to incorporate gradations of customization based on where the learning is performed (e.g. learning performed on the robot should support robot-personalization, learning performed at the infrastructure Spoke should be customizable to the local environment, while AI models at the Hub should learn and relay general global information). Federated learning methodologies (Section 9.4.4) specifically designed for heterogeneous RIDs will be directly applicable in this scenario.

Creating AI techniques for safe human-robot interaction through audio, natural language, gestures, and physical interaction is also a high priority. The primary purpose of such interchanges will be to provide instructions to robots. A major hurdle is that robotic learning with audio or natural language instructions in real world scenarios will need large and diverse multi-modal training sets. Creating such training data using simulated environments may not always serve the purpose [77].

Human-robot interaction (for guidance) can vary from high level task assignments to low-level detailed task instructions, and from one-way communication to a dialog between humans and robots. Techniques developed for learning from demonstration [78, 79] can be employed for achieving this in structured environments [80]. It will also be important, as discussed earlier, to allocate resources to ensure safety in such interactions.

9.5.3 Multi-modal Learning

A unique challenge in IoT-robotics is the noise and variability in sensed information caused by the dynamic nature of the operational environment. For example, if the AI task is to perform inspection of equipment or machinery in the environment and find anything that is out of the ordinary, the performance of the task is adversely affected by the changes in lighting across day and night, changes in noise-levels at various locations in the environment, and movement of equipment during the operations. Clearly, AI pipelines that choose inputs from multiple modalities to achieve robustness will perform better. However, even then, the AI pipeline would suffer from the variations in the environment as the noisy inputs from one or more modalities affect the accuracy. Further, such variations are unpredictable and unknown in advance. Thus, the available training data may not cover all possible variations that are encountered at inference time. Another problem with using all sensor modalities in every situation is the energy consumption, especially as most of RIDs are battery-powered and energy usage grows with usage of multiple modalities. Thus, there is a need for an approach that can dynamically choose the right sensor modalities for a pre-defined AI task.

Prior work in multi-modal sensing is focused on simply leveraging multiple sensors to improve AI task performance [81, 82]. Dynamically choosing the right modalities for a given AI task remains unexplored.

9.5.3.1 Context-based Multi-modal Sensing

In recent years, there has been extensive work on self-supervised multimodal learning in AI literature [65, 66, 83]. Most of these methods focus on learning some joint distribution across multi-modal representations as proxy (without human labeling). These representations are then leveraged for further downstream tasks often in a few-shot learning setup. One research direction is to adapt and extend self-supervised approaches for multi-modal joint learning with RIDs and combine them with the idea of a general purpose OOD technique (e.g. [18]), as previously described, to address the problem of dealing with unpredictable and unforeseen variations in an environment being sensed (Figure 9.13).

Specifically, a self-supervised approach can be used to learn multi-modal representation of the environment. Such learning can happen on unlabelled data collected from the environment using any number of available sensor modalities. Also, an AI task pipeline can be trained based on

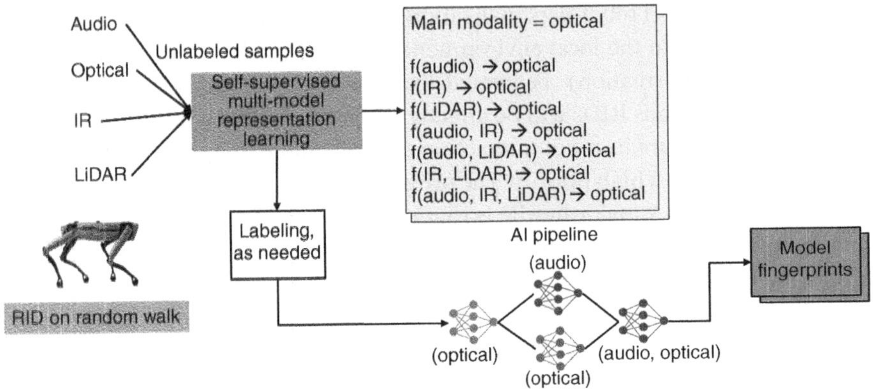

Figure 9.13 Training pipeline for context-based multi-modal sensing.

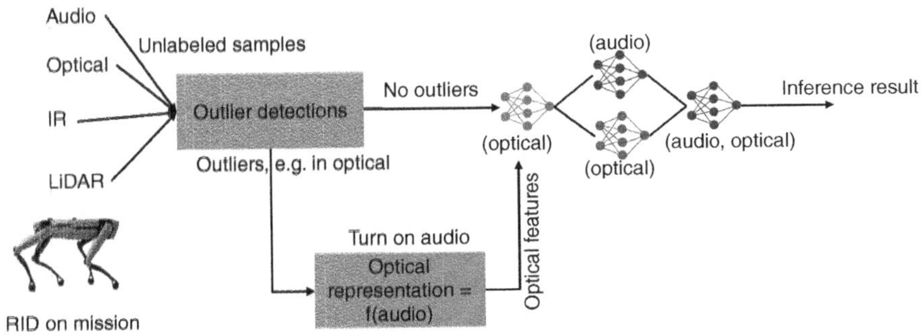

Figure 9.14 Inference pipeline for context-based multi-modal sensing.

one or more fixed modalities, e.g. visual or audio+visual, which we call the *primary modalities* for the task.

At inference time, only the main sensor modalities are turned on by default. An OOD technique is used to determine whether a sample in the main modality is of poor quality. If so, all available sensor modalities are turned on and the self-supervised joint representations are used to infer a representation of the main modalities based on the samples of the rest of the modalities. The generated main modality representation is then used to make inferences. This way, robustness against any variations in the main modality can be covered by other modalities. A limitation of this approach is that if in a situation, most of the modalities show unforeseen variations, then the inferred main modalities may not correspond to ground truth (Figure 9.14).

9.5.3.2 Adaptive Navigation to Optimize Sensing

RID operations can benefit from the fact that they are capable of adaptive navigation and sensing. This intertwining of the two capabilities creates a variety of possibilities for effective completion of a task in low-resource scenarios. An interesting research direction is to develop a method wherein online reinforcement learning is combined with an OOD method (e.g. [18]) as a reward function for adaptive navigation to optimize sensing. The action and state space are pre-defined based on the navigational and sensing capabilities of the RID. However, the reward function is not local to the RID and is in the control of the Spoke or the Hub server where the AI task is deployed.

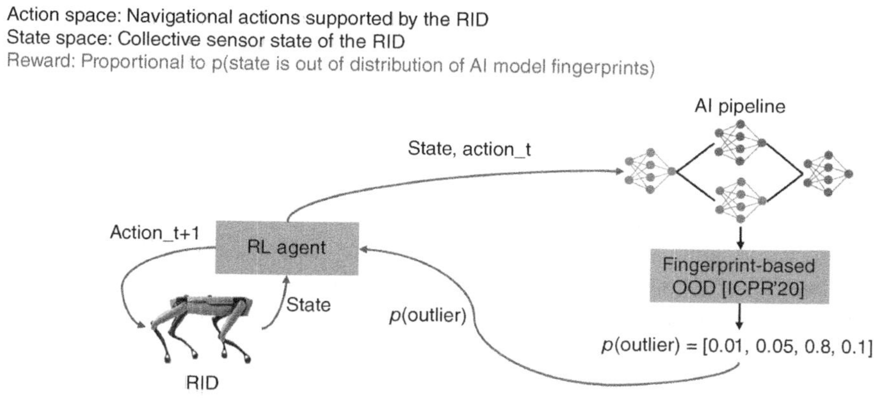

Figure 9.15 Pipeline for adaptive navigation for optimized sensing.

For example, if the RID is the Spot robot, the action space consists of movement in a 3D environment, change in pose, and configuration of each of the sensors. The state space consists of a vector representation of the sensed inputs across the sensor array, e.g. optical camera, thermal camera, microphones, and LiDAR. Given a specific AI task, e.g. inspecting a fire extinguisher, a reward function could be defined to correspond to the quality of the sensed state, e.g. whether or not a fire extinguisher is clearly visible in the optical camera. One approach to defining the reward function that works across a variety of AI tasks is to base it on the likelihood that a given sensed state is out-of-distribution for the AI task (Figure 9.15).

9.6 Conclusions

In this chapter, we discussed key challenges, solutions, and findings from certain IoT industrial inspection applications we have worked on in the context of robotic IoT devices. We introduced the novel Distributed AI paradigm and presented the architecture of Distributed AI to address AI challenges in IoT era. We also did a deep dive into key technologies that enable Distributed AI applications. Finally we discussed some novel research directions in the confluence of IoT, Distributed AI, and robotics which we are actively pursuing. We hope that our discussion will significantly enhance the visibility of this important multidisciplinary research area and foster dialogue and collaboration among academic and industrial researchers in IoT, AI, robotics, and distributed computing.

References

1 Turner, M.J. (2021). Digital business depends on distributed clouds. *IDC Technology Spotlight.*
2 Gudi, S.L.K.C., Ojha, S., Johnston, B. et al. (2018). Fog robotics for efficient, fluent and robust human-robot interaction. *Proceedings of IEEE 17th International Symposium on Network Computing and Applications (NCA).*
3 Gudi, S.L.K.C., Johnston, B., and Williams, M.-A. (2019). Fog robotics: a summary, challenges and future scope. *CoRR* abs/1908.04935. https://arxiv.org/abs/1908.04935.
4 Tanwani, A.K., Mor, N., Kubiatowicz, J. et al. (2019). A fog robotics approach to deep robot learning: application to object recognition and grasp planning in surface decluttering.

Proceedings of 2019 International Conference on Robotics and Automation (ICRA), pp. 4559–4566. IEEE.

5 Tian, N., Tanwani, A.K., Chen, J. et al. (2019). A fog robotic system for dynamic visual servoing. *Proceedings of 2019 International Conference on Robotics and Automation (ICRA)*, pp. 1982–1988. IEEE.

6 Tanwani, A.K., Anand, R., Gonzalez, J.E., and Goldberg, K. (2020). RILaaS: Robot inference and learning as a service. *IEEE Robotics and Automation Letters* 5 (3): 4423–4430.

7 Szegedy, C., Vanhoucke, V., Ioffe, S. et al. (2015). Rethinking the inception architecture for computer vision. *CoRR* abs/1512.00567. http://arxiv.org/abs/1512.00567.

8 Redmon, J. and Farhadi, A. (2018). YOLOv3: An incremental improvement. *CoRR* abs/1804.02767. https://arxiv.org/abs/1804.02767.

9 Rublee, E., Rabaud, V., Konolige, K., and Bradski, G. (2011). ORB: An efficient alternative to SIFT or SURF. *Proceedings of the 2011 IEEE International Conference on Computer Vision*, pp. 2564–2571.

10 Ren, S., He, K., Girshick, R., and Sun, J. (2015). Faster R-CNN: towards real-time object detection with region proposal networks. *Advances in Neural Information Processing Systems 28*, pp. 91–99.

11 Netzer, Y., Wang, T., Coates, A. et al. (2011). Reading digits in natural images with unsupervised feature learning. *Google Research.*

12 Mfula, H., Ylä-Jääski, A., and Nurminen, J.K. (2021). Seamless Kubernetes cluster management in multi-cloud and edge 5G applications. *International Conference on High Performance Computing & Simulation (HPCS 2021).*

13 Yousefpour, A., Fung, C., Nguyen, T. et al. (2019). All one needs to know about fog computing and related edge computing paradigms: a complete survey. *Journal of Systems Architecture* 98: 289–330.

14 IBM (2021). Distributed AI APIs at IBM API hub. https://developer.ibm.com/apis/catalog/edgeai--distributed-ai-apis/Introduction/ (accessed 25 October 2022).

15 Ko, B.J., Wang, S., He, T., and Conway-Jones, D. (2019). On data summarization for machine learning in multi-organization federations. *2019 IEEE International Conference on Smart Computing (SMARTCOMP)*, pp. 63–68. IEEE.

16 Lu, H., Li, M.-J., He, T. et al. (2020). Robust coreset construction for distributed machine learning. *IEEE Journal on Selected Areas in Communications* 38 (10): 2400–2417.

17 Lu, H., He, T., Wang, S. et al. (2022). Communication-efficient k-means for edge-based machine learning. *IEEE Transactions on Parallel and Distributed Systems* 33 (10): 2509–2523.

18 Lee, W.-H., Millman, S., Desai, N. et al. (2021). NeuralFP: Out-of-distribution detection using fingerprints of neural networks. *Proceedings of the 25th International Conference on Pattern Recognition (ICPR)*, pp. 9561–9568. IEEE.

19 LeCun, Y., Bottou, L., Bengio, Y., and Haffner, P. (1998). Gradient-based learning applied to document recognition. *Proceedings of the IEEE* 86 (11): 2278–2324.

20 Xiao, H., Rasul, K., and Vollgraf, R. (2017). Fashion-MNIST: a novel image dataset for benchmarking machine learning algorithms. *arXiv preprint arXiv:1708.07747*. https://arxiv.org/abs/1708.07747.

21 Krizhevsky, A. and Hinton, G. (2009). Learning Multiple Layers of Features from Tiny Images.

22 Reed, R. (1993). Pruning algorithms-a survey. *IEEE Transactions on Neural Networks* 4 (5): 740–747.

23 Huttenlocher, P.R. (1979). Synaptic density in human frontal cortex - developmental changes and effects of aging. *Brain Research* 163 (2): 195–205.

24 Li, H., Kadav, A., Durdanovic, I. et al. (2016). Pruning filters for efficient convnets. *CoRR* abs/1608.08710. http://arxiv.org/abs/1608.08710.

25 He, Y., Zhang, X., and Sun, J. (2017). Channel pruning for accelerating very deep neural networks. *Proceedings of the IEEE International Conference on Computer Vision*, pp. 1389–1397.

26 Liu, Z., Li, J., Shen, Z. et al. (2017). Learning efficient convolutional networks through network slimming. *Proceedings of the IEEE International Conference on Computer Vision*, pp. 2736–2744.

27 Luo, J.-H., Wu, J., and Lin, W. (2017). ThiNet: A filter level pruning method for deep neural network compression. *Proceedings of the IEEE International Conference on Computer Vision*, pp. 5058–5066.

28 Wen, W., Wu, C., Wang, Y. et al. (2016). Learning structured sparsity in deep neural networks. *Advances in Neural Information Processing Systems 29*.

29 Molchanov, P., Tyree, S., Karras, T. et al. (2016). Pruning convolutional neural networks for resource efficient transfer learning. *CoRR* abs/1611.06440. http://arxiv.org/abs/1611.06440.

30 Hu, H., Peng, R., Tai, Y.-W., and Tang, C.-K. (2016). Network trimming: a data-driven neuron pruning approach towards efficient deep architectures. *CoRR* abs/1607.03250. http://arxiv.org/abs/1607.03250.

31 He, Y., Kang, G., Dong, X. et al. (2018). Soft filter pruning for accelerating deep convolutional neural networks. *CoRR* abs/1808.06866. http://arxiv.org/abs/1808.06866.

32 Yu, R., Li, A., Chen, C.-F. et al. (2018). NISP: Pruning networks using neuron importance score propagation. *Proceedings of the IEEE/CVF Conference on Computer Vision and Pattern Recognition (CVPR)*, June 2018.

33 Guo, Y., Yao, A., and Chen, Y. (2016). Dynamic network surgery for efficient DNNs. *CoRR* abs/1608.04493. http://arxiv.org/abs/1608.04493.

34 He, Y., Liu, P., Wang, Z. et al. (2019). Filter pruning via geometric median for deep convolutional neural networks acceleration. *Proceedings of the IEEE/CVF Conference on Computer Vision and Pattern Recognition (CVPR)*, June 2019.

35 Zhuang, Z., Tan, M., Zhuang, B. et al. (2018). Discrimination-aware channel pruning for deep neural networks. *CoRR* abs/1810.11809. http://arxiv.org/abs/1810.11809.

36 Zhu, M. and Gupta, S. (2017). To prune, or not to prune: exploring the efficacy of pruning for model compression. *arXiv preprint arXiv:1710.01878*. https://arxiv.org/abs/1710.01878.

37 Yang, T.-J., Chen, Y.-H., and Sze, V. (2017). Designing energy-efficient convolutional neural networks using energy-aware pruning. *Proceedings of the 2017 IEEE Conference on Computer Vision and Pattern Recognition (CVPR)*, pp. 6071–6079.

38 Han, S., Liu, X., Mao, H. et al. (2016). EIE: Efficient inference engine on compressed deep neural network. *CoRR* abs/1602.01528. http://arxiv.org/abs/1602.01528.

39 Anwar, S., Hwang, K., and Sung, W. (2017). Structured pruning of deep convolutional neural networks. *ACM Journal on Emerging Technologies in Computing Systems (JETC)* 13 (3): 1–18.

40 Han, S., Mao, H., and Dally, W.J. (2015). Deep compression: compressing deep neural networks with pruning, trained quantization and Huffman coding. *arXiv preprint arXiv:1510.00149*. https://arxiv.org/abs/1510.00149.

41 He, K., Zhang, X., Ren, S., and Sun, J. (2015). Deep residual learning for image recognition. *CoRR* abs/1512.03385. http://arxiv.org/abs/1512.03385.

42 Szegedy, C., Liu, W., Jia, Y. et al. (2014). Going deeper with convolutions. *CoRR* abs/1409.4842. http://arxiv.org/abs/1409.4842.

43 Lyons, R.G. (2004). *Understanding Digital Signal Processing*, 2e. Prentice Hall: Hoboken, NJ. ISBN 0131089897.

44 Tee, J. and Taylor, D.P. (2019). A quantized representation of probability in the brain. *IEEE Transactions on Molecular, Biological and Multi-Scale Communications* 5 (1): 19–29.

45 Choudry, F., Fiesler, E., Choudry, A., and Caulfield, H.J. (1990). A weight discretization paradigm for optical neural networks. *Proceedings of the SPIE International Congress on Optical Science and Engineering*, pp. 164–173.

46 Park, E., Yoo, S., and Vajda, P. (2018). Value-aware quantization for training and inference of neural networks. *CoRR* abs/1804.07802. http://arxiv.org/abs/1804.07802.

47 Gholami, A., Kim, S., Dong, Z. et al. (2021). A survey of quantization methods for efficient neural network inference. *CoRR* abs/2103.13630. https://arxiv.org/abs/2103.13630.

48 Chen, W., Wilson, J.T., Tyree, S. et al. (2015). Compressing neural networks with the hashing trick. *CoRR* abs/1504.04788. http://arxiv.org/abs/1504.04788.

49 Jacob, B., Kligys, S., Chen, B. et al. (2018). Quantization and training of neural networks for efficient integer-arithmetic-only inference. *Proceedings of 2018 IEEE/CVF Conference on Computer Vision and Pattern Recognition.*

50 Banner, R., Nahshan, Y., Hoffer, E., and Soudry, D. (2018). ACIQ: Analytical clipping for integer quantization of neural networks. *CoRR* abs/1810.05723. http://arxiv.org/abs/1810.05723.

51 Wang, K., Liu, Z., Lin, Y. et al. (2019). HAQ: Hardware-aware automated quantization with mixed precision. *Proceedings of the IEEE/CVF Conference on Computer Vision and Pattern Recognition*, pp. 8612–8620.

52 Hubara, I., Courbariaux, M., Soudry, D. et al. (2016). Binarized neural networks. *Advances in Neural Information Processing Systems 29.*

53 Courbariaux, M., Hubara, I., Soudry, D. et al. (2016). Binarized neural networks: training deep neural networks with weights and activations constrained to + 1 or −1. *arXiv preprint arXiv:1602.02830.* https://arxiv.org/abs/1602.02830.

54 Umuroglu, Y., Fraser, N.J., Gambardella, G. et al. (2017). FINN: A framework for fast, scalable binarized neural network inference. *Proceedings of the 2017 ACM/SIGDA International Symposium on Field-Programmable Gate Arrays*, pp. 65–74.

55 Cai, H., Zhu, L., and Han, S. (2018). ProxylessNAS: Direct neural architecture search on target task and hardware. *arXiv preprint arXiv:1812.00332.* http://arxiv.org/abs/1812.00332.

56 Cai, H., Gan, C., Wang, T. et al. (2019). Once-for-all: train one network and specialize it for efficient deployment. *arXiv preprint arXiv:1908.09791.* https://arxiv.org/abs/1908.09791.

57 Alwani, M., Chen, H., Ferdman, M., and Milder, P. (2016). Fused-layer CNN accelerators. *2016 49th Annual IEEE/ACM International Symposium on Microarchitecture (MICRO)*, pp. 1–12.

58 Zhang, X., Lu, H., Hao, C. et al. (2020). SkyNet: A hardware-efficient method for object detection and tracking on embedded systems. *Proceedings of Machine Learning and Systems* 2: 216–229.

59 Swaminathan, S., Garg, D., Kannan, R., and Andres, F. (2020). Sparse low rank factorization for deep neural network compression. *Neurocomputing* 398: 185–196.

60 Papadimitriou, D. and Jain, S. (2021). Data-driven low-rank neural network compression. *2021 IEEE International Conference on Image Processing (ICIP)*, pp. 3547–3551. IEEE.

61 Jain, S., Hamidi-Rad, S., and Racapé, F. (2021). Low rank based end-to-end deep neural network compression. *Proceedings of the 2021 Data Compression Conference (DCC)*, pp. 233–242.

62 Lin, T.-Y., Maire, M., Belongie, S. et al. (2014). Microsoft COCO: common objects in context. *European Conference on Computer Vision (ECCV)*, pp. 740–755. Springer.

63 Deng, J., Dong, W., Socher, R. et al. (2009). ImageNet: A large-scale hierarchical image database. *Proceedings of the 2009 IEEE Conference on Computer Vision and Pattern Recognition*, pp. 248–255.

64 Kairouz, P., McMahan, H.B., Avent, B. et al. (2019). Advances and open problems in federated learning. *arXiv preprint arXiv:1912.04977*. https://arxiv.org/abs/1912.04977.

65 Korbar, B., Tran, D., and Torresani, L. (2018). Cooperative learning of audio and video models from self-supervised synchronization. *Advances in Neural Information Processing Systems 31*.

66 Rouditchenko, A., Boggust, A., Harwath, D. et al. (2021). AVLnet: Learning audio-visual language representations from instructional videos. *Proceedings of 2021 INTERSPEECH*, pp. 1584–1588.

67 Wang, Y., Yao, Q., Kwok, J.T., and Ni, L.M. (2020). Generalizing from a few examples: a survey on few-shot learning. *ACM Computing Surveys (CSUR)* 53 (3): 1–34.

68 Motes, J., Sandström, R., Lee, H. et al. (2020). Multi-robot task and motion planning with subtask dependencies. *IEEE Robotics and Automation Letters* 5 (2): 3338–3345.

69 Henkel, C., Abbenseth, J., and Toussaint, M. (2019). An optimal algorithm to solve the combined task allocation and path finding problem. *Proceedings of the 2019 IEEE/RSJ International Conference on Intelligent Robots and Systems (IROS)*, pp. 4140–4146.

70 Solis, I., Motes, J., Sandström, R., and Amato, N.M. (2021). Representation-optimal multi-robot motion planning using conflict-based search. *IEEE Robotics and Automation Letters* 6 (3): 4608–4615.

71 Agha, A., Otsu, K., Morrell, B. et al. (2021). NeBula: Quest for robotic autonomy in challenging environments; team costar at the DARPA subterranean challenge. *arXiv preprint arXiv:2103.11470*. https://arxiv.org/abs/2103.11470.

72 Ghosh, M., Amato, N.M., Lu, Y., and Lien, J.-M. (2013). Fast approximate convex decomposition using relative concavity. *Computer-Aided Design* 45 (2): 494–504.

73 Wu, P., Chen, S., and Metaxas, D.N. (2020). MotionNet: Joint perception and motion prediction for autonomous driving based on bird's eye view maps. *Proceedings of the IEEE/CVF Conference on Computer Vision and Pattern Recognition (CVPR)*, pp. 11385–11395.

74 Liu, X., Qi, C.R., and Guibas, L.J. (2019). FlowNet3D: Learning scene flow in 3D point clouds. *Proceedings of the IEEE/CVF Conference on Computer Vision and Pattern Recognition (CVPR)*, pp. 529–537.

75 Zeng, W., Luo, W., Suo, S. et al. (2019). End-to-end interpretable neural motion planner. *Proceedings of the IEEE/CVF Conference on Computer Vision and Pattern Recognition (CVPR)*, pp. 8660–8669.

76 Gui, L.-Y., Zhang, K., Wang, Y.-X. et al. (2018). Teaching robots to predict human motion. *Proceedings of the 2018 IEEE/RSJ International Conference on Intelligent Robots and Systems (IROS)*, pp. 562–567.

77 Chang, P., Liu, S., Chen, H., and Driggs-Campbell, K. (2020). Robot sound interpretation: combining sight and sound in learning-based control. *Proceedings of the 2020 IEEE/RSJ International Conference on Intelligent Robots and Systems (IROS)*, pp. 5580–5587.

78 Bowen, C. and Alterovitz, R. (2018). Probability-weighted temporal registration for improving robot motion planning and control learned from demonstrations. *Proceedings of the 2018 International Workshop on the Algorithmic Foundations of Robotics*, pp. 246–263.

79 Van Den Berg, J., Miller, S., Duckworth, D. et al. (2010). Superhuman performance of surgical tasks by robots using iterative learning from human-guided demonstrations. *Proceedings of the 2010 IEEE International Conference on Robotics and Automation*.

80 Kelly, M., Sidrane, C., Driggs-Campbell, K., and Kochenderfer, M.J. (2019). HG-DAgger: Interactive imitation learning with human experts. *Proceedings of the 2019 IEEE International Conference on Robotics and Automation (ICRA)*, pp. 8077–8083.

81 Wang, L., Luc, P., Recasens, A. et al. (2021). Multimodal self-supervised learning of general audio representations. *arXiv preprint arXiv:2104.12807*. https://arxiv.org/abs/2104.12807.

82 Nadon, F., Valencia, A.J., and Payeur, P. (2018). Multimodal sensing and robotic manipulation of non-rigid objects: a survey. *Robotics* 7 (4): 74.

83 van den Oord, A., Li, Y., and Vinyals, O. (2018). Representation learning with contrastive predictive coding. *arXiv preprint arXiv:1807.03748*. https://arxiv.org/abs/1807.03748.

10

AI Enabled Processing of Environmental Sounds in Commercial and Defense Environments

David Wood[1], Jae-wook Ahn[1], Seraphin Calo[1], Nancy Greco[1], Keith Grueneberg[1], Tadanobu Inoue[2], Dinesh Verma[1], and Shiqiang Wang[1]

[1] *IBM Thomas J. Watson Research Center, Yorktown Heights, New York, USA*
[2] *IBM Research, IBM Japan, Chuo-ku, Tokyo, Japan*

Abstract

Ambient sounds provide a wealth of information which can be useful in many IoT solutions in commercial and defense use cases. The application of AI based techniques to traditional acoustic signal processing can provide many interesting use cases which include a diverse set such as detecting faults in industrial manufacturing, identifying possible illnesses in chicken farms, efficient process management of naval facility equipment, and detecting possible intruders at borders. The use of AI provides an augmented capability for a data driven understanding of the environment, but also comes with several challenges. AI models need to operate in environments which may be different from the environment within which they are trained. Effective use of AI models in acoustics requires technologies that can enable these models to retrain themselves, or adapt themselves dynamically within the deployed environment. In this chapter, a system for deploying AI based acoustic models in real-world environments, and lessons learned from them is described.

10.1 Introduction

Acoustic data can provide insights concerning what is happening in a particular environment and differentiate among the varied events that may be occurring. Traditionally, the analysis of the acoustic data has focused on developing high-level physical models and using these models for identifying the environmental context. Physical modeling refers to techniques that aim to model, using simple mathematical approximations, the physical processes that give rise to the sound. More recently machine learning (ML) models have been applied for the data interpretation. These can be effective on their own but can also be combined with the physical models which give basic insights about physical systems, while the AI-based approaches can be used to identify more complex phenomena. In this work a system to train acoustic models, run them within a distributed environment and act upon the results obtained is presented.

First, the challenges in applying ML to the analysis of data from acoustic sensors are considered (Section 10.1.1). These include their data requirements, the interpretability of the models, and model drift. Next, an overview of a system for developing and applying the ML models is presented

IoT for Defense and National Security, First Edition. Edited by Robert Douglass, Keith Gremban, Ananthram Swami, and Stephan Gerali.

(Section 10.1.2). In Section 10.1.3 differences between ML models for IoT Acoustics and those for Speech Processing is discussed.

Use cases that exploit the analysis of ambient sounds arise in both military and commercial domains, and can either be exclusively based on acoustic analysis or use sounds with other modalities such as vision, power and seismic data. Defense use cases are described in Section 10.2.1, and include: perimeter defense, vehicle classification, intelligence surveillance and reconnaissance (ISR) operations, and fleet and facilities maintenance. Commercial use cases are discussed in Section 10.2.2. These include: manufacturing, vehicle monitoring, animal husbandry, healthcare, and security.

The architecture of the proposed end-to-end operational system supporting AI on the edge is then presented (Section 10.3), based on a core set of components that are identified and characterized in detail. Subsequently, the key technologies necessary for instantiating the system are discussed (Section 10.4). These include: data management and curation, the pipeline for the development of AI components, the choice of ML models, anomaly detection, and model adaptation. Finally in Section 10.5 future challenges and conclusions are presented.

10.1.1 Challenges

ML has been applied to the analysis of data from acoustic sensors in recent years. These methods have been shown to be able to provide better performance than conventional signal processing methods in many situations that have been studied [1]. There are, however, limitations of ML-based methods in that they are data-driven and thus require large amounts of data for training and testing. This leads to two big challenges: (i) the curation, management, and global dissemination of acoustic datasets; and, (ii) the efficient extraction of critical information and its comparison to data from other data sets in the same context.

To build a predictive analytics model, one begins with a known source of data. Most commonly used is parametric data along the lines of pressure, temperature, power consumption, where a correlation over time can be created to show the influence of these data parameters on the performance, and the predicted performance of the machine and equipment of interest. To a large extent these models work, as long as the data can be consistently collected and analyzed. However, when the models (actually the environments) drift or miss predicting an outage, major malfunction or catastrophe, which then drives investigation and scrutiny into the model's build, performance and reliability. When the investigation moves to identifying how the anomaly got beyond the predictive model's detection, focus is placed on the breadth of the training data set and whether there might be unrepresented environmental conditions.

Technicians, operators and mechanics are all sources of data to predict or diagnose a problem – "it didn't sound right. It was making a grinding sound, or clicking, or banging, that got louder." Sound is a rich data set, that humans have used to survive, diagnose, investigate and repair so why should it not be incorporated into work flows? There are several challenges, many of which are not unique to sound but a few that are. The primary challenge is how to convert a sound to a parameter set in such a way that correlations to other system state can be easily performed to pinpoint root causes and then to use in predictive analytics models.

For any type of predictive or preventive analytics one needs anomaly data, which is not always easy to obtain. Welding defects, for example, in the manufacture of equipment are notoriously difficult to detect as they do not always lie on the surface and currently need destructive analysis to identify. Destructive analysis is not only costly and not timely, but it also introduces error due to the sampling rate a client may choose. To create an effective model, a sufficient amount of anomaly or defect data is required to develop strong correlations, but if the destructive sample rate does not capture the level of defect occurring, the ability to detect or predict the onset of a defect is weak. Humans can often detect anomalies in welding by hearing a different sound than normal, for example, when the gas supply is not sufficient to produce an arc capable of making a solid weld. When they hear that different sound, they can take immediate action. Today robots automate much of the welding and the challenge is how to replicate the human's ability to detect an anomaly, audibly. This is where acoustics and ML can be leveraged looking at the patterns of sounds -training models to understand what is normal and what is not normal. The benefits are potentially huge and go well beyond welding use cases, having broad impact in detecting the onset of problems in all types of machines. Acoustic ML can detect a problem instantly, which may have gone undetected by any other means. This instantaneous detection prevents machine failure and product quality impact from malfunctioning machines and avoids the destructive analysis. It can also capture the deep expertise of the human who labels the training data based on their expert knowledge. That expertise is now captured and retained forever through the AI models. Acoustics as a new data source can reduce the Mean Time To Detect a problem, and build an expert knowledge base not affected by a dynamically changing workforce. To produce highly effective acoustics models, there are a few challenges to address.

There are challenges in the selection of a sensor and its positioning with respect to the source of the sound. For the choice of sensor, considerations as to where the sound is likely to originate from and how close to the origin it can be placed must be taken into account. If distance is a limitation, due to height, width or other factors a highly directional microphone would be recommended. If proximity is easily accomplished, place the microphone near the specific component to be monitored (e.g. pump or engine). An omni-directional microphone can be placed above the machine to capture a wider range of sounds from the larger environment. If the origin of the data could be best acquired when making contact to the component, like a gear shaft, then a contact microphone, similar to a stethoscope, can be mounted to a component. A contact microphone is better at detecting sounds within a component, and will have less noise interference from the environment, but can still pick up some background noise including voices. A part of the microphone decision is also how to retrieve the data from the microphone; will the microphone have compute, network and storage, in a form factor like a mobile phone, or will the microphone be plugged into an edge gateway or can the acoustic device transmit the data wirelessly like some small accelerometers are doing today.

Once the microphone device has been selected, choosing how to label the data is key, as it determines how the model will be used. Will the model be a classification model, where it identifies normal, or abnormal, and if abnormal, what is the fail label associated with it. The labeling is not only key to building the model, but to help the end user react and take action based on the inference result. If a model is created labeling a fan to be normal or abnormal, there is a wealth of information missing from the abnormal data that could be useful when action has to be taken. A label on the data that indicates the type of abnormality, for example the fan has a broken blade, can be used to train a model that can identify the reason for the abnormality. Enterprise Asset Management Systems can initiate work orders immediately to dispense a technician to do the repair. When designing the acoustic model it is recommended to build into the labeling semantics the ability to establish actions based on the inference results.

A classification model uses labeled data to learn the association of labels with patterns in the data. In dynamic environments new failure modes may occur at any time, and a classification model would not detect these because it was not trained to do so. An anomaly detecting model can be designed to expect the unexpected and as such is often more challenging to build depending on the data and environment. Very simplistically, an anomaly model is trained to a normal baseline. Anything outside the normal signature is deemed an outlier and can serve many purposes. First it alerts the operations team that an unknown event has occurred and needs investigation. This can be accomplished in many ways through integration into the information and operational techology systems to issue a work order or send a text message. A person can receive the message and be directed to the location of the captured outlier. With further investigation by the human, the sound could be labeled appropriately and the root cause identified. If the sound turned out to represent normal operation, the anomaly model could be retrained to now recognize this sound as normal. The techniques to produce a classification model and anomaly model vary and are described in detail in subsequent sections.

With the data collected and labeled, the next steps are to curate the data to insure it has the right quality and integrity to build the model. To facilitate that curation, the sound clip is broken up into segments, and those segments can be labeled, or the entire sound clip can be labeled. Depending on the desired outcome of the model, the size of the segment, in seconds, can be optimized to insure the capture of the sound of interest, as well as the performance of the model. A pump making a constant hum may have a different optimization applied than a short squeak occurring inside of a car, and the manner in which the model is created needs to reflect that. In many environments, a common background noise is that of humans speaking and the model needs to be trained to ignore that in most scenarios, and there are techniques and methods to do that.

In most instances one of the biggest challenges in getting models started, is getting enough data for the abnormal or failing conditions. Even for normal conditions this can be a problem. If there is not enough data to achieve the desired performance of the model, data augmentation techniques can be employed to create data from the small amount that one has. Acoustic data is amenable to various data augmentation techniques including frequency and time shifting, addition of expected noise, or frequency filtering to name a few. Such techniques can be applied to the training data to broaden the data sets to avoid over-fitting and/or better adapt the model to the anticipated operational setting.

Once a model is created, the ability to continuously adapt it with new sounds to improve the robustness of the models, is critical. The human always needs to be in the loop to perform and/or validate labeling and to validate the models produced, but it is valuable to minimize the time required to perform these tasks and limit mistakes. Cluster analysis can aid in more efficient labeling and validation techniques like a confusion matrix are tools to help understand where and how to improve the model. Building in feedback loops to automatically save samples of the inference results, by label, for a human to validate can help insure the model is meeting its performance targets or determine the source of drift.

Data distributions almost always shift between model training and deployment. Models that have been trained for use in a particular environment may not perform well in others. Alternatively, while the model may seem to work well when originally deployed, its performance may degrade with time due to environmental drift (e.g. engine wear). Such a degradation may be steady over time, periodic or recurring as with seasonal data. Identifying model drift requires the ongoing monitoring of the accuracy and confidence of the model. The challenge is in identifying systematic changes in data against a background of natural fluctuations and trends.

ML models are also less interpretable than conventional methods for the analysis of acoustic data. Interpretability is useful in that it can aid in trust, and bolster the safety and contestability of model results.

In summary, acoustics is coming to the forefront as a powerful new source of data to help predict and monitor the health of machines and equipment. AI models can be shown to achieve high performance results when they overcome the challenges which accompany their use in acoustics environments.

10.1.2 System Overview

A minimal environment to train acoustic models, run them at the edge and act on their results consists of three primary components, as shown in Figure 10.1.

At the core is a cloud-based infrastructure that provides a number of key capabilities. The first is data storage in which the acoustic data and its metadata (i.e. labeling, timestamps, etc.) are made available to data curation and training facilities. A data curation user-interface provides the ability to refine and adjust the labeling for the acoustic samples.

Today, acoustic models are generally trained in the cloud where powerful GPUs (graphical processor units) are readily available. Model tuning is the process identifying optimal parameters (i.e. frequency range, numbers of features, etc.) to achieve the best accuracy. This is typically an iterative process facilitated by having sufficient compute resources.

Finally, the core cloud service also provides the ability to manage the models and their deployment to the edge devices. Initial assignment of model to devices can be defined, updated models can be made available to the edge devices when available, or a schedule of models can be defined to match operational schedules (e.g. different welds in a manufacturing plant). The edge device and cloud services coordinate to deliver the model as required.

In addition to model management, the edge device provides the ability to capture acoustic data samples. The samples are typically one to ten second clips with which a set of unambiguous labels can be associated. The edge device may allow the capture of training data. If so, it is useful to apply the labeling (i.e. whale, click, etc) during training data capture.

Once the acoustic model is trained and deployed to the edge device, the captured audio clips are provided to the model to identify/classify the sound. The classification results produced by the model for a given sound are made available to an actuator that decides how best to respond to the classification result. The actuator may be running at the edge, in the cloud or both. For example, the actuator at the edge may use classification results to control a local engine operation based on anomaly detection and the cloud may maintain historical edge state.

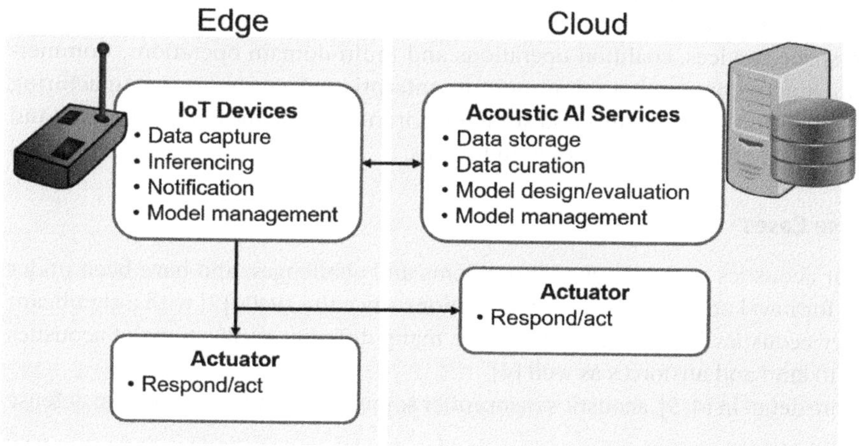

Figure 10.1 System overview – a system for edge data inferencing and actuation with cloud-based training and data management.

Each of these components and their use including model training and management are discussed in subsequent sections.

10.1.3 IoT Acoustics vs. Speech Recognition

Automatic speech recognition (ASR) is a well-developed and capable technology, but differs in a number of ways from that which is needed and/or available for IoT acoustics. First and foremost is that ASR models are used by millions of end-users and as such, large amounts of training time (thousands of GPU hours) can be justified in creating them. In addition, ASR includes both an acoustic model and a natural language understanding (NLU) model. The former identifies discrete word possibilities and the latter selects among multiple options to construct sensible phrases. The NLU model helps to improve accuracy above that which would be available by the acoustic model alone. The use cases in acoustic IoT scenarios are narrower in scope so that large training times are often not justifiable. In addition, use cases most often identify discrete sounds and so do not include anything like an NLU model - in particluar, the sequence and timing of discrete event sounds is usually irregular and follows no set sequence. In addition, ASR models are designed to suppress background sounds, while in IoT use cases these are the sounds of interest. Also unlike ASR, IoT acoustics often requires the identification of rare or unknown sounds.

Finally, another important contrast with much of today's ASR use is that for a typical smart phone, the recognition is primarily performed off device in the cloud. For most IoT use cases, this is not feasible because (i) the IoT device may lack any connectivity to the cloud and (ii) the inference frequency across devices and time is too high to defer to a central location. This generally means that the acoustic models need to be capable of running on the small form-factor edge devices where memory and CPU capacity may be limited.

10.2 Use Cases

There are many use cases that can benefit from analysis of ambient sounds in both defense environments and commercial environments. Some of the use cases are based simply on using the single modality of sound, while others are based on using sounds along with other modalities such as vision or LiDAR (Light Detection and Ranging).

Defense environments include scenarios involving the armed forces of a nation including its land, naval, air and space services, coalition operations and multi-domain operations. Commercial environments include scenarios that are present in enterprise office spaces, manufacturing plants, commercial agriculture, data centers and similar environments. Some of the scenarios and use-cases may be common across the two types of industries.

10.2.1 Defense Use Cases

Defense use cases for acoustics span a variety of problems and challenges, and have been under active consideration for naval and other defense applications since the 1940s [2] with a significant focus on under-water acoustics [3]. However, there are many different applications of acoustics which are pertinent to land and air-forces as well [4].

As discussed in more detail in [4, 5], acoustic sensors offer some significant advantages in defense use cases because:

- acoustic sensors are passive, resulting in sensing that can be clandestine and low power
- sound propagation is not limited by line of sight, and sounds waves can bend around obstacles

- sensors are light-weight, low cost, compact and robust
- military activities are inherently noisy
- different types of military vehicles and equipment often have characteristic identifying sound patterns
- sensors are relatively low-power, resulting in longer battery life

10.2.1.1 Perimeter Defense

The defense of the perimeter of a secure zone or boundary is a basic requirement in many army scenarios. For the physical security of military installations [6], measures such as field-works, facility construction, detection and procedural elements are required. Different type of sensors, including acoustic sensors, play an important role in the detection measures associated with perimeter security. Sounds can identify many intrusions of the physical security, identify possible events of interest and complement other modalities of sensing including visual and seismic sensors.

Events related to perimeter defense which can be detected using acoustics include explosions, gunshots, or footfall from intruders. Other types of anomalous sounds also provide events of interest that ought to be investigated, e.g. tapping against the perimeter fieldwork may be an event worth investigating. Sounds are usually used to complement other modalities as well, e.g. visual to better assist in determining events of interest.

Sounds can also be used to assess the robustness of the fieldwork used for perimeter defense. A moving robot which can tap a robotic hammer lightly against the wall can be used to measure the depth, texture and strength of the wall. When coupled with such robots, acoustics analysis can provide mechanisms to check for robustness of perimeter defense infrastructures in a consistent and regular basis. Manpower is limited in almost all defense facilities, and offloading such periodic validation to a machine results in improved efficiency and lower error rate.

10.2.1.2 Vehicle Classification

Military vehicles are large and noisy, and this noise pattern allows the detection of them from a distance using acoustic and seismic sensors. While seismic sensors are limited to ground vehicles, acoustics can be used for land vehicles as well as for identification of airborne vehicles. This is helpful in identifying approaching enemy vehicles and taking appropriate measures against any threats the vehicles may cause. The ability to track the direction of movement of enemy vehicles can be a very good aid to situational awareness in the battlefield.

Different types of aeroplanes also have a different acoustic signature. ML techniques can help identify such airborne vehicles. Because of the light-weight nature of the acoustic sensors, they can be mounted on drones to extend the range of acoustic sensing.

ML based approaches for vehicle classification have been shown to be effective via a variety of studies in both defense and commercial scenarios [7–9]. Since ML based techniques for classification can be used for classifying other types of assets, acoustics-only systems can also be used to detect, track and classify other types of entities, such as personnel, drones, planes, and specific types of events [10].

10.2.1.3 Activation of Other Modalities

Many modalities of sensing are used in Intelligence Surveillance and Reconnaissance (ISR) operations in defense, and each of these modalities has their own strengths and weaknesses. These modalities include seismic sensors, passive infrared, magnetic sensors, electrostatic or electric field sensors, chemical sensors, and visual sensors [11]. Visual modality is great for surveillance and observation, but visual sensors tend to use a significant amount of bandwidth and processing of the image or video signals cause a substantial drain on the battery available within the sensors. These modalities are also susceptible to line of sight issues, and the sensing field can be blocked by means of other obstacles.

Acoustic sensors use low-power, have no line of sight limitations, have 360° coverage, and hence provide a good mechanism as an always-on trigger for other more power-hungry modalities of ISR sensors. The acoustic sensors are used to watch for conditions when the other sensing modalities ought to be turned on. Such activation results in a more power-efficient solution for comprehensive suite of ISR applications.

It is worth noting that the support of other modalities using acoustics also works in the other direction. Other modalities can be used to supplement a purely acoustics-only approach. Vibration sensors can activate the recording of sounds for perimeter defense. Similarly, when privacy concerns are important, visual detection of people present in a machine room may be used to turn off or turn on the recording of sounds. Several other use-cases deploying multiple modalities can also be envisioned in both military and commercial use cases.

10.2.1.4 Fleet and Facilities Maintenance

Frequently, defense operations involve a substantial number of ground vehicles, airborne assets and/or maritime ships. In order for an effective operation of the fleet of vehicles, acoustics can play an important role. The sounds of vehicles can indicate any problems or anomalies in their operation. Abnormal sounds from vehicles may indicate a need for maintenance. In naval ships, sounds from the engine rooms can be analyzed to detect any issues in the operation of the ship. Similarly, abnormal sounds like drips, or water drops can be used to monitor the conditions in the hold of a ship. In ground vehicles, as they return to a base camp, sounds of vehicles driving by can be used to detect possible problems.

Similarly, sounds can be used to monitor the ambient environment in rooms housing equipment for facility maintenance. Abnormal sounds such as squeaks, knocking or grating of gears can indicate problems in the building equipment. The sound of running engines can be tracked to monitor how long and under what conditions (e.g. idle, max load, shutdown) a machine has been operational and to decide when to schedule a maintenance inspection request.

10.2.2 Commercial Use Cases

There are a wide range of domains to which acoustic modeling can be applied in commercial settings. Any setting in which sounds are produced might be a candidate, but use cases in which acoustics can be used to automate processes for additional efficiency or safety are of particular interest. Some of those key domains are discussed in the following sections.

10.2.2.1 Manufacturing

In manufacturing, using acoustics to detect the onset of a malfunction, like the low gas scenario for welding, is a proven technique. Extensions into numerous other machine scenarios including cavitating chillers, pumps not being able to handle a shift in a material's viscosity, fans starting to malfunction or gear shafts grinding are just a few examples where acoustics is playing a strong role in detection as there is often no other means to detect these malfunctions.

10.2.2.2 Vehicle Monitoring

In aerospace, acoustics is emerging as an opportunity to monitor the operations of the aircraft. Car manufacturers are also beginning to equip cars with a multitude of microphones to pinpoint the source of a squeak, or when a part has begun to wear or an abnormal sound requires a mechanic to investigate. Trains are installing drive-throughs equipped with acoustic sensors used to detect wear and tear. Any vehicle in operation, once equipped, could be monitored for the onset of malfunction.

10.2.2.3 Animal Husbandry

Animal husbandry is exploring the use of acoustics to detect the early onset of a disease in large enclosures where diseases could spread quickly with devastating impact. Microphones positioned in the ceilings can detect coughs in pigs, cows, chickens and turkeys and alert owners quickly to investigate or isolate that region of the pen and the animals affected. Sound can also alert danger to animals being mistreated or in danger from a predator. Acoustics has deep history in application to the identification of whales, which can be identified by their unique song. This can be useful to see if they are in harm's way or simply to identify their migratory paths.

10.2.2.4 Healthcare

Certain medical conditions like a heart murmur can take months and even years to develop the expertise of detecting the problem. Acoustics is being explored in these areas to help capture the expertise of a diagnostician and share it through an acoustic model.

10.2.2.5 Security

Perimeter defense as described in Section 10.2.1.1 is also useful in commercial environments. For example, acoustics is being used for intrusion detection to augment visual surveillance, or as a standalone to detect certain sounds of an intrusion into a building, asset, or secure area. In recent years, unmanned aerial vehicles (UAVs) have become increasingly accessible to the public, but UAVs can be utilized for malicious activities in order to exploit vulnerabilities by spying on private properties, critical areas or to carry dangerous objects. Al-Emadi et al. [12] proposed audio-based drone detection and identification using deep learning to address this physical security concern.

10.3 System Architecture

To realize an end-to-end operational system supporting AI on the edge a core set of components are identified and shown in Figure 10.2.

At the center of the cloud-based components is a core set of Acoustic AI Services providing data and model management services including data curation, model training, model evaluation and model deployment to edge devices. These AI services rely on a flexible audio storage system supporting attached metadata, especially labeling. A graphical user interface workbench provides facilities to utilize the core AI services. Rich facilities for examining audio data both visually and acoustically, applying and modifying labels, and training, evaluating and deploying models are included.

The IoT/Edge device provides access to the physical environment via an audio or accelerometer sensor. The device can be used as part of model development to capture training data. Once a model is trained the device is used during operational environment monitoring to produce classification results. The operational AI model results are published to an edge- or cloud-based message broker allowing a flexible framework to attach actuators to take action based on the edge AI results - for example, raise an alarm when breaking glass is detected.

The above described system is used for two fundamental capabilities. The first is model creation in which data is gathered, curated (i.e. labeled) and models are created, evaluate and tuned using

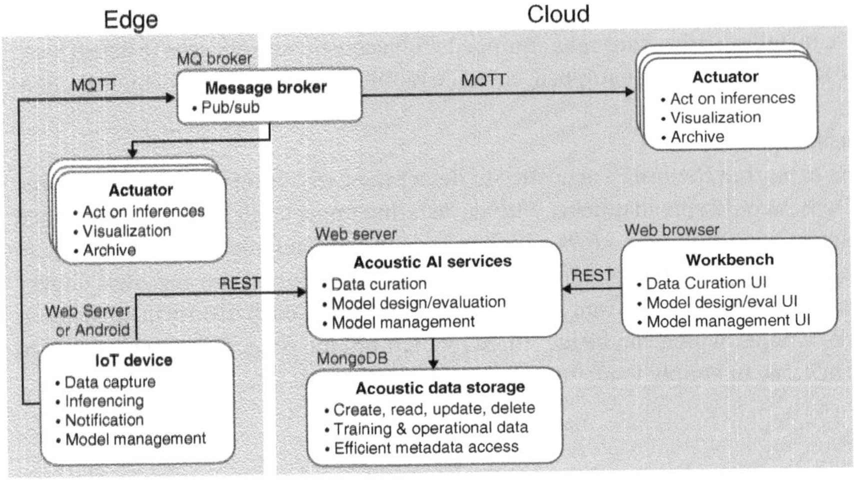

Figure 10.2 System architecture – edge- and cloud-based services to enable acoustic model inferencing at the edge, data capture, and inference response.

the labeled data. The second capability includes deploying the model, capturing sounds and applying them to the model. The model creation capabilities can be outlined in a workflow as shown in Figure 10.3. These steps are largely independent of the data type on which the model is being trained.

- **Data Capture**: Capture include the acquisition of audio (or vibration) data from a sensor and storing it in a form for later use in curation and model training. The sensor might, for example, be an accelerometer, underwater hydrophone, or a standalone or embedded microphone. Data captured for training purposes may have labels associated at recording time or they may be applied later in the curation step. Data captured for monitoring is unlabeled and is provided to a trained model for scoring. Scoring results may be sent to either or both of a storage system or actuator.
- **Data Curation**: This is the process of identify and labeling acoustic features that the model will be trained on. For audio or any data with a high sampling rate, a spectrogram is typically used to show the evolution of the power spectrum through time. This can help visually identify acoustic patterns, for example, a bad bearing at specific pitch or the up swing of a whale call.
- **Model Training**: Is a compute and data intensive process to learn the relationships between labels and the associated acoustic features. There are many types of models that can be

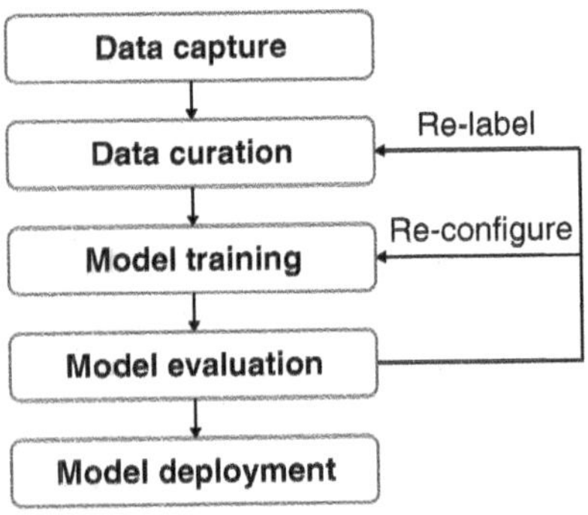

Figure 10.3 Model development – an iterative process to create a robust model.

considered, but two broad sets of models are deep and shallow. The shallow models include nearest-neighbor, Gaussian mixture and decision trees. Deep models are some form of multi-layer convolutional neural network. Both are discussed in Section 10.4.3.
- **Model Evaluation**: Involves running the model on a held out (test) set of labeled data to determine its accuracy on the test data. Evaluation can identify weaknesses in the model, for example, a lack of training data for a bad bearing may result in poor accuracy for that label value.
- **Model/Data Tuning**: This is an iterative process in which model parameters (e.g. frequency ranges, feature type, etc.) may be modified based on model evaluation results.

The above capabilities are generally delivered through a set of command line tools or a graphical user interface (i.e. workbench).

The second set of operation supports the management and use of trained models in an operational setting and consists of the following:

- **Device/Model Management**: Involves the deployment of trained models across a distributed IoT environment. After training a model, it is assigned to one or more edge devices and deployed for use there. The edge device coordinates with the cloud-based model management system to instantiate new models on the edge device. Other configuration or policy, such as scheduling or start/stop monitoring commands, may also be managed from the cloud to provide controls over the edge device operation.
- **Inferencing**: Includes the capture of environmental sounds on which the deployed model is applied to produce inference/classification results. Inference results are optionally communicated to the Message Broker for distribution to registered Actuators.
- **Model Monitoring**: Involves the tracking of model performance during its use on the edge. This can be done using reverse testing, described below, to track relative accuracy over time. This is important to track drift in the environment relative to the original data on which the model was trained.

Each of these two primary capabilities, model training and operational use of the model are detailed in Section 10.4.

10.4 Technology

The core challenges lie in curating acoustic data, training models and monitoring and adapting them once they have been deployed. Approaches to solving these challenges are discussed below.

10.4.1 Data Management and Curation

Data Curation is the process of managing, organizing, and labeling data needed to train effective ML models. For acoustic and other Internet of Things (IoT) datasets, data curation may require a great deal of effort due to the large amount of data generated. As an example, collecting samples of acoustic data needed to train models usually requires hundreds or thousands of sound clips that need to be managed and labeled. This section will discuss advanced features of a workbench, which is a graphical user interface providing basic functionalities for browsing, filtering, reviewing and bulk labeling of the data. Advanced features that optimize the labeling process, including visualization, automatic labeling, and other techniques to manage the proliferation of acoustic and IoT data used by a data scientist or ordinary user are desired.

Splitting Sounds: The volume of sound clips collected from devices needed to train acoustic models can be overwhelming to label at the individual clip level. To help with this burden, bulk labeling should be supported in various ways. Splitting sounds into smaller segments assures that the labels are atomic, meaning the label applies to a majority of the sound and not just a small portion of the clip. Splitting long clips can be done on import in a couple of ways when using a data management workbench. First, when importing a sound, a long sound clip can be split into smaller clips of a fixed length. Then a label can be added to all of the clips uniformly, or played back and selected clips can be labeled individually. Another method is to visualize the sound as a spectogram, playback the sound wave, and select the segments to label and segments to discard.

Recording Session: In some scenarios, sound is recorded for an extended period, but cannot be labeled online in real-time. Another method of bulk labeling is a "recording session", which indicates a temporal grouping of the collected data resulting from an "episode" of data capture. It turns out, from our experience in the real-world experiments, that such a temporal grouping typically corresponds to the semantic grouping of the collected data in IoT scenarios. It thus provides the user not only a simple reminder of when and how the data in a session is recorded, but also, more critically, a convenient and intuitive way to assign the labels across different but strongly related pieces of data within the same session. A tool can be provided in a workbench to label a recording session during playback interactively. Sound clips can be discarded or labeled and saved as the user is listening to the clip.

Assisted Labeling: Given a partially labeled dataset, it is possible to recommend labels using a semi-automatic feature of a workbench. Assisted labeling is the process of training a classifier using a set of partially labeled data, and suggesting the labels for the remaining unlabeled sound recordings selected by a user. The guided labeling algorithm uses a shallow model that is dynamically trained using a selected set of labeled sounds. The suggested labels are shown and can be automatically applied to the sounds, or edited by the user. The process clusters the sounds by similarity, so that the same labels can be applied to like sounds. The interactive assisted labeling is iterative and can be continued until the user finishes or all of the sounds are labeled (Figure 10.4).

Data Visualization: Data visualization helps identify overall properties of the larger dataset, e.g. intuitively and immediately identifying the uneven distribution of data points or even mislabeled items. An unbalanced training dataset may cause inaccurate results in some classes which aren't represented equally as the other classes. If a dataset is associated with an ontology of label categories, the relationship between the dataset and ontology can be explored visually in a hierarchical manner. Using this method, gaps in the labeled data can be easily found and subsets of the dataset

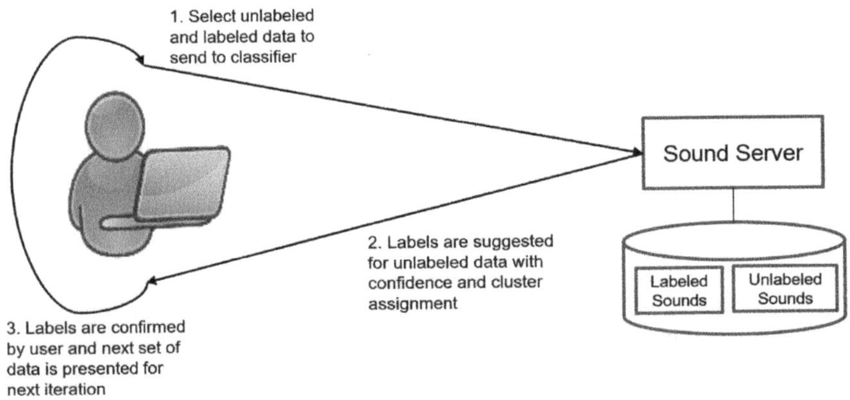

Figure 10.4 Assisted labeling – using existing labeled data to guide labeling of unlabeled data.

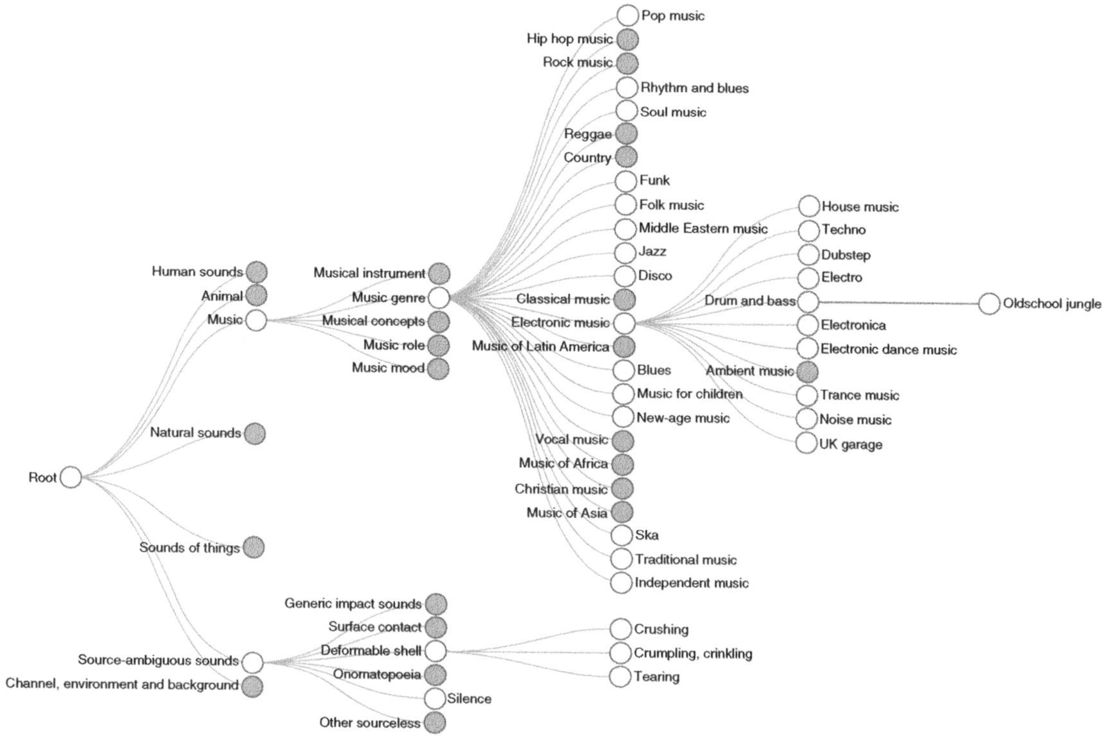

Figure 10.5 Data visualization techniques – ontology browsing.

can be selected for training by selecting a branch of the ontology. As previously mentioned, spectograms of a sound wave can be used to select and label sound clips visually. Another visualization technique is to cluster the label categories as bubble diagrams to easily detect unbalanced regions of the data set (Figure 10.5).

The various Data Curation techniques can help reduce the labeling time and cost. Additionally, a more correctly labeled and balanced training set will produce a more accurate model.

10.4.2 Model Training Pipeline

The main building block of the AI pipeline includes ML techniques to classify sounds in the environment. The classification result is then provided to an automated system or human via an appropriate mechanism so that the receiver can take appropriate actions. For classification, the system generally includes the following steps.

Feature Extraction: Raw acoustic signals are usually represented and stored as waveforms. While these waveforms capture complete information about the signal, it is difficult to train models using them directly since they are highly redundant. It is therefore necessary to extract useful features from the signal. In the acoustics domain, mel-frequency cepstral coefficients (MFCC) and log-Mel are often used as feature extractors, which transform the signal into the frequency domain together with some filtering and additional transformations [13, 14]. In particular, MFCC is obtained by several steps of signal transformation, including taking the Fourier transform of the windowed signal in the time domain, mapping both the frequencies and powers of the resulting spectrum into the log scale, and then applying the discrete cosine transform (DCT).

Log-Mel is similar to MFCC but without the last DCT step. MFCC is slightly more compressed and is often better suited to shallow models, whereas log-Mel maintains higher fidelity and is therefore often used with deep models. The benefits of MFCC and log-Mel are that they resemble the sensitivity of the human ear across different frequencies and energy levels of the signal. It also compresses the waveform into a much smaller vector that significantly reduces the complexity for processing and storage. Depending on characteristics of the sound, MFCC can be replaced with other frequency-based features, such as Fourier coefficients, obtained using fast Fourier transform (FFT), or mel-frequency filter banks (MFFB). Both FFT and MFFB are intermediate steps in computing MFCC.

To capture the temporal variation of the signal, instead of extracting the features from the entire signal directly, the signal is segmented into multiple subwindows, where the feature vector is computed on each subwindow. An example of the full feature extraction and processing pipeline is shown in Figure 10.6.

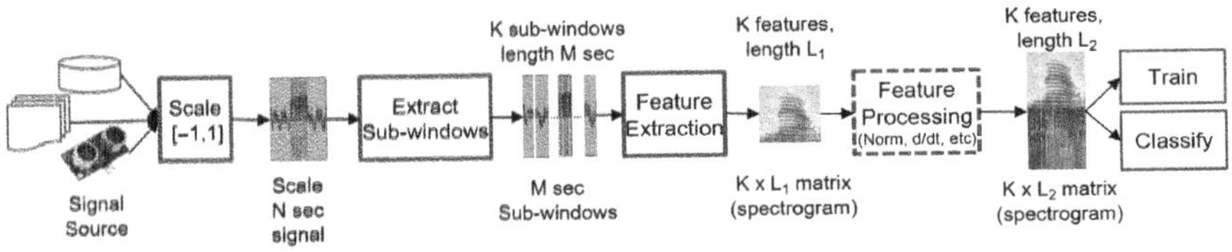

Figure 10.6 Feature extraction and processing pipeline – processing of audio data to produce spectrograms that are used to train a model and inference over. Source: Parallax Inc.

Using this pipeline, a spectrogram of the signal is created that includes information on the variation of frequencies over time, as shown in Figure 10.7.

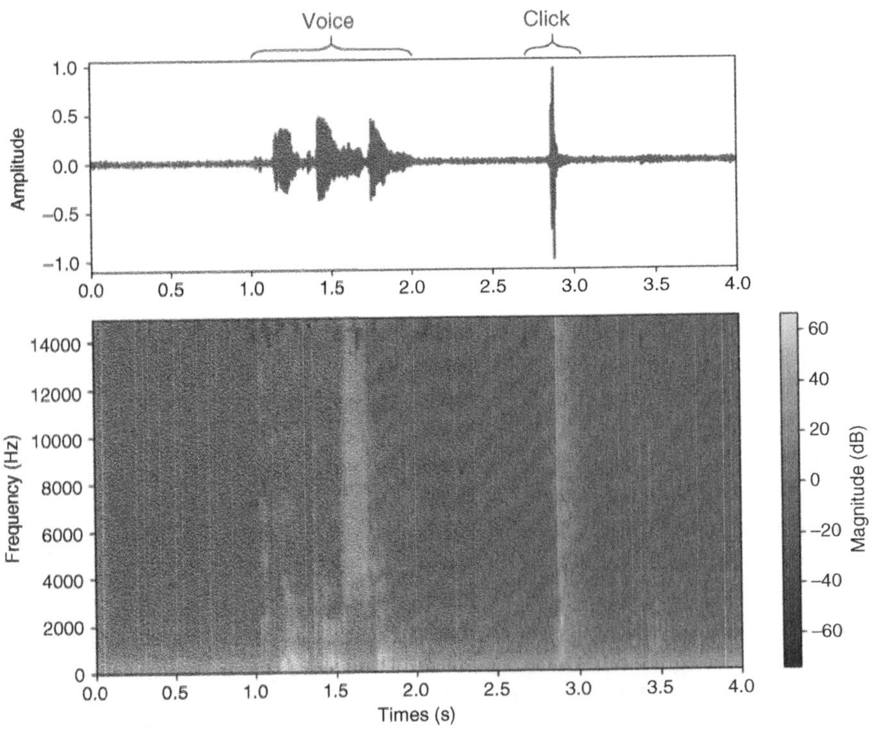

Figure 10.7 Example waveform, its spectrogram and labels. Source: Parallax Inc.

To ensure a smooth transition, two adjacent subwindows can partially overlap in time. To avoid frequency leakage, a filtering window (e.g. Hamming window) is applied on each subwindow. In addition to the original features, delta features that capture the "velocity" of change between neighboring subwindows may also be computed. For coefficients $\{c_t\}$ where t is the subwindow index, the delta feature d_t is computed as

$$d_t = \frac{\sum_{n=1}^{N} n\left(c_{t+n} - c_{t-n}\right)}{2\sum_{n=1}^{N} n^2}, \tag{10.1}$$

for a neighborhood of length N. The "acceleration" (i.e. second-order delta) can also be obtained by computing Eq. (10.1) again on the delta features.

Detection and Classification Model: The system trains a model using the features of input data and the target predicted class (label), in the case of supervised learning. For unsupervised learning, the goal can be to group the sound data into clusters in which each cluster has similar data samples, for instance. Unsupervised learning can be useful for detecting unknown sounds that are outside the distribution of the training dataset. In many practical use cases, supervised and unsupervised models can be combined, wherein the unsupervised model first determines whether the input sound is something that the system knows about, and if yes, the supervised model gives the predicted class label of the sound. Details of model design are discuss in Section 10.4.3.

Post-processing of Results: The results produced by the model may need to be post-processed. For example, it maybe useful to normalize the model outputs in order to compute a unified score for comparing the confidence of different predictions. To make a more accurate prediction than using a single model, results from multiple models may be combined using ensemble techniques.

10.4.3 Models

The choice of models depends on the characteristics of the task and also the amount and complexity of training data. As a rule of thumb, shallow models may be beneficial in cases with a small amount of training data (as little as a few minutes) or when the efficiency of training and inference is important. Deep models may be beneficial where the training data has a complex underlying structure that cannot be captured by shallow models, and there is a significant amount of data (many minutes to hours) available to train deep models without overfitting.

Figure 10.8 illustrates basic recipe for ML [15].

First, a model is trained and checked for accuracy on the training data. If the performance on training data is poor, it means that the model is "underfitting" and a more complex (i.e. deeper) model is needed. If needed, a deeper model is trained and checked for accuracy on training data again. If the accuracy on training data is good, then the accuracy of the model on held out set of test data, which is not used for training, is identified. If the accuracy on test data is not good, it

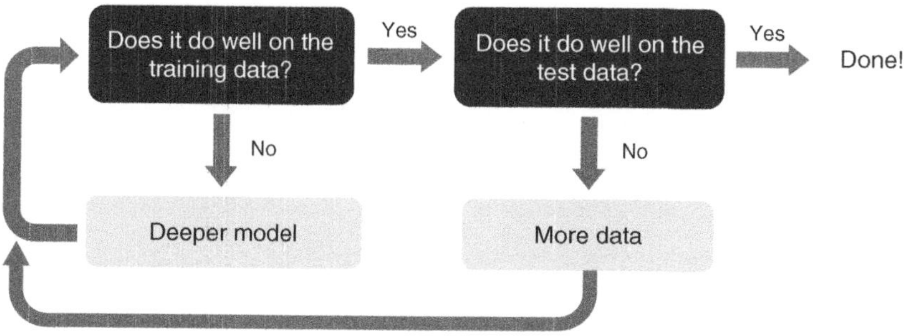

Figure 10.8 Illustration of basic recipe for machine learning.

means the model is "overfitting" and more training data is needed. This process is repeated until the model produces good accuracy on both training and test data.

10.4.3.1 Shallow Models

Classic shallow models such as clustering, nearest neighbor, and support vector machine (SVM) may be applied in the feature space for classification purpose. For acoustic data, Gaussian mixture models (GMMs) have been proven to give good performance in many practical scenarios. In essence, GMMs learn a separate Gaussian mixture distribution for each sound class, as shown in Figure 10.9. For classification, the predicted class is identified by finding the label of the mixture distribution in which the input data has the maximum likelihood.

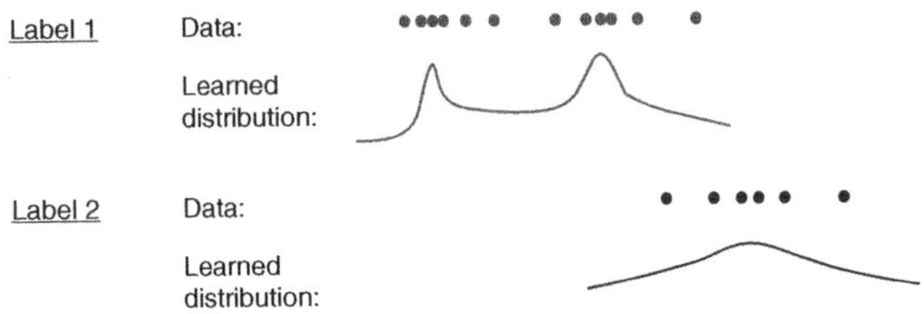

Figure 10.9 Illustration of Gaussian mixture models (GMMs).

Some of these models can be also used for detecting known vs. unknown sounds. For example, the distance to known feature points can be used to determine whether a data sample is known or unknown. Similarly, the likelihood of Gaussian mixture distributions can be used to make this determination. Other techniques for outlier/novelty detection, such as one-class SVM, can also be used.

10.4.3.2 Deep Models

In addition to shallow models, deep models, especially convolutional neural networks (CNNs), are used for acoustic ML. CNN is a structure inspired by biological vision processes and can optimize the filters through automated learning. Figure 10.10 shows a typical CNN model [16]. CNN has two parts, feature extractor and classifier. The first feature extractor consists of repeated convolutional layers and pooling layers to enable feature extraction from 2D data. The later classifier consists of fully connected layers and provides classification from the extracted features.

CNNs have been broadly used for image recognition, but they can also be applied in acoustic scene classification [17–21]. However, widely available pretrained CNN models for images applied to acoustic classification have not performed as well as acoustics-specific CNN models as proposed by Inoue et al. [22, 23].

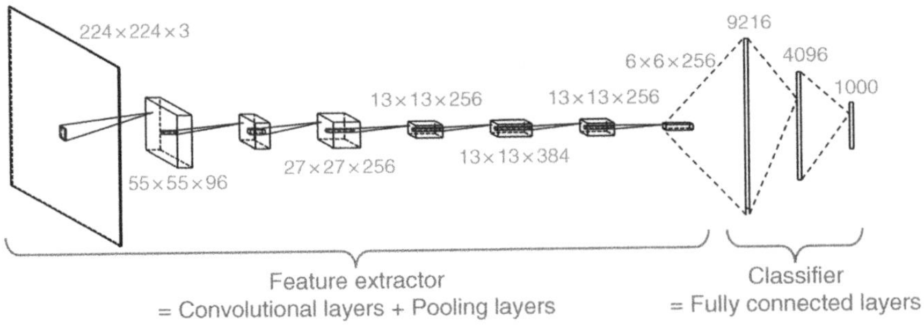

Figure 10.10 Illustration of typical CNN model.

It is well-known that deep learning requires a large amount of data to train an accurate model. To increase the amount of training data and reduce overfitting, numerous data augmentation methods have been studied in the acoustic literature, ranging from simple noise injection to more sophisticated approaches [18]. Some musically inspired deformations such as pitch shifting and time stretching are adopted to augment training sound data [18, 24]. Jaitly and Hinton [25] showed that the data augmentation based on vocal tract length perturbation is effective to improve the performance of ASR. Takahashi et al. [26] mixed two sound sources within the same class to generate a new sound. Tokozume et al. [27] proposed a method to mix two sound sources from different classes. Both labels and sounds are mixed and referred to as between-class data. Room simulation can simulate recordings of arbitrary microphone arrays within an echoic room. It supports research related to developing and experimenting with multichannel microphone arrays and higher order ambisonic playback. It models both specular and diffuse reflections in a shoebox type environment [28].

They train the model solely using the generated data without using the original data. Zhang et al. [29] proposed a similar approach to use between-class data, but they also use mixing of sounds from the same class in the training. Inoue et al. [22, 23] proposed a method to shuffle and mix two sound sources from the same classes to increase the variation in the training samples on both the temporal sequence and event density of the sound events.

These data augmentations can increase the amount of training data artificially, which can mitigate "overfitting" and improve the performance on test data, that is not included in training data.

10.4.3.3 Inference Performance on the Edge

Many of the IoT use cases require execution of the model directly on the IoT device to provide local inferencing services. As such it is important to understand the feasibility of running acoustic models on edge-class IoT devices for which there may be limited memory and computational resources. To characterize the feasibility of running models locally, inferencing performance is captured for two classes of Raspberry Pi devices, which can serve to represent an edge platform. Only inferencing performance is considered as training is generally done in the cloud. As such, GPUs are generally not required on the edge for the models considered here.

Figure 10.11 shows the execution time of inference on Raspberry Pi 3B (quad core 1.2GHz 64-bit ARM CPU and 1GB RAM) and Raspberry Pi 4B (quad core 1.5GHz 64-bit ARM CPU and 4GB

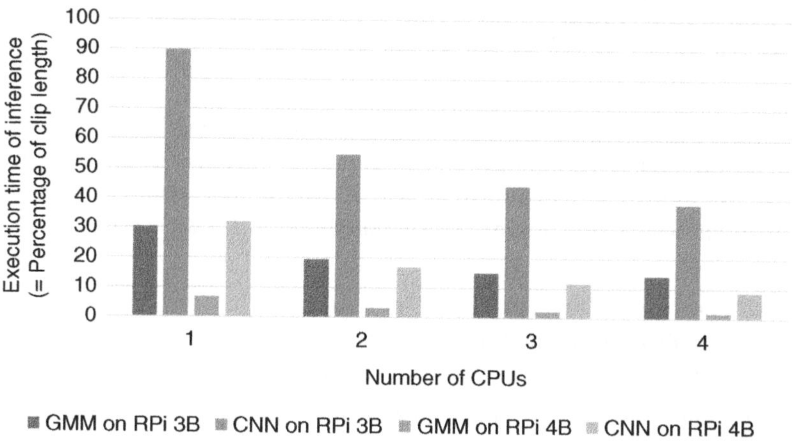

Figure 10.11 Execution time of inference on RaspberryPi 3B and 4B for 1 second audio.

RAM) [30] for 1 second audio clips. Inferencing was run in a container to allow control over the number of available CPUs.

Both the shallow (GMM) and deep (CNN) models were run with different numbers of CPUs as shown in the horizontal axis. The vertical axis shows the execution time of inference relative to the length of the sound clip being classified. If less than 100% then the inferencer can maintain real-time continuous performance. As shown in Figure 10.11, both models are able to maintain real-time performance at all CPU levels. Performance improves with more CPUs, which is primarily due to the fact that feature extraction is a highly-parallelizable operation. More important is the fact that the Raspberry Pi, as an exemplar IoT platform, demonstrates the feasibility of running acoustic models on small IoT platforms.

10.4.4 Anomaly Detection

Anomaly detection is the task of finding unusual samples in a set of data. The "unusual samples" may indicate abnormal states and can be identified by comparing with a set of known normal states. In the setting known as "unsupervised" anomaly detection, the training data consists of only "normal" data; namely, anomalous samples are not known a priori. Algorithms for anomaly detection can be used in many applications such as quality inspection of products, maintenance of equipment, detection of network intrusions, and detection of fraud.

Anomaly detection is an important topic in ML. Many approaches to solve this task have been proposed [31, 32]. Deep learning is used in this work due to the type (i.e. sound) and amount of data. Approaches using deep learning for unsupervised anomaly detection can be broadly categorized as reconstruction-based methods and feature-learning-based methods. As for reconstruction-based methods, the model is trained to learn the distribution of normal samples, and anomalies can be detected by analyzing reconstruction errors as an anomaly score [33–35]. The reconstruction error is usually higher for anomalies as the model is only trained to reconstruct the normal samples. As for feature-learning-based methods, a feature-extraction model is trained to map normal data into a small region in the feature space. Anomalies can be detected by analyzing the distance from normal samples in the feature space [36, 37]. As a variation of feature-learning-based methods, classifier confidence, specifically a maximum softmax probability (MSP), can be utilized. As for this method, an anomalous sample is considered to be outside of the distributions that the classifier learned, and it generally has lower MSP [38]. Inoue et al. [39] proposed a method to transform normal sounds to generated pseudo classes. The generated pseudo classes are then used to learn a classifier that predicts which transformation was applied for each sound sample from normal sounds. In the inference phase, anomaly scores are calculated using the confidence values produced by the trained model for given sounds.

In some cases of real-world applications, one can use a small amount of anomalous sounds in addition to normal sounds for training. Even in this problem setting, anomaly detector can show advantages over binary classifier as shown in Figure 10.12 [40]. As an example, consider a set of training and test data containing normal and anomalous sounds. (Figure 10.12a). Three models can be trained:

- binary classifier trained with normal and anomaly data (Figure 10.12b),
- anomaly detector trained with only normal data (Figure 10.12c), and
- anomaly detector trained with normal and anomaly data (Figure 10.12d).

Since the binary classifier (Figure 10.12b) is trained to find the boundary between normal and anomaly data during training, unseen anomaly data during training (in the lower right) can be mis-classified as "normal". Since the anomaly detector trained with only normal data

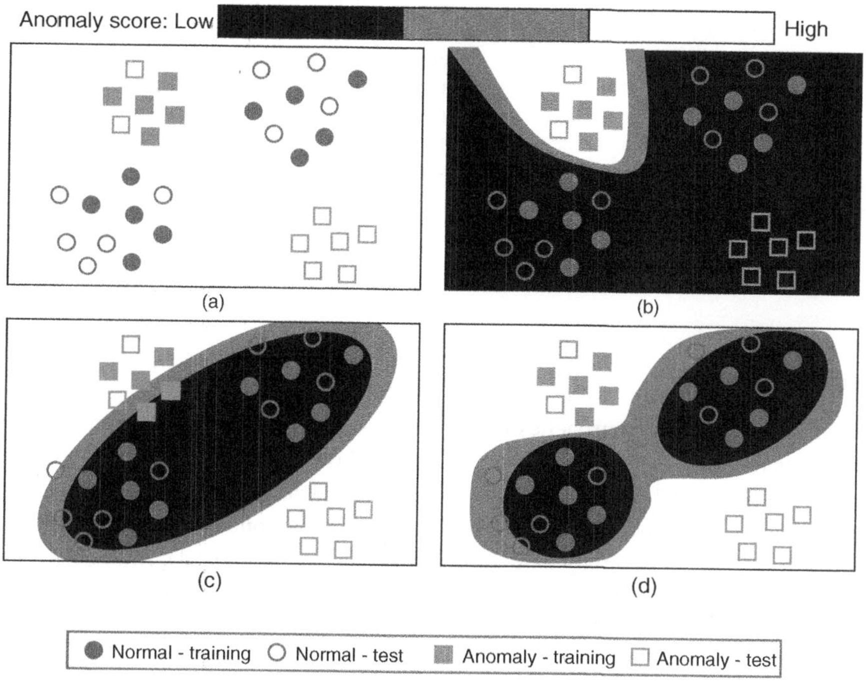

Figure 10.12 Illustration of anomaly detector advantage. (a) Training and test data; (b) Binary classifier; (c) Anomaly detector trained with only normal data; (d) Anomaly detector trained with normal + anomaly data.

(Figure 10.12c) does not have any information about anomaly data, anomaly data (in upper left) may have a low anomaly score. If anomaly data can be added to the training (Figure 10.12d), even if it is a small amount, a better anomaly detection performance can be enabled.

10.4.5 Model Drift

In the assessment of the prediction quality of a ML model, it is typically assumed that there exists a set of labelled testing data. This assumption is not always true, especially if a model is being used in an environment other than the one in which it was trained. Model adjustment or adaptation may be required to maintain effectiveness in the new environment.

Such a situation also occurs when the relationship between the variable being predicted (labels) and the variables being used for prediction (features) changes with time. This is called model drift and can arise when the statistical properties of the target variable itself change (concept drift) or the statistical properties of the predictors change (data drift). An example of the latter is when the patterns in the data change due to seasonality. In order to detect model drift or the need for model adjustment, a method for predicting the effectiveness of a model without the existence of labeled testing data is required. One approach that has been used to estimate the quality of predictions on unlabeled testing samples is based on reverse testing. In this method it is assumed that a labeled training set and an unlabeled testing set are available. First a model is learned from the training set and is applied to label the testing set. Then the model is retrained on the pseudo-labeled testing set and its prediction qualities are evaluated on the training set, as shown in Figure 10.13. It has been shown that the reverse testing qualities are more accurate estimates of the testing qualities than the traditional training qualities in a number of examples using real world data sets [41, 42].

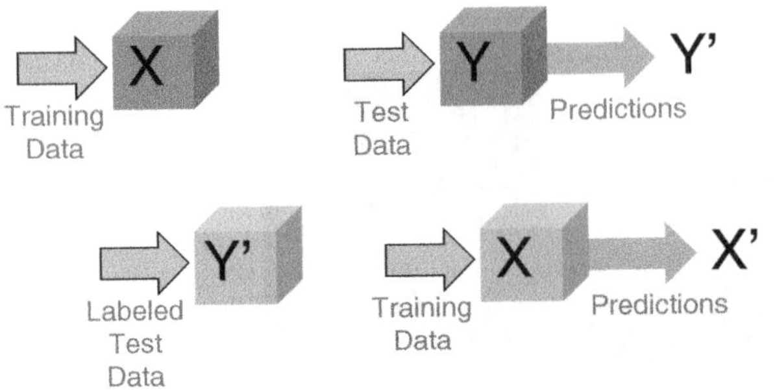

Figure 10.13 Reverse testing accuracy on training data – using training data (X) to a create model that is applied to test data (Y) to produce predictions (Y'), then using the predictions to train a second model that is evaluated using the original training data (X).

10.4.6 Model Update/Evolution

When model drift detection, e.g. using approaches like reverse testing described in Section 10.4.5, indicates that the model is no longer valid, models need to be updated. In this section, the task of model update and evolving the model which needs to be done periodically is considered.

In order to update the model, the process to follow depends on whether one is using a classification model (see Section 10.4.3) or an anomaly detection model (see Section 10.4.4). In the case of anomaly detection, one would be using training data that does not require any explicit labeling of the input sounds. In such environments, it is sufficient to collect the training data that is generated in the environment and to use that training data periodically to retrain the model and to update it. The update process can be periodic and set to occur at predetermined intervals or it can be triggered by means of some specific events (e.g. when the number of reported anomalies or the rate at which anomalies are reported exceeds a specific rate). In some cases, the retraining may be triggered by sensors in some other modality. As a specific example, suppose acoustics are being used to detect abnormal sounds in an engine room. The readings from an electric meter could indicate the subset of engines that are active in the room, and different intervals of electric power reading may indicate the need to change the model that detects anomalous sounds.

In the classification model use cases sounds are defined into one or more labels. In deployment, the data that is collected is not labeled. However reverse testing described in Section 10.4.5 could indicate that the model needs to be updated. A new model can be retrieved from a model catalogue and one can use heuristics such as one described in [43] to determine which model is the best fit for the data seen during operations. The updated model can then be deployed.

Another alternative is to rely on a back-end process to determine if new data classes have been introduced into the environment. This two-tier process has an acoustic module classifying sounds. A selected subset of sounds is sent to a human or slower-loop process for manual classification. This manual classification technique can then be used to create new data for training sounds.

An example of this can be seen in Figure 10.14, where an AI-based acoustic model is used to monitor sounds of a machine. Initially, the training data only consists of two types of sounds, normal and abnormal, and the machine can use it to detect and report anomalous sounds. However, when new anomalous sounds are generated, they are passed over to a trouble ticketing system. A technician addressing those tickets could record the type of problems that caused the sound to happen (e.g. broken belt, lack of lubrication etc.). As more of the newly such classified sound samples are available, one can train the model to recognize these new classes of sounds.

Another specific case of model adaptation is the adjustment of models using techniques like style transfer. This approach is described in Section 10.4.7.

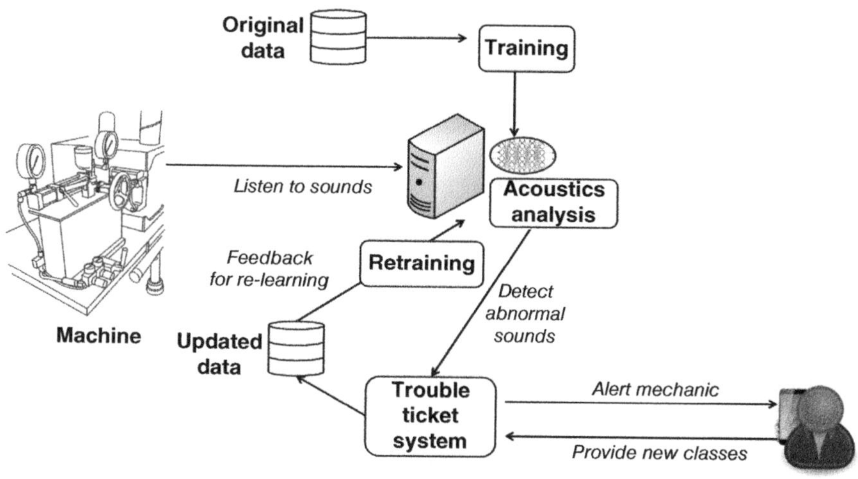

Figure 10.14 Model adaptation using a second tier for labeling.

10.4.7 Model Adaptation

In distributed acoustic ML, insuring the generalizability of the models across different environments and making them adaptive to the environmental changes is an important task. The distributed nature makes the generic ML problem even more difficult, as the individual environments may have different data distribution to be included in the models, while they share common attributes at the same time.

The training data for acoustic ML models are frequently sensitive to environmental changes. The same event can create different acoustic signatures in different environments. Even in the same location, the direction or distance from sound capturing devices (i.e. microphones) can result in different acoustic signatures and even can change over time depending on the gradual or seasonal change of the environmental attributes. If one wants to detect anomalies of motors by observing their sounds, one needs to consider all these possible variations. It would be virtually impossible to manually collect all possible training data that can meet these variations, because it is often not feasible to capture audio samples in multiple locations over a long period of time. In the case of anomaly sounds, it is even more difficult given the lower rate of occurrence of such events.

In order to address this problem and adapt the models to different environments, an efficient method to generate acoustic data adapted to new environments is needed. Traditionally, the ML community has devised various methods such as regularization [44], transfer learning [45, 46], and data augmentation [47]. Transfer learning has been acknowledged to be an effective method to expand the amount of training data by reusing a pre-trained model and transferring knowledge learned from one environment to another as a starting point [48, 49]. Data augmentation is a suite of techniques that enhance the size and quality of training data sets such that better deep learning models can be built using them [47]. See Section 10.4.3.2 for more details.

Aside from the rather simpler techniques, style transfer attempts to adapt newly generated data to a new environment while keeping the nature of the acoustic features of events of interest by transferring environmental features to generate new acoustic data. Style transfer for images was originally developed for generating a new image [50] that resembles the texture of a "style" image while maintaining the structure of objects in a "content" image. A well-known example is to convert a photo (content) with the brush touch of Vincent van Gogh (style). The resulting image looks like a painting of the content object drawn with the brush and the technique used in the style image.

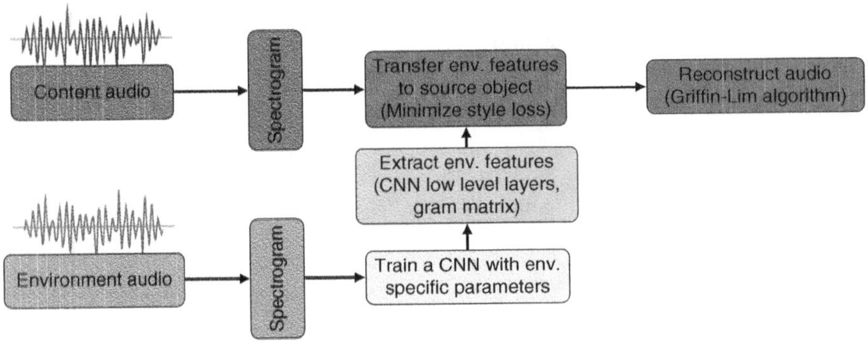

Figure 10.15 Environment transfer for acoustic model adaptation: the environment audio and content audio are transformed to a spectrogram and each feature is extracted using a CNN. The environment feature is transferred to the content audio so that the resulting audio can represent the change of the content audio in a different environment. The new strategy (the new convolutional filter configuration) to enhance the method for environment audio takes place in the lower row of the diagram.

Recently, this technique has been used for transferring styles of a source audio [51] and generating a new audio that resembles styles or textures of the original style audio data. However, the attempts mostly aimed to generate new audio samples that copy the acoustic texture of style audio and were limited to utilities such as musical instruments timbre variations. Acoustic style transfer can be effectively used for adapting distributed environmental features by transferring acoustic environment style to another environment. Figure 10.15 shows the architecture that generates new adapted data using the environmental style transfer technique.

10.5 Summary

In this chapter the use of acoustics data to characterize events occurring in a sensor monitored environment has been presented. In particular, the application of acoustic ML models for data interpretation has been put forward as an effective way to derive insights into the state of the environment and its evolution. A number of use cases have been described, both in the military and commercial domains.

To realize an end-to-end operational system a hybrid cloud approach has been proposed and the core set of necessary components identified. These span both cloud and edge environments, and provide for the training of acoustic models, running them at the edge, and taking actions on the results obtained. Technologies for data management and curation, feature extraction, and detection and classification of data have been described. The different AI models that can be used and their inference performance have also been considered, as have model adaptation, model drift, and model update and evolution.

Future investigations would address the challenges identified in applying acoustic ML models to the analysis of data from acoustic sensors. These include obtaining sufficient data for model training, especially for the characterization of abnormal events. Then once a model is created, the ability to continuously adapt it to new sounds or new environments with different combinations of sounds is a key problem.

This is a rich and important research area that can provide powerful analysis techniques. It is an area of investigation that is just starting to show its potential.

References

1 Bianco, M.J., Gerstoft, P., Traer, J. et al. (2019). Machine learning in acoustics: theory and applications. *The Journal of the Acoustical Society of America* 146 (5): 3590–3628. https://doi.org/10.1121/1.5133944.

2 Wenz, G.M. (1972). Review of underwater acoustics research: noise. *The Journal of the Acoustical Society of America* 51 (3B): 1010–1024.

3 Muir, T.G. and Bradley, D.L. (2016). Underwater acoustics: a brief historical overview through world war II. *Acoustics Today* 12 (3): 40–48.

4 Damarla, T. (2015). *Battlefield Acoustics*. Springer.

5 Ferguson, B.G. (2019). Defense applications of acoustic signal processing. *Acoustics Today* 15 (3): 10–18.

6 Zoltán, K. (2016). Physical perimeter security of military facilities. https://tudasportal.uni-nke.hu/xmlui/handle/20.500.12944/14497 (accessed 25 October 2022).

7 George, J., Mary, L., and Riyas, K.S. (2013). Vehicle detection and classification from acoustic signal using ANN and KNN. *International Conference on Control Communication and Computing (ICCC)*, pp. 436–439. IEEE.

8 Bansal, A., Aggarwal, N., Vij, D., and Sharma, A. (2018). An off the shelf CNN features based approach for vehicle classification using acoustics. *International Conference IoT in Social, Mobile, Analytics and Cloud in Computational Vision and Bio-Engineering*, pp. 1163–1170. Springer.

9 Altmann, J., Linev, S., and Weiß, A. (2002). Acoustic–seismic detection and classification of military vehicles–developing tools for disarmament and peace-keeping. *Applied Acoustics* 63 (10): 1085–1107.

10 Pham, T. (2010). Acoustic sensing for urban battlefield applications. *The Journal of the Acoustical Society of America* 127 (3): 1779.

11 Damarla, T. (2010). Sensor fusion for ISR assets. *Ground/Air Multi-Sensor Interoperability, Integration, and Networking for Persistent ISR*, volume 7694, p. 76941C. International Society for Optics and Photonics.

12 Al-Emadi, S., Al-Ali, A., Mohammad, A., and Al-Ali, A. (2019). Audio based drone detection and identification using deep learning. *15th International Wireless Communications & Mobile Computing Conference (IWCMC)*.

13 Huang, X., Acero, A., Hon, H.-W., and Reddy, R. (2001). *Spoken Language Processing: A Guide to Theory, Algorithm, and System Development*, vol. 1. Upper Saddle River, NJ: Prentice Hall PTR.

14 Lyons, J. (2015). Mel frequency cepstral coefficient (MFCC) tutorial. *Practical Cryptography*. http://practicalcryptography.com/miscellaneous/machine-learning/guide-mel-frequency-cepstral-coefficients-mfccs/.

15 Ng, A. (2015). Keynote speech: deep learning. *2015 Nvidia GPU Technology Conference*. https://video.ibm.com/recorded/60113824/highlight/619422 (accessed 25 October 2022).

16 Krizhevsky, A., Sutskever, I., and Hinton, G.E. (2012). ImageNet classification with deep convolutional neural networks. *Proceedings of the 25th International Conference on Neural Information Processing Systems (NIPS)*.

17 Piczak, K.J. (2015). Environmental sound classification with convolutional neural networks. *Machine Learning for Signal Processing (MLSP)*.

18 Salamon, J. and Bello, J.P. (2017). Deep convolutional neural networks and data augmentation for environmental sound classification. *IEEE Signal Processing Letters* 24: 279–283.

19 Han, Y., Park, J., and Lee, K. (2017). Convolutional neural networks with binaural representations and background subtraction for acoustic scene classification. *Detection and Classification of Acoustic Scenes and Events (DCASE).*

20 Adavanne, S., Pertilä, P., and Virtanen, T. (2017). Sound event detection using spatial features and convolutional recurrent neural network. *International Conference on Acoustics, Speech and Signal Processing (ICASSP).*

21 Dai, W., Dai, C., Qu, S. et al. (2017). Very deep convolutional neural networks for raw waveforms. *International Conference on Acoustics, Speech and Signal Processing (ICASSP).*

22 Inoue, T., Vinayavekhin, P., Wang, S. et al. (2018). Domestic activities classification based on CNN using shuffling and mixing data augmentation. *Detection and Classification of Acoustic Scenes and Events (DCASE) Technical Report.*

23 Inoue, T., Vinayavekhin, P., Wang, S. et al. (2019). Shuffling and mixing data augmentation for environmental sound classification. *Detection and Classification of Acoustic Scenes and Events (DCASE) Workshop.*

24 McFee, B., Humphrey, E.J., and Bello, J.P. (2015). A software framework for musical data augmentation. *International Society for Music Information Retrieval (ISMIR)*, pp. 248–254.

25 Jaitly, N. and Hinton, G.E. (2013). Vocal tract length perturbation (VTLP) improves speech recognition. *International Conference on Machine Learning (ICML).*

26 Takahashi, N., Gygli, M., Pfister, B., and Van Gool, L. (2016). Deep convolutional neural networks and data augmentation for acoustic event recognition. *INTERSPEECH*, September 2016.

27 Tokozume, Y., Ushiku, Y., and Harada, T. (2018). Learning from between-class examples for deep sound recognition. *International Conference on Learning Representations (ICLR).*

28 Schimmel, S.M., Muller, M.F., and Dillier, N. (2009). A fast and accurate "shoebox" room acoustics simulator. *IEEE International Conference on Acoustics, Speech and Signal Processing*, pp. 241–244.

29 Zhang, H., Cisse, M., Dauphin, Y.N., and Lopez-Paz, D. (2018). mixup: Beyond empirical risk minimization. *International Conference on Learning Representations (ICLR).*

30 Raspberry Pi. Raspberry Pi products. https://www.raspberrypi.com/products/ (accessed 25 October 2022).

31 Aggarwal, C.C. (2016). *Outlier Analysis*, 2e. Springer Publishing Company, Incorporated ISBN 3319475770.

32 Chalapathy, R. and Chawla, S. (2019). Deep learning for anomaly detection: a survey. *arXiv:1901.03407.*

33 Zhou, C. and Paffenroth, R.C. (2017). Anomaly detection with robust deep autoencoders. *Proceedings of the 23rd ACM SIGKDD International Conference on Knowledge Discovery and Data Mining*, pp. 665—674. New York, NY, USA: Knowledge Discovery and Data Mining (KDD). ISBN 9781450348874.

34 Schlegl, T., Seeböck, P., Waldstein, S.M. et al. (2017). Unsupervised anomaly detection with generative adversarial networks to guide marker discovery. *International Conference on Information Processing in Medical Imaging*, pp. 146–157. Springer.

35 Kimura, D., Chaudhury, S., Narita, M. et al. (2020). Adversarial discriminative attention for robust anomaly detection. *2020 IEEE Winter Conference on Applications of Computer Vision (WACV)*, pp. 2161–2170.

36 Ruff, L., Vandermeulen, R., Goernitz, N. et al. (2018). Deep one-class classification. *Proceedings of the 35th International Conference on Machine Learning*, pp. 4393–4402 (10–15 Jul 2018).

37 Chong, P., Ruff, L., Kloft, M., and Binder, A. (2020). Simple and effective prevention of mode collapse in deep one-class classification. *Proceedings of International Joint Conference on Neural Networks (IJCNN).*

38 Hendrycks, D. and Gimpel, K. (2017). A baseline for detecting misclassified and out-of-distribution examples in neural networks. *Proceedings of International Conference on Learning Representations.*

39 Inoue, T., Vinayavekhin, P., Morikuni, S. et al. (2020). Detection of anomalous sounds for machine condition monitoring using classifcation confidence. *Detection and Classification of Acoustic Scenes and Events (DCASE) Workshop.*

40 Ruff, L., Vandermeulen, R.A., Goernitz, N. et al. (2020). Deep semi-supervised anomaly detection. *International Conference on Learning Representations (ICLR).*

41 Fan, W. and Davidson, I. (2006). Reverse testing: an efficient framework to select amongst classifiers under sample selection bias. *ACM SIGKDD International Conference on Knowledge Discovery and Data Mining*, pp. 147–156.

42 Bhaskaruni, D., Moss, F.P., and Lan, C. (2018). Estimating prediction qualities without ground truth: a revisit of the reverse testing framework. *24th International Conference on Pattern Recognition (ICPR)*, pp. 49–54.

43 Desai, N., Ganti, R.K., Kwon, H. et al. (2018). Unsupervised estimation of domain applicability of models. *IEEE Military Communications Conference (MILCOM)*, pp. 34–39. IEEE.

44 Kukacka, J., Golkov, V., and Cremers, D. (2017). Regularization for deep learning: a taxonomy. http://arxiv.org/abs/1710.10686.

45 Pratt, L.Y. (1993). Discriminability-based transfer between neural networks. *Advances in Neural Information Processing Systems*, pp. 204–211.

46 Pan, S.J. and Yang, Q. (2009). A survey on transfer learning. *IEEE Transactions on Knowledge and Data Engineering* 22 (10): 1345–1359.

47 Shorten, C. and Khoshgoftaar, T.M. (2019). A survey on image data augmentation for deep learning. *Journal of Big Data* 6 (1): 60. https://doi.org/10.1186/s40537-019-0197-0.

48 Olivas, E.S., Guerrero, J.D.M., Martinez-Sober, M. et al. (2009). *Handbook of Research on Machine Learning Applications and Trends: Algorithms, Methods, and Techniques: Algorithms, Methods, and Techniques.* IGI Global.

49 Goodfellow, I., Bengio, Y., and Courville, A. (2016). *Deep Learning.* MIT press.

50 Gatys, L.A., Ecker, A.S., and Bethge, M. (2015). A neural algorithm of artistic style. *CoRR*, abs/1508.06576. http://arxiv.org/abs/1508.06576.

51 Grinstein, E., Duong, N.Q.K., Ozerov, A., and Pérez, P. (2018). Audio style transfer. *International Conference on Acoustics, Speech and Signal Processing (ICASSP)*, pp. 586–590, April 2018.

Section 3

Introduction: Security, Resiliency, and Technology for Adversarial Environments

Ananthram Swami

U.S. Army DEVCOM Army Research Laboratory, U.S. Army Futures Command, Adelphi, MD, USA

Internet of Things (IoT) devices, platforms and networks deployed for critical national infrastructure or military applications face unique security challenges compared to commercial IoT devices or to traditional computing systems. Key challenges here include the scale of the network, the severe resource constraints under which IoT [and in particular, Internet of Battlefield Things (IoBT)] devices must operate. These constraints include limited computing power, CPU capacity, memory, and battery power. Further, these devices must operate in an adversarial environment and cope with denied/disrupted, intermittent and limited (bandwidth) communication environments. Thus, the traditional notions of host vs. network-based defenses need to be re-examined. Vulnerabilities emerge from heterogeneity – in devices, network protocols, computing architectures, and operating systems. Connecting things is critical to making them smarter and resilient to the loss of a fraction of the nodes; indeed, any single node or sensor will have only a limited view of the operating conditions and hostile forces, limited by its modality and sensing power. Inter-connectivity and interactivity make the IoBT/IoT smarter (and an IoT/IoBT without a network is just a bunch of isolated dumb things), but also significantly increase the attack surface. Indeed, the hyper-connectivity in IoT/IoBT obviously increases opportunities for cyberattack, but it also offers new defensive approaches as multiple devices could potentially guard themselves collectively and collaboratively. Indeed, scale and multi-modality offer multiple vantages and thus collaborative defense against deception. Given the large-scale availability of commercial IoT solutions, DoD and other national security organizations will leverage commercial IoT to keep pace with the state of the art and must deal with challenges that are not critical to commercial deployments. Military operations are increasingly multi-national, and often involve non-government organizations (NGOs). Thus, IoBTs are, perhaps, best considered as network of networks, and security-resiliency guarantees in any one network do not translate to guarantees for the overall network of networks. The six chapters in Section 3 describe these challenges in detail, discuss existing solutions, and propose new avenues of attack and defense.

While identifying and fixing security flaws and practices described in this section are good practice for commercial IoT systems, they are essential for military and security IoT systems that can count on being attacked and where the impact of their compromise can lead directly to destruction of property and lives. Defense and security organizations must have a process for detecting the sorts of vulnerabilities that the chapters in this section outline and a procedure for fixing them before incorporating commercial IoT devices and technology.

IoT for Defense and National Security, First Edition. Edited by Robert Douglass, Keith Gremban, Ananthram Swami, and Stephan Gerali.

Chapter 11, *Assuring Resilience by Design for Cyber Physical Data-Driven Systems (CPDDS)*, examines safety and security issues in the context of CPDDS – Cyber Physical Data Driven System, a generalization of an embedded system in which computing devices communicate with each other and the physical world through sensors and actuators in a feedback loop. A CPDDS is a specific example of IoT. CPDDS finds applications in critical infrastructure such as transportation, power grid and water and fuel distribution. Applications within DoD include autonomous unmanned systems (UXS). Thus, assurance of physical and cyber health of such national infrastructure and DoD assets is paramount. Assurance guarantees should include provable safety, reliability, and trustworthiness. CPDDS will rely on a variety of vendors and developers that provide the components such as commercial cloud computing services, the hardware, trained Machine Learning (ML) models, etc. Assurance guarantees for each of the components (hardware, system, software, data, application) are necessary but not sufficient; assurance guarantees are also required for the interconnectivity of the networks, which must often communicate over disadvantaged communication links, and for the system as a whole so as to ensure correctness and safety of ensuing decisions. The authors discuss S&T gaps that need to be addressed and several possible approaches. Formal methods (FM) could provide provable guarantees, but they need to be adapted to data-driven models. Lack of workforce trained in FM highlights the need for automated but explainable methods. Network science concepts (such as network-of-networks, multi-graph representations, percolation) have been useful for analysis and have yielded some insights but have not yet been translated into practice. Concepts from Age of information are potentially critical for network control and data assurance. The chapter thus argues that the all-domain nature of the problem of assuring the design of CPDDSs requires a multi-disciplinary perspective.

Chapter 12, *Vulnerabilities in Tactical IoT Systems*, discusses vulnerabilities related to different components of a typical IoT system. An IoT system consists of several devices and components that interact both over the cyber and the physical medium. Thus, ensuring the device or component-level security is not adequate, as integration and interaction introduce a large range of vulnerabilities. IoT systems have been built with functionality, rather than security, as the goal. IoT systems are finding increased deployment both in commercial and tactical systems in applications related to sensing and actuation of control systems. This chapter provides a detailed view of the vulnerabilities in IoT systems that are deployed in tactical applications. Discussed vulnerabilities include weak authentication for network services, insecure firmware and software update mechanisms (e.g., the SolarWinds hack), over-privileging, interconnectedness and interaction between components – both cyber and physical, Apps using hard-coded keys (passwords), or sending credentials over plaintext, and vulnerabilities in all layers of commonly used communication protocols. The chapter provides several examples of recent attacks that have exploited these vulnerabilities. It concludes with a discussion of potential countermeasures and healthy cyber practices.

Continuing the theme of the previous two chapters, Chapter 13, *Intrusion Detection/Prevention Systems for IoT*, addresses the question of detection of attacks on IoT devices. There are several challenges due to scale and resource constraints of IoT devices, which preclude direct application of intrusion detection and mitigation systems designed for conventional computer systems. For example, scale hinders quick application of patches. Limited memory and CPU are not conducive to standalone endpoint intrusion detection algorithms: it is difficult to install firewalls and anti-virus protection schemes which are typically memory and CPU hungry. Other challenges arise from the diversity of network protocols, CPU and operating system architectures. The chapter discusses the pros and cons of host vs. network-based intrusion detection in the IoT context. It describes several IoT-specific attacks at the network and application layers. This chapter provides a comprehensive comparative analysis of several intrusion detection systems for IoTs, based on the location of collectors and analyzers, detection algorithms, and specific threats. Vulnerabilities in IoT devices may be used as a stepping-stone to attacks on host systems to which they are connected. Further research is needed on multi-step attack detection, as well as on detection of advanced types of stealthy attacks. A challenge here is tradeoff between improved accuracy with a network-based detector vs. latency and communication overhead.

Chapter 14, *Bringing Intelligence at the Network Data Plane for Internet of Things Security*, discusses opportunities for enhancing security of IoT devices by taking advantage of a programmable data plane offered by the Software-Defined Networking (SDN) architecture. Timely deployment of updated security policies is crucial to cope with novel threats, but approaches proposed for standard computer architectures cannot be applied to IoTs due to CPU and memory constraints, and diversity of IoT network protocols. To this end, the chapter describes a two-stage data-driven deep learning solution. A convolutional neural network (CNN) is trained to classify traffic; the abstract features learned by this CNN are then converted, by a second CNN, to packet header byte strings that are amenable to the match-and-action mechanism of SDN. A second aspect is that the use of a binarized neural network considerably reduces the memory and computational load such that it can be deployed on edge devices.

An earlier chapter discussed the use of anomaly-based detection of intrusion threats to IoT/IoBT. Such data-driven approaches must rely on continuous learning to cope with the dynamics of the environment. These techniques will be computationally expensive and will likely need distributed computing approaches. To cope with the vast amounts of data that a large-scale IoT/IoBT produces, there will be a need for in-network processing of the data. Such processing also offers flexible multi-vantage multi-modal confirmation/verification of the IoBT's health and threats. The increasing complexity of many applications requires in-network processing of data from many sources, leveraging hybrid edge and cloud computing resources, subject to the availability of communication links. With the above context, Chapter 15, *Distributed Computing for Internet of Things under Adversarial Environments*, discusses the challenges in orchestrating distributed computing, rapidly identifying optimal placement of computing tasks, while coping with the dynamics in resource availability and network conditions. Another daunting challenge is to cope with adversarial manipulation of both the data as well as the meta-data regarding resource and network availability, as well as compromise of the computing and networking resources. The chapter describes a comprehensive threat model, existing distributed computing frameworks and schemes for verifiable computing, and discuss open research issues and potential approaches. It also discusses the use of non-sympathetic ("gray") nodes, and issues related to trust.

Earlier chapters discussed the vulnerabilities arising from the interconnectedness of IoT/IoBT devices; compromise of one node could serve as a stepping-stone for attacks on the rest of the network. Chapter 16, *Ensuring the Security of Defense IoT Through Automatic Code Generation*, discusses the state-of-the-art approaches to controlling the interfaces between a node and the network, and it describes the limitations of these approaches. The chapter proposes the use of weakness-free interfaces between an IoT node and the network. The underlying concept is that interface code can be automatically generated starting from high-level specifications using an automatic code generator that has been verified by formal methods to create weaknesses-free code. A detailed description of the auto-code generation is provided, and an extended example – automatic generation of router software – is discussed in detail. For IoBT applications, weakness-free interface software would be auto-generated by a trusted authority and augmented with anti-tamper elements to guarantee that it stays secure and unmodified. A detailed discussion of the interface code issuing authority is included, along with a recommendation for consolidating this authority in one DoD-wide anti-tamper authority to provide uniformity of solutions.

The future battlefield will see the IoBT pitted against the adversary's IoBT. How will the IoBT cope in battle against a determined peer adversary? The adversary will likely attack the IoBT on all three fronts of confidentiality, integrity, and availability. The chapters in this section largely discuss how an adversary might attack the IoBT's integrity by injecting malware, and by exploiting vulnerabilities within the things and in their connections. Equally important is the notion of confidentiality. An important challenge is to thwart the adversary's ability and opportunities to acquire information about our IoBT – its topology, composition, and functionality, going beyond the traditional notion of information usage. The adversary will try to gain "stimulative intelligence" by probing the IoBT both physically as well as in the cyber domain. As discussed earlier, the key characteristics of an IoBT that differentiate it from civilian IoT or traditional networked computing systems are the extreme scale, heterogeneity, dynamics and adversarial environment. The scale and density of the IoBT potentially provide a solution for dealing with compromised elements and mitigating their effects. The scale also provides a potential advantage against traditional traffic analysis – but it comes with the price that IoBT will glow (in the RF spectrum). Honeynets and honeypots may provide forms of deception, camouflaging the IoBT's functionality and structure.

The chapters in this section provide perspectives on the state-of-the-art and potential research directions on security and resilience of IoT/IoBT. Further information may be obtained from the public-domain websites of the academic authors, from DEVCOM ARL's IoBT CRA program (https://iobt.illinois.edu) and in DEVCOM ARL's Cyber Security CRA (https://www.arl.army.mil/business/collaborative-alliances/current-cras/cyber-security-cra/). Other programs in this space include: DARPA's Ocean of Things (OoT) program that seeks to enable persistent maritime situational awareness over large ocean areas by deploying thousands of intelligent sensors as a distributed intelligent sensor network, DARPA's CHARIOT program that is investing in developing revolutionary security technologies for IoT, and the European Union's Alliance for Internet of Things Innovation.

11

Assurance by Design for Cyber-physical Data-driven Systems

Satish Chikkagoudar[1], Samrat Chatterjee[2], Ramesh Bharadwaj[1], Auroop Ganguly[3], Sastry Kompella[1], and Darlene Thorsen[2]

[1] *Information Technology Division, U.S. Naval Research Laboratory, Washington, DC, USA*
[2] *Data Sciences & Machine Intelligence Group, Pacific Northwest National Laboratory, Richland, WA, USA*
[3] *Department of Civil & Environmental Engineering, Northeastern University, Boston, MA, USA*

Abstract

Currently, Cyber-Physical Data-Driven Systems (CPDDS) employ machine learning for the classification, data fusion, and control of our nation's infrastructure, such as the power grid, transportation networks (e.g. fuel distribution, air traffic control), and DoD long-duration collaborative autonomous platforms including unmanned underwater, ground, surface, space, and aerial systems. Many CPDDSs are system-of-systems that should be designed to communicate over disadvantaged networks. It is important to assure that the CPDDSs are resilient against physical and cyber threats by design. Additionally, their design should tolerate misclassification errors resulting from natural and/or adversarial distribution shifts within their data driven components. The all-domain nature of the problem of assuring the design of CPDDSs requires a multi-disciplinary perspective as outlined in this chapter.

11.1 Introduction

U.S. critical infrastructure systems increasingly rely on process automation enabled by the seamless integration of information flow in cyberspace and system operations in physical space. A cyber-physical data driven system (CPDDS) is a generalization of an embedded system where computing devices communicate with each other and the physical world through sensors and actuators in a feedback loop. CPDDSs comprise hardware, software, and data-driven applications embedded into the physical world. The data driven applications rely on machine learning (ML) and reasoning to achieve system objectives. Many CPDDSs are system-of-systems that should be designed to communicate over disadvantaged networks. System operations enabled by flow of information lead to enhanced efficiency but may also result in additional vulnerabilities. These vulnerabilities may exist in the cyberspace or physical space and could be exploited/targeted by natural and/or man-made threat vectors potentially leading to catastrophic losses. Essential properties of a CPDDS include [1]:

- **Reactive Computation**: Continuous interaction with the environment in the form of inputs and outputs, where correctness refers to input/output sequences that correspond to acceptable behaviors.

IoT for Defense and National Security, First Edition. Edited by Robert Douglass, Keith Gremban, Ananthram Swami, and Stephan Gerali.

- **Concurrency**: Execution of multiple threads of information in parallel including information exchange to achieve computational goals, e.g. autonomous mobile robots. Formal models may be synchronous (components execute in lock-step) or asynchronous (components execute at independent speeds).
- **Feedback Control**: Measurement via sensors and influence via actuators. Dynamical control systems have the mathematical tools for design and analysis. In cyber-physical systems, a controller consists of discrete concurrent components operating at multiple modes interacting with continuous dynamics in the physical environment.
- **Real-Time Computation**: Includes timing delays and timing-dependent coordination protocols to ensure system predictability.
- **Safety-Critical Operations**: Involves assurance for detecting design errors and ensuring high reliability in operations.

CPDDSs are used in critical infrastructure including multi-modal transportation, communication, electric power grid, and water distribution. Examples of DoD relevant CPDDSs include long-duration collaborative autonomous systems such as unmanned underwater, ground, surface, and aerial systems (UUVs, UGVs. USVs, and UASs). Within the DoD, CPDDSs address advanced and persistent adversary threats from a safe standoff distance. Therefore, assurance of physical and cyber health of such DoD assets is paramount. Assurance encompasses provable safety, reliability, and trustworthiness guarantees. Currently, these objectives are not being achieved in practice within the design space of CPDDSs.

The problem of assuring CPDDSs encompass the following research challenges:

- **Computing Hardware Assurance**: Use of Commercial Off-The-Shelf (COTS) hardware for CPUs and GPUs introduces vulnerabilities such as Spectre and Meltdown [2], leading to data exfiltration and lack of assured model fidelity.
- **System Software Assurance**: The increasing reliance on COTS for software infrastructure and processes – DevSecOps – introduces unprecedented cyber-vulnerabilities due to frequent changes to the codebase. The CPDDS community is reliant on large, open-source codebases with attendant supply chain vulnerabilities. For example, DoD developed formal methods were used to assure the trusted kernel seL4 [3], which is currently being pursued for acquisition by China, making us aware of the vulnerability of such systems falling into the hands of adversaries.
- **Application Assurance**: The complexity and variability in mission requirements and design of fielded CPDDSs raise additional challenges. The operational context of these systems ranges from fully autonomous to non-autonomous, AI-dependent to non-AI based, and a continuum of human involvement on unmanned vs. manned platforms. Clearly, there is a great degree of design tradeoffs and assurance objectives that have to be addressed to successfully develop and field such disparate systems.
- **Data Assurance**: CPDDSs need to have an Information Barrier for filtration and data transduction to ensure confidentiality and integrity of data in motion [4]. Additionally, there needs to be a tamper-proof information repository shielded from adversaries (despite gaining physical access) to ensure confidentiality and integrity for data at rest.
- **System Assurance**: System assurance research challenges include [5, 6]: How do attacks and disruptions affect the system state estimation and control algorithms? How to consider closed-loop dynamics of controlled systems (typically modeled by differential or difference equations) in conjunction with discrete actions associated with networked components? Can the system also operate using open-loop control when limited sensor information is available? How to integrate temporal properties of physical systems with computing and networking systems to

facilitate concurrency and real-time computations? What are new theoretical formalisms that couple continuous dynamics of physical systems with discrete dynamics of cyber systems under uncertainty? How to conduct system-wide verification through formal methods and algorithms for large-scale CPDDSs?

- **Decision Assurance**: Decision assurance is achieved by providing explainability with compound or high consequence scenarios as well as generating robust risk-informed defense options under uncertainty and within multi-agent settings.
- **Assuring Interconnected Networked CPDDSs**: Modern infrastructure forms a network of assets from a network science perspective. The London Rail Network, for example, consists of the Underground subway, the Overground passenger trains, and the Dockland Light Rail. These three networks interconnect at shared nodes or rail stations, forming a so-called *network-of-networks*. Figure 11.1 presents an overview of a network-of-networks system along with the challenges associated with complexity, uncertainty, heterogeneity, and dynamics. Similarly, various modern infrastructure networks such as national airspace system, railroad, communication, fuel pipelines, water pipelines, and so on, are interconnected and interdependent. Hence, any changes or threats to one of the networks can adversely affect other networks.

In the path to achieve the above-listed research goals for assuring CPDDSs, one needs to overcome the following operational challenges:

- **Safety**: The traditional high-assurance approach for enforcing safety properties is by a combination of formal methods and run-time detection/ enforcement. However, the technical challenge is covering all real-world conditions, which results in state spaces that are huge, and therefore intractable. Symbolic methods somewhat mitigate this problem, but the assurance challenge for safety remains a research problem.
- **Reliability**: In order to prove system robustness under conditions that include natural or adversarial shifts, we're confronted with the same situation as above. Additionally, in order to defend against adversarial attacks on autonomous systems reliant on deep learning (DL), we need to exhaustively and stochastically craft adversarial examples with specific statistical characteristics.

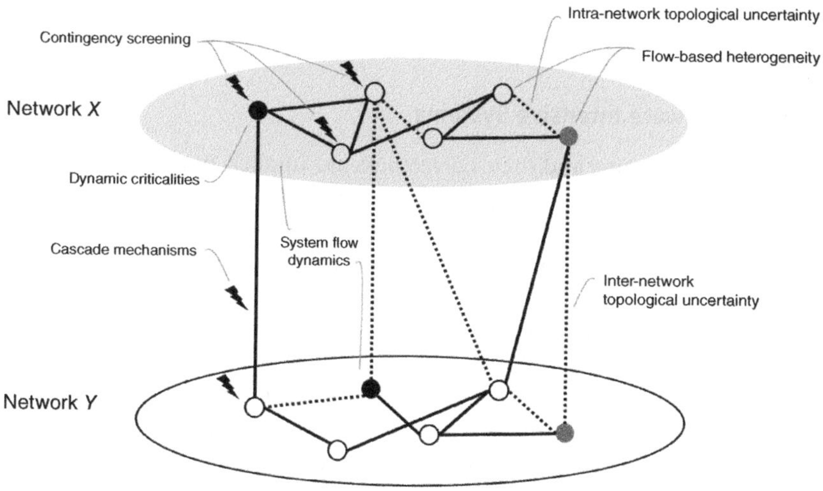

Figure 11.1 An overview of a Network-of-networks CPDDS.

Once these adversarial examples have been generated, we search for *"minimal"* adversarial perturbations which lead to misclassification, thereby allowing us to institute measures that mitigate misclassification of adversarial examples.

- **Trust**: In order to build trust in ML, one needs to extract human-readable descriptions of decisions made by ML models based on white-box interpretation. However, the current state-of-practice is to use explainability, which attempts to provide explanations using adjunct models (such as LIME [7]) rather than inspect the mechanisms behind such decisions.
- **Scalability and Flexibility**: CPDDSs have a high dimensional state space and need to be reliably operated under varying environments. For operational modeling of large-scale systems, including dependencies and interdependencies, a potential approach may include study and application of Hybrid Automata principles [8, 9]. This approach addresses scalability issues in model building (i.e. model structure and function) and computational effort needed for running these models. In terms of flexibility, both higher and lower fidelity models may be incorporated in a unified setting. The modular structure also allows for composing multiple models to develop higher-level models that can be used further, thereby eliminating the need to build models from a single starting component every time. Open hybrid automata (OHA) can couple continuous dynamics of physical systems with discrete dynamics of networked components [8, 9]. This includes integrating state machines and differential equations in a unified mathematical formalism, with the ability to include inputs and outputs to model subsystems and combine them to develop more complex system models.

The solution for any assurance problem within the high assurance community is the application of formal methods. However, current state-of-the-art in formal methods for CPDDSs is deficient due to imprecise specifications, inadequate environmental modeling, and poor scalability of existing methods and tools. Moreover, the brittle nature of data-driven algorithms has compelled researchers such as Rodney Brooks, the founder of iRobot, to declare: *"The reality is that just about every successful deployment [of data-driven algorithms in the field] has either one of two expedients: It has a person somewhere in the loop, or the cost of failure, should the system blunder, is very low"* [10]. However, fielding CPDDSs in mission- and safety-critical contexts proceeds at a rapid pace, in spite of the inability of extant formal methods to catch up with the state of practice. Clearly, there is much need for fundamental and applied research towards formulating and using formal methods for CPDDSs.

11.1.1 Formal Methods for Software Intensive Systems

The following assumptions about programs and their correctness are implicit in applying formal methods to practical systems: Associated with each program are properties (such as types, assertions, and other annotations), which are readily available, unassailable, inviolable, and invariant. Moreover, the correctness of these properties is both necessary and sufficient for the correctness of the programs they annotate.

However, none of these assumptions necessarily hold for systems such as CPDDSs that are intended to solve real-world problems. An example of such a program is one that implements the 4-coloring algorithm for planar graphs. If we assume that program annotations can characterize the function being computed by the program, the proof of its correctness is probably derivable from the proof of correctness of the 4-coloring problem. However, even for such programs, the correctness of its annotations is often predicated upon extraneous factors in the program's execution environment, such as the word length of the processor, the size of the address space, or the amount of available memory. This is because program code is generically written for an abstract machine; the program may execute on a real machine that may not correctly implement some of these abstractions. In such an event, the program will fail in unexpected ways. Also, program annotations may never be able to capture quantitative aspects such as the space and time requirements of the program. Such properties are central to the program's "correctness" since correctness often entails user expectations about the time and space requirements for successful execution on specific data sets. Even if we assume that it is feasible to precisely characterize such machine requirements and non-functional properties, it is not clear how their correctness could possibly be established by a verification system. The situation becomes hopeless for programs the correctness of whose annotations depends upon extraneous factors.

11.1.2 Adapting Formal Methods for Data Driven Systems

ML, in particular DL, has received wide press in the recent past. Although requiring massive computational power to train, these algorithms show great promise by being able to recognize objects with human level precision [11, 12] and translating human speech in real-time [13]. However, in addition to coming to grips with recent advances in deep learning, we must also understand its limitations. For instance, data sparsity and data poisoning attacks may lead to classification errors, producing incorrect results which can be both embarrassing and damaging. Two recent examples are Google's image classifier misidentifying humans as gorillas[14] and Microsoft's chatbot Tay learning to spew racist and misogynistic hate speech minutes after being turned on [15]. Generative adversarial networks (GAN) demonstrate that deep learning algorithms can be deliberately tricked by adversarial examples [16]. A trained neural network can be tricked into grossly misclassifying objects with extremely high confidence, by mere manipulation of input images not discernible to the human eye or even by images that look like noise to the human viewer [17]. The dangers of adversarial attacks can have a profound impact on society – self-driving vehicles can be hijacked or misdirected with seemingly innocuous signage [18], and security can be compromised with tampered data.

A popular misconception about DL, which provides an illusion of success, is to test a neural network on its training data. However, what many people do not realize is that the fundamental goal of ML is to generalize beyond the examples in the training set [19]. This is because no matter how much training data we provide, it is unlikely that the same data will be encountered during testing. This corresponds to the **"no free lunch"** theorem of Wolpert and Macready [20], which states that no learner can beat random guessing over all possible functions to be learned. Consider for example learning a Boolean function over 100 variables from a million examples. Of the 2^{100} possible classes to be learned, we have only provided 10^6 examples. There are additionally $2^{100} - 10^6 \approx 1.3 \times 10^{30}$ possible inputs whose classes are yet unknown. Clearly, there is no way to do this that beats flipping a coin. Or is there?

Experts in DL algorithms such as John Launchbury (formerly at DARPA and current Chief Scientist at Galois) contend that their phenomenal success is due to what is termed the manifold hypothesis [21]. High-dimensional natural data tend to clump and be shaped differently when visualized in lower dimensions, known as manifolds.

In order to assure DL based systems, one needs to adapt extant formal methods and tools to reason about manifolds and their relationship to relevant system properties. One hypothesis that is currently being explored is that each manifold in a deep neural network represents a unique functional entity and they lead to insight/understanding of input data classification. Unlike extant approaches such as Reluplex [22], which try to reason about the entire network, this insight gives us the ability to map decision boundaries of a feed-forward neural network a layer at a time, thereby mitigating the state explosion problem encountered by extant approaches.

11.2 Methods for Assurance

An approach to assuring CPDDSs comprises development of tools and methods to achieve the following objectives:

- Guaranteeing application independent system properties by proofs of absence of hardware and software system vulnerabilities, including absence of data exfiltration. This can be achieved by exploiting advances in automated theorem provers[23], lightweight formal methods [24], runtime verification [25], and combinations thereof [26].
- Guaranteeing application specific system properties by exploring and exploiting recent and historical contexts in formal verification of systems [27–30].
- Proof of data and model fidelity in the face of cyber and physical attacks [31, 32].
- Full stack design and assured safety and security in cyber and physical domains [33].
- *Platform Integrity and Availability*: Anti-tamper and denial of data access to adversaries is necessary for fielded systems.
- *Information Freshness*: CPDDSs need to operate on fresh data to meet their operational objectives. Age of Information (AoI) [34, 35] is a measure of information freshness.
- *Assuring Decisions*: Even though most of the decisions are going to be made in an operational setting, one will need to ensure that there are tools to assure any decision making by techniques inspired from the risk science and operations research community.
- *Assuring Interconnected Networked CPDDSs*: It is important to address ways to assure interconnected networked CPDDSs from a system-of-systems perspective. We can achieve this objective using methods that are inspired by the risk science and resilience communities.

In the subsections below, we elaborate on some of the above objectives.

11.2.1 Tools for Information Freshness

The concept of AoI is useful in any communications system where the receiver has an interest in fresh information. This is indeed the case in numerous applications of wireless systems that require the transmission of status updates between nodes, such as sensor networks, situational awareness applications, and environment monitoring. Traditional network metrics of throughput and delay are inadequate for describing the performance of these status monitoring applications. This is because the delay for a particular packet may be small, but if its transmission occurred a long time ago, the information, as observed at the current time, is no longer fresh. On the other hand, throughput may be high, such that packets arrive very frequently at the receiver. However, since the queues are usually saturated at high throughput, the packets that arrive at the destination are typically generated a long time ago but delayed in a queue (at the source or relay nodes). This results in even recently received packets no longer being fresh. Therefore, a different metric such as AoI is needed to convey the freshness of information at the receiver.

The age metric can be defined as $\Delta(t) = t - u(t)$, where $u(t)$ is the time stamp of the most recently received packet by the monitor. Considering a simple system of a first-come, first-served queue, Figure 11.2 shows how the age metric typically evolves over time. We see from Figure 11.2 that a packet is received at the monitor at time τ_0, and at time 0, the age is equal to $-t_0$. As time progresses,

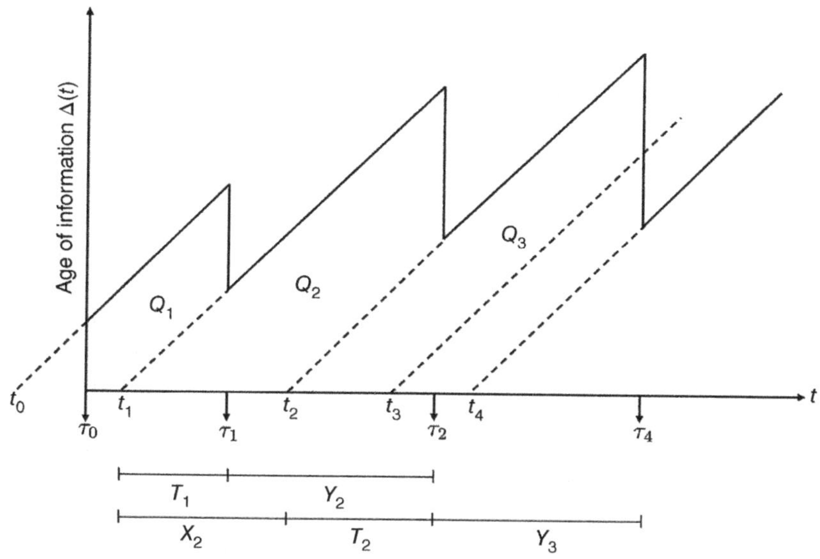

Figure 11.2 Evolution of the Age metric. Here X is the inter-arrival time, Y is the inter-departure time, and T is the total time spent in the system. The time-average Age can be obtained by computing the average area under the trapezoids denoted by Q.

the information ages linearly until the next packet is received at time τ_1, which has an age equal to $\tau_1 - t_1$. Continuing this process, we observe that the function exhibits a "saw-tooth" pattern that is characteristic of the age function.

Typical AoI studies model the system as memoryless and have a source that transmits packets to a monitor through an M/M/1/K queue, where the last entry in the Kendall notation [36] describes a total capacity of $(K - 1)$ packets in the queue and one packet in service. In most cases, an arriving packet that encounters a full capacity system never enters the system and is dropped. We refer to the time between packet generations as the inter-arrival time $Xi = (t_i - t_{i-1})$, $i = 1, 2, 3, \ldots$. The inter-arrival times are modeled as random if the source does not have control over the exact times at which it can transmit updates. In that case, typically, the X_i's are modeled as i.i.d. exponential random variables with rate λ. The total time spent in the system from arrival to service is given by T_k, $k = 1, 2, \ldots$. We also define the inter-departure time Y_k as the time between the instants of complete service for the $(k-1)^{st}$ packet served and the kth packet served. This will be useful in the computation of the average age.

The time-average age can be obtained using a graphical approach to compute the area under the saw-tooth curve, specifically by computing the average area of a trapezoids Q_i, $i = 1, 2, 3, \ldots$ associated with each packet updating the age. This is a non-trivial computation even for simple queueing systems (e.g. M/M/1 queue [34, 37]), since it involves computing the expected value of the product of the packet inter-arrival time at the source and the system time, which are not independent. A closed form expression for the average age has the following form:

$$\Delta = \lambda_e E[Q_k] = \lambda_e \left(\frac{1}{2}E[(T_{k-1} + Y_k)^2] - \frac{1}{2}E[T_k^2]\right) = \lambda_e \left(\frac{1}{2}E[Y_k^2] + E[T_{k-1}Y_k]\right). \tag{11.1}$$

Here λ_e is the average rate of the transmitted packets and $E[\ .\ ]$ is the expectation.

This expression (Eq. 11.1) provides us the ability to analyze the impact of the timeliness of information exchange on application performance and resilience over a varied set of network instantiations.

Research on the age metric has focused on optimizing the performance of systems that are modeled by different types of queues, with various arrival/departure processes, number of servers, and queue capacities. Analyzing age shows that when the number of servers is finite, there is an optimum packet generation rate for which the average age is minimized. Also, the availability of more network resources results in the lowering of the average age, i.e. the average age decreases as the number of servers increases [34, 35]. Additionally, deterministic arrival and departure processes achieve a lower average age than memoryless processes. There has been other work that suggests that the age decreases as the queue capacity decreases, when packets in the queue are replaced with newer packets or when other side information is available. Specifically, age with a system capacity of one or two can be much lower than that of a system with infinite capacity, and the ability to replace packets in the buffer when newer packets arrive does even better. Furthermore, AoI exhibits interesting behaviors in the presence of competing sources as well as when updates have to travel beyond a single hop.

11.2.2 Methods for Decision Assurance

A recent report by The President's National Infrastructure Advisory Council (NIAC) [38] indicates that catastrophic power outages may occur with little or no notice and may result from myriad types of scenarios - for example, a sophisticated cyber-physical attack resulting in severe physical infrastructure damage; or attacks timed to follow and exacerbate a major natural disaster. The NIAC study focused not on cause but consequences categorized as severe, widespread, and long-lasting. One of the major recommendations was to improve our understanding of how cascading failures across critical infrastructure will impact restoration and survival. As summarized in a recent news report [39], *"a cyberattack timed to coincide with a natural disaster could be especially problematic,"* and while a simultaneous natural disaster and major cyberattack directed at the U.S. power grid has not yet happened, their potential for *"happening at the same time is enough for NIAC to encourage the government to take steps to mitigate potential consequences following such an event."* Moreover, the National Science & Technology Council's (NSTC) FY2019 federal cybersecurity Research & Development plan [40] includes the strategic defensive element of dynamic adaptation by efficiently reacting to disruption, recovering from damage, maintaining operations while completing restoration. Developing a scalable modeling and simulation capability to predict with high fidelity consequences of all-hazard threats to CPDDSs, will enable the ability to: (i) reason over possible responses to disruptions; and (ii) inform autonomic cyber response that will shorten the time to recover.

11.2.2.1 Scenario Generation for CPDDSs

In the context of CPDDSs, preparing and safeguarding against Black Swan (highly improbable and extreme) and Perfect Storm (rare combination of circumstances leading to drastic impacts) type events on an ongoing basis is a significant modeling and computational challenge. Multi-hazard natural and man-made threats are typically categorized as low probability and high consequence events. Complex and interdependent processes in CPDDS operations that may be perturbed by dynamic human, environmental, and organizational factors may contribute to the frequency and/or consequences of such events. Adverse scenarios are traditionally developed using a combination of the following techniques [41–43]: (i) failure mode and effects analysis, (ii) hazard and operations analysis, (iii) logic trees (i.e. probability, decision, and event trees; and fault, attack, and success trees), (iv) influence diagrams, (v) hierarchical holographic modeling, and (vi) bow-tie analysis.

Typically, a scenario is characterized by its expected frequency or probability and its consequence or measure of damage. Using the techniques listed above, a risk analyst supported by subject matter experts may generate a list of mutually exclusive and collectively exhaustive (complete, finite, and disjoint) scenarios or scenario categories with their corresponding likelihoods and consequences. However, Black Swan and Perfect Storm type events are especially challenging to enumerate and quantify using traditional methods listed above. For example, recent literature in the complex offshore drilling environment [44–48] suggests the incorporation of human and organizational factors in systems risk analysis using Bayesian networks and simulation modeling within hybrid settings. However, most analyses are geared toward quantifying risk from a specific hazard or scenario using a collection of ad hoc analysis and computational tools that precludes a rigorous scenario generation process.

Advancing the state-of-the-art in scenario planning may be achieved by defining scenarios as a set of critical systemic operational features; and focusing on the development of an integrated computational engine (addressing systemic and human uncertainties and dynamics) that generates a suite of high-consequence scenarios for CPDDS operations using: (i) systems modeling and dynamic simulation, (ii) logic trees, (iii) human reliability modeling, and (iv) Dynamic Bayesian Networks (DBNs). These methods may be informed by large-scale ML, Monte Carlo simulation, and rule-based artificial intelligence techniques to address relationships among operational features and expose patterns among feature combinations. Such a dynamic modeling and simulation setting may help: (i) conduct a comprehensive exploration of the scenario space, (ii) better identify critical operational vulnerabilities and potentially high-consequence scenarios, and (iii) conduct robust what-if analyses; thereby enhancing safe operations through scenario planning within high-consequence CPDDS operations. These high-consequence scenarios can then be used to: (i) develop mitigation strategies and priorities, (ii) detect areas where safety training is required, and (iii) provide inputs to quantitative risk analysis.

A modular approach to implement a multi-method modeling framework for high consequence scenario generation is presented in Figure 11.3. The modeling phases in this figure are explained below and consist of: (i) System Decomposition, (ii) Feature Selection using event trees, fault trees, and human and organizational error (HOE) Analysis, (iii) DBNs including structure and parameter learning; and inference using ML and simulation, (iv) System dynamics (SD) Simulation, and (v) Scenario Generation utilizing rule-based artificial intelligence engines.

The integrated modeling approach may yield novel insights that may not otherwise be readily retrieved using traditional methods and relying solely on expert judgments. A brief description of the modeling phases is below.

11.2.2.1.1 System Decomposition

The key goal of system decomposition is to understand composition of a CPDDS and characterize its key functional relationships. An additional key goal is to identify and characterize mechanisms where HOE can affect CPDDS functions and safety. Systems and their sub-systems may be identified and parameterized using CPDDS subject matter experts. The elicitation of system composition knowledge may be conducted in a semi-automated fashion. Ontologies and semantic/ontological tools may be used to ensure the ease and quality of knowledge transfer. System decomposition methods (including functional analysis, hierarchical functional analysis, hierarchical task analysis, functional decomposition and allocation matrix, design structure matrix, and functional hazard analysis) may also be used to portray relationships between functional activities within the system and subsystems. Relationships at this level may then be used to help inform the event tree, fault tree and human and organizational effects models.

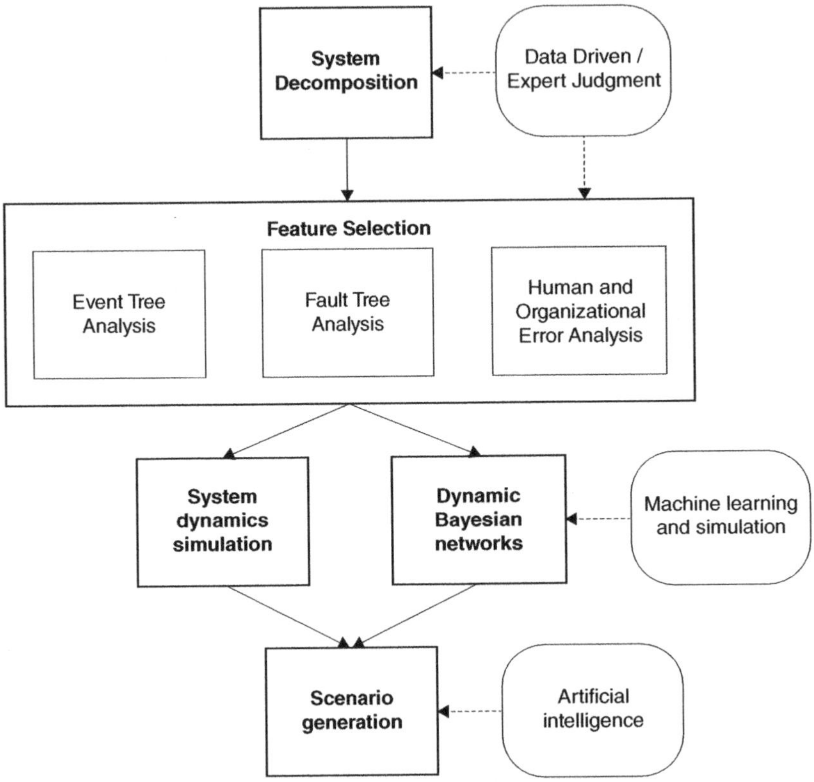

Figure 11.3 High Consequence Scenario Generation Modeling Framework for CPDDS. The analytical steps of this framework begin with system decomposition. The output of this framework is scenario generation.

System decomposition traditionally relies on system expert knowledge and concludes once sufficient detail is provided to inform construction of subsequent models. The constructs derived from the decomposition process will carry through downstream scenario generation algorithms providing traceability of scenarios. Knowledge can also be extracted from data sources to help with knowledge elicitation. To this end, we envision being able to incorporate existing data, lessons learned, safety standards, policy documents, system datasheets, and other sources of information (e.g. CPDDS simulator generated data) to help with this system decomposition task. In addition, stakeholder workshops may yield additional data and information related to the system decomposition task and help drive initial parameter sets for the feature extraction component of the CPDDS scenario generation modeling framework.

11.2.2.1.2 Feature Selection

The system decomposition phase may be followed by the identification of critical features that incorporate systemic and human operational vulnerabilities. This step may include adopting multiple probabilistic modeling techniques, including event trees, fault trees, and HOE analysis in conjunction with data-driven expert judgments. Following subsections contain brief descriptions of these methods.

Event Tree Analysis Event tree analysis is a forward, bottom-up, Boolean logic-based risk quantification technique for investigating system outcomes caused by an initiating event [49, 50]. Event trees are widely used for analyzing system consequences within chemical, nuclear, and aerospace industries. The paths created by these trees contain branch probabilities that are combined to produce path probabilities. The tree structure contains a sequential progression, over time, of the effects of initiating events on a system, and an end state or outcome is associated with each path. For CPDDSs, event trees may be generated and analyzed for multiple hazards to identify critical systemic vulnerabilities. These system features may serve as inputs to the DBNs and SD simulation.

Fault Tree Analysis Fault tree analysis is a top-down, Boolean logic-based failure analysis technique for investigating system failures/faults [49, 50]. Fault trees are also widely used for failure analysis within high-hazard industries (e.g. chemical, nuclear, and aerospace). Fault trees combine events with logic gates (AND, OR) to identify factors that contribute to various types of system failures. The probabilities of top failure events are calculated through combinations of probabilities of minimum-cut sets and their intersections. For CPDDSs, event trees may be generated and analyzed for multiple hazards to identify critical systemic vulnerabilities. These system features may serve as inputs to the DBNs and SD simulation.

Human and Organizational Error (HOE) Analysis Traditional quantitative risk analyses focus mainly on equipment and structure failure, omitting the historical evidence pointing to human and organization error as the leading cause for multi-hazard incidents [48]. Additionally, several standardized human reliability analysis methods exist that are aimed at estimating likelihoods for human errors, some of which are general enough to apply to diverse CPDDS environments. A mixture of the aforementioned methods may be used to help enumerate the feature/factor space [51].

11.2.2.1.3 Dynamic Bayesian Networks (DBN)

A Bayesian network (BN) is a graphical model that represents uncertainties (as probabilities) associated with discrete or continuous random variables (nodes) and their conditional dependencies (edges) within a directed acyclic graph [52–54]. BNs model the relationships among variables and may be updated as additional information about these variables becomes available. DBNs extend BNs with model features related over time. The network topology across time remains consistent and additional arcs are introduced to represent causal relationships among features over time. BNs and DBNs support both diagnostic (root cause) and predictive (future forecasting) analysis capabilities. Moreover, propagation of probabilistic information supports what-if analyses where the effects on downstream parameters are sought because of perturbations to upstream parameters. BNs and DBNs may be generated using critical systemic operational features through relationships identified by industry experts and ML approaches (including structure and parameter learning). Monte Carlo simulation methods will also be adopted to generate an ensemble of scenarios for further refinement.

11.2.2.1.4 System Dynamics (SD) Simulation

SD is a dynamic simulation modeling technique used to understand the dynamic behavior of complex systems [55]. SD modeling comprises of causal loops, stock and flow diagrams, feedback mechanisms, and time delays that help address nonlinearities among system features. Mathematically, SD modeling involves a system of nonlinear differential and integral equations. For example, if a feature varies due to changes in its causal features, SDs help determine the extent of the variation and the rate at which variation occurs.

Selected features derived from event tree, fault tree, and HOE analyses may inform the SD simulation environment and drive a system-level simulation over time representing CPDDS operational behaviors. The SD simulation may evaluate primacy of specific features, and feature interactions that would also generate an ensemble of scenarios that will be refined and compared with those produced using DBNs.

11.2.2.2 Consequence Assessment for CPDDSs

A discussion of multi-hazard risk assessment under uncertainty with focus on consequence estimation follows. The CPDDS challenges and system representation approaches above apply to the risk formulations below in terms of how various natural and/or man-made threat intensities impact system performance.

An overarching CPDDS consequence assessment analytical framework [56–58] consists of five phases/modules: (i) cyber-physical system (CPS) modeling, (ii) hazard intensity analysis, (iii) engineering parameter analysis, (iv) system damage analysis, and (v) loss exceedance analysis. CPS modeling may focus on novel abstractions of an embedded system (S) where computing devices communicate with each other and the physical world through sensors and actuators in a feedback loop. This consists of discrete concurrent components operating at multiple modes interacting with continuous dynamics in the physical environment. Hazard intensity analysis generates measures of intensity (IM) associated with varying hazard levels for a given CPDDS model abstraction (S); engineering parameter analysis calculates the response of a system, in terms of an engineering parameter (EP) variability, for a given IM; system damage analysis produces measures of damage (DM) to system elements using EP variability and fragility functions; and finally loss exceedance analysis entails using these DMs to develop probabilistic estimates of loss (e.g. system performance degradation) that will serve as decision variables/functions (DV) for autonomic resilience decision-making. This analytical approach characterizes uncertainty at each analysis phase and propagates it probabilistically (Eq. 11.2).

$$g\,[DV|\,S] = \iiint p\,[DV|DM,S]\,p\,[DM|EP,S]\,p\,[EP|IM,S]\,g\,[IM|S]\,dIMdEPdDM, \tag{11.2}$$

where, $g[X|Y]$ is $P(X > x|Y)$ and $p[X|Y]$ refers to the probability density of X given Y.

11.2.3 Assurance of Interconnected Networked CPDDSs

The U.S. National Academies defines resilience as *"the ability to prepare and plan for, absorb, recover from, or more successfully adapt to actual or potential adverse events"* [59]. Figure 11.4 presents an illustration of CPDDS resilience framing with varying system functionality, $f(t)$, over time, t, based on the phases of plan, absorb, recover, and adapt. An adverse event e (natural or man-made) at time, t_e, may lead to degradation in system functionality followed by system recovery. However, compound hazard events may entail additional failure events such as e' at time $t_{e'}$, that may lead to further loss of functionality resulting in a sawtooth shaped dynamic system functionality profile - with shaded region representing area of the functionality function or the overall systemic impact, a measure of system resilience. Recent literature [60] also suggests that this resilience measure is informed by the measure of risk associated with an adverse event and quantified as a combination of threat, vulnerability, and consequence. As a result, the risk measure contributes to the drop in system functionality and is a part of the resilience measure. Typically, the operational goal of a CPDDS would be to minimize the impact area (or increase resilience) through robust and proactive design, mitigation, and response/recovery decision options.

CPDDSs are complex interdependent engineered systems often represented as networks (i.e. graph G=(V,E) where, V is vertex or node set and E is edge or link set) with dynamic properties. In light of potential multi-hazard threats to CPDDSs, it is imperative for stakeholders to identify, test,

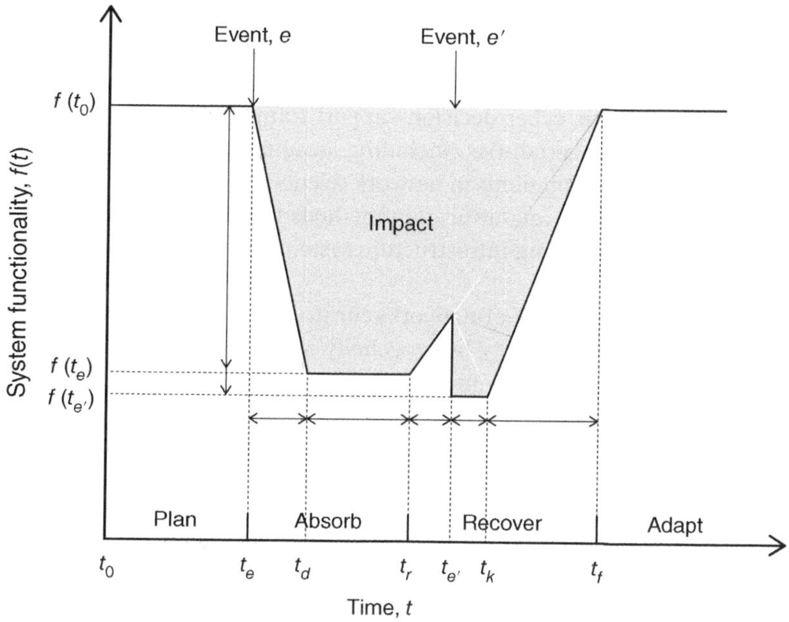

Figure 11.4 CPDDS Resilience Framing Illustration. Four phases of system resilience comprise of plan, absorb, recover, and adapt. In the plan phase, system functionality is typically optimized to operate at a desired level $f(t_0)$. Following a disruptive event, e in the absorb phase, the functionality of a system, $f(t)$ at time t degrades to $f(t_e)$. This is followed by the recover phase where system functionality begins to improve but drops again due to new event e', and eventually recovers to functionality at the start. Finally, in the adapt phase, the system maintains original functionality or in some cases may even surpass prior levels due to adaptation mechanisms.

and incorporate adaptive resilience strategies that prevent and/or mitigate impact to operations while addressing operational system complexities, dependencies, heterogeneities, and dynamics. However, translating and generalizing theoretical advances in network resilience (typically based on graph-theoretic, optimization, and percolation-based methods) to operational real-world engineering practice is a significant research challenge primarily due to incomplete and uncertain network structure information, flow-based network component heterogeneities, and dynamic system criticalities.

A well-grounded modeling approach for handling uncertainty and heterogeneity in interdependent networks is through a network-of-networks construct. Real-world networked systems interact through dependency connections, and failures originating in one network cascade to another through these connections [61]. This cascading failure process is typically modeled based on percolation theory, whereby nodes belonging to the giant mutually connected cluster remain operational and other nodes become nonoperational, as removal of nodes occur based on an initial disruptive event [62]. Theoretical advances typically categorize coupling among networks as: (i) Assortative: nodes in networks A and B are sorted in descending order of load followed by connections between them until all links are created; (ii) Disassortative: nodes in network A are sorted in descending and in B are sorted as ascending followed by connections between them until all links are created; or (iii) Random: random connection between nodes in A and B [63]. These connectivity assumptions, however, may not hold in real-world systems. Nonetheless, these theoretical advances have led to deep insights about the robustness and fragility of complex networks against random and targeted attacks, typically originating from a single point of failure. Recently, using a network science approach, Bhatia et al. [64] demonstrated the failure and recovery of a real-world rail transportation network. Halappanavar et al. [65] have been studying the combinatorial explosion problem with contingency selection applied to the real-world electric power grid and have developed efficient graph-theoretic algorithms to prune this contingency space. Also, Chatterjee et al. [66, 67] have developed a probabilistic cyber decision-support framework for quantifying mixed types of system and attack-related uncertainties (including aleatory – inherent variability and epistemic – incomplete data). These developments in network science theories and increasing applications of graph-theoretic and uncertainty quantification methods for real-world infrastructure security form the foundation for investigating infrastructure system structure, behavior, and impacts.

Developing resilience approaches with a network-of-networks construct, using real-world operational data and modeling assumptions (i.e. uncertainty, heterogeneity, and dynamic criticalities), to generate loss of functionality estimates in CPDDS will inform consequence-based decision support, risk assessments, and contingency analysis using mitigation strategies. Increasing the accuracy and precision in estimating loss of functionality in critical infrastructure systems due to cyber-enabled disruptions will fundamentally improve pre- and post-event decision making for infrastructure stakeholders thereby enhancing national security and economic prosperity. To achieve this goal, key research areas include network representation, dynamic cascade modeling, and multi-agent decision optimization.

11.2.3.1 Network Representation

Network science-based approaches may be blended with the dynamical aspects of CPDDS networks to assess robustness and recovery characteristics. Recently, Gao et al. [68] developed an analytical approach to identify control and state parameters of a multi-dimensional system for predicting behavior while separating the roles of dynamics and topology onto a universal resilience function:

$$\frac{dx_{eff}}{dt} = F\left(x_{eff}\right) + \beta_{eff} G\left(x_{eff}, x_{eff}\right). \tag{11.3}$$

Here, functions F and G represent the non-linear dynamical laws that govern the system, and β_{eff} is the single microscopic parameter that summarizes network interactions. While resilience characteristics of isolated networks have been studied from both structural and dynamical perspectives [64], resilience properties of network-of-networks have either been studied using statistically generated networks (e.g. Erd?s-Rényi models and scale-free networks), or using stylized simplified models. A key research problem is to develop an analytical approach that combines the structural and dynamical characteristics of network-of-networks as initially presented in Bhatia et al. [69] and further explored in Yadav et al. [70]. Assuming for a system with n components, the coupled non-linear dynamical equation can be written as:

$$\frac{dx_i}{dt} = F\left(x_i\right) + \sum_{j=1}^{n} A_{ij} G\left(x_i, x_j\right). \tag{11.4}$$

Here $F\left(x_i\right)$ is the internal dynamics (or self-dynamics) of each component and $G\left(x_i, x_j\right)$ represents the interaction between i^{th} and j^{th} component. For network-of-networks, equation (11.4) can be written as simultaneous differential equations to account for inter-network interactions:

$$\frac{dx_i}{dt} = F\left(x_i\right) + \sum_{j=1}^{n} A_{ij} G_1\left(x_i, x_j\right) + \sum_{k=1}^{m} A_{ik} G_2(x_i, y_k), \tag{11.5}$$

$$\frac{dy_k}{dt} = F\left(y_k\right) + \sum_{j=1}^{m} A_{kj} G_3\left(y_k, y_j\right) + \sum_{l=1}^{n} A_{kl} G_4(y_k, x_l). \tag{11.6}$$

Here, G_1 and G_3 represent the non-linear functions within the system whereas G_2 and G_4 are the functions that capture the dynamics across the system. For N-dimensional systems, the number of parameters, M, is proportional to the square of number of components making the analysis of linear stability infeasible with available tools. While the applicability of the universal resilience function in equation (11.4) [68] has been demonstrated on both built and natural systems, no formalism has been proposed yet to address the dimensionality curse and system state-space explosion for network-of-networks. To account for inter-network interactions, equations 11.5 and 11.6 above can be reformulated as:

$$\frac{dx_{eff}}{dt} = F'\left(x_{eff}\right) + \beta'_{eff} G'(x_{eff}, y_{eff}), \tag{11.7}$$

$$\frac{dy_{eff}}{dt} = F'\left(y_{eff}\right) + \beta''_{eff} G'(x_{eff}, y_{eff}). \tag{11.8}$$

where β' and β'' are macroscopic resilience parameters used to collapse the information about $N^2 + M^2 + N * M$ parameters into low dimensional parameters.

11.2.3.2 Dynamic Cascade Modeling

The notion of importance of a node in a network is typically captured using the concept of centrality. In addition to centrality-based metrics, further generalization to an influence maximization concept may enhance modeling of dynamic cascades via a percolation theory-based modeling construct. This approach leads to a generalized model for sub-modular optimization to address dynamic system criticalities. Identifying key nodes in a complex network has been a central research question spanning several disciplines ranging from power grids to epidemiology to viral marketing. Frequently used metrics such as degree and betweenness centralities do not accurately capture the underlying dynamics of a given complex network. The influence maximization approach popularized by Kempe et al. [71] provides a rigorous computational framework that builds on different models of local interaction to identify top-k influential nodes for a network by simulating the dynamics in a Monte Carlo setting. Given that it is possible to activate k nodes in a network (directed graph G=(V,E)), the influence maximization problem aims to find that particular set of k vertices called the seed set, that when activated results in maximal activations on the network among all possible such sets of k vertices. The sub-modularity property for identifying top-k influential nodes relates to the phenomenon of diminishing gains as more candidates are added in the decreasing order of the returns (for example, influence or reachability of a node). The key advantage of working with sub-modular functions is that greedy algorithms provide solutions that are $(1 - 1/e)$ or 63% optimal.

Such an influence maximization framework captures uncertainty by generating a series of realizations (samples) of the given network. For example, in the independent-cascade model, each edge (s,t) comes with a probability that represents the one-shot chance for node s to activate node t, and in the linear-threshold model a vertex can get activated if the number of neighboring vertices that are active exceeds a threshold. An important goal here is to generalize the concept of influence maximization to include multi-layer networks. Traditional infrastructure system risk assessment practice focuses on component level analysis, which relies on fragility estimates of various components against a particular hazard. However, given the inherent complexity of interdependent infrastructure systems, component-based fragility approaches have become cost and time prohibitive. To assess multi-hazard risks targeted toward network-of-networks, tailoring the optimal percolation approach using a scalable, collective influence algorithm for isolated networks may be explored [18]. This will yield the minimal set of network attributes (nodes, edges and weights) such that if the members of this set are removed, the impact on system performance can be evaluated in terms of network fragmentation. Percolation modeling will lead to understanding of network fragmentation and the size of the giant mutually connected clusters may be translated into loss of functionality (defined as number of components in the giant component) as: state of critical functionality (SCF) = *fragmented functionality* (FF)/*total functionality* (TF) in CPDDS.

11.2.3.3 Multi-Agent Decision Optimization

Following the network representation and dynamic cascade modeling for CPDDSs, a next step is to generate assured decision options via defense optimization using broad concepts of game theory and network interdiction. Game theory-based mathematical modeling approaches, involving strategic decision-makers within non-cooperative settings, are increasingly being adopted for addressing cybersecurity challenges [72–75]. Since game-theoretic approaches for cybersecurity involve adversarial agents, with unknown and/or partially-known objectives, joint attack and defense policy identification may not be possible in certain conditions. Various taxonomies for classifying game-based modeling approaches have been proposed [66]. These game formulations contain assumptions about rounds of game plays, past player actions, types of players, number of cyber-system states, number of player actions in each system state, and payoff (reward or penalty) functions associated with player actions.

Game-based attack-defense models consider complex scenarios and represent dynamic interactions effectively but could be further enhanced by increased focus on uncertainties in attacker payoff functions [76]. In a realistic setting, a defender cannot assume availability of all the necessary information - not only about the attackers, but also about their own system. Since a cyber-attacker's payoff generation mechanism is largely unknown, appropriate representation and propagation of uncertainty is a critical task. In addition, one must account for the lack or absence of perfect cyber-system state information. These uncertainties may arise due to inherent randomness or incomplete knowledge about the behavior of the system or the events affecting the system. For example, a cyber-system's state over time may be uncertain possibly due to partial observability. Moreover, there may be multiple types of attackers targeting a system at a given point in time.

Advances in state-space modeling of cyber-systems and reinforcement learning approaches for Markov decision processes have inspired the development of partially observable stochastic games (POSG) and their potential applications for cybersecurity [67, 72, 77–81]. A POSG comprises of multiple players where each player independently chooses actions, makes observations, and receives payoffs while the system state transitions based on player action combinations [67, 80].

POSGs are very general formulations, and as a result become intractable. The decision-making goal is to identify joint policies (that map from observation history and system states to actions) of players that form a Nash equilibrium. Under equilibrium conditions, no player gains by unilaterally changing their policy. Typically, these problems may be categorized into: (i) Planning - where complete specification of the cyber-system environment is known and optimal joint policies are desired; and (ii) Learning - where players need to interact with the cyber-system environment to learn about the system and one another, while updating their policies based on these interactions. Solving such problems involve iteratively finding policies that achieve high rewards on average over the long run. The objective in a POSG is typically to maximize the expected cumulative value (i.e. a function of payoffs) for each player [78, 82].

Various approaches for solving POSGs have been proposed, including: (i) dynamic programming with iterative elimination of weakly dominated strategies [80] and (ii) transformations of POSG to a series of Bayesian games (with incomplete information about other player payoffs) that have similar properties as the original POSG [81]. Recently, Tipireddy et al. [83] described as agent-centric approach for decomposing the POSG into a distribution of Partially Observable Markov Decision Process (POMDPs) where the defense agent generated optimal policies against a collection of adversarial agents. As a result, a distribution of optimal policies and recommended best options were identified. Heuristic based algorithms were used to efficiently solve the POMDPs and can be applied even for large-scale state and action settings. These distributed POMDP-based methods can also be paired with formal methods to assure safety-critical CPDDS operations. More recently, Dutta et al. [84] developed a constrained reinforcement learning (RL) approach for autonomous cyber defense that combines RL with constraints verification via satisfiability modulo theory (SMT). Simulation results indicate that a hybrid RL-SMT based cyber defense agent can learn optimal policies rapidly and defeat diversified attack strategies a majority of the time. Moreover, in realistic cybersecurity settings, insufficient and uncertain information about system properties and attacker goals may be available to a defender. Chatterjee et al. [66, 67, 76] have also proposed a probabilistic framework for quantifying attacker payoff uncertainty within a stochastic game setup that accounts for dependencies among a cyber-system's state, attacker type, player actions, and state transitions.

This approach adopts conditional probabilistic reasoning to characterize dependencies between these modeling elements. Probabilistic theories (such as total probability theorem) and functions (such as marginal and conditional) may then be applied to simulate attacker payoff probability distributions under various system states and operational actions. The framework is flexible and accounts for multiple types of uncertainties (aleatory - statistical variability and epistemic - insufficient information) in attacker payoffs within an integrated probabilistic framework [66].

Uncertainty from randomness (aleatory) is typically addressed using statistical probability distributions, while incomplete knowledge (epistemic) may be represented with mathematical intervals. Furthermore, depending on these representations, uncertainty propagation methods may include: (i) Monte Carlo sampling analysis, (ii) Interval analysis and (iii) Probability bounds analysis. Application of uncertainty propagation techniques result in the generation of probability distributions, intervals, or intervals of distributions associated with attacker payoffs, that serve as critical inputs within stochastic cybersecurity games. Note that these probabilities may be informed and updated based on empirical event and system data, simulation experiments, and/or informed judgments of subject matter domain experts. The game-theoretic and uncertainty quantification methods outlined above can be enhanced further for complex interdependent networks to model the dynamics between attackers (at different layers/networks) and defenders (acting at a given layer/network or at the system level).

11.3 Discussion and Conclusion

Current state-of-the-art in system design for CPDDSs is deficient due to imprecise specifications, inadequate environmental modeling, and poor scalability. However, fielding CPDDSs in mission- and safety-critical contexts proceeds at a rapid pace, in spite of the inability of extant formal methods to catch up with the state of practice. Therefore, for the interim, assurance practitioners pursue a process-based verification and certification approach because some assurance is better than no assurance at all. The dearth of formal method workforce means that more automated techniques need to be developed to reduce the workload of existing professionals and to reduce the barrier to entry for new professionals. Fundamental research being carried out to improve ML model explanability will help with assurance of CPDDSs. Clearly, there is much need for fundamental and applied research towards formal methods for CPDDSs.

While the Age of Information work has established a clear need for network control that is influenced by information freshness, such as in monitoring for situational awareness, there has been no clear deployed approach yet that effectively integrates the concepts from information freshness into existing cyber-physical system architectures and scenarios.

The use of commercial cloud computing resources increases the operational risk and complexity by introducing more vulnerabilities (e.g. supply chain vulnerabilities like with the Solarwinds hack) and timeliness issues, thus making the assurance of CPDDSs that use such resources challenging. Additionally, interconnected networked CPDDSs pose research challenges associated with multi-hazard risk assessment, system resilience characterization, and multi-agent decision optimization. More R&D for scalable and flexible tools and methods is needed to address these challenges in operational settings.

References

1 Alur, R. (2015). *Principles of Cyber-Physical Systems*. MIT press.

2 Abu-Ghazaleh, N., Ponomarev, D., and Evtyushkin, D. (2019). How the spectre and meltdown hacks really worked. *IEEE Spectrum* 56 (3): 42–49.

3 Brookes, J. (2021). China, Singapore line up for dumped CSIRO seL4 team. https://www.innovationaus.com/china-singapore-line-up-for-dumped-csiro-sel4-team/ (accessed 25 October 2022).

4 Chikkagoudar, S., Hagge, T., McDonald, B. et al. (2016). Information barriers for imaging: scale-invariant feature transformations and homomorphic encryption. *The 57th Annual Meeting of Institute of Nuclear Materials Management*. Institute of Nuclear Materials Management.

5 Cardenas, A., Amin, S., Sinopoli, B. et al. (2009). Challenges for securing cyber physical systems. *Workshop on Future Directions in Cyber-physical Systems Security*. DHS, July 009. http://chess.eecs.berkeley.edu/pubs/601.html (accessed 25 October 2022).

6 Kim, K.D. and Kumar, P.R. (2013). An overview and some challenges in cyber-physical systems. *Journal of the Indian Institute of Science* 93 (3): 341–352.

7 Dieber, J. and Kirrane, S. (2020). Why model why? Assessing the strengths and limitations of LIME.

8 Heracleous, C., Kolios, P., Panayiotou, C.G. et al. (2017). Hybrid systems modeling for critical infrastructures interdependency analysis. *Reliability Engineering & System Safety* 165: 89–101.

9 Lygeros, J., Tomlin, C., and Sastry, S. (2008). Hybrid Systems: Modeling, Analysis and Control. *Electronic Research Laboratory, University of California, Berkeley, CA, Tech. Rep. UCB/ERL M*, 99.

10 Brooks, R. (2021). A human in the loop: AI won't surpass human intelligence anytime soon. *IEEE Spectrum* 58 (10): 48–49. doi: 10.1109/MSPEC.2021.9563963.

11 Karpathy, A. (2014). What I learned from competing against a ConvNet on ImageNet. http://karpathy.github.io/2014/09/02/what-i-learned-from-competing-against-a-convnet-on-imagenet/ (accessed 25 October 2022).

12 Steinberg, R. (2017). Areas where artificial neural networks outperform humans. *Venturebeat*. https://bit.ly/2pFBxGk (accessed 25 October 2022).

13 Hsu, J. (2018). Starkey's AI transforms hearing aids into smart wearables. *IEEE Spectrum* 27. Available: https://spectrum.ieee.org/starkeys-ai-transforms-hearing-aid-into-smart-wearables.

14 Simonite, T. (2018). When it comes to gorillas, Google photos remains blind, *Wired*, January 2018.

15 Hunt, E. (2016). Tay, Microsoft's AI chatbot, gets a crash course in racism from Twitter. *The Guardian* 24 (3). Available: https://www.theguardian.com/technology/2016/mar/24/tay-microsofts-ai-chatbot-gets-a-crash-course-in-racism-from-twitter.

16 Nguyen, A., Yosinski, J., and Clune, J. (2015). Deep neural networks are easily fooled: high confidence predictions for unrecognizable images. *Proceedings of the IEEE Conference on Computer Vision and Pattern Recognition*, pp. 427–436.

17 Goodfellow, I.J., Shlens, J., and Szegedy, C. (2014). Explaining and harnessing adversarial examples. *arXiv preprint arXiv:1412.6572*.

18 Garfinkel, S. (2017). Hackers are the real obstacle for self-driving vehicles. *MIT Technology Review*.

19 Domingos, P. (2012). A few useful things to know about machine learning. *Communications of the ACM* 55 (10): 78–87.

20 Wolpert, D.H. and Macready, W.G. (1997). No free lunch theorems for optimization. *IEEE Transactions on Evolutionary Computation* 1 (1): 67–82.

21 Cayton, L. (2005). Algorithms for Manifold Learning. *Univ. of California at San Diego Tech. Rep*, 12 (1–17): 1.

22 Katz, G., Barrett, C., Dill, D.L. et al. (2017). Reluplex: An efficient SMT solver for verifying deep neural networks. *International Conference on Computer Aided Verification*, pp. 97–117. Springer.

23 Rodhe, I. and Karresand, M. (2015). *Overview of Formal Methods in Software Engineering*. Totalförsvarets forskningsinstitut (FOI), Swedish Defence Research Agency.

24 Gilliam, D.P., Powell, J.D., and Bishop, M. (2005). Application of lightweight formal methods to software security. *Proceedings of the Workshop on Enabling Technologies: Infrastructure for Collaborative Enterprises, WETICE*, volume 2005. https://doi.org/10.1109/WETICE.2005.19.

25 Espada, A.R., del Mar Gallardo, M., Salmerón, A. et al. (2019). A formal approach to automatically analyse extra-functional properties in mobile applications. *Software Testing, Verification and Reliability* 29 (4–5): e1699.

26 Bharadwaj, R. (1996). Tools to support a formal verification method for systems with concurrency and nondeterminism. PhD thesis. McMaster University.

27 Garavel, H., ter Beek, M.H., and de Pol, J. (2020). The 2020 expert survey on formal methods. *International Conference on Formal Methods for Industrial Critical Systems*, pp. 3–69. Springer.

28 Brat, G., Drusinsky, D., Giannakopoulou, D. et al. (2004). Experimental evaluation of verification and validation tools on Martian rover software. *Formal Methods in System Design*, volume 25. https://doi.org/10.1023/B:FORM.0000040027.28662.a4.

29 D'silva, V., Kroening, D., and Weissenbacher, G. (2008). A survey of automated techniques for formal software verification. *IEEE Transactions on Computer-Aided Design of Integrated Circuits and Systems* 27 (7): 1165–1178.

30 Cordeiro, L.C., de Lima Filho, E.B., and Bessa, I.V. (2020). Survey on automated symbolic verification and its application for synthesising cyber-physical systems. *IET Cyber-Physical Systems: Theory and Applications* 5. https://doi.org/10.1049/iet-cps.2018.5006.

31 Pinar, A. (2020). Rigorous cyber experimentation for security of cyber physical systems. *Technical report*. Livermore, CA (United States): Sandia National Lab.(SNL-CA).

32 NIST (2012). Proceedings of the Cybersecurity in Cyber-Physical Systems Workshop. https://csrc.nist.gov/publications/detail/nistir/7916/final (accessed 25 October 2022).

33 Giraldo, J., Urbina, D., Cardenas, A. et al. (2018). A survey of physics-based attack detection in cyber-physical systems. *ACM Computing Surveys (CSUR)* 51 (4): 1–36.

34 Kam, C., Kompella, S., and Ephremides, A. (2013). Age of information under random updates. *IEEE International Symposium on Information Theory - Proceedings.* https://doi.org/10.1109/ISIT.2013.6620189.

35 Kam, C., Kompella, S., and Ephremides, A. (2014). Effect of message transmission diversity on status age. *IEEE International Symposium on Information Theory - Proceedings.* https://doi.org/10.1109/ISIT.2014.6875266.

36 Kendall, D.G. (1953). Stochastic processes occurring in the theory of queues and their analysis by the method of the imbedded Markov chain. *The Annals of Mathematical Statistics* 24 (3): 338–354.

37 Newell, G. (1982). *Applications of Queueing Theory.* Springer.

38 The President's National Infrastructure Advisory Council (2018). Surviving a catastrophic power outage: how to strengthen the capabilities of the nation. https://www.dhs.gov/sites/default/files/publications/NIAC%20Catastrophic%20Power%20Outage%20Study_508%20FINAL.pdf (accessed 25 October 2022).

39 Stone, J. (2018). U.S. must prep for a cyberattack that coincides with a natural disaster, industry council says. https://www.cyberscoop.com/national-infrastructure-advisory-council-cyberattack-natural-disaster/ (accessed 25 October 2022).

40 National Science & Technology Council (2018). FY2019 Federal Cybersecurity R&D Strategic Plan Implementation Roadmap. https://www.nitrd.gov/pubs/FY2019-Cybersecurity-RD-Roadmap.pdf (accessed 25 October 2022).

41 Chatterjee, S., Brigantic, R.T., and Waterworth, A.M. (2021). *Applied Risk Analysis for Guiding Homeland Security Policy and Decisions.* Wiley.

42 Sage, A.P. (2015). *Risk Modeling, Assessment, and Management.* Wiley.

43 John Garrick, B. (2008). *Quantifying and Controlling Catastrophic Risks.* Academic Press.

44 Aven, T., Hauge, S., Sklet, S., and Vinnem, J.E. (2006). Methodology for incorporating human and organizational factors in risk analysis for offshore installations. *International Journal of Materials and Structural Reliability* 4 (1): 1–14.

45 Massoud Azizi, S. (2014). PRA application to offshore drilling critical systems. *PSAM 2014 - Probabilistic Safety Assessment and Management.*

46 Ahmad, M., Pontiggia, M., and Demichela, M. (2014). Human and organizational factor risk assessment in process industry and a risk assessment methodology (MEDIA) to incorporate human and organizational factors. *Chemical Engineering Transactions* 36. https://doi.org/10.3303/CET1436095.

47 Arnhus, M. (2014). Modeling of technical, human and organizational factors and barriers in marine systems failure risk: modeling of stability operations on a semi-submersible unit with the use of Bayesian belief networks. Master's thesis. Trondheim, Norway: Institutt for marin teknikk.

48 Wang, Y.F., Li, Y.L., Zhang, B. et al. (2015). Quantitative risk analysis of offshore fire and explosion based on the analysis of human and organizational factors. *Mathematical Problems in Engineering* 2015. https://doi.org/10.1155/2015/537362.

49 Haldar, A. and Mahadevan, S. (2000). *Reliability Assessment Using Stochastic Finite Element Analysis.* Wiley.

50 Paté-Cornell, M.E. (1984). Fault trees vs. event trees in reliability analysis. *Risk Analysis* 4 (3): 177–186.

51 Kirwan, B. (1996). The validation of three human reliability quantification techniques THERP, HEART and JHEDI: Part 1 - Technique descriptions and validation issues. *Applied Ergonomics* 27. https://doi.org/10.1016/S0003-6870(96)00044-0.

52 Jensen, F.V. and Nielsen, T.D. (2007). *Bayesian Networks and Decision Graphs*, vol. 2. Springer.

53 Murphy, K.P. (2012). *Machine Learning - A Probabilistic Perspective*. The MIT Press.

54 Koller, D. and Friedman, N. (2009). *Probabilistic Graphical Models: Principles and Techniques*. MIT Press.

55 Forrester, J. (2009). *Some Basic Concepts in System Dynamics*. Sloan School of Management, MIT.

56 Audigier, M.A., Kiremidjian, A.S., Chiu, S.S., and King, S.A. (2000). Risk analysis of port facilities. *12WCEE* (2311).

57 Porter, K.A. (2003). An overview of PEER's performance-based earthquake engineering methodology. *Proceedings of 9th International Conference on Applications of Statistics and Probability in Civil Engineering*, pp. 1–8.

58 Pant, R., Hall, J., Thacker, S. et al. (2014). TRC/University of Oxford. *National Scale Risk Analysis of Interdependent Infrastructure Network Failures Due to Extreme Hazards*. Newcastle University.

59 National Research Council (2012). Disaster Resilience: A National Imperative. https://www.nap.edu/catalog/13457/disaster-resilience-a-national-imperative (accessed 25 October 2022).

60 Linkov, I., Bridges, T., Creutzig, F. et al. (2014). Changing the resilience paradigm. *Nature Climate Change* 4 (6): 407–409.

61 Liu, X., Stanley, H.E., and Gao, J. (2016). Breakdown of interdependent directed networks. *Proceedings of the National Academy of Sciences of the United States of America* 113. https://doi.org/10.1073/pnas.1523412113.

62 Buldyrev, S.V., Parshani, R., Paul, G. et al. (2010). Catastrophic cascade of failures in interdependent networks. *Nature* 464. https://doi.org/10.1038/nature08932.

63 Tan, F., Xia, Y., and Wei, Z. (2015). Robust-yet-fragile nature of interdependent networks. *Physical Review E - Statistical, Nonlinear, and Soft Matter Physics* 91. https://doi.org/10.1103/PhysRevE.91.052809.

64 Bhatia, U., Kumar, D., Kodra, E., and Ganguly, A.R. (2015). Network science based quantification of resilience demonstrated on the Indian railways network. *PLoS ONE* 10. https://doi.org/10.1371/journal.pone.0141890.

65 Halappanavar, M., Chen, Y., Adolf, R. et al. (2012). Towards efficient N-x contingency selection using group betweenness centrality. *Proceedings - 2012 SC Companion: High Performance Computing, Networking Storage and Analysis, SCC 2012*. https://doi.org/10.1109/SC.Companion.2012.45.

66 Chatterjee, S., Halappanavar, M., Tipireddy, R., and Oster, M. (2016). Game theory and uncertainty quantification for cyber defense applications. *SIAM News* 49 (6), 1–5.

67 Chatterjee, S., Halappanavar, M., Tipireddy, R. et al., (2015). Quantifying mixed uncertainties in cyber attacker payoffs. *2015 IEEE International Symposium on Technologies for Homeland Security, HST 2015*. https://doi.org/10.1109/THS.2015.7225287.

68 Gao, J., Barzel, B., and Barabási, A.L. (2016). Universal resilience patterns in complex networks. *Nature* 530. https://doi.org/10.1038/nature16948.

69 Bhatia, U., Chatterjee, S., Ganguly, A.R. et al. (2018). Aviation transportation, cyber threats, and network-of-networks: modeling perspectives for translating theory to practice. *2018 IEEE International Symposium on Technologies for Homeland Security, HST 2018*. https://doi.org/10.1109/THS.2018.8574123.

70 Yadav, N., Chatterjee, S., and Ganguly, A.R. (2020). Resilience of urban transport network-of-networks under intense flood hazards exacerbated by targeted attacks. *Scientific Reports* 10. https://doi.org/10.1038/s41598-020-66049-y.

71 Kempe, D., Kleinberg, J., and Tardos, É. (2003). Maximizing the spread of influence through a social network. *Proceedings of the ACM SIGKDD International Conference on Knowledge Discovery and Data Mining*. https://doi.org/10.1145/956750.956769.

72 Lye, K.W. and Wing, J.M. (2005). Game strategies in network security. *International Journal of Information Security* 4. https://doi.org/10.1007/s10207-004-0060-x.

73 Roy, S., Ellis, C., Shiva, S. et al. (2010). A survey of game theory as applied to network security. *Proceedings of the Annual Hawaii International Conference on System Sciences*. https://doi.org/10.1109/HICSS.2010.35.

74 Liang, X. and Xiao, Y. (2013). Game theory for network security. *IEEE Communication Surveys and Tutorials* 15. https://doi.org/10.1109/SURV.2012.062612.00056.

75 Do, C.T., Tran, N.H., Hong, C. et al. (2017). Game theory for cyber security and privacy. *ACM Computing Surveys (CSUR)* 50 (2): 1–37.

76 Chatterjee, S., Tipireddy, R., Oster, M.R., and Halappanavar, M. (2015). A probabilistic framework for quantifying mixed uncertainties in cyber attacker payoffs. *National Cybersecurity Institute Journal* 2 (PNNL-SA-114140).

77 Ramuhalli, P., Halappanavar, M., Coble, J., and Dixit, M. (2013). Towards a theory of autonomous reconstitution of compromised cyber-systems. *2013 IEEE International Conference on Technologies for Homeland Security (HST)*, pp. 577–583. IEEE.

78 Oliehoek, F., Spaan, M., Robbel, P., and Messias, J.V. (2009). MADP Toolbox 0.2. *Technical report*. Informatics Institute, Amsterdam University.

79 MacDermed, L., Isbell, C., and Weiss, L. (2011). Markov games of incomplete information for multi-agent reinforcement learning. *Workshops at the Twenty-Fifth AAAI Conference on Artificial Intelligence*.

80 Hansen, E.A., Bernstein, D.S., and Zilberstein, S. (2004). Dynamic programming for partially observable stochastic games. *AAAI Workshop - Technical Report*, WS-04-08.

81 Chatterjee, S., Tipireddy, R., Oster, M., and Halappanavar, M. (2016). Propagating mixed uncertainties in cyber attacker payoffs: exploration of two-phase Monte Carlo sampling and probability bounds analysis. *2016 IEEE Symposium on Technologies for Homeland Security, HST 2016*. https://doi.org/10.1109/THS.2016.7568967.

82 Sutton, R.S. and Barto, A.G. (2018). *Reinforcement Learning: An Introduction*. MIT Press.

83 Tipireddy, R., Chatterjee, S., Paulson, P. et al. (2017). Agent-centric approach for cybersecurity decision-support with partial observability. *2017 IEEE International Symposium on Technologies for Homeland Security, HST 2017*. https://doi.org/10.1109/THS.2017.7943478.

84 Dutta, A., Al-Shaer, E., and Chatterjee, S. (2021). Constraints satisfiability driven reinforcement learning for autonomous cyber defense. *arXiv preprint arXiv:2104.08994*.

12

Vulnerabilities in IoT Systems

Zheng Fang and Prasant Mohapatra

Department of Computer Science, University of California, Davis, CA, USA

Abstract

IoT systems are finding increasing use in tactical systems where they help in applications related to sensing and actuation to aid various control systems. Since the IoT systems are built using several devices and components, ensuring device or component-level security is not adequate. Integration of these devices to build a system introduces a broad range of vulnerabilities that are very critical for tactical applications. In this Chapter, we provide a detailed view of these vulnerabilities in IoT systems that are deployed in tactical applications.

12.1 Introduction

Internet of Things (IoT) is one of the most booming trends in recent years. We are seeing an increasing number of smart home platforms, such as Samsung SmartThings [1], Google Home [2], Philips Hue [3], and so on. IoT systems have also being applied to many other scenarios, such as healthcare [4, 5], industrial manufacturing [6, 7], and battlefield [8, 9], etc. It is predicted that IoT market size will reach $1.6 trillion by 2025 [10]. The number of IoT devices connected to the internet is projected to be 30.9 billion by 2025 [11].

However, numerous vulnerabilities have been reported in IoT systems, threatening system security and even users' safety. This is largely due to the fact that most of the IoT manufacturers focus on functionality to gain market share and tend to ignore security issues. Another reason could be that IoT developers lack a sense of security when designing systems or implementing IoT functionalities.

The first common vulnerability is **unprotected network services**, meaning that the services do not do authentication, or are authenticated using weak or hard-coded credentials. Many IoT devices' firmware runs unprotected network APIs, such as thermostats [12], smart speakers [13],and Wi-Fi bulbs [14], etc. The APIs may be exploited by attackers on the same LAN. The Mirai malware [15] utilized such a vulnerability and created botnets to launch distributed denial-of-service (DDoS) attacks on well-known sites like OVH [16] and Dyn [17].

Moreover, some IoT devices have **insecure firmware update mechanism**. For example, HP printers do not verify the validity of the firmware before performing firmware update [18]. Essentially, the printers' remote firmware update (RFU) feature is enabled by default, and they accept any firmware sent remotely to port 9100 without authentication. The impact of such a vulnerability

IoT for Defense and National Security, First Edition. Edited by Robert Douglass, Keith Gremban, Ananthram Swami, and Stephan Gerali.

is huge because, after replacing the firmware with a malicious one, the attacker gets full control of the device. He can further use the compromised printer to attack other devices in the IoT system.

Last but certainly not least, a popular IoT platform is found to have **over-privileging** issues. Fernandes et al. [19] analyzed the SmartThings platform [1] and discovered design flaws in its access control and event subsystem. They applied static analysis, runtime testing, and manual analysis on 499 SmartThings IoT apps (called SmartApps) and 132 device handlers, and found that over 55% of the SmartApps are over-privileged because of the coarse-grained access control model. Moreover, if a SmartApp gains access to a device, it is granted all the IoT events sent by the device. This leads to information leakage. The authors also identify an event spoofing vulnerability in the platform.

Sometimes, even though a newer version of the firmware with a patch is released, the firmware's upgrade rate is low. Cui et al. [18] conducted an internet-wide scanning for IPv4 IPs to compute the firmware update rate for HP printers. The results show that two months after the announcement of the vulnerability [20] and the release of firmware updates, only 1.08% of vulnerable HP LaserJet printers were patched worldwide. What's worse, some of the firmware developers ignore the vulnerability reports [12, 14].

A unique feature of the IoT systems is that devices may interact with each other via physical channels. For example, a heater can change the room temperature, and the temperature increase will be detected by a temperature sensor in the IoT system. This can potentially incur unexpected chains of device actions or even be employed by an attacker. IoTMon is a static framework which identifies hidden inter-app chains based on physical channels using static analysis and natural language processs (NLP) techniques. In [21], Ding et al. identified four features of physical interactions of IoT devices: spatial context, temporal context, implicit effect, and joint effect, which are further discussed in Section 12.5. They designed and implemented a dynamic safety and security policy enforcement framework to detect risky physical interactions.

IoT apps are unique to IoT systems. They allow users to customize IoT devices' behaviors and usually introduce interaction between different devices. Researchers are paying increasing attention to IoT app security as IoT apps can be chained via physical channels or device actions, causing the system to enter unexpected states. IoT apps can also be leveraged by adversaries to launch complicated IoT attacks.

The interactive nature of different IoT components makes it critical to consider IoT system security in a holistic way, including both cyber and physical attacks. Since IoT systems involve different components which can be vulnerable, usually there are multiple attack paths to a certain resource in an IoT system. Hence, system defenders should evaluate IoT security at the system level to uncover as many threats as possible and to provide complete guidance on system hardening.

In the following sections of this chapter, we discuss vulnerabilities and threats discovered on various components of IoT systems, and the IoT platform itself. We also discuss countermeasures and design principles to make IoT systems more secure.

12.1.1 IoT System Components

Most of the existing IoT platforms, e.g. [1, 2, 22–24] etc., share similar architecture. Figure 12.1 illustrates a typical IoT system, which consists of 6 components: devices, communication protocols, IoT cloud, IoT apps, physical environment, and mobile apps. IoT *devices* include sensor, actuator, and infrastructure devices such as Wi-Fi router and various gateways. The devices communicate with each other and the corresponding gateways using wireless, low-power, and short-range *communication protocols*. IoT *applications* are programmed using trigger-action paradigm, where triggers are some cyber/physical events or device status change and actions are devices' actions,

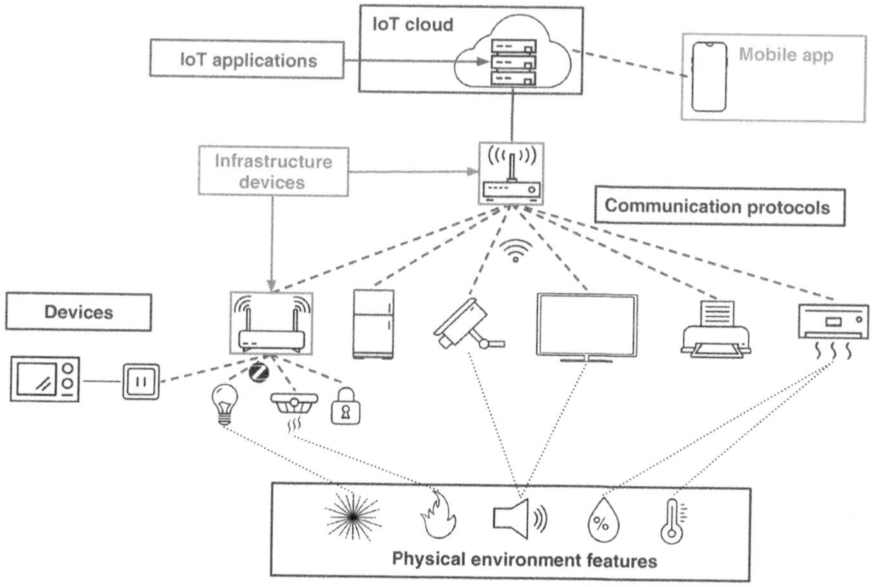

Figure 12.1 Typical IoT system and its components. IoT devices can sense environment events or change environment features, causing a chain of effects.

such as turning on the AC and locking the door, etc. The *mobile apps* are used for device setup, remote monitoring and control. IoT *cloud* hosts IoT apps and provides many other features like device management and analytics. The *physical environment* is a necessary component of IoT systems because they sense and modify physical environment features. For example, if a humidifier is turned on, the smart home's humidity will increase, which will be captured by the humidity sensor.

12.1.2 Vulnerabilities and Threats

Due to the rapid expansion of IoT market and a lack of security consciousness, vulnerabilities in IoT systems are ubiquitous. And interactive nature of IoT components can escalate the impact of a security breach in unanticipated ways. In this subsection, we explain the functionality of each IoT component, outline the potential vulnerabilities and threats, and discuss the potential impact on other components.

12.1.2.1 Devices

We use the term "IoT devices" to refer to both end devices connected to the IoT network and network infrastructure devices such as routers and gateways. Most of the device vulnerabilities are rooted in the firmware [25–28]. However, some vulnerabilities are found in the device's physical components such as gyroscopic sensors [29] and microphone [30].

A critical but often neglected threat to IoT devices relates to vulnerabilities of companion mobile apps, such as hard-coded keys. For example, the issue with August lock mobile app enables the attacker to decrypt communication between the lock base station and the mobile phone, resulting in Wi-Fi credential leakage if the attacker is physically adjacent to the smart home [31]. Moreover, the hard-coded key of the Eques plug mobile app makes it possible for the attacker to send spoofed commands to the remote server, allowing him to control end devices through the internet [32].

Once a device is compromised, it can be used to attack other components of the IoT system in three different ways. First of all, if the attacker gets root privilege to a fully-fledged device such as a router or camera, he can send malicious commands to other devices on the same network. Second, the attacker can utilize the compromised device to inject physical events, such as changing the room's illuminance or injecting audio/video. Moreover, the attacker can take advantage of the devices' physical dependency to compromise other devices. For example, he can disable air conditioner by simply launching a DoS attack to the smart outlet it plugs in, or he can stop the smart sprinkler by compromising the water valve.

12.1.2.2 Communication Protocols

IoT systems utilize short-range, low-power protocols such as Zigbee and Bluetooth to communicate with end devices. As a result, these end devices are first connected to a gateway (also called base station, bridge or hub) in order to communicate with the remote cloud. Notice that in an IoT system each protocol can have *multiple network instances*. For example, a smart home user can have both SmartThings [1] and Xiaomi [33] devices; even if they both use Zigbee protocol, there are two different Zigbee networks, one for the Smartthings platform and the other for Xiaomi. For system security analysis, we should consider multiple wireless network instances individually, because compromise of the first network instance does not necessarily lead to compromise of the second one.

Some devices use plaintext or hard-coded key [34] to transmit Wi-Fi credentials during the device setup phase. As a result, an attacker can easily steal the credentials and access the Wi-Fi network. Since there is no firewall or medium access control MAC address filtering in most of the smart home networks, if the attacker gains access to the network, he can freely send packets to other devices on the same network. To make things worse, many IoT devices, such as router, camera, or thermostat, expose unprotected network services to the home network, making it possible for the attacker to further compromise these devices after he joins the home network.

12.1.2.3 IoT Applications

Typically, IoT applications can be designed using *if-this-then-that* paradigm, where *this* represents a trigger and *that* means some action. A trigger can be either an individual IoT event, such as device state change, or sensor input. An action is some IoT device behavior, such as turning on the AC.

IoT apps automate device behavior and this results in device dependency, which exposes extra attack surface to the adversary. Consider "If smoke is detected, sound the alarm and open the window." as an example. To make the victim's window open, the attacker does not have to directly attack the window opener; instead, he can just compromise the smoke detector, and the IoT app will do the rest of the attack for him. Even though the attacker has to know the existence of such an app before he launches the attack, researchers have shown this can be done via wireless sniffing [35, 36].

For an IoT system, when multiple IoT apps are installed, they can issue conflicting commands, or create infinite loops, making it necessary to analyze all of the installed IoT apps together, which is detailed in Section 12.4.

12.1.2.4 Physical Medium

One of the unique features of IoT systems compared with other networked systems is that IoT devices can interact with each other via the physical medium. For example, if the door lock is locked, then the door opener cannot make the door open. There are multiple physical channels, seven of which were identified in [37] and are explained in Section 12.5. Different from IoT apps-based device dependency which only exist if the app is installed by the user, physical device dependencies always exist in an IoT system.

Table 12.1 IoT components and typical vulnerabilities/threats.

IoT system component	Vulnerabilities/Threats
IoT app	Conflicting commands Repeated commands Sniffing
Mobile app	Malware Cryptographic misuse
Device firmware	Unprotected network service Unprotected firmware updating process Buffer overflow
Device hardware	Spoofing DoS Unprotected buttons/pins
Communication protocol	Sniffing Spoofing Jamming
Physical medium	Event injection Device chaining

An attacker can utilize various physical dependencies to compromise IoT devices. For instance, he can turn off the AC or light bulb plugged into a smart outlet by launching a denial-of-service (DoS) attack on the outlet.

12.1.2.5 Mobile Apps

IoT mobile apps are used to setup the corresponding IoT devices. Besides, both IoT devices and mobile apps have connections to the remote cloud server so that users can utilize mobile apps to remotely control IoT devices. The most common type of vulnerability in mobile apps is cryptographic misuse. In particular, there are several reports [31, 32] about hard-coded keys in different apps, resulting in Wi-Fi credential leak or remote device control.

Table 12.1 summarizes common vulnerabilities/threats in all of the core IoT components. In the following sections, we survey recent research on these vulnerabilities, discuss the root causes, and describe some mitigating methods.

12.2 Firmware

IoT devices' firmware is the software providing all control, monitoring, and networking functionalities. Nowadays, most of the IoT devices support firmware update to fix bugs or to add new features to the device. However, numerous vulnerabilities have been discovered and reported about firmware. Even the firmware update process itself can be vulnerable [18].

12.2.1 Unprotected Network Services

Many IoT devices host network services, such as Telnet, SSH, HTTP, etc. These services are used for different purposes. For example, HTTP provides some Web APIs for the user to control the device.

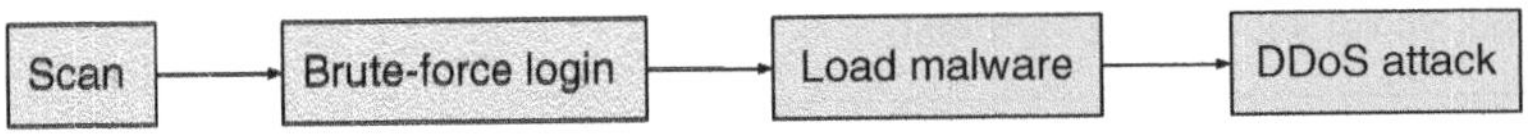

Figure 12.2 Mirai botnet formation and DDoS attack.

And the Telnet or SSH service can be used for debugging or interactive device configuration. However, these services should be protected with proper authentication. Otherwise, once the attack joins the IoT LAN, he can easily control these devices. According to OWASP [38], weak, guessable, or hard-coded passwords is the top 1 threat in IoT systems, which means many IoT devices' firmware fail to implement proper authentication.

Even though this kind of vulnerability seems naive, due to the ubiquity of IoT devices, these vulnerable devices can wreak havoc on large sites. A recent prominent example is the Mirai botnet [15], which first started in September 2016. Attackers have utilized Mirai botnets to launch distributed denial-of-service (DDoS) attacks on websites like Krebs on Security [39], OVH [16], and Dyn [17]. The formation of Mirai botnet and the DDoS attack to a victim server is shown in Figure 12.2.

The first step is rapid scanning, where the Mirai malware sends TCP SYN probes to random public IPv4 addresses on Telnet ports, i.e. TCP port 23 and port 2323. If Mirai finds a potential victim, it then enters the brute-force login step, in which it tries to log into the Telnet shell using 10 username-password pairs randomly chosen from the pool of 62 pre-set credentials. If the login is successful, the malware reports the victim device information and the corresponding credentials to the command and control (C&C) server. Then the loader program infects the victim by logging into the victim, downloading and executing the malware. Now the victim becomes a bot and listens to the commands from the C&C server. On receiving the attack command, all of the bots will launch DDoS attack on the target server.

In addition to forming botnets, IoT devices running unprotected network services can also be exploited individually by attackers to change device status or environment features of the IoT systems. For example, Dan Crowley [40] uncovers unprotected HTTP APIs on Radio Thermostat CT80 and CT50. The APIs are not encrypted and do not require authentication. Any attacker on the same LAN as the thermostat can send spoofed HTTP requests to fetch device information or to change the status of the device. The complete list of APIs can be found on [41]. As another example, CVE-2019-18980 [42] describes the unprotected APIs discovered on Philips Wiz A19 Wi-Fi smart bulbs. The bulb listens to UDP port 38899. The command is formatted as JSON string and there is no authentication or encryption required. Hence, as long as the attacker is on the IoT LAN, he can completely control the bulb via these unprotected APIs. For example, he can turn the bulb on or off, or change its color or brightness, etc.

Brannon Dorsey finds unprotected HTTP APIs on Roku TV, Google Home, and Sonos WiFi Speakers. They allow attackers on the LAN to control the devices without authentication. Brannon designed a proof-of-concept DNS rebinding attack to show that such vulnerabilities even allows adversaries on the internet to control these devices. Details can be found in Section 12.3.6.

12.2.2 Unprotected Firmware Updating

Some IoT devices fail to verify the validity of the firmware image before updating. Cui et al. [18] demonstrate how vulnerable firmware updating process can be exploited to allow attackers to inject malicious firmware into HP LaserJet printers and get root privilege on them.

CVE-2011-4161 [43] reveals that the RFU feature is enabled by default, which allows attackers to send malicious firmware remotely to port 9100 without authentication. Cui implemented a proof-of-concept malware, which can be injected into the printer via standard Printer Job

Language (PJL) commands or be embedded in PostScript files. The malware can have various capabilities, such as sniffing the network traffic, building a reverse IP tunnel to penetrate perimeter firewalls, or allowing the attacker to control the device interactively.

The authors conducted exhaustive scans of IPv4 IPs and identified over 90,000 unique vulnerable printers inside government, educational, military, and enterprise organizations, etc. They also found that the firmware update rate is alarmingly slow. Two months after the release of the official patch, the overall patch rate was only 1.08%. This phenomenon indicates that firmware developers should design mechanisms to prompt users to update device firmware or implement automatic firmware update so that discovered firmware vulnerabilities can be patched in a timely manner. Last but not least, developers should integrate code signing into the firmware so that the validity of the newer firmware is verified before the updating process.

12.2.3 Buffer Overflow

Buffer overflow is another type of threat rooted in IoT firmware. Many IoT devices are found to be running network services vulnerable to buffer overflow attack. The result of such an attack could be denial-of-service or full device control.

Researchers from Tactical Network Solutions [44] demonstrate buffer overflow vulnerability on D-Link DCS series cameras. The vulnerability exists in `alphapd`, the cameras' web server. The attacker on the LAN can launch a buffer overflow attack by providing a string longer than 0× 28 bytes as the value of the `WEPEncryption` parameter when requesting `wireless.htm` file using HTTP API. As a result, the attacker can execute arbitrary code on the camera.

As another example, researchers from Cisco Talos identify a buffer overflow vulnerability, CVE-2018-3902 [45], on Samsung SmartThings Hub STH-ETH-250 devices with firmware version 0.20.17. The firmware is Linux-based and runs a series of daemons managing communication in different protocols, including Ethernet, ZigBee, Z-Wave and Bluetooth. In particular, the `video-core` process incorrectly parses the user-controlled JSON payload sent as HTTP POST data, leading to a stack buffer overflow.

The `hubCore` process is used to communicate with the remote SmartThings cloud via a persistent TLS connection. The `video-core` process handles live streaming by connecting to the smart camera on the same local network via RTSP protocol. The `video-core` hosts an HTTP server listening to the `localhost` address with port 3000. The remote cloud may communicate with the `video-core` by sending HTTP requests to the `hubCore`, which will relay the requests to the HTTP server of the `video-core`. Therefore, an adversary can impersonate the remote cloud and send HTTP requests to the `hubCore` process, which forwards the message to the vulnerable `video-core` process. As a result, the attacker gets root privilege on the SmartThings hub.

12.3 Communication Protocols

There are many wireless communication protocols developed for low-cost, low-power, low-data-rate wireless IoT networks. The most popular ones are Wi-Fi, Zigbee, Z-Wave, and Bluetooth. Design flaws in communication protocols have huge impacts on IoT security because they are the industry standards — If a certain protocol is found vulnerable, all of the devices using such a protocol will be affected.

Table 12.2 compares the data rate, range, and power consumption of the common wireless protocols used in IoT systems. From the table, we can see that Wi-Fi has the highest data rate and

Table 12.2 Specs of common wireless protocols for IoT devices.

Protocol	Data rate	Range	Power draw
Wi-Fi (802.11ax)	Up to 3.5 Gbps	Up to 92 m	High
Zigbee	Up to 250 kbps	Up to 100 m	Low
ZWave	Up to 40 kbps	Up to 100 m	Low
Bluetooth low energy (BLE)	Up to 2 Mbps	Up to 60 m	Low
Bluetooth 5	Up to 3 Mbps	Up to 240 m	Medium

highest power consumption, so devices requiring heavy data transmission such as cameras and smart speakers use Wi-Fi protocol, and are connected to an AC outlet. Simple sensors such as motion sensor or temperature sensor usually use low power consumption protocols such as Zigbee or ZWave. In the following subsections, we discuss various vulnerabilities in these protocols.

12.3.1 Wi-Fi

Wi-Fi is a suite of wireless network protocols based on the IEEE 802.11 family of standards. Many IoT devices communicate using Wi-Fi for its high data rate. Typical examples are cameras, smart speakers, TVs, gaming consoles, etc. Wi-Fi security is very important in smart home IoT systems because Wi-Fi devices usually have multiple functionalities and oftentimes home routers are the entry point of the attacks.

Vanhoef and Ronen [46] present several attacks against the Dragonfly handshake [47] of WPA3 [48] and EAP-pwd [49]. This is really alarming because many of the vulnerabilities identified in [46] are related to protocol design. Below are some of the novel attacks described in that paper.

Downgrade & Dictionary Attack Against WPA3-transition: In WPA3 there is a transition mode so that a WPA3 access point (AP) can be connected to clients that only support WPA2. In this mode, a Wi-Fi network supports both WPA3 and WPA2 with an identical password. In WPA2, the attacker only needs to capture a single authenticated 4-way handshake to launch a dictionary attack. Therefore, the adversary can set up a rogue AP to broadcast a WPA2 network with the same SSID. As a result, the victim will transmit an authenticated message to the rogue AP, and the attacker can now launch a dictionary attack.

Security Group Downgrade Attack: There is another design flaw that the WPA3's Dragonfly handshake lets the client choose different elliptic curve or security groups. So the adversary can send spoofed commit frames to Wi-Fi clients to force the client to choose a weak security group.

Cache-based Side-channel Attack: If the malicious process is running on the same device as the benign process, then the adversary can monitor cache access patterns on the victim machine. They can further use the leaked patterns to launch a dictionary attack by comparing the pattern with the one measured when running the handshake protocol with a guessed password. The authors of [46] also considers this vulnerability as a WPA3's design flaw.

Denial-of-service Attack: The designers of Dragonfly try to prevent attackers from abusing the high overhead of its hash-to-curve method by requiring the client to reflect a secret cookie sent by the AP before the AP processes the client's commit frame. However, the adversary can easily capture and replay secret cookies. The authors describe a proof-of-concept attack using a Raspberry Pi device with a WNDA3200 Wi-Fi dongle and a professional AP. They successfully launch a DoS attack to the AP by spoofing 8 commit exchanges per second using curve P-521.

12.3.2 Zigbee

Zigbee is a suite of communication protocols built on top of the IEEE 802.15.4 standard. It is widely used in low-power, low-bandwidth IoT systems such as smart home and smart healthcare. Figure 12.3 shows the Zigbee protocol stack. Zigbee has its own network layer and application layer. The application (APL) layer consists of the application support sub-layer (APS), ZigBee device objects (ZDO), and application objects which are defined by the manufacturer.

Huang and Lin [50] discovered that several Xiaomi Zigbee devices (e.g. Xiaomi DGNWG03LM, ZNCZ03LM, MCCGQ01LM, etc.) use the default trust center link key, which has been leaked, to encrypt the network key when they join the Zigbee network created by the ZigBee coordinator (i.e. Xiaomi DGNWG03LM). After getting the network key, the attacker can send spoofed messages to the device to get device information or take over the device. In [51], Huang et al. found an issue in Xiaomi DGNWG03LM and ZNCZ03LM devices. The attacker can repeatedly send spoofed ZigBee trust center rejoin requests to make ZNCZ03LM irresponsive to the user's requests.

ZigBee Light Link (ZLL) differs from the Zigbee protocol as there is no Coordinator or Trust Centre in ZLL. All of the devices in the ZLL network share the same network key to encrypt/decrypt communications between them. When a new device is trying to join the ZLL network, the network key is encrypted using the ZLL master key which is pre-installed in all ZLL-certified nodes during manufacture.

Ronen et al. [52] describe a threat in ZLL standard by creating a worm targeting Philips Hue smart bulbs. The worm will spread over large areas if the density of the vulnerable devices exceeds a certain value. The vulnerability in the Atmel's implementation of the ZLL Touchlink protocol makes it possible for the attacker to bypass proximity check to reset the light bulbs and to join the attacker's Zigbee network. Another implementation flaw is that the code does not verify the destination address field properly. This allows the attacker to factory reset all the bulbs in the Zigbee range in batch by sending the reset command as a broadcast message.

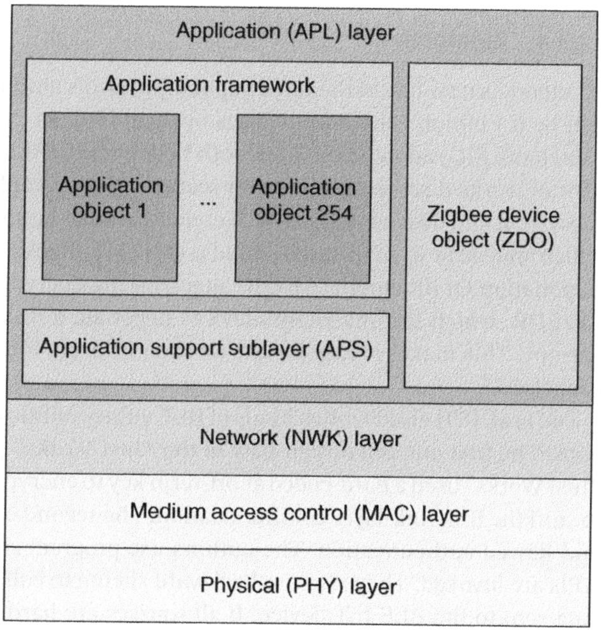

Figure 12.3 Zigbee protocol stack. The blue boxes are defined by Zigbee standard while the red boxes are defined by Zigbee device manufacturers.

The worm can have far-reaching impact. For example, by updating the firmware of the light bulbs, the attacker can brick the device, turn on/off the compromised bulbs, or even trigger epileptic seizures. Moreover, the attacker can cause wireless network jamming by setting the devices in "test mode", which transmits wave signals on any of th 2.4 GHz 802.11 channels.

12.3.3 Z-Wave

Z-Wave is another low-power protocol suite widely used in IoT systems. Its protocol stack consists of Physical layer, MAC layer, Network layer, and Application layer. Z-Wave's physical and MAC layer are defined following the ITU-T G.9959 standard [53].

Fouladi and Ghanoun [54] discover two implementation vulnerabilities of the Z-Wave protocol in a Z-Wave smart lock. The first one encrypts the network key using a hard-coded key of 16 bytes of zeros, making it easy for the attacker to steal the Z-Wave network key. The second vulnerability is due to a lack of state validation in the key exchange protocol. More specifically, upon receiving a key exchange start packet, the device does not check whether a previously provisioned network key already exists before proceeding. As a result, the attacker can overwrite the doorlock's network key and take full control of the device.

The most up-to-date Z-Wave security paring process, S2, seeks to resolve the network key encryption vulnerability by using Diffie-Hellman key exchange and adding a 5-digit authentication code to the controller (optional). However, to retain backward compatibility with older Z-Wave devices, the controller will change from S2 to S0 if the device only supports S0 pairing protocol, which effectively defeats the enhanced security brought by S2. This is a design choice so all Z-Wave devices are susceptible to this vulnerability.

Researchers from Pen Test Partners present a Z-Wave downgrading attack on Yale Conexis L1 smart lock [55]. Any active attacker present at the time of device pairing can send spoofed packets to the controller. As a result, the controller will pair with the device using S0 protocol. This not only allows the attacker to sniff the network key and control the device, but also puts all of the devices on the Z-Wave network paired using S0 protocol at risk.

12.3.4 Bluetooth

Bluetooth is now one of the most popular protocols among IoT devices. According to Statista, there will be 6.4 billion Bluetooth devices by 2025 [56]. In 2010, the Bluetooth Special Interest Group (Bluetooth SIG) adopted the Bluetooth Low Energy (BLE) as a subset of Bluetooth v4.0. There exist a lot of works discussing Bluetooth security. For example, Cyr et al. [57] analyzed the Fitbit Flex ecosystem and discovered that BLE credentials can be sniffed during the pairing process over TLS, which may lead to man-in-the-middle (MITM) attacks. Antonioli et al. [58] present a novel Key Negotiation Of Bluetooth (KNOB) attack on the encryption key negotiation protocol of Bluetooth BR/EDR, which lets the victim devices negotiate a link-layer encryption key with only 1 byte of entropy. This makes it easy for the attacker to brute force the encryption key and manipulate the packets.

Zuo et al. [59] discover two kinds of BLE vulnerabilities in IoT devices and the companion mobile apps. The first one is a design flaw of the "Just Works" pairing protocol. Before BLE versions 4.2, "Just Works" used a hard-coded short-term key to encrypt the long-term key, enabling the adversary to sniff the BLE link-layer communication. The second one is the absence of encryption-decryption and flawed authentication. The authors use program slicing to check whether the cryptographic APIs are invoked. They also use backward slicing to collect all of the sources that contribute to the data sent to the BLE IoT device. If all sources are hard-coded, then the authentication is flawed.

They also collected BLE UUIDs to fingerprint BLE IoT devices by conducting static analysis of mobile apps.

Recently, von Tschirschnitz et al. [60] describe a design flaw in the pairing mechanism of Bluetooth v5.2, which allows two devices to pair using different methods. They exploit this method confusion to launch a MITM attack. Since BLE devices have different input-output capabilities (IOCap)s, there are two variations of the attack: Passkey on Numeric (PoN) and Numeric on Passkey (NoP). They evaluated their attack using several smartphones, a smartwatch, headphones, and a banking device. After establishing the MITM position, the authors were able to relay, eavesdrop, and manipulate all the communication between the two BLE devices.

12.3.5 Physical Layer

Jamming and side channel attack are common threats to the physical layer of IoT communication protocols. The jamming attack can be used alone to make certain devices or channels irresponsive. Moreover, it can also be used as an intermediary step of some complicated attacks, such as the MITM attack described in [60]. The side channel attacks are often used by the attacker to infer critical information such as encryption algorithms and cryptographic keys.

12.3.5.1 Jamming Attack

In jamming attack, the adversary makes IoT devices unavailable or irresponsive by injecting unwanted wireless signals into the communication channel [61]. Since a typical IoT system involves more than one wireless protocols and different protocols utilize overlapped channel bands, IoT devices are especially vulnerable to jamming attacks.

In [52], the authors describe a possible jamming attack scenario where they can cause a malicious firmware update to Philips Hue Zigbee bulbs to jam the 802.11 Wi-Fi traffic. This is because Zigbee devices operate at 2.4 GHz frequency band, which is widely used by many standards such as IEEE 802.11 b/g. The malware can change the bulbs to "test mode", which transmits continuous wave signals without first checking whether the channel is clear. The test signal can be used to jam Wi-Fi or other protocols using 2.4 GHz bands, such as Thread or 6LoWPAN.

In [62], the authors describe three different jamming attacks to Wi-Fi devices. First, the authors present a continuous jamming attack to make the channel unusable by other devices. The second type is selective jamming attack, where the attacker targets a specific Wi-Fi device and jams specific frames already in the air. The last one is a novel channel-based MITM attack, achieved by using continuous jamming to force the client to connect to the rogue AP.

12.3.5.2 Side Channel Attack

When IoT devices do computation and communicate with each other, they are generating a lot of physical signals, such as electromagnetic signals, sound and heat. This makes them vulnerable to side channel attacks. Most of the reported side channel analysis of IoT systems are on inferring the encryption algorithms and extracting cryptographic keys from the devices.

For example, Camurati et al. [63] present a side channel attack on mixed-signal chips (i.e. chips having both analog and digital circuits on the same die), which are commonly used in wireless communication protocols, such as Bluetooth and Wi-Fi. The paper identified a new type of threat called *screaming channels*, i.e. electromagnetic leakage in digital circuits picked up and transmitted by radio transceivers. To exploit such a vulnerability, the authors proposed two novel side channel analysis techniques: correlation radio analysis (CRA) and template radio analysis (TRA). Using only the radio signal the chip emits and the knowledge of plaintexts, they recovered the AES key from a distance of up to 10 m.

In the work done by Ronen et al. [52] described in Section 12.3.2, the authors employ correlation power analysis (CPA) to break Philips' cryptographic bootloader, which uses counter with CBC-MAC (CCM) authenticated encryption mechanism. The authors launched the CPA attack to retrieve the CBC MAC state and utilized a differential power analysis (DPA) attack to retrieve all of the round keys. As a result, they can create and re-encrypt any malicious firmware, and sign it with the newly calculated MAC.

The side channel attack described in [63] can be performed at a distance of 10 m, while the one described in [52] involves correlation power analysis and differential power analysis and requires hands-on access to the device.

12.3.6 TCP/IP Suite & Application Layer

Though typically most of the end devices in an IoT system communicate using low-power protocols, there are devices running the internet protocol suite directly, such as cameras, routers, TVs, and various gateways, etc., which connects to the system's local area networks (LAN)s using Wi-Fi or Ethernet. Worse still, in many LANs, there is no firewall; and these devices may expose unprotected ports to the LAN. Hence, many IoT systems are vulnerable to common network or web security threats such as sniffing, DNS rebinding, cross-site scripting (XSS), or command injection attack, etc.

CVE-2018-13022 [64] reports a reflected cross-site scripting vulnerability in the Xiaomi Mi Router 3 device version 2.22.15. When some malformed requests sent to the router's web server trigger a 404 Not Found error, the request URL is reflected in the error body. However, due to the lack of an `X-Content-Type-Options: nosniff` header, certain web browsers, such as Internet Explorer, may interpret the page as HTML. As a result, the attacker can execute arbitrary JavaScript code on the victim's browser by tricking the victim into clicking the malformed request URL.

Unprotected network services, i.e. services without proper authentication, encryption, or input/output filtering, are among the top IoT vulnerabilities according to OWASP [38]. While unprotected ports are directly exploitable when the attack is on the IoT LAN, DNS rebinding attack enables the adversary on the internet to compromise the vulnerable IoT devices on the LAN.

Brannon Dorsey describes a DNS Rebinding attack on several IP-based smart home devices [13], including smart speaker, smart TV, thermostat, and a gateway. The root cause is that these devices expose unprotected APIs to the private network. In this attack, the adversary sets up a malicious DNS server and tricks the user into clicking a malicious webpage. The DNS response's expiration time is set to a very small value, e.g. 2/,s. The malicious JavaScript code repeatedly make HTTP POST requests to the malicious webpage. Soon the DNS entry becomes stale and the browser makes another DNS lookup. When the malicious DNS server receives the browser's second DNS request, it responds with the IP address of the vulnerable device. Hence, the HTTP POST request is sent to the device.

Chau et al. [65] utilizes symbolic execution to analyze real implementations of SSL/TLS libraries used mainly in IoT systems for *noncompliance*, i.e. the X.509 certificate chain validation logic (CCVL) is over-permissive, over-restrictive, or both. If a CCVL is overly permissive, the corresponding IoT device is vulnerable to the man-in-the-middle attack. They analyzed 9 implementations and uncovered 48 noncompliances.

12.4 IoT Apps

Different from other wireless networking systems, IoT systems allow users to install IoT apps to automatically control the behavior of the IoT devices based on sensor inputs and IoT devices' states. IoT apps are designed in *trigger-action* paradigm, which can be described in natural language

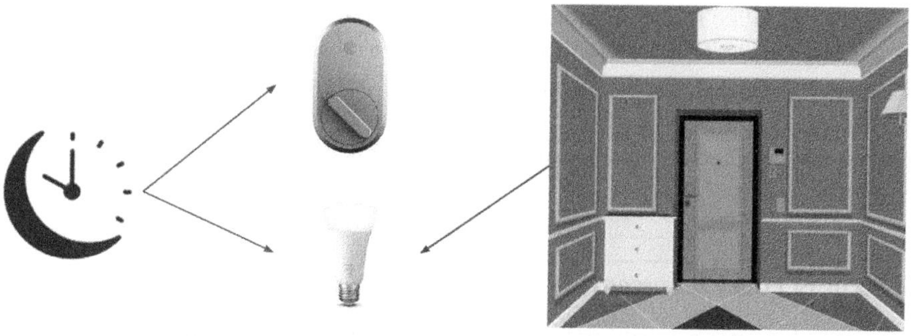

Figure 12.4 App1: If it's 11 p.m., and the hallway light is off, then turn on the hallway light. App2: If no one is at home, and the hallway light is on, then turn off the hallway light. Source: zuich/Adobe Stock.

using *if-this-then-that* format where *this* represents physical or cyber events and *that* represents device actions.

However, there are multiple threats introduced by IoT apps. First, different apps can issue conflicting commands to a certain device. Suppose there are two IoT apps installed in a smart home with the first one specifying "When there is smoke, the door must be unlocked", and the second one saying "When a camera does not recognize a face, the door must be locked." If the smart home is on fire and firefighters arrive at the home for rescue, the first app will try to unlock the door, but the second app will keep the door locked.

Second, two IoT apps can be chained if the first app's action is the input of the second app. When multiple IoT apps are installed, there can be infinite loops due to app chaining. Consider Figure 12.4 as an example. The semantics of the two apps are described in the caption below the figure. If it is after 11 p.m. and no one is at home, then the hallway light will be turned on and off indefinitely.

Moreover, IoT apps can be exploited by the attacker. For example, if the user installs an app which opens the window when room temperature is high, then the attacker whose goal is to break into the home can just try to compromise the heater or the thermostat to increase the room temperature, and the IoT app will trigger a window opening command.

Finally, the attacker can design and publish malicious IoT apps, which may steal sensitive information, such as the doorlock's pin-code, or trigger malicious behavior such as unlocking the door when no one is at home. IoTBench [66] collects 19 hand-crafted malicious SmartThings' SmartApps [1] that contain data leaks. IoTCom [67] also created one malicious IoT app which changes the `location.mode` system status variable.

12.4.1 Checking Safety and Security Properties

Nguyen et al. [68], Celik et al. [69], Wang et al. [70], and Trimananda et al. [71] utilize model checking to detect app conflicts and violation of pre-defined safety and security properties. They come up with dozens of safety and security properties for IoT systems. They also translate IoT apps' source code and the properties into formats amenable to model checking.

To alleviate state space explosion problem, Nguyen et al. [68] designs two optimizations: only considering apps that interact with each other, and removing unnecessary interleaving unlikely to yield useful assessment of unsafe behaviors. In addition, the authors propose a method to attribute the cause of property violations. Celik et al. [69] also addresses the state explosion problem by discretizing numerical-valued variables representing system states, such as temperature, humidity, and battery level. Evaluations show that the number of states of the compressed model is orders of magnitude smaller than that of the original model.

Alhanahnah et al. [67] introduce IoTCom, a static analysis framework to automatically discover hidden and unsafe IoT app interactions. The authors identify and formally define seven types of multi-app coordination threats and develop a formal model of IoT systems based on relational logic. For evaluation, they collected 3732 IoT apps and created 622 non-overlapping app bundles, each consisting of 6 apps. IoTCom detected 2883 violations and it runs much faster than [68], consuming 92.1% less time on average.

12.4.2 Dynamic Security Policy Enforcement

Celik et al. [72] present a dynamic security policy enforcement framework for IoT systems. The framework consists of three components: a *code instrumentor* which collects apps' information at runtime, a *data collector* which stores the apps in a dynamic model representing app's execution, and a security service which enforces the security policies on the dynamic model of individual apps or sets of IoT apps using an information flow analysis algorithm. Experiments on both hand-crafted and real-world IoT apps show that the framework can accurately identify policy violations.

Zhang et al. [36] designs a framework to monitor IoT apps' behaviors. It first analyzes the IoT app source code and the app descriptions to build deterministic finite automaton (DFA) model for the app. Then it leverages wireless fingerprinting to infer the state transition of the DFA. If the inferred DFA transition is different from the DFA generated from the IoT app's source code, then an unexpected app behavior is detected.

12.4.3 IoT App Sniffing

In order for adversaries to utilize IoT apps to attack a certain IoT system, they have to know the IoT apps installed on the system. HoMonit [36] is a dynamic system which monitors the execution of the IoT apps with the help of encrypted wireless traffic. In this work, the authors describe a side-channel attack which enables the adversary to infer IoT events by sniffing encrypted wireless traffic. To launch such an attack, the adversary can first fingerprint IoT events using the encrypted packet sequences sent by devices at hand. With the fingerprints, the adversary can sniff the encrypted wireless traffic of a smart home and compare the sniffed packets with the fingerprints.

Gu et al. [73] propose an algorithm to automatically discover IoT event dependencies based on long sniffed packet sequence. The key observation is that if two IoT events belong to the same IoT app, then they usually appear together with small latency. Thus, if two events are always temporally correlated in the long packet sequence, they are from the same app or app chain. After identifying all the app-based event pairs, the adversary can infer event sequences from the same app or chained apps using the algorithm presented in the paper. This work demonstrates that IoT system defenders cannot rely on security by obfuscation and assume the attacker has no knowledge about the IoT apps installed.

12.5 Physical Dependencies

Another important feature of IoT is that devices can impact and sense the physical environment. For example, an AC can change the temperature of a smart home, and the temperature drop will be sensed by a temperature sensor. Such physical dependencies may generate unexpected IoT app chaining or even be exploited by attackers to compromise an IoT system.

Ding and Hu [37] create a framework, IoTMon, to automatically discover potential physical interactions across different IoT apps. First, they conduct static program analysis to find triggers and actions for the IoT app. Then, they utilize NLP techniques to analyze IoT app descriptions by extracting entity keywords, computing similarities between each pair of keywords, and clustering keywords based on the similarities. The authors analyzed 185 open-source SmartThings SmartApps and identified seven physical channels: temperature, humidity, illumination, location, motion, smoke, and leakage. They further discovered 162 hidden interaction chains among different SmartApps, 37 of which are highly risky. Notice that these seven physical channels are not exhaustive — there could be other physical channels, such as radiation, chemicals, and biotoxins, etc.

In 2021, Ding et al. [21] propose a dynamic safety and security enforcement framework called IoTSafe. This is the first work on dynamic testing for IoT physical interaction discovery. In the paper, they identify four critical features of IoT physical interaction. The *spatial context* refers to the location information of IoT devices, which affects the physical dependencies between different devices. For example, if the heater and the temperature are in different rooms, then these two devices may not be chained via the room temperature. The *temporal context* relates to the time interval for the physical interaction to happen. For instance, turning on the light bulb will immediately increase the illuminance, while it may take dozens of minutes for the temperature sensor to capture the temperature change after a heater is turned on.

The *implicit effect* means that one device's action may have unintended physical impacts. For example, a heater may implicitly reduce the room's humidity; a fan or a cleaning robot may cause the motion detector to report motion. *Joint effect* describes implied physical influence when multiple devices are functioning together. Consider a thermostat's heating app which turns on the heater if the temperature is 68 °F, the heater's behavior which will eventually increase the room temperature by 10 °F, and another app which opens the window when room temperature is above 75 °F. The joint effect of these two IoT apps is window opening.

Given an IoT system instance, IoTSafe utilizes static analysis based on deployed apps to construct static interaction graphs. Then it generates test cases for each group of devices, tests devices' actions in parallel, and collects the sensors' readings. After the testing process, the framework generates real physical interaction paths by removing invalid paths in the static interaction graph and adding implicit interactions. After that, the runtime prediction module generates models for physical channels and maintains the models as users change the trigger condition of an IoT app. Finally, the policy enforcement module utilizes a control server to compare the users' specified policy and the current IoT system state, and identify policy violations.

12.6 Companion Mobile Apps

Most of the existing IoT devices have companion mobile apps for device setup, and remote monitoring and control. In a typical IoT infrastructure, a companion mobile app only connects to the device's remote cloud, and the device also maintains a persistent connection with the cloud. However, such design creates additional attack vectors enabling remote adversaries to exploit IoT devices from the internet.

In 2019, a researcher identified a vulnerability in the mobile app of a smart plug [32], which uses a hard-coded AES-256 key to encrypt the communication between the plug and the mobile app, making it possible for the attacker to sniff the wireless traffic or to inject commands to the victim plug.

The smart plug has a local network service listening to UDP port 27431. The vendor's cloud server hosts an Apache MINA server listening to UDP port 9123, and an Extensible Messaging and Presence Protocol (XMPP) server listening to port 5222. After wireless sniffing and analyzing the mobile app's source code, the author finds that the AES encryption function is imported from a shared object (`.so`) file for 32-bit ARM binaries. Using the reverse engineering tool Ghidra, the author extracts the hard-coded key from the `.so` file. With the hard-coded AES key, the attacker can send commands to the remote Apache MINA server to query the smart plug's information and status. He can also send spoofed commands to the XMPP server to control the smart plug over the internet.

The hard-coded AES key is a specific instance of cryptographic misuse. By *cryptographic misuse* we mean the app developers do not invoke the cryptographic APIs with proper arguments. For example, as the use of RSA algorithm with 1024-bit keys is likely to be brute-forced, when invoking RSA cipher the `key length` argument should not be smaller than 1024. A violation of this rule is considered as a cryptographic misuse. Such vulnerabilities are common in IoT mobile apps, allowing attackers to sniff IoT devices' information and potentially control victim devices. Rahaman et al. [74] design and implement Cryptoguard, a high-precision cryptographic misuse detection tool for Java and Android projects based on program slicing.

12.7 Hardware

The hardware of an IoT device involves its electronic circuits, mechanical components, micro-electromechanical system (MEMS), ports, buttons, and pins, etc. However, numerous attacks on IoT devices have been reported due to a lack of physical protection; some ports, buttons, or even debugging pins are easily accessible to the adversaries. Recently, researchers have found vulnerabilities on MEMS components.

Sugawara et al. [30] discover a novel type of vulnerability on MEMS microphones of smart speakers, such as Google Home, Echo Dot, and Facebook Portal Mini, etc. The attacker can inject arbitrary audio signals to the smart speakers by aiming laser beams at the vulnerable microphones. As the intensity of the laser beam is proportional to the supplied current, the authors convert the intensity of the sound signal to that of the light signal by utilizing the laser driver to perform amplitude modulation (AM) of the laser diode's current. Experimental results show that it is possible to launch the attack from a distance up to 110 m. The impact of this vulnerability can be huge, because nowadays many IoT systems are equipped with smart speakers, and the voice commands can be used to control other IoT devices, such as doorlocks or heaters, etc.

Son et al. [29] present a DoS attack on drones by generating acoustic noise at the resonant frequencies of the dones' MEMS gyroscopes. A drone's flight attitude control system takes gyroscope data as one of the inputs. However, the MEMS gyroscope may generate unexpected data at its resonance frequency, causing the attitude control system to malfunction. The authors tested 15 kinds of MEMS gyroscopes from four vendors and found the resonance frequencies for 7 of them. In real-world attack experiments on two target drones, the authors successfully made one of the drones to fall down, using a small Bluetooth speaker as the sound source at a a distance of 10 cm from the target gyroscope.

There are many attacks on electronic circuits. For example, Kune et al. [75] demonstrated that specially crafted electromagnetic interference (EMI) signals can be injected into the analog-to-digital converters (ADC)s of implantable medical devices and consumer electronics to generate fake sensing signals. Shoukry et al. [76] showed that it is possible to spoof the measurements of anti-lock braking systems (ABS) by attacking magnetic wheel speed sensors using EMI.

In 2015, researchers from Pen Test Partners [77] discovered a hardware vulnerability in Ring doorbell, which allows the attacker outside the smart home to gain access to the home's Wi-Fi network. The Ring doorbell is mounted on the door with two Torx T4 screws. So it is very easy for the attacker to take off the doorbell, flip it over, and press the setup button. This triggers the doorbell to enter AP mode and creates a temporary Wi-Fi network. Since there is no authentication required, the attacker can connect to the AP and steal the home Wi-Fi's credentials by sending a REST API to the doorbell's vulnerable webserver.

12.8 IoT Platforms

There are dozens of commodity IoT platforms, such as Samsung SmartThings [1], Google Home [2], Apple HomeKit [22], and Philips Hue [3], etc. These platforms share similar architecture as shown in Figure 12.1 in Section 12.1. In addition, they have similar access control mechanism — using the abstraction of *capabilities* to represent commands or device statuses. Each IoT app is associated with a set of capabilities for it to function. However, researchers have discovered serious design flaws on popular IoT platforms. It is difficult to fix platforms' design flaws as all of the IoT apps developed specifically for a certain platform have to be re-written to adapt to the changes.

Fernandes et al. [19] analyzed the Samsung SmartThings platform and discovered two critical design flaws in two aspects: (i) the SmartThings capability model and (ii) the event subsystem. In addition, the authors constructed four proof-of-concept attacks to demonstrate how these design flaws could be exploited by attackers.

12.8.1 Over-privileging

Fernandes et al. [19] found that SmartApps (IoT apps for the Samsung SmartThings platform) are significantly over-privileged. They identified two types of over-privileging due to the granularity of the access control model.

Coarse-grained Capabilities: A capability consists of commands and attributes, representing the device's functionality and status, respectively. A SmartApp is over-privileged if the set of the requested commands and attributes is larger than the set of the commands and attributes needed. The authors discovered that 55% of the SmartApps are over-privileged due to the capabilities being too coarse-grained. For example, if an app requests the `lock` command of `capability.lock`, the app also gets access to the `lock` command.

Coarse SmartApp–SmartDevice Binding: In SmartThings, a SmartApp may get access to capabilities it does not explicitly request — it will automatically gets access to all commands and attributes of all the capabilities associated with the device handler if the user selects a certain device to be used by the SmartApp. By comparing the set of the capabilities granted to an SmartApp and the set of the capabilities used, the authors conclude that 42% of the SmartApps end up gaining unnecessary capabilities.

12.8.2 Data Leakage

Event sniffing: Fernandes et al. [19] discovered that the SmartThings platform fails to properly protect device events. If a SmartApp is granted access to a device, the SmartApp is allowed to subscribe to all of the events published by that device. This leads to serious data leakage as devices usually use events to communicate sensitive data. For example, some smart locks use `codeReport` events to transmit locks' pin-codes in plaintext. In the paper the authors designed a proof-of-concept attack to sniff the smart lock's pin-code using a malicious SmartApp claiming to only monitor the lock's battery life. With the sniffed pin-code, the attacker can break into the smart home.

Event spoofing: In SmartThings, there is no access control for raising events, so any SmartApp can spoof device events and location-related events. An event object contains various state information, a location ID, and a hub ID, which can be easily obtained by the attacker. Fernandes et al. [19] demonstrate this vulnerability by designing a proof-of-concept attack to trigger a fake alarm. In the attack scenario, there is a CO alarm app subscribing to the events sent by a CO detector, and a malicious SmartApp. The malicious app spoofs an event for CO detection, and the alarm app will catch this event and trigger the alarm.

The authors reported the vulnerabilities to the SmartThings team in 2015. Since then, the team started a new SmartThings platform whose capability model has been improved dramatically. In 2020, the platform discussed in [19] was shut down completely [78].

12.9 Countermeasures

In this section, we try to give some coutermeasures to help IoT system designers to reduce the chance of introducing vulnerabilities or design flaws. We also discuss existing challenges and call for novel tools to improve IoT system or component security.

We should analyze IoT security in a holistic view. Due to ubiquitous component interaction, to analyze the potential impacts of an individual vulnerability, we need to consider all of the components of an IoT system. Fang et al. [79] come up with a novel IoT hypothesis graph to detect vulnerabilities at the system level.

The design of access control in IoT apps and devices should follow the least privileging principle. This is because the effects of access control violations are asymmetric. For example, it may cause inconvenience if an IoT app is denied access to door unlock operation, but the results might be dangerous if the app is accidentally granted access to such an operation.

IoT devices should not expose unprotected network service. There are numerous vulnerability reports based on unprotected network services. The IoT firmware developer should follow this principle to reduce attack vectors. It is desirable to run simple tests for this type of vulnerability before delivering firmware.

The firmware update process should be protected. New firmware images should be validated before the update process. Otherwise, the attacker can inject arbitrary malicious firmware and gain full control of the device.

Implementing automatic firmware update or prompting the user to update firmware. As firmware and protocol vulnerabilities are being discovered, it is ideal to keep the firmware up-to-date so that vulnerabilities are patched in a timely manner.

There is a need for mobile app and IoT app sanitization. To prevent attackers from publishing malicious IoT apps on the app market, we need tools which can automatically detect malicious apps with high precision. Similarly, tools are needed for cryptographic API misuse detection in IoT mobile apps.

Do not expose unnecessary physical buttons or debugging pins. This can help reduce physical attacks.

12.10 Conclusions

In this chapter, we identified core components of modern IoT systems, and gave a comprehensive view of the IoT vulnerabilities/threats in different components by surveying recent research papers and vulnerability reports. From these concrete examples, we can see that vulnerabilities are ubiquitous in each IoT component. Most of the discovered vulnerabilities may also occur in various IoT scenarios, including healthcare, battlefield, industry, or agriculture, etc.

Given that IoT systems are distributed, consisting of multiple highly interactive components with numerous vulnerabilities, hardening IoT systems is extremely challenging. This is because there can be multiple attack paths leading to the same effect. Besides, since the impact of a vulnerability can propagate via different components, the attack paths can also be very deep, making it difficult to anticipate by IoT system designers. As a result, it is necessary to view IoT security in a holistic way, even if the goal is to harden a specific device.

Finally, we discussed some security principles in IoT system design and present countermeasures to help harden IoT systems.

References

1 SmartThings. https://smartthings.developer.samsung.com (accessed 1 November 2022).

2 Google Nest. https://developers.google.com/home?hl=en (accessed 1 November 2022).

3 Philips Hue. https://developers.meethue.com/develop (accessed 1 November 2022).

4 Catarinucci, L., De Donno, D., Mainetti, L. et al. (2015). An IoT-aware architecture for smart healthcare systems. *IEEE Internet of Things Journal* 2 (6): 515–526.

5 Baker, S.B., Xiang, W., and Atkinson, I. (2017). Internet of Things for smart healthcare: technologies, challenges, and opportunities. *IEEE Access* 5: 26521–26544.

6 Sanchez-Iborra, R. and Cano, M.-D. (2016). State of the art in LP-WAN solutions for industrial IoT services. *Sensors* 16 (5): 708.

7 Chen, B., Wan, J., Shu, L. et al. (2017). Smart factory of industry 4.0: key technologies, application case, and challenges. *IEEE Access* 6: 6505–6519.

8 Abdelzaher, T., Ayanian, N., Basar, T. et al. (2018). Will distributed computing revolutionize peace? The emergence of battlefield IoT. In *2018 IEEE 38th International Conference on Distributed Computing Systems (ICDCS)*, pp. 1129–1138. IEEE.

9 Farooq, M.J. and Zhu, Q. (2018). On the secure and reconfigurable multi-layer network design for critical information dissemination in the internet of battlefield things (IoBT). *IEEE Transactions on Wireless Communications* 17 (4): 2618–2632.

10 Forecast end-user spending on IoT solutions worldwide from 2017 to 2025. https://www.statista.com/statistics/976313/global-iot-market-size (accessed 1 November 2022).

11 Internet of Things (IoT) and non-IoT active device connections worldwide from 2010 to 2025. https://www.statista.com/statistics/1101442/iot-number-of-connected-devices-worldwide (accessed 1 November 2022).

12 No Authentication Vulnerability in Radio Thermostat. https://www.trustwave.com/en-us/resources/security-resources/security-advisories/?fid=18874 (accessed 1 November 2022).

13 Brannon Dorsey. Attacking private networks from the internet with DNS rebinding. https://medium.com/@brannondorsey/attacking-private-networks-from-the-internet-with-dns-rebinding-ea7098a2d325 (accessed 1 November 2022).

14 Eric Pendergrass. Cheap, Hackable IoT Light Bulbs (or, Philips Bulbs Have No Security). https://blog.dammitly.net/2019/10/cheap-hackable-wifi-light-bulbs-or-iot.html (accessed 1 November 2022).

15 Antonakakis, M., April, T., Bailey, M. et al. (2017). Understanding the Mirai botnet. *26th USENIX Security Symposium (USENIX Security 17)*, pp. 1093–1110.

16 Klaba, O. Octave Klaba Twitter. https://twitter.com/olesovhcom/status/778830571677978624 (accessed 1 November 2022).

17 Hilton, S. DYN analysis summary of Friday October 21 attack. http://hub.dyn.com/dyn-blog/dyn-analysis-summary-offriday-october-21-attack (accessed 1 November 2022).

18 Cui, A., Costello, M., and Stolfo, S. (2013). When firmware modifications attack: a case study of embedded exploitation. *Network and Distributed System Security Symposium (NDSS).*

19 Fernandes, E., Jung, J., and Prakash, A. (2016). Security analysis of emerging smart home applications. *2016 IEEE Symposium on Security and Privacy (SP)*, pp. 636–654. IEEE.

20 HP. HPSBPI02728 SSRT100692 rev.7 - Certain HP Printers and HP Digital Senders, Remote Firmware Update Enabled by Default. https://support.hp.com/us-en/document/c03102449 (accessed 1 November 2022).

21 Ding, W., Hu, H., and Cheng, L. (2021). IoTSafe: Enforcing safety and security policy with real IoT physical interaction discovery. *Network and Distributed System Security Symposium (NDSS).*

22 Apple HomeKit. https://developer.apple.com/homekit (accessed 1 November 2022).

23 OpenHAB. OpenHAB. https://www.openhab.org (accessed 1 November 2022).

24 Alexa. https://developer.amazon.com (accessed 1 November 2022).

25 Costin, A., Zaddach, J., Francillon, A., and Balzarotti, D. (2014). A large-scale analysis of the security of embedded firmwares. *USENIX Security.*

26 Costin, A., Zarras, A., and Francillon, A. (2016). Automated dynamic firmware analysis at scale: a case study on embedded web interfaces. *Asia CCS.*

27 Hernandez, G., Fowze, F., Tian, D. et al. (2017). FirmUSB: Vetting usb device firmware using domain informed symbolic execution. *ACM Conference on Computer and Communications Security (CCS).*

28 Shoshitaishvili, Y., Wang, R., Hauser, C. et al. (2015). Firmalice-automatic detection of authentication bypass vulnerabilities in binary firmware. *NDSS.*

29 Son, Y., Shin, H., Kim, D. et al. (2015). Rocking drones with intentional sound noise on gyroscopic sensors. *USENIX Security.*

30 Sugawara, T., Cyr, B., Rampazzi, S. et al. (2020). Light commands: laser-based audio injection attacks on voice-controllable systems. *USENIX Security.*

31 Molly Price. August smart locks could be giving hackers your Wi-Fi credentials. https://www.cnet.com/home/security/august-smart-locks-could-be-giving-hackers-your-wi-fi-credentials/ (accessed 1 November 2022).

32 CVE-2019-15745, Eques Elf Smart Plug. https://nvd.nist.gov/vuln/detail/CVE-2019-15745 (accessed 1 November 2022).

33 Xiaomi. https://global.developer.mi.com/ (accessed 1 November 2022).

34 CVE-2019-17098, August Lock. https://nvd.nist.gov/vuln/detail/CVE-2019-17098 (accessed 1 November 2022).

35 Gu, T., Fang, Z., Abhishek, A. et al. (2020). IoTGaze: Iot security enforcement via wireless context analysis. *IEEE INFOCOM.*

36 Zhang, W., Meng, Y., Liu, Y. et al. (2018). HoMonit: Monitoring smart home apps from encrypted traffic. *CCS.*

37 Ding, W. and Hu, H. (2018). On the safety of IoT device physical interaction control. *ACM Conference on Computer and Communications Security (CCS).*

38 OWASP. OWASP IoT top 10. https://wiki.owasp.org/index.php/OWASP_Internet_of_Things_Project#tab=IoT_Top_10 (accessed 1 November 2022).

39 Krebs, B. Krebs on security. https://krebsonsecurity.com/ (accessed 1 November 2022).

40 CVE-2013-4860, Radio Thermostat. https://nvd.nist.gov/vuln/detail/CVE-2013-4860 (accessed 1 November 2022).

41 Radio Thermostat Company of America. Radio Thermostat APIs. https://github.com/brannondorsey/radio-thermostat/blob/master/RTCOAWiFIAPIV1_3.pdf (accessed 1 November 2022).

42 CVE-2019-18980, Philips Wiz A19 Wi-Fi Smart Bulb. https://nvd.nist.gov/vuln/detail/CVE-2019-18980 (accessed 1 November 2022).

43 CVE-2011-4161, HP Printers. https://nvd.nist.gov/vuln/detail/CVE-2011-4161 (accessed 1 November 2022).

44 Tactical Network Solutions. D-link camera exploit. https://github.com/tacnetsol/CVE-2019-10999 (accessed 1 November 2022).

45 CVE-2018-3902, Samsung SmartThings Hub. https://nvd.nist.gov/vuln/detail/CVE-2018-3902 (accessed 1 November 2022).

46 Vanhoef, M. and Ronen, E. (2020). Dragonblood: Analyzing the dragonfly handshake of WPA3 and EAP-PWD. *IEEE Symposium on Security and Privacy.*

47 Harkins, D. (2008). Simultaneous authentication of equals: a secure, password-based key exchange for mesh networks. *2008 Second International Conference on Sensor Technologies and Applications (sensorcomm 2008)*, pp. 839–844. IEEE.

48 Wi-Fi Alliance. WPA3 specification version 3.0. https://www.wi-fi.org/download.php?file=/sites/default/files/private/WPA3_Specification_v3.0.pdf (accessed 1 November 2022).

49 Network Working Group. Extensible authentication protocol (EAP). https://datatracker.ietf.org/doc/html/rfc3748 (accessed 1 November 2022).

50 Huang, Y.-C. and Lin, H.-Y. CVE-2019-15913. https://github.com/chengcheng227/CVE-POC/blob/master/CVE-2019-15913.md (accessed 1 November 2022).

51 Huang, Y.-C. and Lin, H.-Y. CVE-2019-15914. https://github.com/chengcheng227/CVE-POC/blob/master/CVE-2019-15914_1.md (accessed 1 November 2022)

52 Ronen, E., Shamir, A., Weingarten, A.-O., and Flynn, C.O. (2017). IoT goes nuclear: creating a ZigBee chain reaction. *IEEE Symposium on Security and Privacy (SP).*

53 International Telecommunication Union. G.9959: Short range narrow-band digital radiocommunication transceivers - PHY, MAC, SAR and LLC layer specifications. https://www.itu.int/rec/T-REC-G.9959 (accessed 1 November 2022).

54 Fouladi, B. and Ghanoun, S. (2013). Security evaluation of the Z-Wave wireless protocol. *Black Hat USA* 24: 1–2.

55 Tierney, A. Z-Shave. Exploiting Z-Wave downgrade attacks. https://www.pentestpartners.com/security-blog/z-shave-exploiting-z-wave-downgrade-attacks/ (accessed 1 November 2022).

56 Vailshery, L.S. Bluetooth device shipments worldwide from 2015 to 2025. https://www.statista.com/statistics/1220933/global-bluetooth-device-shipment-forecast/ (accessed 1 November 2022).

57 Cyr, B., Horn, W., Miao, D., and Specter, M. (2014). *Security Analysis of Wearable Fitness Devices (Fitbit)*, vol. 1. Massachusetts Institute of Technology.

58 Antonioli, D., Tippenhauer, N.O., and Rasmussen, K.B. (2019). The KNOB is broken: exploiting low entropy in the encryption key negotiation of bluetooth BR/EDR. *28th USENIX Security Symposium (USENIX Security 19)*, pp. 1047–1061.

59 Zuo, C., Wen, H., Lin, Z., and Zhang, Y. (2019). Automatic fingerprinting of vulnerable BLE IoT devices with static UUIDs from mobile apps. *Proceedings of the 2019 ACM SIGSAC Conference on Computer and Communications Security*, pp. 1469–1483.

60 von Tschirschnitz, M., Peuckert, L., Franzen, F., and Grossklags, J. (2021). Method confusion attack on bluetooth pairing. *Proceedings of the IEEE Symposium on Security and Privacy (S&P).*

61 Xu, W., Trappe, W., Zhang, Y., and Wood, T. (2005). The feasibility of launching and detecting jamming attacks in wireless networks. *Proceedings of the 6th ACM International Symposium on Mobile Ad Hoc Networking and Computing*, pp. 46–57.

62 Vanhoef, M. and Piessens, F. (2014). Advanced Wi-Fi attacks using commodity hardware. *Proceedings of the 30th Annual Computer Security Applications Conference*, pp. 256–265.

63 Camurati, G., Poeplau, S., Muench, M. et al. (2018). Screaming channels: when electromagnetic side channels meet radio transceivers. *Proceedings of the 2018 ACM SIGSAC Conference on Computer and Communications Security*, pp. 163–177.

64 CVE-2018-13022, Xiaomi Mi Router. https://nvd.nist.gov/vuln/detail/CVE-2018-13022 (accessed 1 November 2022).

65 Chau, S.Y., Chowdhury, O., Hoque, E. et al. (2017). SymCerts: Practical symbolic execution for exposing noncompliance in X.509 certificate validation implementations. *2017 IEEE Symposium on Security and Privacy (SP)*, pp. 503–520. IEEE.

66 Celik, Z.B. IoTBench test-suite. https://github.com/IoTBench/IoTBench-test-suite (accessed 1 November 2022).

67 Alhanahnah, M., Stevens, C., and Bagheri, H. (2020). Scalable analysis of interaction threats in IoT systems. *Proceedings of the 29th ACM SIGSOFT International Symposium on Software Testing and Analysis*, pp. 272–285.

68 Nguyen, D.T., Song, C., Qian, Z. et al. (2018). IotSan: Fortifying the safety of IoT systems. *International Conference on Emerging Networking EXperiments and Technologies (CoNEXT).*

69 Celik, Z.B., McDaniel, P., and Tan, G. (2018). Soteria: Automated IoT safety and security analysis. *USENIX ATC 18.*

70 Wang, Q., Datta, P., Yang, W. et al. (2019). Charting the attack surface of trigger-action IoT platforms. *CCS.*

71 Trimananda, R., Aqajari, S.A.H., Chuang, J. et al. (2020). Understanding and automatically detecting conflicting interactions between smart home IoT applications. *Proceedings of the 28th ACM Joint Meeting on European Software Engineering Conference and Symposium on the Foundations of Software Engineering*, pp. 1215–1227.

72 Celik, Z.B., Tan, G., and McDaniel, P.D. (2019). IoTGuard: Dynamic enforcement of security and safety policy in commodity IoT. *NDSS.*

73 Gu, T., Fang, Z., Abhishek, A., and Mohapatra, P. (2020). IoTSpy: Uncovering human privacy leakage in IoT networks via mining wireless context. *2020 IEEE 31st Annual International Symposium on Personal, Indoor and Mobile Radio Communications*, pp. 1–7. IEEE.

74 Rahaman, S., Xiao, Y., Afrose, S. et al. (2019). CryptoGuard: High precision detection of cryptographic vulnerabilities in massive-sized java projects. *Proceedings of the 2019 ACM SIGSAC Conference on Computer and Communications Security*, pp. 2455–2472.

75 Kune, D.F., Backes, J., Clark, S.S. et al. (2013). Ghost talk: mitigating EMI signal injection attacks against analog sensors. *2013 IEEE Symposium on Security and Privacy*, pp. 145–159. IEEE.

76 Shoukry, Y., Martin, P., Tabuada, P., and Srivastava, M. (2013). Non-invasive spoofing attacks for anti-lock braking systems. *International Conference on Cryptographic Hardware and Embedded Systems*, pp. 55–72. Springer.

77 Pen Test Partners. Steal your Wi-Fi key from your doorbell? IoT WTF! https://www.pentestpartners.com/security-blog/steal-your-wi-fi-key-from-your-doorbell-iot-wtf/ (accessed 1 November 2022).

78 SmartThings. Changes to the Legacy SmartThings Platform. https://community.smartthings.com/t/announcement-changes-to-our-legacy-smartthings-platform/197958 (accessed 1 November 2022).

79 Fang, Z., Fu, H., Gu, T. et al. (2019). ForeSee: A cross-layer vulnerability detection framework for the Internet of Things. *2019 IEEE 16th International Conference on Mobile Ad Hoc and Sensor Systems (MASS)*, pp. 236–244. IEEE.

13

Intrusion Detection Systems for IoT

Hyunwoo Lee, Anand Mudgerikar, Ninghui Li, and Elisa Bertino

Department of Computer Science, Purdue University, West Lafayette, IN, USA

Abstract

This chapter focuses on techniques to detect attacks on IoT devices. The chapter first motivates the use of intrusion detection systems (IDS) for IoT environments and in which aspects IDS for IoT differ from IDS for conventional computer systems. Then the chapter covers the design dimensions for IDS for IoT and approaches/systems that have been proposed, especially focusing on their research challenges. Finally, the chapter discusses research directions in the design, deployment, and management of IDS with concluding remarks.

13.1 Introduction

The internet of things (IoT) are the network of physical objects or "things" including embedding electronics, software, and network communication capabilities, which enable these objects to collect and exchange data. IoT also allows objects to be controlled remotely across network infrastructures, creating opportunities for direct integration between the physical world and computer-based systems.

However, IoT devices raise security concerns as they are at higher risk than conventional computer systems [1]. Managing the security of IoT networks can be challenging as some of those networks may contain thousands of IoT devices. As an example, it is difficult to patch such a large number of devices to remove vulnerabilities [2, 3]. Furthermore, some IoT devices have limited functions and be resource-constrained, thus lacking the capability of running standalone endpoint protection. As a result, they can be easily exploited for malicious activities. For example, IoT devices were exploited by the Mirai botnet [4], which produced the largest distributed denial-of-service (DDoS) attack on dynamic domain name system (DNS) services in 2016. IoT devices can also be misused; an example is that of a refrigerator sending spam emails [5]. Or worst, vulnerabilities of IoT devices (see for example Ref. [6]) can result in safety risks. Effective defenses and comprehensive approaches for IoT security are thus critical.

A fundamental building block for system security is represented by intrusion detection systems (IDS). An IDS is a specialized software or hardware that collects network or host logs and analyzes them to prevent and/or detect intrusions. Because they have proven to be effective, they have been extended for use in the protection of IoT devices.

IoT for Defense and National Security, First Edition. Edited by Robert Douglass, Keith Gremban, Ananthram Swami, and Stephan Gerali.

The main objective of this chapter is to review IDSes proposed for IoT devices and analyze them based on the following three dimensions. First, we categorize the IDSes according to the research challenges they aim to address and their core techniques. For instance, many IDSes have been designed based on enhanced machine learning (ML) algorithms for anomaly detection. These IDSes differ in terms of the features they consider and the algorithms they use for the classifiers. Second, we categorize the IDSes based on the threats that they aim to prevent, such as routing attacks in IPv6 over low-power wireless personal area networks (6LoWPAN). However, not all IDSes are designed assuming specific threats. Some have been designed to enhance the detection performance in a general way. We review them as well. Third, we categorize the IDSes based on their design. Specifically, we survey the IDSes concerning: (i) from where the IDS collects logs to be analyzed (i.e. host-based or network-based); (ii) the type of architectures the IDS uses (i.e. centralized, decentralized, or distributed); and (iii) the type of detection mechanism that the IDS relies on (i.e. signature-based, anomaly-based, or hybrid).

The rest of this chapter is organized as follows. In Section 13.2, we introduce some background on IDS and IoT-specific network protocols. Then, in Section 13.3 we present details about the design dimensions of IDS. Next, in Section 13.4 we analyze 18 IDS proposed between 2016 and 2021 and we carry out a comparative analysis. Finally, in Section 13.5 we outline some future research directions.

13.2 Background

This section provides background about IDSes and their design dimensions. Then, we describe the characteristics of IoT environments and an overview of IoT-specific protocols. Finally, we discuss why IDSes are suitable for IoT.

13.2.1 Intrusion Detection Systems

An *intrusion* is a malicious activity from an adversary to disrupt a target system. To avoid such disruptions, the IDS detects intrusions by analyzing *logs* collected from hosts or networks and raises alarms if an intrusion is detected. To this end, the IDS typically consists of two modules: a *collector* and an *analyzer*. The former is responsible for collecting logs that are sent to the latter for analysis in order to detect attacks or anomalies by using a *detection mechanism*. Note that a collector and an analyzer do not necessarily have to be on the same machine. For example, there can be several collectors located across multiple machines, with only one analyzer.

In general, IDSes are categorized based on the placement of collectors, the architecture of analyzers, and the type of detection mechanisms as described below. Note that these three aspects are important factors in designing and implementing IDSes.

13.2.1.1 Placement of Collectors

There are two types of IDS: host-based and network-based.

- **Host-based IDS (HIDS).** An HIDS deals with host logs recording information such as lists of running processes or data about CPU/memory usage, as well as network packets that a host sends/receives, from a collector installed on a host. The collector can be at the operating system (OS) level or the application level. IDSes have typically higher accuracy in detecting anomalies if the collector has the OS-level privilege. One drawback of HIDSes is that the collector should be installed individually in each device, which may incur management overhead to a network administrator of a network with a large number of devices. Based on the logs from the collector, the HIDS can check file integrity, enforce some security policies, and detect rootkits. Examples of open-source HIDSes are open source HIDS SECurity (OSSEC) [7] and Tripwire [8].

- **Network-based IDS (NIDS).** An NIDS aims to detect malicious network activities. It analyzes the network traffic and raises alarms upon detecting anomalies. To this end, the collector is usually located at the edge of the network, such as at a router, to monitor all the packets passing through between the IoT network and the Internet. Since collectors are not located at the devices, NIDSes do not require changes in IoT devices. Therefore, one important advantage of NIDSes is that they can be deployed regardless of IoT devices. Examples of open-source NIDSes widely used are Snort [9] and Suricata [10].

13.2.1.2 Architecture of Analyzers

IDSes can be classified into three different types based on where analyzers are located: centralized, decentralized, and distributed.

- **Centralized IDS.** In a centralized IDS, there is a single analyzer detecting attacks based on logs from collectors. Note that one can place the analyzer on various locations. For example, the analyzer can be on a specific host, on a fog/edge computing platform, or in a cloud infrastructure. If the analyzer is in a host, one drawback is that the analyzer uses host resources. Another is that it only takes information limited to the internal data. Therefore, the analyzer can only detect anomalies with a single and narrow point of view. For example, the IDS may easily detect transmission control protocol (TCP) synchronize (SYN) packets used for port scanning targeting multiple devices from a network perspective, but the packets may look like simple TCP SYN packets from a host perspective. On the other hand, sensitive logs (e.g. from a health care device) do not need to be transmitted over a network; thus, there is no chance for a network adversary to learn or manipulate any logs. Also, it guarantees an instant response to threats. The analyzer in a cloud infrastructure can use sufficient resources and process lots of logs from a broader range of devices. Therefore, the IDS can make the most optimal decision. However, massive (and possibly sensitive) logs should be transmitted over a network, so an adversary can learn any information from the logs and manipulate them if the transmission is not secured. Furthermore, if the IDS is used in delay-sensitive domains such as in vehicular ad hoc networks, network latency due to the analyzer in the cloud infrastructure may not meet the requirements [11].
- **Decentralized IDS.** A decentralized IDS uses a hierarchical organization of analyzers. For example, there can be a central analyzer in a cloud infrastructure and several analyzers in edge computing platforms. Each of the latter may be responsible for some networks and may receive logs from the networks. Within the hierarchy, the lower-layer analyzer can aggregate or anonymize data and send them to the upper-layer analyzer, thereby the detail of the logs are not exposed to a potential adversary. Also, an analyzer in an edge computing platform can detect anomalies faster than an analyzer in a cloud infrastructure.
- **Distributed IDS.** A distributed IDS consists of several IDSes, each of which is a centralized IDS and collaborates with others. It has the same advantages as the decentralized IDS regarding privacy and fast response. The difference from the decentralized IDS is that the analyzers in the distributed IDS comprise a point-to-point network, rather than a hierarchy, to share information since they are all standalone.

13.2.1.3 Detection Mechanisms

There are two types of IDSes based on their detection mechanisms. They are signature-based and anomaly-based.

- **Signature-based IDS.** A signature-based IDS maintains a list of pre-defined attack patterns. It monitors the network or host logs, and alerts users when it finds matching patterns. It is simple to implement and has a high true positive rate for known attacks. However, it is unable to detect unknown attacks and always requires manual effort to generate detection rules.
- **Anomaly-based IDS.** An anomaly-based IDS maintains a model that captures the normal behavior of the system. It monitors the network or host logs, and alerts users when it finds the patterns deviated from the normal behavior. Unlike a signature-based detection mechanism, it is able to detect unknown attacks without any manual effort. However, it generally suffers from a high false-positive rate.

13.2.2 Characteristics of IoT Environments

There are several factors of IoT environments that should be considered in designing and implementing the IDSes in IoT networks.

13.2.2.1 Simple Networking Patterns

Networking patterns of IoT devices are simple because the devices usually have specific functionality and run one or a few applications. For instance, a temperature/humidity monitor or a motion detector in a smart home periodically (more than 100 s) sends a group of consecutive packets in a short time (less than 1 s) [12]. Such a pattern is different from network patterns of conventional systems, which are complicated as the patterns are associated with a user's diverse behavior. Therefore, modeling the normal behavior does not require complex statistical or probabilistic inference modules, compared with the IDS for traditional networks.

13.2.2.2 Diverse Network Protocols

There are many new protocols designed for IoT devices across diverse protocol layers. For example, constrained application protocol (CoAP) [13] and message queuing telemetry transport (MQTT) [14] are application layer protocols designed for resource-constrained devices. 6LoWPAN [15] has been designed to use IPv6 in lossy networks. Also, the IPv6 routing protocol has been designed for low power and lossy networks (RPL) [16]. These new protocols are typically used in environments where there are many resource-constrained devices. One cannot use existing IDSes in such an environment [17]. For example, existing signature-based IDSes, such as Snort [9], mainly work with rules for application protocols running over TCP/IP (e.g. hypertext transfer protocol (HTTP)), so they become useless when the IDS is deployed in such an environment. Therefore, new rules for attacks against new protocols should be generated, and also new decoders for new protocols should be developed to apply the new rules to packets.

13.2.2.3 Small Number of Threads

Unlike a general-purpose machine, an IoT device runs a single or a small number of threads. Furthermore, spawning a new thread is unusual. The reason is that tasks running in an IoT device are usually simple and repetitive. For instance, a temperature sensor in a room periodically only measures air temperature and reports the temperature to the cloud server. Therefore, modeling the normal host behavior does not require any complex approach.

13.2.2.4 Various Types of CPU Architectures and Operating Systems

Many CPU architectures and operating systems are tailored for IoT devices. CPU architectures for IoT devices include as ARM [18] or RISC-V [19]. Furthermore, operating systems used in IoT devices, including Contiki [20, 21], RIOT [22], TinyOS [23], and FreeRTOS [24], are different in many aspects, such as CPU scheduling, multi-processing, and memory management. Therefore, if a collector resides in a device, it should support diverse architectures and operating systems.

13.2.2.5 Resource Constraints

IoT devices are usually more resource-constrained than general-purpose devices. Some IoT devices have low-power CPUs and do not have much storage or memory capabilities. For example, Tmote Sky, a platform for low power and high data-rate sensor network applications, is equipped with T1 MSP430 that is a 16-bit RISC based 8 MHz microcontroller, and has only 10K RAM and 48K flash memory. Furthermore, some IoT devices are battery-powered; thus, power consumption should be an important factor.

13.2.2.6 Large Numbers of Devices

A large number of IoT devices can be deployed in smart cities, farms, or factories, where the number of devices could be more than hundreds of thousands [25]. Therefore, the management of the IoT domain may be complicated. For example, maintaining the firmware of devices up-to-date is challenging if the devices are from different vendors, and managing a myriad of secrets for IoT devices would be also difficult.

13.2.2.7 Dynamics and Autonomy

In some cases, there are many dynamics in IoT networks for defense or military applications. For example, some sensor devices are distributed for surveillance and reconnaissance. To be practical, these sensor devices should be easily added to or removed from networks. However, it is not guaranteed that all the devices maintain their connectivity. To be connected to the Internet, these sensor devices typically perform autonomous networking, such as retrieving neighbors or advertising themselves, and multi-hop routing to a gateway. In other words, they usually perform distributed and ad hoc networking rather than centralized and infrastructural networking. Therefore, we should consider that sometimes there is no centralized device in the network and the devices can be easily exposed to malicious devices.

13.2.3 IoT-Specific Protocols

13.2.3.1 IoT Network-layer Protocols

The IoT network-layer protocols are designed to allow IoT devices to communicate with entities across the Internet. To support low-data-rate and low-power devices, the protocols usually run over IEEE 802.15.4 which specifies the physical layer and the media access control layer for small devices. In what follows, we describe two such protocols: 6LoWPAN and RPL.

13.2.3.1.1 6LoWPAN

6LoWPAN [15] has been designed to connect resource-constrained devices to the Internet by enabling the transmission of IPv6 packets on top of IEEE 802.15.4 networks. 6LoWPAN is an adaptation layer between IPv6 and IEEE 802.15.4. Since the maximum transmission unit of the network is smaller than the header size of IPv6, 6LoWPAN allows the IPv6 packet delivery in a compressed or fragmented form. In the 6LoWPAN network, there is a gateway between the network and the Internet, called the 6LoWPAN border router (6BR). All the devices in the network are connected to the 6BR directly or indirectly via other devices and finally, they are connected to the Internet.

13.2.3.1.2 RPL

RPL [16] is a source routing and distance-vector routing protocol based on IPv6 designed for the IoT network. It constructs a destination-oriented directed acyclic graph (DODAG) in which each device has its own ID (i.e. an IPv6 address), one or more parents, and a list of neighbors. Also, it has a rank that defines the relative position with respect to the DODAG root. The rank of a device increases based on the number of devices between the device and the DODAG root. In other words, parent devices always have a rank smaller than the ranks of their children's devices. To build DODAG, a device uses DODAG information solicitation (DIS) messages to probe neighbor devices and request information about their views of the network including their ranks or parent nodes. The recipients respond with DODAG information objects (DIOs) that contain such information. An IPv6 packet from a device is sent to the Internet via parent devices and the 6BR. In the case of 6LoWPAN, the 6BR become the DODAG root.

13.2.3.2 IoT Application-layer Protocols

The IoT application-layer protocols are designed for resource-constrained IoT devices to support the lightweight exchange of application data. There are two representative protocols, namely the MQTT which utilizes a broker, and CoAP, a simplified version of HTTP.

13.2.3.2.1 MQTT

MQTT is based on a publisher/subscriber mechanism by which a publisher publishes a message with a topic on a broker and a subscriber receives the message from the broker if the subscriber subscribes to the topic. It is useful for IoT devices that have sleep duty cycles; thus, the device-to-device communications do not require both devices to be awake simultaneously. In MQTT, a publisher or a subscriber registers itself to a broker by sending the CONNECT message. Once it is registered, the subscriber sends a SUBSCRIBE message with a topic. Then, it receives data related to the topic when the publisher sends a PUBLISH message to a broker with the topic and a payload.

13.2.3.2.2 CoAP

CoAP is a specialized application protocol for resource-constrained devices. It enables communications between devices in the same network, between devices and other parties on the Internet, and between devices on different IoT networks. It also meets several requirements for communication among resource-constrained devices, including multicast support, low overhead, and simplicity.

13.2.4 IDS in IoT Environment

13.2.4.1 Relevance of IDS in IoT Environment

There are two reasons why IDSes are suitable for IoT environments.

First, it is difficult to deploy defense tools, such as anti-virus software, on some IoT devices. Such tools are usually computationally heavy and thus, it is challenging for the devices to run them because of their limited resources. Equipping those devices with powerful CPUs and sufficient memory to run defense tools would not be viable because it would increase the device costs. Therefore, instead of installing standalone systems on the devices, one can consider deploying the NIDSes at the edge of the networks, thus delegating intrusion detection tasks from the IoT devices to the edge.

Second, the simple behavior patterns of the IoT devices make anomaly detection effective. In conventional systems, anomaly-detection mechanisms may have high false-positive rates [26]. The reason is that general-purpose systems run many different applications, which have often complex behaviors difficult to profile and predict. In addition, network behavior patterns are affected by users' unpredictable behavior. On the other hand, the IoT devices often run a small number of applications responsible for periodic and repetitive tasks (e.g. sensing), and thus, new behavior patterns do not occur frequently.

13.2.4.2 Challenges for IDSes in IoT Dynamic and Autonomous Environment

There are three challenges in IoT networks where a dynamic level is high and connectivity is not always guaranteed.

First, a distributed "security" architecture is the obvious choice as connectivity to central devices is intermittent. In other words, each device must have a local analyzer to detect anomalies. Since devices in these networks are generally resource-constrained, how to deploy a distributed system where each device takes its own decision on anomalies is challenging. At the same time, it is also necessary to consider what individual devices should do when they have connections to central devices.

Second, designing lightweight detection mechanisms for analyzers is a critical issue. If a signature-based detection mechanism is used in the analyzer, one must address the scalability problem with respect to the number of signatures. On the other hand, if an anomaly-based detection mechanism is used in the analyzer, one should address the problem of keeping the model size small while at the same time minimizing the false-positive rate of the model. Alternatively, one can consider a split architecture in which each individual device takes decisions based on some heuristics, whereas final decisions are taken when connected to some central device.

Third, attacks to autonomous systems should also be considered. Typically, in the conventional network, the network itself is assumed to be secure; thus, one usually focuses on attacks from the external Internet. However, as IoT devices could be easily compromised or exposed to malicious devices in highly dynamic and autonomous scenarios, one should consider not only the attacks from the Internet but also attacks on the autonomous routing protocols.

13.3 IoT Attack Scenarios

In designing and implementing an IDS, one should understand what attacks the IDS aims to detect or prevent for what scenario. There are many kinds of IoT attacks across diverse layers. First, IoT devices are exposed to the typical application layer attacks from the Internet. Many botnet campaigns (e.g. Mirai [4] or IoTroop [27]) are usually included in this category since they rely on attacks based on the protocols used on the Internet such as user datagram protocol (UDP), HTTP, or secure shell (SSH). Next, there are IoT attacks from an internal adversary in the IoT network as the devices can be compromised or exposed to malicious devices in autonomous routing protocols. This category is divided into two classes: (i) attacks on the IoT-specific network-layer protocols and (ii) attacks on the IoT-specific application-layer protocols.

13.3.1 Attacks from the Internet

These attacks are launched from the Internet. That is, an external adversary attempts to compromise IoT devices to use them as backdoors to a target network. Many botnet campaigns often have multi-steps including the following attacks.

13.3.1.1 Port Scanning

An adversary performs port scanning on a target system. It is a reconnaissance attack to acquire useful information about the victim. For example, the adversary may obtain information about the SSH port used by the victim and perform an SSH bruteforce attack as the next step.

13.3.1.2 Telnet/SSH/HTTP Bruteforce

An IoT gateway or a device may open the telnet/SSH port for maintenance purposes, or it may provide a web interface for a user to monitor the network or control devices within the network. Those telnet/SSH/web services are generally protected with users' credentials that consist of usernames and passwords. The attack is used by an adversary to get unauthorized access to the victim. In botnet campaigns, once the adversary acquires the credential, the victim becomes a bot.

13.3.1.3 SYN/ACK/UDP/HTTP Flooding

There are many kinds of DDoS attacks by flooding a target system with a variety of packet types including SYN, acknowledgment (ACK), UDP, or HTTP. An adversary launches such an attack by directing its bots to send a number of packets toward the victim system.

13.3.2 IoT-specific Network-layer Attacks

There are several attacks on 6LoWPAN/RPL networks. Most of the attacks are from an internal adversary aiming to disrupt the routing protocols.

13.3.2.1 Hello Flood Attack

An internal adversary sends the hello message to legitimate nodes to make them believe the adversary is their neighbor. The attack aims to make the legitimate nodes send data packets to the adversary. Thereby, the adversary can read and control the packets.

13.3.2.2 Neighbor Attack

This attack is launched by an internal adversary that broadcasts DIO messages to neighbors with incorrect information (e.g. illegitimate devices) about the network. If a device that receives the messages finds any new node from the messages, the recipient adds the new node to its parent list or selects the new node as the parent, which may disrupt the network.

13.3.2.3 DIS Attack

This attack aims to increase the power consumption of resource-constrained devices by flooding the number of DIO messages. To this end, an internal adversary sends DIS messages to victim devices. Once the devices receive the DIS messages, they reply with the DIO messages, leading to a number of DIO messages in the network.

13.3.2.4 Sinkhole Attack

A sinkhole attack is an internal attack by which an adversary with a compromised device attracts all the traffic from the neighbors of the device by advertising a beneficial routing path. As a result of the attack, the adversary may receive information from other devices since many routing paths may include the adversary's device as if it is a sink. In RPL, an adversary can advertise a better rank to victim devices making them select the compromised device as their parent device.

13.3.2.5 Wormhole Attack

With a wormhole attack, an adversary can disrupt the network by transmitting packets between two colluding devices that are distant. It results in latency or a packet loss in the network.

13.3.2.6 Grayhole (or Selective Forwarding) Attack

A grayhole attack (a.k.a. a selective forwarding attack) is an internal attack by which an adversary selectively drops packets. The adversary may use this attack to deliver packets with non-critical data but discard packets that contain important data such as reporting an enemy in a military scenario.

13.3.3 IoT-specific Application-layer Attacks

There are several proposed IDSes for MQTT or CoAP networks. The attacks considered in this category are mostly denial-of-service (DoS) attacks launched by an internal adversary to deplete resources in the victim devices.

13.3.3.1 CONNECT/CONNACK Flooding

This attack targets an MQTT broker. An adversary sends the CONNECT message to the broker. Then, the broker opens the state for the connection for a certain period, which may result in DoS.

13.3.3.2 CoAP Request/ACK Flooding

The target system that this attack aims to disrupt is a CoAP server. An adversary can send a CoAP request or a CoAP ACK to the server to make it unavailable.

13.4 Proposed IDSes for IoT

This section provides a review of the IDSes for IoT. We first classify 18 existing IDS proposals tailored for IoT based on their research topics. For each topic, we list research challenges and describe IDSes that deal with the topic focusing on what they are and how they achieve goals. Finally, we present an analysis per topic. To provide a systematic view of IDSes, we consider the following indicators to understand envisioned IDSes. Table 13.1 summarizes the indicators for each IDS.

- **Type.** We describe the type of the IDS based on (i) the placement of collectors (host-based (H) or network-based (N)), (ii) the architecture of analyzers (centralized (CE), decentralized (DE), or distributed (DI)), and (iii) the detection mechanism (signature-based (S), anomaly-based (A), or hybrid-based (SA)).
- **Core Technique.** We describe the core technique of an IDS.
- **Evaluation Method.** We list what kind of evaluation methods, such as simulation or testbed evaluation, is used to validate an IDS technique.
- **Validated Detection.** We list all the attacks that an IDS detects in the evaluation.

13.4.1 Definition of Normal/Abnormal Behavior

To detect anomalies from the logs, normal, or abnormal behavior should be accurately defined. For example, all the signature-based IDS consider activities that match signatures as abnormal behavior. We introduce IDS techniques related to defining normal or abnormal behavior and list prominent issues for each technique.

13.4.1.1 Legitimate IP Addresses

The simplest solution to define the normal behavior is to use whitelisting or blacklisting of IP addresses. Due to its simplicity, this solution is preferred in resource-constrained environments where heavy computation is undesirable. In detail, the IDS defines activities from legitimate IP addresses as normal behavior. When the IDS finds an activity from an illegitimate IP address, it reports the activity as an anomaly. The main challenge in this IDS is how to tell legitimate IP addresses from illegitimate IP addresses.

13.4.1.1.1 Heimdall

Habibi et al. [28] propose a network-based, anomaly-based, and centralized IDS with an IP access control approach. In detail, the IDS is located at the edge of the network where all the communications are passing through, aiming to prevent botnet campaigns. As botnet malware spreads by connecting other victims like a worm, the IDS allows communications only toward the legitimate destinations listed in its whitelist. To this end, it learns legitimate destinations per device by profiling the normal behavior of the device. Whenever the IDS captures a DNS query packet, it validates the IP address by referring to VirusTotal[1], an online service that analyzes suspicious files and uniform resource locators (URLs). If the IP address does not turn out to be malicious, the address is added to the whitelist; otherwise, it is added to the blacklist. Since the IP addresses in the blacklist can be legitimate later on, HEIMDALL periodically validates the IP addresses in the blacklist to remove valid ones from the list. The authors demonstrate the feasibility of HEIMDALL on their testbed against botnet campaigns.

13.4.1.1.2 Combining MUD Policies for IoT IDS

Hamza et al. [29] design a network-based, signature-based, and centralized IDS with the manufacturer usage descriptions (MUD) policies [30] and software-defined networking (SDN) (COMBINING MUD in 13.1). The IDS utilizes MUD profiles that describe security policies for devices, such as access control lists, written by manufacturers. When an SDN switch receives new traffic that does not match any rule, it forwards the traffic to the IDS. The IDS maintains a list of devices in the network and retrieves the corresponding MUD profile from a MUD server. From the MUD profile, the IDS extracts the access control entries and translates them into rules. The resultant rules are inserted into the SDN switches to block unauthorized access. The authors evaluate the IDS with the simulation-based approach (SDN pcap simulator [31]) against unauthorized access.

13.4.1.1.3 Discussion

Both approaches are similar in that they validate IP addresses by referring to trusted entities. HEIMDALL relies on VirusTotal and COMBINING MUD refers to manufacturers and MUD servers. Therefore, the performance of these IDSes depends on how much such trusted entities quickly and correctly update information about the IP addresses. Despite their similarity, we classify HEIMDALL as an anomaly-based IDS and COMBINING MUD as a signature-based IDS. The reason is that HEIMDALL learns by itself the legitimate IP addresses per device while COMBINING MUD gets a list of legitimate IP addresses from MUD files and uses them as if they are rules in the signature-based mechanism.

13.4.1.2 Threshold

Another approach to defining normal behavior is to use threshold values. In this approach, a threshold value is set to the number of particular events occurring in a fixed time. If the number of occurring events is higher than the threshold value, the IDS reports the events as an anomaly. We introduce three IDSes in this category – the first two IDSes are for RPL and the other is for MQTT.

13.4.1.2.1 Machine learning-based IDS (ML-IDS)

Shukla [32] develops a network-based, anomaly-based, and centralized IDS based on the K-means algorithm, decision tree, and a hybrid of these. The author suggests three IDSes, referred to as K-Means (KM) clustering-based IDS (KM-IDS) based on the K-means algorithm, Decision Tree

1 https://www.virustotal.com

(DT)-based IDS (DT-IDS) based on the decision tree algorithm, and Hybrid IDS utilizing both KM-IDS and DT-IDS. Those IDSes aim to detect the victims of wormhole attacks by setting the distance threshold value between any two devices. That is, if two devices agree that they are neighbors but they are far apart from each other, the IDSes determine that a wormhole attack occurs and these devices are victims. The ML algorithms are used to set the threshold values. In detail, KM-IDS clusters the devices based on their distance from the 6BR. The anomaly is detected when the 6BR receives requests to update neighbors from two devices in different clusters. In DT-IDS, the threshold value is set as the averaged value of the sum of distances between any two devices. Then, if two devices for which the distance is above the threshold send requests to update neighbors to the 6BR, DT-IDS determines an attack is ongoing. In Hybrid-IDS, when KM-IDS reports an anomaly because of two devices from different clusters, Hybrid-IDS delays its decision and checks if the distance between the two devices is less than the threshold set by DT-IDS. If it is the case, Hybrid-IDS determines that this is not an anomalous event since the actual distance is not far even if they are in different clusters; otherwise, the Hybrid-IDS reports that the anomalous event is occurring. In this way, it tries to reduce false positives. Although the author implements the IDSes and conducts experiments on random networks, the detection rate without breaking into true-positive and false-positive is reported; thus, the effect of the approach is unclear.

13.4.1.2.2 Anomaly-based IDS in RPL-based IoT

Farzaneh et al. [33] propose a host-based, anomaly-based, and distributed IDS based on the threshold values (Anomaly in 13.1). The IDS is installed in all the nodes and each node monitors the others, aiming to detect the neighbor attack and the DIS attack. To this end, the IDS sets the threshold values for the attacks based on the network settings (e.g. the number of nodes in the network or their distances). The authors find that the number of messages that a device receives from its neighbors follows a normal distribution. After the threshold values are determined, the IDS reports the detected attack if a node receives the related messages (DIO messages or DIS messages) more than the threshold values. The authors conduct simulation-based evaluation using Cooja [34] and show the acceptable performance.

13.4.1.2.3 Secure-MQTT

Haripriya and Kulothungan [35] propose a network-based, anomaly-based, and centralized IDS based on a fuzzy logic-based approach. The IDS aims to detect DoS attacks based on CONNECT and CONNACK messages. The fuzzy system in Secure-MQTT gets a fraction of connection requests from a publisher and a fraction of requests from a subscriber as inputs, which are fuzzified by a fuzzy membership function. Based on such a function a fuzzy classifier is derived and rules are generated. When Secure-MQTT receives packets the related rule is activated and used to determine whether those are malicious or not after defuzzification. Secure-MQTT is validated with the Cooja simulation tool [34].

13.4.1.2.4 Discussion

The main challenge of these approaches is how to set reasonable threshold values. The three IDSes use different ways to decide the threshold values. ML-IDS relies on the ML algorithm itself, Anomaly uses the normal distribution, and Secure-MQTT utilizes the fuzzy logic.

13.4.1.3 Automata

There is one IDS that describes devices behavior by using automata to perform anomaly detection.

13.4.1.3.1 Automata-based Intrusion Detection Method

Fu et al. [36] design a network-based, signature-based, and centralized IDS based on automata learning (Automata in 13.1). The IDS is located in an IoT gateway and gets a view of the network topology from the collected network packets. Then, it learns the network structure that consists of an automaton per device, in which the transitions are triggered by input and output messages of the device. From the network packets, the IDS extracts automata transition sequences within a fixed time window and checks if any of the sequences is an intrusion. For each sequence, it checks with the known malicious sequences and reports an intrusion if any of the sequences match a malicious sequence. Otherwise, it checks with the known normal sequences and sends the sequence to experts for manual inspection if it does not match any of them. Once the experts assess the sequence to be malicious, the sequence is added to the known malicious sequences for future use. The authors build their testbed and evaluate their IDS against the replay attack, the jamming attack, and the fake attack.

13.4.1.3.2 Discussion

This approach leverages automata (or a finite state machine) to detect anomalies. The rationale behind this choice is that automata are widely used to model or implement the behavior of a protocol [37] and anomalies are detected if a transition sequence deviates from the automata. The main challenge of this approach is, thus, how to accurately develop such automata for protocols. Automata can be manually predefined, but an automatic automata generation would be also helpful to make this approach scalable with respect to the number of supported protocols.

13.4.1.4 Federated Learning

One IDS uses federated learning to generalize the normal behavior of the devices from multiple networks.

13.4.1.4.1 DÏoT

Nguyen et al. [38] propose a network-based, anomaly-based, and decentralized IDS with device-type-specific anomaly detection and federated learning. IoT devices typically show simple patterns in networking but the patterns are varied depending on device types such as a camera or a sensor. In this regard, the authors propose a device-type-specific anomaly detection that shows higher accuracy compared with the detection without considering the device types. As profiling the normal behavior of IoT devices is difficult due to the scarcity of communications in IoT networks, the IDS uses a generalized model for a specific type learned from diverse networks via federated learning. To this end, there is a module called *IoT security service* that maintains global models aggregated with the locally collected data for device types. In detail, DÏoT monitors packets at the edge of the network and identifies device types from the networking patterns. Once it identifies the device types, it fetches models from the IoT security service and uses them to detect anomalies. The authors set up their testbed and evaluate the system with the Mirai botnet. They show high accuracy of the system and efficiency of the federated learning.

13.4.1.4.2 Discussion

One interesting point of this approach is that it generates device-type-specific models by leveraging federated learning. Note that this approach is especially useful for small networks where the IoT devices generate small numbers of packets typically insufficient to develop a model of device behavior.

13.4.2 Enhancements of ML-based Detectors

Classification is one of the main tasks that ML algorithms perform. IDSes use classification to detect anomalies. Therefore, one important challenge is how to enhance accuracy of ML-based detectors. We introduce four proposals below.

13.4.2.1 Compression Header Analyzer Intrusion Detection System (CHA-IDS)

Napiah et al. [39] propose a network-based, hybrid-based, and centralized IDS based on the 6LoWPAN compression header. Unlike prior IDSes that usually use rank devices to identify routing attacks, CHA-IDS uses the 6LoWPAN compression header as a feature. The rationale behind the use of such a feature is that the compression header includes routing information; thus, it improves accuracy in detecting routing attacks. Once CHA-IDS collects the 6LoWPAN packets, it runs a correlation-based feature selection algorithm to choose the significant features to identify abnormal activities. Then, it generates a model to detect anomalies. The authors perform experiments with the Cooja simulator [34] and demonstrate that CHA-IDS works well against hello flood, sinkhole, and wormhole attacks.

13.4.2.2 E-Spion

Mudgerikar et al. [40] design a host-based, anomaly-based, and centralized IDS with a device-edge split architecture. In the architecture, the collectors (*SysMon* in E-Spion) are installed in the devices, while the analyzer runs at the network edge. There are three variants concerning what information the collectors collect. They are called Process White listing Module, Process Behavior Module, and System-call Behavior Module. The collectors periodically send the collected logs to the analyzer to detect anomalies from them. To ensure the integrity of the logs, the collectors also transmit the hash chains of the logs and the analyzer uses the logs only if the hash chains are verified. The authors evaluate the system on their testbed with running malware binaries and show a high accuracy detection rate and the least amount of CPU and storage usage.

13.4.2.3 Deep learning-based IDS (DL-IDS)

Otoum et al. [41] propose a network-based, anomaly-based, and centralized IDS with an optimal feature selection and stacked-deep polynomial network. DL-IDS begins with preprocessing the dataset, followed by the optimal feature selection phase to remove all uncertainties. To this end, the spider monkey optimization (SMO) algorithm [42] is used. The resultant features are learned in each layer of the stacked-deep polynomial network. The authors perform simulation-based evaluation with the NSL-KDD dataset.

13.4.2.4 Multiclass Classification Procedure

Alaiz-Moreton et al. [43] propose a network-based, anomaly-based, and centralized IDS based on the gradient boosting classifier and recurrent neural networks (RNNs) (Multiclass in 13.1). The authors create three MQTT datasets that contain DoS attacks, man-in-the-middle attacks, and active topic extraction attacks from their testbed with sensors, actuators, and a broker. They use the gradient boosting classifier and the Long Short Term Memory (LSTM) / Gated Recurrent Unit (GRU) RNN networks as multiclass classifiers to detect three different attacks. The authors evaluate the IDS on their testbed.

13.4.2.5 Discussion

The IDSes generally perform the four steps – data collection, feature selection, model generation, and anomaly detection. The four IDSes aim to enhance the performance of the ML-based detectors in different steps. E-Spion collects host information to detect anomalies; CHA-IDS introduces a new feature compared with prior work; DL-IDS uses a new feature selection algorithm called SMO; and Multiclass adopts a multiclass classifier instead of a binary classifier. From this analysis, we can see that selecting the proper features is an important direction for enhancing ML-based detectors.

13.4.3 Lightweight Detector Implementation

Lightweight IDSes have also been proposed for resource-constrained environments. The following two projects aim to implement a signature-based IDS and an anomaly-based IDS over Raspberry Pis, respectively, which model routers at the edge of the network.

13.4.3.1 Raspberry Pi IDS (RPiDS)

Sforzin et al. [44] developed a network-based, signature-based, and centralized IDS with Snort [9] running over Raspberry Pi 2 (RPi2). The work aims to deploy the general-purpose open-source Snort IDS on the RPi2 and evaluate its performance. They succeed in running Snort over RPi2 and evaluate it by simulating packet traces. As RPiDS runs Snort, it detects attacks covered by the Snort rules. Their experimental result shows that capturing packets incurs high usage of CPU and the number of IDS rules negatively affects the performance of the IDS.

13.4.3.2 Passban IDS

Eskandari et al. [45] propose a network-based, anomaly-based, and centralized IDS with one-class classification techniques. The motivation for the design of this IDS is that it is difficult to use a signature-based IDS on an IoT device. Therefore, the authors aim to provide a lightweight software implementation that can be deployed on typical resource-constrained IoT gateways. They set the Raspberry Pi 3 Model B (RPi) as their reference deployment board and show that their lightweight anomaly-based IDS works without depleting CPU and memory resources. Passban employs one-class classification techniques that are useful to detect unknown attacks by using the isolation forest [46] and the local outlier factor (LOF) [47] algorithms. They show on their testbed that their implementation works well against the port scanning, the HTTP login bruteforce, the SSH login bruteforce, and the SYN flooding attacks with low false-positive rates.

13.4.3.3 Discussion

The above two approaches utilize two different detection mechanisms. RPiDS runs a signature-based detection mechanism over RPi. Without any special lightweight algorithms, they run Snort with enough storage on RPi, while Passban relies on one-class classification algorithms to perform anomaly-based detection. Considering anomaly detection, it is unavoidable for RPiDS to increase the number of signatures to achieve high recall, while Passban can detect unknown anomalies without any change to the system. However, it will be prone to the issue of high numbers of false positives. Therefore, as mentioned before, one must design solutions able to support increasing numbers of signatures to perform signature-based detection in resource-constrained environments. On the other hand, the issue of high false positives should be addressed in order to effectively use anomaly-based detection mechanisms. To assess the feasibility of the idea, one should set up a target resource-constrained device (e.g. Raspberry Pi in these two examples) and prove the idea by running a lightweight IDS on the target device.

13.4.4 Combination of Diverse Detectors

To cover diverse attacks, some IDSes take advantage of various detection techniques. We introduce three such IDSes. Two of them aim to combine signature-based and anomaly-based detection. The other IDS is based on a multi-tier architecture where each tier implements different techniques.

13.4.4.1 IDS with Game-theoretic Methodology

Sedjelmaci et al. [48] propose a host-based, hybrid-based, and distributed IDS based on a game-theoretic methodology (Game in 13.1). The IDS aims to achieve high accuracy by using both signature-based and anomaly-based detection and to reduce energy consumption. To this end, the IDS running at a device activates the anomaly-based detection only when it expects a new signature to occur (i.e. an unknown attack pattern). This approach is implemented by modeling a security game between an IDS and an adversary, and the IDS enables anomaly-based detection when the equilibrium state is expected by Nash equilibrium [49]. The evaluation is conducted against flooding attacks using the TOSSIM [50] simulation tool.

13.4.4.2 Hybrid Intrusion Detection and Prevention System (IDPS)

Shurman et al. [51] develop a network-based, hybrid-based, and centralized IDS based on the combination of signature and anomaly-based detection. Once the IDS receives packets to be analyzed, it first checks if the source addresses of the packets are in the IP lists. If the packets turn out to be anomalous, the IDS blocks the corresponding IP address. Otherwise, the signature-based detector checks the packets if they match any signatures. If there is no matched signature, the anomaly-based detector analyzes the packets. Finally, the IDS determines whether the given packets are anomalous or not. The core of this IDS is to combine an IP address-based access control, a signature-based detection, and anomaly-based detection. Hybrid IDPS benefits from fast matching of an IP address-based access control, high true positive rates of signature-based detection, and a capability of unknown attack detection of anomaly-based detection. Although the authors have implemented the IDS and argue that they have evaluated it, it is unclear how they conduct experiments. They only show the number of IP addresses blocked by the IDS, which is less than three addresses. Moreover, they do not show whether the combination of both mechanisms is effective and do not present the performance of anomaly detection for unknown attacks.

13.4.4.3 IDPS

Ali and Yousaf [52] propose a network-based, anomaly-based, and centralized IDS with a three-tier architecture in SDN. The system aims to detect illegitimate users that generate malicious packets. To this end, the authors suggest a system with three tiers, which include: (i) user authentication in Tier 1; (ii) packet validation by fuzzy filtering in Tier 2 where SDN switches are located; and (iii) flow validation by the convolutional neural network in Tier 3 where SDN controllers are placed. With user authentication, only an authenticated user can join the network. To detect a compromised user, the system extracts features of packets and classifies them with fuzzy filtering into normal packets, suspicious packets, or malicious packets. The suspicious packets and new normal packets are sent to Tier 3. For normal packets, rules for SDN switches are generated to make these packets forwarded at Tier 2. The suspicious packets are again analyzed with convolution neural networks and if they turn out to be malicious, the related rules are sent to the switches for future prevention. The authors perform simulations based on OMnet++ [53].

13.4.4.4 Discussion

The above approaches perform anomaly detection effectively by integrating several techniques. The important design choice is *how* to integrate *what* techniques. In detail, Game and Hybrid IDPS integrate signature-based detection and anomaly-based detection. Note that both approaches do not always use signature-based detection and anomaly-based detection simultaneously. Game uses signature-based detection by default and applies anomaly-based detection only when the equilibrium state is expected, while Hybrid IDPS uses anomaly-based detection only when there is no matching signature. Also both Hybrid IDPS and IDPS integrate additional techniques as well as detection mechanisms. Hybrid IDPS leverages the IP address-based access control mechanism and

IDPS utilizes user authentication. These additional techniques are responsible for identifying packets to be analyzed with detection mechanisms that are relatively heavy in computation; thus, these techniques make IDSes resource-efficient.

13.4.5 Optimal Detector Selection

Since the IDSes use diverse techniques to detect various attacks, how to coordinate multiple detectors to achieve good accuracy become an important topic. We introduce the following two IDSes proposed to select optimal detectors in specific cases.

13.4.5.1 Kalis

Midi et al. [54] propose Kalis, a network-based, hybrid-based, distributed IDS that selectively leverages different detection techniques based on the knowledge of the network. As the IDSes typically use several detection techniques to cover lots of attacks at the cost of inaccuracy, Kalis uses knowledge about the network to disable unnecessary detection techniques, thus reducing the number of false positives. This knowledge is autonomously acquired by the IDS by monitoring the network. Furthermore, each Kalis IDS advertises itself to other Kalis IDSes for collaboration. Once a Kalis IDS discovers peers, it exchanges information with them to detect anomalies from a global view of the network. The authors evaluate Kalis on their testbed and demonstrate that the knowledge-driven approach is beneficial and it is reactive to changes of the environment.

13.4.5.2 Reinforcement learning-based IDS (RL-IDS)

Pasikhani et al. [55] propose a network-based, hybrid-based, and distributed IDS based on reinforcement learning. The IDS employs several lightweight signature-based or anomaly-based detectors, each of which is built from a subset of the training dataset. The subsets have different patterns of attacks; thus, the detectors have different strengths and weaknesses in analyzing RPL attacks. Over the detectors, RL-IDS uses reinforcement learning to select an appropriate detector to analyze the current input packets. The agent gets a reward when it successfully selects the detector in particular scenarios. Simulation-based evaluation with NS-3 [56] is performed and the results show that the IDS is effective against the various RPL attacks.

13.4.5.3 Discussion

The above approaches aim to solve the problem of optimal detection selection for a given scenario. This topic is very important in practice as using many detection techniques may result in high false positives and inefficiency of resource use. To address this challenge, Kalis uses information about the network and RL-IDS utilizes reinforcement learning. These approaches enable IDSes to integrate many detection techniques to make an IDS adaptive to diverse environments.

13.5 Research Directions

This chapter has covered 18 IDS systems for IoT, which have been proposed between 2016 and 2021. Although the research community has attempted to propose many IDS systems with diverse techniques, much more remains to be done. The IDSes that deal with complex attacks should enable the protection of IoT devices from advanced threats. As seen in Table 13.1, most of the systems that cover application-layer attacks deal with simple DoS attacks. However, other

Table 13.1 IDS/IPS systems proposed in between 2016 and 2021.

Name	Type	Core technique	Evaluation method	Validated detection
			Definition of normal/abnormal behavior	
Heimdall [28]	N/A/CE	IP access control using DNS	Testbed	Botnet Campaign
Combining MUD [29]	N/S/CE	IP access control with MUD	Simulation (NSL-KDD)	Unauthorized access
ML-IDS [32]	N/A/CE	K-means, decision tree, and their combination	Testbed	Wormhole
Anomaly [33]	H/A/DI	Threshold values	Simulation (Cooja [34])	o Neighbor attack o DIS attack
Secure-MQTT [35]	N/A/CE	Fuzzy logic	Simulation (Cooja [34])	o CONNECT flooding
Automata [36]	N/A/CE	Automata learning	Testbed	o Replay attack o Jamming attack o Fake attack
DÏoT [38]	N/A/DE	Device-type-specific anomaly detection and federated learning	Testbed	DoS attacks
			Enhancements of ML-based detectors	
CHA-IDS [39]	N/SA/CE	6LoWPAN compression header	Simulation (Cooja [34])	o Hello flood o Sinkhole o Wormhole
E-Spion [40]	H/A/CE	Device-edge split architecture	Testbed	Malware binaries
DL-IDS [41]	N/A/CE	Optimal feature selection and stacked-deep polynomial network	Simulation (NSL-KDD)	DoS attacks Port scanning
Multiclass [43]	N/A/CE	Gradient boosting and recurrent neural networks	Testbed	o Flooding o Man-in-the-middle o Active topic extraction
			Lightweight detector implementation	
RPiDS [44]	N/S/CE	Snort over RPi2	Testbed	Attacks related to Snort rules
Passban [45]	N/A/CE	One-class classification	Testbed	o Port scanning o SSH Bruteforce o HTTP bruteforce o SYN flooding
			Combination of diverse detectors	
Game [48]	H/SA/DI	Game-theoretical methodology	Simulation (TOSSIM [50])	Flooding

(Continued)

Table 13.1 (Continued)

Name	Type	Core technique	Evaluation method	Validated detection
Hybrid IDPS [51]	N/H/CE	Combination of signature and anomaly-based detection	Simulation	- (Unclear)
IDPS [52]	N/A/CE	Combination of signature and anomaly-based detection	Simulation (OmNet++ [53])	DoS attacks
Kalis [54]	N/SA/DI	Knowledge of networks	Testbed	Optimal detector selection o ICMP flooding o SYN flooding o Grayhole o Smurf o Wormhole o Hello flooding o Data modification o Replication
RL-IDS [55]	N/SA/DI	Reinforcement learning	Simulation (NS-3 [56] / KDD Cup [59])	o DoS attacks o Port scanning

advanced attacks should be covered. For instance, many botnet campaigns have multiple steps such as reconnaissance, infection, and action. Multi-step attack detection has not been largely investigated in the context of IoT. Once we can detect a step of the attack, we can also predict the next step of the attack, which makes it easier and faster to contain and counter the attack. Also, there are many types of advanced stealthy attacks, such as stealthy SSH login attacks [57], which are hard to detect. Detecting such attacks may require obtaining much more intelligence, such as more information about the network and smarter algorithms; thus, it could be more challenging for resource-constrained environments. Finally, intrusion detection activities become even more challenging when dealing with rapidly changing IoT systems, such as mobile IoT systems, where different communication technologies may be used at different times, such as WiFi and 4GLTE/5G, and communication can be fragmented. As IDS relies on preexisting knowledge, such as attack signatures and/or normal behavior profiles, we need mechanisms to quickly adapt such preexisting knowledge, such as transfer learning techniques [58], and mechanisms to enhance the robustness of the underlying ML models used by IDSes.

Acknowledgement

The work reported in this paper has been funded by NSF under Grants CNS-2112471 and DGE-2114680.

References

1 Bertino, E. and Islam, N. (2017). Botnets and Internet of Things security. *IEEE Computer* 50 (2): 76–79. https://ieeexplore.ieee.org/stamp/stamp.jsp?arnumber=7842850.

2 Paul, F. (2017). Fixing, upgrading, and patching IoT devices can be a real nightmare. *NETWORKWORLD*. https://www.networkworld.com/article/3222651/fixing-upgrading-and-patching-iot-devices-can-be-a-real-nightmare.html (accessed 14 July 2022).

3 Tshuva, N. (2020). Patching vulnerabilities in IoT devices is a losing game. *Info Security*. https://www.infosecurity-magazine.com/opinions/patching-vulnerabilities-iot/ (accessed 14 July 2022).

4 Antonakakis, M., April, T., Bailey, M. et al. (2017). Understanding the Mirai botnet. *Proceeding of 26th USENIX Security Symposium (USENIX Security '17)*, pp. 1093–1110. https://www.usenix.org/system/files/conference/usenixsecurity17/sec17-antonakakis.pdf (accessed 14 July 2022).

5 Starr, M. (2014). Fridge caught sending spam emails in botnet attack. *CNET*. https://www.cnet.com/home/kitchen-and-household/fridge-caught-sending-spam-emails-in-botnet-attack/ (accessed 14 July 2022).

6 Thomson, I. (2016). Wi-Fi baby heart monitor may have the worst IoT security of 2016. *The Register*. https://www.theregister.com/2016/10/13/possibly˙worst˙iot˙security˙failure˙yet (accessed 14 July 2022).

7 OSSEC. OSSEC - World's most widely used host intrusion detection system - HIDS. https://www.ossec.net/ (accessed 14 July 2022).

8 Tripwire. Cybersecurity and Compliance Solutions — Tripwire. *Help/Systems*. https://www.tripwire.com/ (accessed 14 July 2022).

9 Snort (1998). Snort - network intrusion detection & prevention system. *Cisco*. https://www.snort.org/ (accessed 14 July 2022).

10 Suricata (2009). Home - Suricata. *The Open Information Security Foundation*. https://suricata.io/ (accessed 14 July 2022).

11 Rahman, S.A., Tout, H., Talhi, C., and Mourad, A. (2020). Internet of Things intrusion detection: centralized, on-device, or federated learning? *IEEE Network* 34 (6): 310–317. https://ieeexplore.ieee.org/document/9183799 (accessed 14 July 2022).

12 Cho, E., Park, M., Lee, H. et al. (2019). D2TLS: Delegation-based DTLS for cloud-based IoT services. *IoTDI '19: Proceedings of the International Conference on Internet of Things Design and Implementation*, pp. 190–201. https://dl.acm.org/doi/10.1145/3302505.3310081 (accessed 14 July 2022).

13 Shelby, Z., Hartke, K., and Bormann, C. (2014). The constrained application protocol (CoAP). *Internet Engineering Task Force* "RFC 7252. https://datatracker.ietf.org/doc/html/rfc7252 (accessed 14 July 2022).

14 OASIS Standard (2015). MQTT Version 3.1.1 Plus Errata 01. *OASIS*. http://docs.oasis-open.org/mqtt/mqtt/v3.1.1/mqtt-v3.1.1.pdf (accessed 14 July 2022).

15 Kushalnagar, N., Montenegro, G., and Schumacher, C. (2007). IPv6 over low-power wireless personal area networks (6LoWPANs): overview, assumptions, problem statement, and goals. RFC 4919. https://datatracker.ietf.org/doc/html/rfc4919 (accessed 14 July 2022).

16 Brandt, A., Hui, J., Kelsey, R. et al. (2012). RPL: IPv6 routing protocol for low-power and lossy networks. RFC 6550. https://datatracker.ietf.org/doc/html/rfc6550 (accessed 14 July 2022).

17 Arnaboldi, L. and Morisset, C. (2021). A review of intrusion detection systems and their evaluation in the IoT. *arXiv preprint*, arXiv:2105.08096. https://arxiv.org/pdf/2105.08096.pdf (accessed 14 July 2022).

18 ARM. ARM architecture: a foundation for computing everywhere. *ARM*. https://www.arm.com/architecture/cpu (accessed 14 July 2022).

19 RISC-V. RISC-V International. *RISC-V*. https://riscv.org/ (accessed 14 July 2022).

20 Dunkels, A., Gronvall, B., and Voigt, T. (2004). Contiki - A lightweight and flexible operating system for tiny networked sensors. *In 29th Annual IEEE International Conference on Local Computer Networks*, pp. 455–462. https://ieeexplore.ieee.org/document/1367266 (accessed 14 July 2022).

21 Contiki. Contiki-NG, the OS for Next Generation IoT Devices. https://www.contiki-ng.org/ (accessed 14 July 2022).

22 RIOT. RIOT - The friendly operating system for the Internet of Things. *RIOT*. https://www.riot-os.org/ (accessed 14 July 2022).

23 TinyOS. TinyOS home page. *TinyOS*. http://www.tinyos.net/ (accessed 14 July 2022).

24 FreeRTOS. FreeRTOS - Market leading RTOS (Real Time Operating System) for embedded systems with Internet of Things extensions. *FreeRTOS*. https://www.freertos.org/ (accessed 14 July 2022).

25 Khokale, S. (2019). Importance of remote device management for smart city initiatives. *eInfochips*, 10 December 2019. https://www.einfochips.com/blog/importance-of-remote-device-management-for-smart-city-initiatives/ (accessed 14 July 2022).

26 Jallad, K.A., Aljnidi, M., and Desouki, M.S. (2020). Anomaly detection optimization using big data and deep learning to reduce false-positive. *Journal of Big Data* 7 (1): 1–20. https://journalofbigdata.springeropen.com/articles/10.1186/s40537-020-00346-1.

27 Check Point Research (2017). IoTroop botnet: the full investigation. *cp*, 29 October 2017. https://research.checkpoint.com/2017/iotroop-botnet-full-investigation/ (accessed 14 July 2022).

28 Habibi, J., Midi, D., Mudgerikar, A., and Bertino, E. (2017). Heimdall: Mitigating the Internet of Insecure Things. *IEEE Internet of Things Journal* 4 (4): 968–978. https://ieeexplore.ieee.org/document/7930378.

29 Hamza, A., Gharakheili, H.H., and Sivaraman, V. (2018). Combining MUD policies with SDN for IoT intrusion detection. *IoT S&P '18: Proceedings of the 2018 Workshop on IoT Security and Privacy*, pp. 1–7. https://dl.acm.org/doi/10.1145/3229565.3229571 (accessed 14 July 2022).

30 Lear, E., Droms, R., and Romascanu, D. (2019). Manufacturer usage description specification. RFC 8520. https://datatracker.ietf.org/doc/html/rfc8520 (accessed 14 July 2022).

31 Hamza, A. SDN PCAP simulator. *sdn-pcap-simulator*. https://github.com/ayyoob/sdn-pcap-simulator (accessed 14 July 2022).

32 Shukla, P. (2017). ML-IDS: A machine learning approach to detect wormhole attacks in Internet of Things. *2017 Intelligent Systems Conference (IntelliSys)*, pp. 234–240. https://ieeexplore.ieee.org/document/8324298 (accessed 14 July 2022).

33 Farzaneh, B., Montazeri, M.A., and Jamali, S. (2019). An anomaly-based IDS for detecting attacks in RPL-based Internet of Things. *2019 5th International Conference on Web Research (ICWR)*, pp. 61–66. https://ieeexplore.ieee.org/abstract/document/8765272 (accessed 14 July 2022).

34 Gomes, P.H., Gugri, M., and Aggarwal, S. Cooja Simulator. *Cooja Simulator*. https://anrg.usc.edu/contiki/index.php/Cooja˙Simulator (accessed 14 July 2022).

35 Haripriya, A.P. and Kulothungan, K. (2019). Secure-MQTT: An efficient fuzzy logic-based approach to detect DoS attack in MQTT protocol for Internet of Things. *Journal on Wireless*

Communications and Networking 2019 (1): 1–15. https://jwcn-eurasipjournals.springeropen.com/articles/10.1186/s13638-019-1402-8 (accessed 14 July 2022).

36 Fu, Y., Yan, Z., Cao, J. et al. (2017). An automata based intrusion detection method for Internet of Things. *Mobile Information Systems*. https://doi.org/10.1155/2017/1750637.

37 Lee, D. and Su, D. (1997). Modeling and testing of protocol systems. *Testing of Communicating Systems*, 339–364. Springer, (Kim, M., Kang, S., and Hong, K., Eds.).

38 Nguyen, T.D., Marchal, S., Miettinen, M. et al. (2019). D"IoT: A federated self-learning anomaly detection system for IoT. *2019 IEEE 39th International Conference on Distributed Computing Systems (ICDCS)*. https://ieeexplore.ieee.org/stamp/stamp.jsp?arnumber=8884802 (accessed 14 July 2022).

39 Napiah, M.N., Idris, M.Y.I.B., Ramli, R., and Ahmedy, I. (2018). Compression header analyzer intrusion detection system (CHA-IDS) for 6LoWPAN communication protocol. *IEEE Access* 6: 16623–16638. https://ieeexplore.ieee.org/document/8270652.

40 Mudgerikar, A., Sharma, P., and Bertino, E. (2019). E-Spion: A system-level intrusion detection system for IoT devices. *Proceedings of the 2019 ACM ASIA Conference on Computer and Communications Security*, pp. 493–500. https://dl.acm.org/doi/10.1145/3321705.3329857 (accessed 14 July 2022).

41 Otoum, Y., Liu, D., and Nayak, A. (2019). DL-IDS: a Deep learning-based intrusion detection framework for securing IoT. *Transactions on Emerging Telecommunications Technologies* 33 (3). https://onlinelibrary.wiley.com/doi/full/10.1002/ett.3803 (accessed 14 July 2022).

42 Bansal, J.C., Sharma, H., Jadon, S.S., and Clerc, M. (2014). Spider monkey optimization algorithm for numerical optimization. *Memetic Computing* 6 (1): 31–47. https://jcbansal.scrs.in/uploads/3-Spider˙Monkey˙Optimization.pdf (accessed 14 July 2022).

43 Alaiz-Moreton, H., Aveleira-Mata, J., Ondicol-Garcia, J. et al. (2019). Multiclass classification procedure for detecting attacks on MQTT-IoT protocol. *Complexity*. https://doi.org/10.1155/2019/6516253.

44 Sforzin, A., Mármol, F.G., Conti, M., and Bohli, J.M. (2016). RPiDS: Raspberry Pi IDS – a fruitful intrusion detection system for IoT. *2016 International IEEE Conferences on Ubiquitous Intelligence & Computing, Advanced and Trusted Computing, Scalable Computing and Communications, Cloud and Big Data Computing, Internet of People, and Smart World Congress (UIC/ATC/ScalCom/CBDCom/IoP/SmartWorld)*, pp. 440–448. https://ieeexplore.ieee.org/document/7816876 (accessed 14 July 2022).

45 Eskandari, M., Janjua, Z.H., Vecchio, M., and Antonelli, F. (2020). Passban IDS: An intelligent anomaly-based intrusion detection system for IoT edge devices. *IEEE Internet of Things Journal* 7 (8): 6882–6897. https://ieeexplore.ieee.org/abstract/document/8976157.

46 Liu, F.T., Ting, K.M., and Zhou, Z. (2008). Isolation forest. *2008 8th IEEE International Conference on Data Mining*, pp. 413–422. https://ieeexplore.ieee.org/abstract/document/4781136 (accessed 14 July 2022).

47 Breunig, M.M., Kriegel, H., Ng, R.T., and Sander, J. (2000). LOF: Identifying density-based local outliers. *Proceedings of the 2000 ACM SIGMOD International Conference on Management of Data*, pp. 93–104. https://dl.acm.org/doi/10.1145/335191.335388 (accessed 14 July 2022).

48 Sedjelmaci, H., Senouci, S.M., and Al-Bahri, M. (2016). A lightweight anomaly detection technique for low-resource IoT devices: a game-theoretic methodology. *2016 IEEE International Conference on Communications (ICC)*, pp. 1–6. https://ieeexplore.ieee.org/document/7510811 (accessed 14 July 2022).

49 Kreps, D.M. (1989). Nash equilibrium. In: *Game Theory*, 167–177. Springer.

50 TOSSIM. TOSSIM - TinyOS Wiki. *TOSSIM*. http://tinyos.stanford.edu/tinyos-wiki/index.php/TOSSIM (accessed 14 July 2022).

51 Shurman, M.M., Khrais, R.M., and Yateem, A.A. (2019). IoT denial-of-service attack detection and prevention using hybrid IDS. *2019 International Arab Conference on Information Technology (ACIT)*, pp. 252–254. https://ieeexplore.ieee.org/document/8991097 (accessed 14 July 2022).

52 Ali, A. and Yousaf, M.M. (2020). Novel three-tier intrusion detection and prevention system in software defined network. *IEEE Access* 8: 109662–109676. https://ieeexplore.ieee.org/document/9117020.

53 Varga, A. (2010). OMNeT++. *In Modeling and Tools for Network Simulation*, pp. 35–59. http://src.gnu-darwin.org/ports/science/omnetpp/work/omnetpp-2.3p1/doc/usman.pdf (accessed 14 July 2022).

54 Midi, D., Rullo, A., Mudgerikar, A., and Bertino, E. (2017). Kalis – A system for knowledge-driven adaptable intrusion detection for the Internet of Things. *2017 IEEE 37th International Conference on Distributed Computing Systems (ICDCS)*, pp. 656–666. https://ieeexplore.ieee.org/document/7980009 (accessed 14 July 2022).

55 Pasikhani, A.M., Clark, A.J., and Gope, P. (2021). Reinforcement-learning-based IDS for 6LoWPAN. *The 20th IEEE International Conference on Trust, Security, and Privacy in Computing and Communications*. https://ieeexplore.ieee.org/document/9724461 (accessed 14 July 2022).

56 Riley, G.F. and Henderson, T.R. (2010). The ns-3 network simulator. *Modeling and Tools for Network Simulation* 15–34. https://link.springer.com/chapter/10.1007/978-3-642-12331-3˙2 (accessed 14 July 2022).

57 Javed, M. and Paxson, V. (2013). Detecting stealthy, distributed SSH brute-forcing. *Proceedings of the 2013 ACM SIGSAC Conference on Computer & Communications Security*, pp. 85–96. http://www.icir.org/vern/papers/dist-ssh-det.ccs13.pdf (accessed 14 July 2022).

58 Singla, A., Bertino, E., and Verma, D.C. (2020). Preparing network intrusion detection deep learning models with minimal data using adversarial domain adaptation. *Proceedings of ASIA CCS '20: The 15th ACM ASIA Conference on Computer and Communications Security*, pp. 127–140. https://dl.acm.org/doi/abs/10.1145/3320269.3384718 (accessed 14 July 2022).

59 KDD Cup (1999). Data, Available: http://kdd.ics.uci.edu/databases/kddcup99/kddcup99.html, [Accessed November 30, 2021].

14

Bringing Intelligence at the Network Data Plane for Internet of Things Security

Qiaofeng Qin, Konstantinos Poularakis, and Leandros Tassiulas

Department of Electrical Engineering & Institute for Network Science, Yale University, New Haven, CT, USA

Abstract

Internet of Things (IoT) can aid modern military operations in various ways; from immersive virtual simulations for soldiers' training to autonomous vehicles and environmental sensors for situation awareness and distributed decision making. Yet, security threats arising in massively connected IoT devices continue to challenge their widespread adoption by the Army. It is necessary to equip IoT gateways with firewalls to prevent hacked devices from infecting a larger number of network nodes. Meanwhile, cutting-edge Software Defined Network (SDN) technologies open the door for greater innovation to network control and data planes. The match-and-action mechanism of SDN provides the means to differentiate malicious traffic flows from normal ones, which mirrors the past firewall mechanisms but with a new flexible and dynamically re-configurable twist. However, vulnerabilities of IoT devices and heterogeneous protocols coexisting in the same network challenge the extension of SDN into the IoT domain. To overcome these challenges, we leverage data-plane programming languages that enable intelligent packet processing, and propose two novel data-driven approaches for attack detection. First, we design a two-stage deep learning method that generates flow rules for classifying and separating malicious from normal packets. Our method is tailored to the P4 programming language so as to be adaptive to arbitrary protocols while maintaining high performance of attack detection. Second, we develop a binarized neural network (BNN) based method that offloads the security functionality from a remote server (control plane) to an IoT gateway (data plane) thereby reducing the packet classification latency and flow rule storage demand. Evaluations using network traces of various IoT protocols show significant benefits in accuracy, efficiency and universality over state-of-the-art methods.

14.1 Introduction

Modern military operations have complex communication and computing requirements that cannot be supported by today's network systems. To this end, Internet of Things (IoT) technology has the potential to enable the redesign and successful deployment of these systems. IoT interconnects a multitude of devices interfacing with the physical world as sensors and actuators, facilitating their communication towards accomplishing assigned tasks. In a practical military scenario, these devices can collect real-time information about the environment conditions and push this to designated command and control nodes so that the latter can monitor and adjust if needed the tactical operations.

IoT for Defense and National Security, First Edition. Edited by Robert Douglass, Keith Gremban, Ananthram Swami, and Stephan Gerali.

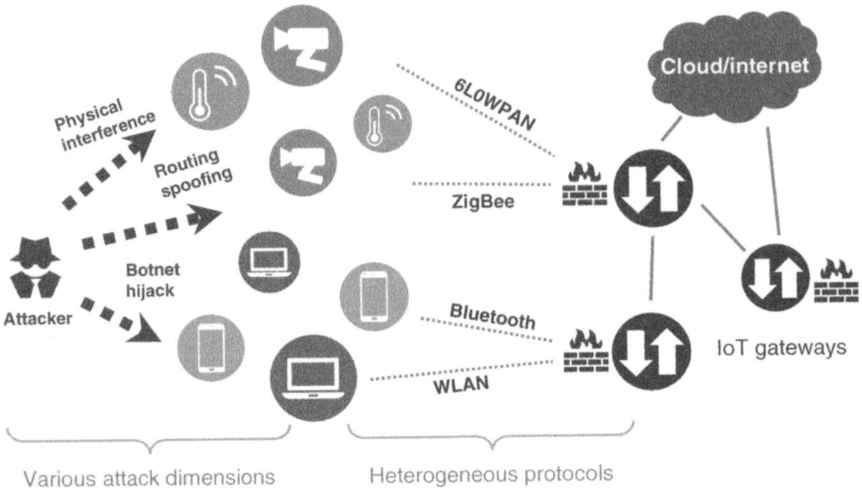

Figure 14.1 Network-layer security approaches such as firewalls deployed at IoT gateways can target various types of attacks in heterogeneous protocols.

In IoT networks with massively interconnected devices, security is a major concern. A large amount of insecure IoT devices have become targets of botnet attacks [1], leading to some of the most potent DDoS attacks in history. IoT devices are vulnerable to more types of attacks compared with other devices [2], such as network attacks in different protocols (e.g. RFID, Zigbee, 6LoWPAN) and even physical attacks. Therefore, it has been a big challenge to guarantee the security of an IoT network.

Traditional methods to secure an IoT device require the deployment of *physical* and *application layer* protection in it, e.g. by strengthening the authentication and encryption during data transmission. However, such approaches usually involve firmware and even hardware modifications, taking a relatively long time period. Devices in which security policies are not updated in time will increase the risk of being hacked and becoming sources of infection to other devices. To prevent malware from spreading, *network layer* security approaches are also necessary. For example, firewalls can be deployed at IoT gateways, monitoring and separating malicious from normal traffic, as depicted in Figure 14.1.

Software Defined Networking (SDN) provides a flexible framework for network management and is widely adopted in IoT networks. This flexibility can be exploited for the development and dynamic reconfiguration of network layer security mechanisms. By separating control and data planes, SDN protocols such as OpenFlow [3] make it possible to develop such mechanisms in a logically centralized and programmable manner. OpenFlow-enabled switches process incoming packets through match-and-action flow rules received from the controller checking specific header fields (e.g. MAC and IP addresses, TCP port, etc.) and performing actions such as forwarding or dropping accordingly.

A firewall can be developed by generating flow rules through *machine learning* algorithms, which have been demonstrated as a promising method for identifying attacks from even unknown or encrypted traffic flows [4]. However, this method presents several limitations. Specifically:

1. **Limitations in Machine Learning Models**: The training features used by the machine learning algorithm are often the specific header fields of the packet. However, heterogeneous IoT protocols may have distinct packet header structures, leading to a problem that the feature extraction process and even the whole learning algorithm should be specifically *redesigned* for

every different protocol. Besides, the *manual* feature extraction adds difficulty to achieve optimal performance.

2. **Limitations in OpenFlow**: The match fields of OpenFlow are *predefined* and *fixed*. Many IoT headers cannot be parsed by it, e.g. compressed IPv6 headers in 6LoWPAN packets, or application layer protocols such as MQTT and RESTful API. As a result, no proper flow rules can be created in these cases. Although OpenFlow can be extended with user-defined headers by OpenFlow extensible match (OXM), it has limited functionality and hardware support in the above scenarios.

Novel network data plane devices, including programmable switches and SmartNICs (or Smart network interface cards), provide possible solutions to the above challenges. Different from OpenFlow which only focuses on the control plane, these devices also bring programmability to the data plane. P4 [5] is a language designed for data plane programming supported by multiple SmartNIC models. Specifically, in P4, the packet headers are customizable by operators with the position and width provided, and table lookup can be conducted on these newly defined headers by the switches. This feature is especially meaningful in IoT scenarios, where support of different IoT protocols can be added by defining their headers [6].

Motivated by the features brought by the programmable data plane device and P4, we propose our first approach, a novel *flow rule generation (FRG)* framework for IoT security and a corresponding learning algorithm which takes advantage of the P4 language. The proposed method operates in two stages. In Stage 1, a learning algorithm trains *a dilated Convolutional Neural Network (Dilated CNN)* with raw packet bytes, skipping the step of manual feature extraction. In Stage 2, *a proper set of header field definitions* is inferred from the trained neural network, based on which flow rules for blocking traffic (dropping packets) are generated and installed in the IoT gateway (data plane switch). This method is applicable to heterogeneous IoT protocols. Besides, it is designed to take the constraints of switch memory cost and packet processing speed into consideration, realizing a trade-off between accuracy and efficiency. Figure 14.2 illustrates its differences compared with the existing OpenFlow-based methods.

In addition to the ability of handling heterogeneous protocols, we also consider reducing the latency and storage cost of existing SDN approaches. On the one hand, additional latency is

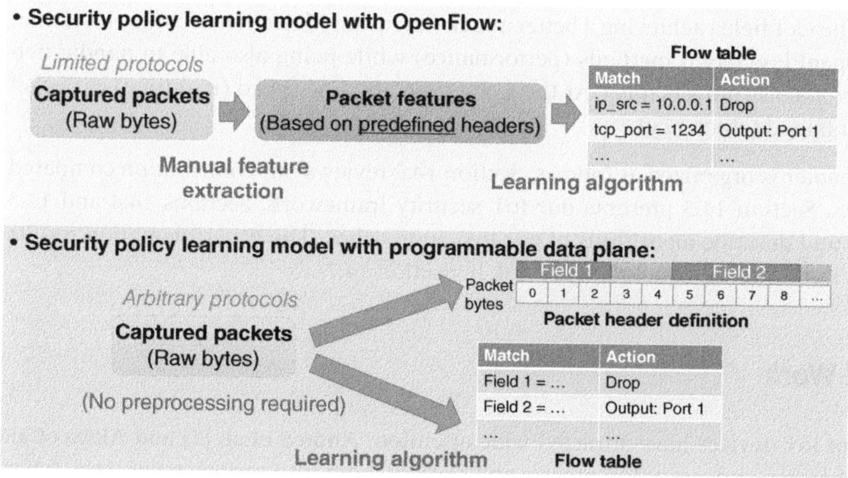

Figure 14.2 A learning model based on P4 can be more flexible to handle heterogeneous protocols than OpenFlow-based ones.

incurred at the data plane device due to waiting for instructions from the SDN controller which is usually hosted in a remote server or host. On the other hand, the storage of match-and-action flow rules consumes memory resources at the data plane device. Compared with traditional networking devices, SmartNICs also guarantee higher capability for executing more complicated logic by supporting P4 and other languages, rather than simple packet forwarding. Network services that are traditionally run in remote servers can be offloaded to SmartNICs, in order to reduce the network overhead and latency. Hence, we propose our second approach that presents the possibility of executing packet classification models locally instead of at a remote controller. We investigate methods based on binarized neural network (BNN) [7], which compresses weights of a neural network model into single bits. In this way, the computation and memory requirements are significantly reduced to a level that SmartNICs can afford.

The contributions of this work can be summarized as follows:

- **IoT Security Framework**: We propose a new framework for securing IoT networks and devices. Taking advantages of the programmable data plane of SmartNIC devices and the P4 language, we aim at developing universal, highly accurate and efficient solutions to identify malicious traffic flows of multiple IoT protocols.
- **Flow Rule Generation (FRG) Approach**: We propose a two-stage algorithm for classifying IoT packets. In the first stage, we train a Dilated CNN with raw packet bytes to set up a traffic classifier, which skips the step of manual feature extraction of OpenFlow based methods and thus requires minimum data preprocessing. In the second stage, we convert the abstract features learned in the trained CNN into a particular set of header fields (byte substrings), so that a proper set of flow rules can be installed at the IoT gateway. This way, the classification can be realized as a switch function at the IoT gateway for lower memory cost and faster processing speed.
- **Binarized Neural Network (BNN) Approach**: We also propose a lightweight packet classification algorithm based on BNN, achieving high accuracy with low memory costs. We discuss how our approach offloads the network security service from the remote controller to the local data plane for improved latency and scalability.
- **Experimental Datasets**: We conduct experiments to create our own new datasets of IoT traffic and multiple types of attacks. With these as well as publicly available datasets, we evaluate the performance of the proposed methods in all aspects. The results show that our method makes proper choices of header fields achieving a better attack (intrusion) detection accuracy level than state-of-the-art OpenFlow based methods (performance) while being also able to handle heterogeneous IoT protocols (universality). At the same time, the line speed (real-time) of packet processing is maintained (efficiency).

The rest of this chapter is organized as follows. Section 14.2 reviews our contribution compared to the related works. Section 14.3 presents our IoT security framework. Sections 14.4 and 14.5 define the problem and describe algorithms of our two approaches. The experimentation results are presented in Section 14.6. We conclude our work in Section 14.7.

14.2 Related Work

Security problems of IoT devices have attracted wide attention. Andrea et al. [2] and Alaba et al. [8] provide comprehensive surveys of IoT attacks and classify them into various types. New types of attacks different from traditional networks threaten IoT security, including a variety of attack methods in IoT protocols such as Zigbee and 6LoWPAN [9–11], as well as physical attacks targeting the sensors and actuators [12, 13]. These works suggest adding authentication mechanisms to the devices. However, a network-level security solution is also necessary for preventing malware from spreading among vulnerable IoT devices, such as botnets [14]. Our firewall implementation at the IoT gateway complements the device-level authentication for a more powerful security guarantee.

Network-level security approaches can be grouped into two categories. The first category applies machine learning methods on specific packet headers [15]. For example, a learning-based method is applied on 6LoWPAN headers [16]. Kalis [17] provides a knowledge-driven solution for detecting IoT attacks, while DÏoT [18] and IoT Sentinel [19] identify the IoT device types by learning. Though these methods are effective, they usually require pre-knowledge from protocol definitions or device manufacturers. Due to the large diversity of IoT devices and protocols, we explore another direction leading to a more universal solution for heterogeneous IoT systems in case that such pre-knowledge is not available.

The second category classifies packets based on raw packet bytes rather than header fields. Machine learning methods, especially neural networks are also widely applied for it [4, 20, 21]. These approaches have high accuracy and are not limited to specific protocol or device types. However, they can only be deployed in a remote server/host rather than a switch (IoT gateway). The reason is that the switch typically lacks the computation power to perform inference of the neural network model. Therefore, packets cannot be processed at the line speed of the switch and the latency is much higher.

Our FRG approach focuses on combining the merits of the two approaches above, developing intrusion detection as a switch function at the IoT gateway and at the same time not relying on assumptions of device and protocol types. Our BNN approach belongs to the second category but offloads the computation from the remote server to the local data plane for higher efficiency. This in-network computation concept has been discussed for various network applications [22]. BNN [7] is a type of neural network with only binary weights and activation functions. It is regarded as a suitable method for deploying services in embedded devices [23]. Attempts are also made to implement BNN in smart network devices [24, 25]. We make similar attempts while performing realistic networking tasks, i.e. packet classification.

Benefiting from their programmable, flexible and efficient packet processing capabilities, recent developments in SDN make the implementation of such a switch function possible. For example, Sensor OpenFlow [26] and SDN-Wise [27] extend the OpenFlow protocol in this direction. Besides, there is an increasing research interest in deploying and managing P4-enabled devices. Sensor data from multiple packets can be aggregated by P4 header operations [16]. Moreover, multi-protocol switching of IoT services can be achieved by deploying P4-enabled switches [6]. Our proposed security framework is also based on P4, which will be described in the next sections.

14.3 System Design

The proposed system has two components. The first part is the control plane, an SDN controller which is a software entity hosted in a node with sufficient computation capacity, e.g. a conventional cloud server or an edge cloud node. The second part is the data plane, which can be an IoT gateway. We consider the case that the IoT gateway is based on programmable switches or SmartNICs that support the P4 language.

14.3.1 Architecture of the FRG Approach

We firstly describe the system design of our FRG approach. P4, or Programming Protocol-independent Packet Processors language is designed for reconfigurability and protocol independence. More specifically, the control plane (controller) is able to define how a data plane device

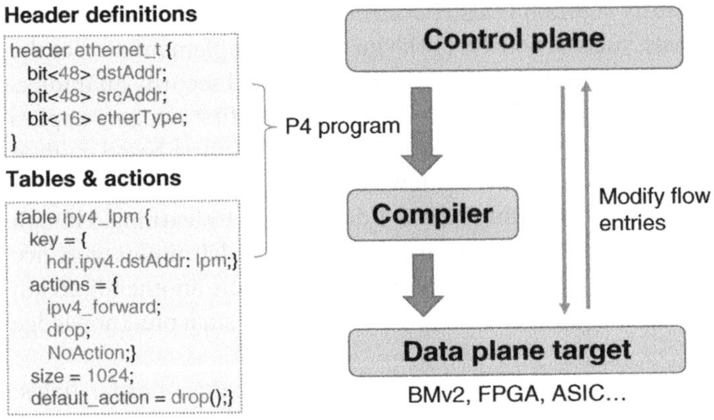

Figure 14.3 P4 language has protocol independence and reconfigurability.

(switch or gateway) parses a packet in a programmable and automated way (reconfigurability). First, one or more headers are defined as a list of fields given their positions and widths in bits. Then, a parser works as a state machine to extract headers, following a series of match-and-action tables, which is similar to OpenFlow, except that header fields are not predefined (protocol independence). The whole workflow is depicted in Figure 14.3.

A P4-enabled gateway is capable of serving IoT devices of heterogeneous network protocols [6]. Our aim is to use the IoT gateway to identify malicious incoming traffic flows (e.g., from a hijacked IoT device) before they are routed to other domains and devices. We program the IoT gateway to execute a firewall function before the routing function. The firewall keeps a match-and-action table recording the features of known packets, which are the values of certain packet header fields. These fields will be checked inside the incoming packets and marked as normal or malicious based on the flow rules installed in the table. Normal packets will be passed to the routing function without modifications. On the other hand, actions can be defined to handle the malicious packets, e.g. blocking them or forwarding them to a honeypot. The flow rules are generated by the SDN controller, where a classifier is deployed and responsible for judging whether a flow is malicious or not. The controller is able to convert classification results into header field definitions and flow rules to install them in the firewall at the IoT gateway either reactively or proactively. The whole architecture is depicted in Figure 14.4.

Two key problems are required to be solved in the proposed system. First, we need to find algorithms for classifying packets with high accuracy. Second, P4 match-and-action tables should be generated, making classification a set of data plane flow rules which achieves line-speed packet processing. Besides, the solution we expect should be universal for heterogeneous IoT protocols, i.e. neither algorithm redesign nor protocol-dependent data preprocessing is required. In the next two sections, we will formally propose a formulation of the packet classification problem and provide algorithms corresponding to these key problems.

14.3.2 Architecture of the BNN Approach

In our first approach above, the security service will incur either a latency (in the case that we install flow rules reactively when receiving an unknown packet) or a memory cost of the match-and-action table (in the case we install flow rules proactively). Although our solution contains methods making such costs acceptable, which will be discussed in later sections, we

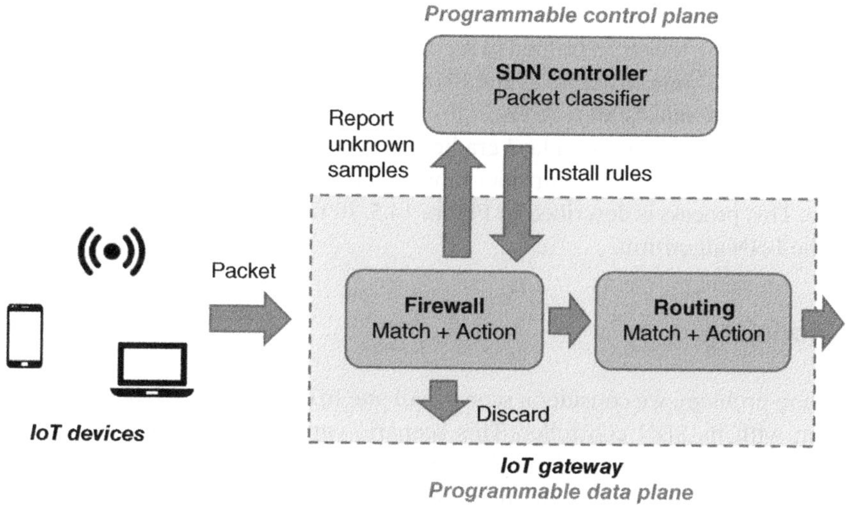

Figure 14.4 Both the control and data planes are programmable in the proposed FRG framework.

also investigate a different approach based on a similar architecture, which offloads the security service to the data plane. In addition to the regular match-and-action table that passively receives rules from the SDN controller, we propose that the IoT gateway locally executes a packet classification model that requires minimal costs. Specifically, we deploy BNN as the model. Even with programmability, most data plane devices are not capable enough for complicated calculations such as the dot product of floating-point numbers, which is a common operation of most neural network algorithms. However, a binarization process turns all weights of BNN into single bits, and therefore only bitwise operations are required for the model inference. SmartNICs such as Netronome Agilio CX supporting P4 and C language are able to execute BNN models.

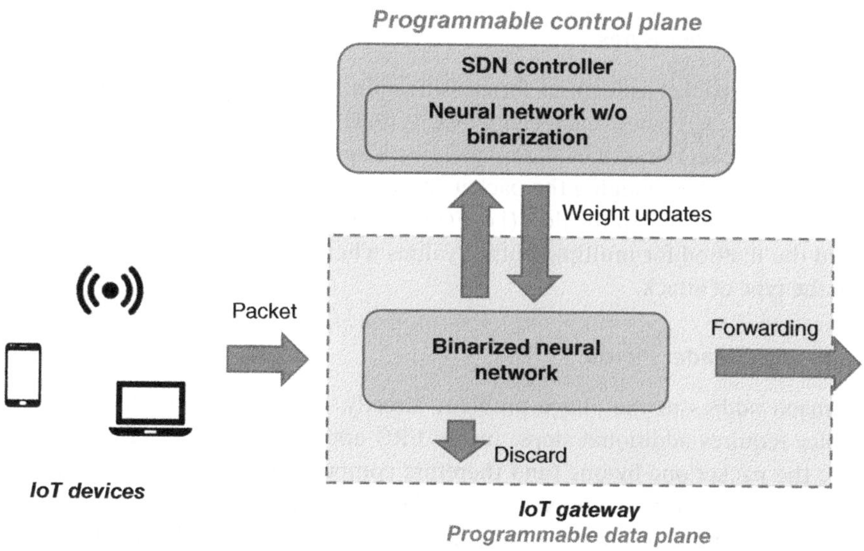

Figure 14.5 Compared with the FRG framework, the BNN approach further offloads the security service to the data plane by deploying a binarized model.

The task of the SDN controller in this approach is assisting the training of the model. After binarization, the neural network can no longer be trained in regular ways based on gradients. Therefore, another neural network with the same structure as the BNN but without binarization is deployed in the controller. When the BNN makes an inference, it can forward the classification result to the controller. The controller performs the standard gradient descent operation to train its neural network. When BNN requires updating, the controller binarizes the weights again and sends the results to the data plane. This process is described in Figure 14.5. In the next two sections we will describe the details of the BNN algorithm.

14.4 Problem Modeling

To model the classification problem, we consider a scenario of one IoT network domain equipped with one gateway along with its SDN controller. This scenario can be easily extended into a multi-domain or multi-gateway topology by deploying the same solution in each domain. The gateway is responsible for identifying attacks among all traffic flows going through it, so that it can block current and future packets of the attack flow to prevent it from spreading, e.g. a hijacked device outside the domain infecting devices inside the domain, and vice versa. We assume that the security of the gateway itself and its SDN controller is not compromised.

The features that can be used for classifying network traffic can be divided into two types, the *packet-level features* (e.g. IP address, TCP port, payload length), and the *flow statistics* (e.g. packet count, duration). The programmable data plane brings opportunities for defining new packet-level features, not restricted to OpenFlow's pre-defined collection, which is particularly important for the IoT network where heterogeneous protocols coexist. Besides, previous studies claim several other merits of learning directly from packet bytes, including the higher accuracy and the ability to classify encrypted traffic [20]. Therefore, our work is focused on the packet-level features type of classification. Nevertheless, we also perform several evaluations which indicate that our methods can be easily applied to the flow-level classification without much modification in later sections.

14.4.1 Classification with Header Bytes

We use the first N bytes of the packet as features for classification. The packet can be thus represented by a vector $\mathbf{x} = (x_1, x_2, \dots, x_N)$ where each element $x_i \in [0, 1]\ \forall i \leq N$ is a number converted from a byte. If the length of a packet is less than N, zero padding is applied. A classifier in the control plane should provide a function $F(\mathbf{x})$ judging the packet. We consider a binary output indicating whether the packet belongs to a normal traffic flow (i.e. $F(\mathbf{x}) = 0$) or a malicious one (i.e. $F(\mathbf{x}) = 1$). We can directly extend the method for multiple output values where the gateway takes different actions depending on the type of attack.

14.4.2 Classification with Header Fields

While our BNN approach addresses the above problem directly, the FRG approach based on match-and-action tables requires additional steps. In the FRG approach, the control plane can check the bytes inside the packet one-by-one (and therefore compute the $F(\mathbf{x})$ value). However, such fine-grained classification may not be possible in the data plane (IoT gateway) as this would require one to install a huge number of flow rules for all possible combinations of the N bytes. This is not feasible since it would lead to unrealistic memory cost and latency of lookup and processing packets.

Taking advantage of P4, any *substring* of packet bytes can be regarded as a *header field* by the gateway, based on which flow rules will be generated. Therefore, we can effectively limit the number and length of flow rules, as well as the associated packet processing latency, by carefully defining a small number of packet byte substrings as header fields at the gateway.

Formally, we define the *Header Fields Definition* $H = \{h_k, k = 1, 2, ..., K\}$ which is a set of K substrings of bytes. Investigation of various P4-enabled devices shows that the number of header fields has an impact on the performance [28]. Therefore, we require that $K \leq K_{\max}$ where $K_{\max} << N$ so that a maximum memory cost and packet processing latency requirement is met. Each element $h_k = (a_k, a_k + L_k)$ is a substring starting from the a_k-th byte of the packet and ending at the $(a_k + L_k - 1)$-th byte, with its length L_k. These substrings should not overlap with each other, i.e. $a_{k+1} \geq a_k + L_k$ for any k, to avoid wasting memory. Unlike the traditional definition of header fields, each of which contains a specific type of information (e.g. network address or port number), we do not restrict that every substring defined by our method corresponds to a clear entity. Instead, *we aim for an algorithm capable of learning the meaning and importance of different substrings*, so that it can minimize the requirement of data preprocessing and be applicable to heterogeneous IoT protocols.

Based on the Header Fields Definition H, the information actually extracted from a packet $\mathbf{x}$ is $\mathbf{x}^H = (x_{a_1}, ..., x_{a_1+L_1-1}, ..., x_{a_K}, ..., x_{a_K+L_K-1})$. Therefore, the packet classification executed at the gateway follows a function different from $F(\mathbf{x})$, which depends on the definition of header fields H. We denote this function by $F^H(\mathbf{x}^H)$. Our goal is to find proper H and $F^H(\mathbf{x}^H)$ functions which satisfy the constraints mentioned above and are able to predict the packet classification at a high accuracy.

14.5 Algorithms and Learning Models

14.5.1 FRG Approach: Overview

We solve the two problems specified in the previous section in *two stages* as depicted in Figure 14.6. In Stage 1, we build and train a *neural network (NN)* as the packet classifier. The training is based on raw packet bytes without considering the definition of header fields. This classifier will be deployed at the control plane. In Stage 2, we calculate *importance scores* for each possible substring of packet bytes using the information from the trained NN (neuron weights), and then select non-overlapping substrings with the largest scores to be included in the header field definition, which will be installed at the gateway (data plane) along with a match-and-action flow table.

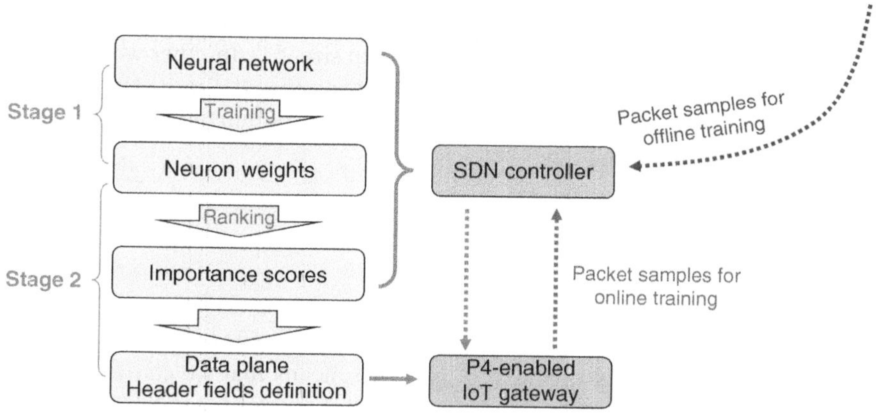

Figure 14.6 Illustration of the proposed two-stage learning approach. Packet classification is realized by the SDN control plane in Stage 1, followed by header field definition and implementation at the IoT gateway in Stage 2.

Initially, the NN is trained offline with captured network traces. The trained NN is then deployed at the controller as the packet classifier. For the data plane, both proactive and reactive operating modes are available according to different scenarios. In the first mode, the controller installs both header field definitions and corresponding flow rules from training data proactively at the gateway. The gateway can therefore process new incoming packets at line speed without forwarding them to the controller. In the second mode, the controller can proactively install header field definitions only, and install flow rules in a reactive way by replying to the gateway's queries. This mode incurs less memory cost in the gateway but increases latency due to the controller-gateway communication each time when the gateway receives unknown packets.

After the initial offline training, with the gateway sampling new packets and sending them to the controller, the two-stage process can be repeated in an online manner optionally, as long as the labels of packets can be acquired by the controller as well. The controller can also dynamically update the header field definition by compiling a new P4 program. All these operations are supported by the P4 specification.

14.5.2 FRG Stage 1: Neural Network Structure

We apply methods of supervised learning for the packet classification. In particular, trained with a labeled dataset (i.e. large amount of packets marked as either malicious or normal), the classifier should be able to infer the expected output of a new input (the function $F(\mathbf{x})$). A NN [29] is a computing system for supervised learning. It consists of several hidden layers and an output layer. Each layer is constructed by building blocks called neurons. For example, if we arrange the neurons of each layer in an array with index n (corresponding to the byte index of the packet), assign another index $i = 1, 2, \ldots, I_t$ for each layer t and take the packet byte vector $\boldsymbol{x}$ as the input, the output of a neuron in the first hidden layer is:

$$c_{ni}^{1} = f(\boldsymbol{w}^{1;ni} \cdot \boldsymbol{x} + \boldsymbol{b}^{1;ni}). \tag{14.1}$$

The output of each layer is the input of the next layer. For the neuron in the t-th hidden layer ($t > 1$), the output is:

$$c_{ni}^{t} = f(\boldsymbol{w}^{t;ni} \cdot \boldsymbol{c}^{t-1} + \boldsymbol{b}^{t;ni}), \tag{14.2}$$

where $\boldsymbol{w}^{t;ni}$ is a 2D vector of trainable weights, $\boldsymbol{b}^{t;ni}$ is a bias term, and f is a non-linear activation function.

Among various NN structures, we adopt the 1D Dilated CNN [30], as depicted in Figure 14.7. In each hidden layer t, connections are local and dilated with step size 2^{t-1}. In other words, each neuron with index i only takes two rows of neurons with indices i and $i + 2^{t-1}$ in its last layer as the inputs. Neurons in the same layer share the same weight values. The output of the hidden layer neurons can be represented in the following way:

$$c_{ni}^{1} = f(w_{\alpha}^{1} \cdot x_n + w_{\beta}^{1} \cdot x_{n+1} + b^1), \tag{14.3}$$

$$c_{ni}^{t} = f(\boldsymbol{w}_{\alpha}^{t} \cdot \boldsymbol{c}_{n}^{t-1} + \boldsymbol{w}_{\beta}^{t} \cdot \boldsymbol{c}_{n+2^{t-1}}^{t-1} + \boldsymbol{b}^{t}), \forall t > 1, \tag{14.4}$$

where $\boldsymbol{w}_{\alpha}^{t}$ and $\boldsymbol{w}_{\beta}^{t}$ are two 1D vectors of trainable weights.

This structure brings two major benefits. First, for any hidden layer neuron c_{ni}^{t}, its inputs are limited in the range between packet bytes x_n and x_{n+2^t-1}, which means that we can establish a correspondence between a neuron c_{ni}^{t} and a substring $(n, n + 2^t)$ following the denotation in the last section. Second, the neuron receptive field is 2^t, increasing exponentially with the network depth.

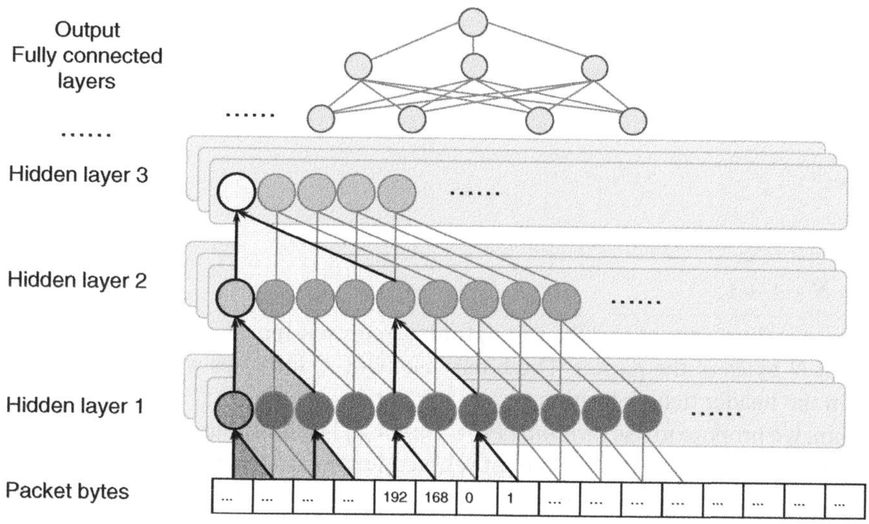

Figure 14.7 Structure of the dilated convolutional neural network (Dilated CNN) is adopted for packet classification, establishing the correspondence between neurons and packet substrings.

With T hidden layers, we can find neurons corresponding to any potential header field of length $2, 4, 8, \ldots,$ up to 2^T bytes. In other words, with a limited amount of layers, we are able to cover a wider range of packet substrings. This is beneficial in both representing the packet structure better and training the neural network more efficiently. After convolutional layers, we have fully-connected layers, the last of which has a single neuron taking the weighted sum of the last hidden layer outputs as the final result. This structure can be easily extended to multi-class classification, as long as we set up more neurons in the output layer.

14.5.3 FRG Stage 2: Header Field Definition

In the next stage, we adopt a neural network pruning technique [31] to the trained network. Pruning compresses the neural network by reducing the number of neurons. With smaller memory and calculating costs, pruning facilitates the processing of NN in IoT scenarios [32], where the capacity of devices may be limited. However, besides this benefit, our main purpose is to deduce an optimal set of header field definition based on the results of pruning, therefore enabling the line-speed packet processing in a P4-enabled gateway.

Pruning leads to an *importance score* of each neuron. Neurons with higher importance scores play a more crucial role in the classification. According to Yu et al. [31], we apply the Inf-FS [33] algorithm to calculate the importance scores of neurons in the last hidden layer. Then, the importance scores are calculated for the remaining layers in a backpropagation manner.

Leveraging the one-to-one correspondence between neurons and header fields in the proposed CNN structure, we extend the notion of importance score from neurons to header fields. Unlike the approach that suggests to greedily select neurons with highest importance scores [31], our problem has additional constraints, e.g. that the header fields should not overlap with each other. Therefore, we propose a new problem formulation.

The input of the problem includes the importance scores of all neurons in each hidden layer t. We denote the importance score of neuron c^t_{ni} by s^t_{ni}. By summing these values, we denote the importance score of a potential header field $(n, n + 2^t)$ by $S_n = \sum_i s^t_{ni}$. Then, we obtain the following optimization problem:

$$\max_{y} \sum_{n=1}^{N^t} y_n * S_n, \tag{14.5}$$

$$\text{s.t.} \sum_{n=1}^{N^t} y_n \leq K_{\max}, \tag{14.6}$$

$$y_n * y_{n+j} = 0, \quad \forall n < N^t, j < L, \tag{14.7}$$

$$L = 2^t, \quad N^t = N - L + 1, \tag{14.8}$$

where $\mathbf{y} = (y_1, y_2, \ldots, y_{N^t})$ is the vector of variables to optimize, representing all possible substrings of length 2^t in the first N bytes of the packet. The binary element y_n indicates whether to select substring $(n, n + 2^t)$ in the header field definition ($y_n = 1$) or not ($y_n = 0$).

To solve this problem, we propose to use *Dynamic Programming* [34]. A *Bellman equation* can be easily defined based on two states; K as the number of selected header fields and n_0 as the starting byte of the latest selected header field. We then have the following equations:

$$V(1, n_0) = S_{n_0}, \qquad \forall n_0 \leq N^t,$$

$$V(K, n_0) = \max_{n+L \leq n_0} V(K-1, n) + S_{n_0}, \qquad \forall n_0 \leq N^t, K > 1.$$

Based on the above equations, any $V(K, n_0)$ value can be calculated by *recursion*. The maximum of our objective function is therefore $\max_{n_0 \leq N^t} V(K_{\max}, n_0)$. As described in Algorithm 1, an optimal set of header fields H can be selected with reasonable $O(K_{\max} * N)$ time complexity.

The parameters $K_{\max}$ (i.e. maximum number of header fields) and $L = 2^t$ (i.e. length of one header field) can be determined according to the capacity of different types of P4-enabled devices [28]. In general, a tradeoff between accuracy and cost can be achieved by adjusting these parameters. With fewer or shorter header fields, some different traffic flows may be regarded as the same one by the gateway, negatively affecting classification accuracy. With more or longer header fields, however, it takes larger memory cost to store flow rules, and may slow down packet processing in some implementations. In the next section, we will evaluate the exact impact of these parameters on different performance metrics.

Algorithm 1: Optimal Header Fields Selection

Input: $S_1, S_2, \ldots, S_{N^t}, K_{\max}, L$

for $n_0 \leq N^t$ **do**
    $V(1, n_0) = S_{n_0}$;
    $H(1, n_0) = \{(n_0, n_0 + L)\}$;
end
for $K = 2, 3, \ldots, K_{\max}$ **do**
    for $n_0 \leq N^t$ **do**
        $n^* = \arg\max_{n+L \leq n_0} V(K-1, n)$;
        $V(K, n_0) = V(K-1, n^*) + S_{n_0}$;
        $H(K, n_0) = H(K-1, n^*) \cup \{(n_0, n_0 + L)\}$;
    end
end
$n^* = \arg\max_{n \leq N^t} V(K_{\max}, n)$;

Output: $H = H(K_{\max}, n^*)$

14.5.4 BNN Approach

For the BNN approach that offloads the classification to the data plane, the header field definition is not necessary, and it only needs to achieve the goal of the FRG approach's first stage. The challenge here is to make the neural network model lightweight enough to be executed by a network data plane device such as the SmartNIC. Following the methods proposed in [7], we constrain a neural network's weights to binary values (either +1 or −1) and choose sign function as the activation function. More specifically, consider a neural network with T fully-connected layers. In a similar fashion as in the first approach, we denote the neuron weights of layer t by a 2D vector W_B^t and denote the input of this layer by c^{t-1}. Then, the output of layer t is:

$$c^t = \text{sign}(c^{t-1} \cdot W_B^t). \tag{14.9}$$

We can acquire the input vector of the first layer c^0 by treating each byte of the original packet vector x as 8 separate bits. In this way, both c^t and W_B^t are binary vectors in all layers. The operation is equivalent to the Hamming weight of two bit strings' XNOR. Algorithms have been developed enabling fast calculations [35], which make the BNN executable in multiple models of SmartNICs. For instance, we implement such algorithms in Netronome Agilio CX. In the next section, we will demonstrate that such lightweight model can also achieve a high accuracy in the packet classification application.

14.6 Evaluation Results

To demonstrate the benefits of our IoT security methods, we perform evaluations using various real traffic datasets. The evaluation results indicate that both the FRG and BNN methods are able to classify packets accurately with an acceptable cost.

14.6.1 Performance of FRG Approach: Setup and Metrics

We start with evaluating the FRG approach by training the accompanying neural network model with labeled traffic datasets. Specifically, we use the following two publicly-available datasets of IoT network traffic:

- **ISCX Botnet 2014 Dataset [36]**: This is a collection of botnet traffic traces from multiple well-known datasets. The types of traffic are mainly HTTP, P2P and IRC. This dataset is already divided into the training set and test set. The test set has more diversity than the training set, in order to evaluate the detection of unknown attacks. It is originally gathered for statistics-based classification and contains a huge amount of packets, therefore we sample 10% of the packets from each flow for packet-level training. We also randomly modify the IP fields because all malicious flows are remapped to fixed IP addresses in the original data.
- **CICAAGM Android Dataset [37]**: This dataset captures the traffic of Android applications in real smartphones, including 250 adware, 150 malware and 1500 benign applications. Besides HTTP, there are also massive HTTPS traces, a large portion of which is SSL/TLS-encrypted. The raw packet bytes are available through PCAP files. We sample 1000 successive packets from each class of the trace for packet-level training and testing.

We also make our *own* efforts to create two *new* datasets using network simulators and real IoT devices we deploy, containing unique threats to IoT devices[1]. These datasets contain protocols that

1 For the benefit of the research community, we make our captured traffic traces publicly available at https://www.kaggle.com/datasets/qfqin0/iot-attack-traffic-traces/.

Table 14.1 We evaluate our solution using four different datasets, representing a large variety of networking scenarios.

Dataset	Scenario	Protocols
ISCX Botnet 2014 [36]	16 types of botnets	HTTP, SSH, SMTP, P2P
CICAAGM [37]	Android applications	HTTP, HTTPS
Cooja [11]	Simulation of IoT networks	RPL
Waspmote [39]	IoT sensors	LR-WPAN

OpenFlow cannot handle. On the contrary, we will demonstrate that P4 and our algorithm work well on them. Specifically, we create the following two new datasets:

- **Cooja Network Simulator Dataset**: Le et al. [38] and Mayzaud et al. [11] analyze different types of attacks in 6LoWPAN networks through the RPL routing protocol with the help of Contiki operating system and its Cooja simulator. Adopting similar methods, we run simulations of 10-node IoT networks with random topologies, and set up a malicious node conducting Version Number Attack and Increased Rank Attack. We collect packet bytes of both malicious and normal traffic flows to generate our dataset.
- **Waspmote IoT Sensor Dataset**: We also create a new dataset with measurements on real IoT devices (not simulator) we deploy. Specifically, we install temperature, humidity and luminosity sensors on a Waspmote [39] Smart Cities Pro sensor board. It periodically sends 802.15.4 low-rate wireless personal area network (LR-WPAN) frames to the gateway containing sensor data. If the electrical connection from a sensor to the board is impeded, the device will still send packets in the same format but with the wrong values. This is indeed categorized as a physical attack on sensors rather than network attack. However, we will demonstrate that our method is also effective in detecting such unconventional attacks.

The information of the datasets described above is summarized in Table 14.1. In each dataset except the first one which has already been split, we randomly pick 80% of the samples for training, and the remaining 20% for testing. We implement several state-of-the-art algorithms and make comparisons with our method. In particular:

- **Proposed P4-based Method**: In Stage 1, we build the deep neural network of the proposed structure with 4 convolutional layers each with 64 filters and the Rectified Linear Unit (ReLU) activation function [40], followed by two fully-connected layers with 100 and 50 neurons. At each hidden layer, a 0.05 dropout rate is set to avoid over-fitting. We keep the hyperameters unchanged when training with different datasets. In Stage 2, we produce the header field definition and install the corresponding flow rules to the IoT gateway.
- **OpenFlow-based Methods**: As a comparison, we consider classification methods based on OpenFlow protocol, representing SDN without programmable data plane. We limit the features of classification within the predefined header fields of MAC, IP, TCP and UDP protocols according to the OpenFlow specification. As stated by Nanda et al. [41], multiple machine learning techniques can be applied to these features, among which we choose two representative methods, decision tree (DT) and support vector machine (SVM).
- **1D Convolutional Neural Networks (1D-CNN)**: We also consider other deep learning approaches for packet classification which (similar to our method) take packet bytes rather than some specific header fields as the input. We implement two 1D-CNN imitating the structures

and hyperameters as methods proposed in [20] and [4], denoted by CNN-1 and CNN-2. These CNNs provide the same type of output as our Stage 1 output. However, they are not capable in producing a header field definition as Stage 2 of our method does. In other words, the classification cannot be executed as a switch function for line-speed packet processing.

We implement DT and SVM models using scikit-learn [42] library, and implement NNs in TensorFlow [43]. To verify the header field definition calculated by our algorithm, we also conduct emulations with Mininet [44] and P4 behavioral model software switch (BMv2) [45]. The experiments are conducted on a desktop computer with Intel Core i7-7700 Processor, 16 GB RAM and GeForce GTX 1060 graphics card.

We evaluate the performance of the classification algorithms using as metric not only accuracy, but also precision and recall. We denote the number of correctly identified malicious packets by TP and incorrectly identified ones by FP. We denote the number of correctly identified normal packets by TN and incorrectly identified ones by FN. The metrics are calculated as follows:

$$\text{accuracy} = \frac{TP + TN}{TP + TN + FP + FN}, \tag{14.10}$$

$$\text{precision} = \frac{TP}{TP + FP}, \quad \text{recall} = \frac{TP}{TP + FN}. \tag{14.11}$$

Considering that the datasets have uneven class distributions (where malicious samples account for around 30% in each dataset, except the Cooja dataset with around 10% malicious samples), we also calculate the F1 score defined as the harmonic mean of precision and recall:

$$F_1 = 2 * \frac{\text{precision} * \text{recall}}{\text{precision} + \text{recall}}. \tag{14.12}$$

14.6.2 Performance of FRG Stage 1 (Classification)

In this subsection, we evaluate Stage 1 of the FRG method. We compare the classification performance of the proposed dilated convolutional neural network with the two other CNN structures as well as with the DT and SVM OpenFlow-based methods.

ISCX Botnet: We train and test all the algorithms on the ISCX dataset. Table 14.2 shows the accuracy of each algorithm. Compared with methods based on OpenFlow headers, the CNNs (including our method) that take raw bytes as the input have significantly better performance. We also find that CNN-based methods outperform other algorithms in both precision and recall rates, leading to higher F1 score.

CICAAGM Dataset: We perform similar training and testing on the CICAAGM Android dataset, which contains a larger diversity of traffic flows including SSL/TLS encrypted ones. The results are

Table 14.2 Performance metrics of the Dilated CNN on ISCX dataset.

Method	Accuracy	Precision	Recall	F_1
DT	0.790	0.694	0.659	0.676
SVM	0.773	0.706	0.544	0.615
CNN 1	0.907	0.897	0.816	0.854
CNN 2	0.909	0.903	0.816	0.857
Proposed	**0.911**	**0.904**	**0.822**	**0.861**

Table 14.3 Performance metrics of the Dilated CNN on CICAAFM dataset.

Method	Accuracy	Precision	Recall	F_1
DT	0.890	0.833	0.771	0.801
SVM	0.895	**0.933**	0.646	0.780
CNN 1	0.882	0.833	0.738	0.782
CNN 2	0.898	0.870	**0.760**	0.811
Proposed	**0.908**	0.927	0.736	**0.820**

Table 14.4 Performance metrics of the Dilated CNN on other datasets.

Dataset	Cooja		Waspmote	
Method	Accuracy	F_1	Accuracy	F_1
CNN 1	0.998	0.991	0.995	0.993
CNN 2	0.994	0.971	0.998	0.996
Proposed	**0.995**	**0.973**	**1.00**	**1.00**

depicted in Table 14.3. Although the performance difference is not as large as in the ISCX dataset, our algorithm still achieves highest accuracy than the other algorithms. We note that while the SVM OpenFlow-based method reaches higher precision, it severely degrades the recall value, leading to a lower F1 score.

Cooja Dataset and Waspmote Dataset: The Cooja and Waspmote datasets are relatively simple, each with smaller amount of samples and only two types of attacks. However, the former contains compressed 6LoWPAN headers, and the latter has abnormalities which can only be identified from the packet payload rather than the headers. Therefore, *the packets are not readable and can no longer be classified by the OpenFlow-based methods (i.e. DT and SVM).*

As shown in Table 14.4, all three CNNs are capable of identifying the RPL routing attacks and sensor physical attacks with accuracy higher than 99%. The performance metrics of different methods are generally at the same level. Except being slightly worse than the CNN-1 in the Cooja dataset, our proposed network has superior performance in accuracy and F1 score. Especially, it achieves perfect prediction in the Waspmote dataset.

Performance Tradeoff: We are also interested in the tradeoff between the different performance metrics. In some cases, it is of major importance to limit the number of false alarms. To achieve this, we can apply a threshold to the CNN output. We depict the respective precision-recall curves for different thresholds in Figure 14.8 for all datasets except the Waspmote dataset where perfect predictions have been reached. We notice that in all datasets there is a space to increase precision further at a cost to recall.

Main Takeaways: (i) P4-based methods with packet bytes as the input can achieve better classification performance compared with OpenFlow-based methods that take as input predefined header fields. They can also handle heterogeneous protocols and application layer contents of packets, where OpenFlow-based methods are not applicable. (ii) Our proposed Dilated CNN structure achieves similar or better performance than other state-of-the-art CNN approaches that take the same input (packet bytes).

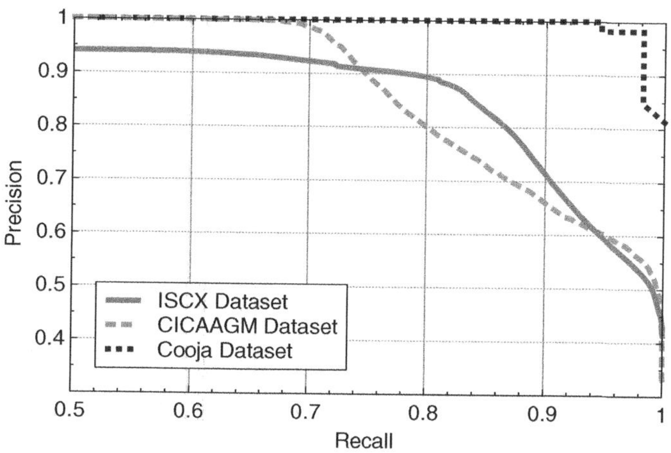

Figure 14.8 A tradeoff between precision and recall rates can be achieved by adjusting the output threshold, characterized by precision-recall curves.

14.6.3 Performance of FRG Stage 2 (Header Field Definition)

The classification performance benefits in the previous subsection are important but not surprising. It was expected that taking packet bytes rather than predefined headers as input to the learning algorithm would achieve superior classification performance as the classifier design space is larger. Still, the above results quantified the exact performance improvement we can achieve and verified the suitability of our proposed Dilated CNN structure compared to other CNN structures.

The main contribution of our work, however, lies in the implementation of the intrusion detection function directly inside the data plane (P4-enabled IoT gateway). This is important because it enables line-speed packet processing that is not available in the other learning methods like CNN-1 and CNN-2. To achieve this, Stage 2 of our learning method uses the trained Dilated CNN to define a particular set of packet byte substrings as header fields that will be used by the gateway to install flow rules. Therefore, matched packets will be directly handled by the gateway without requiring to be forwarded to the SDN controller or another remote firewall function. In the sequel, we elaborate on the header field definition and corresponding classification performance achieved by Stage 2 of our algorithm.

14.6.3.1 Profiles of Importance Scores

Following the procedure described in Section 14.5.3, we calculate the importance scores for all substrings of length $1, 2, 4, 8$ and 16 in the first $N = 128$ byte positions. For example, Figure 14.9 depicts the results of importance scores (after normalization) for every single byte.

The profiles of the datasets show different and complicated tendencies. However, there are also some intuitive results:

- **ISCX Dataset** (with IP addresses masked): The algorithm highly scores both TCP/UDP fields and some positions in the application layer.
- **CICAAGM Android Dataset**: The curve has three peaks in the IP address field, the TCP port field and application layer. This distribution implies that the classifier makes predictions based on information from headers of multiple network layers, which is an advantage from adopting SDN and P4. For example, in the case where the packet payload is SSL/TLS encrypted, even if the classifier is not able to parse application-layer information, it is able to make predictions based

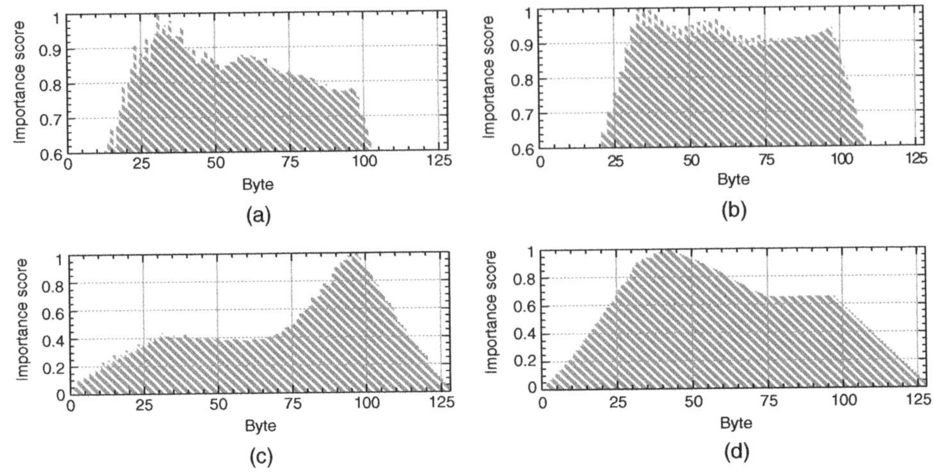

Figure 14.9 Importance scores have distinct distributions in different protocols. As an example, single-byte importance scores of each dataset are shown in the figure. (a) ISCX dataset. (b) CICAAGM dataset. (c) Cooja dataset. (d) Waspmote dataset.

on TCP/IP headers with a high accuracy. On the other hand, the application-layer headers reveal much information in those packets without encryption.

- **Cooja Dataset**: High importance score is given to 97-th byte. It is reasonable because all attacks occur through DODAG Information Object (DIO) messages of 96 bytes [38]. The algorithm takes the packet length into account when making classification.
- **Waspmote Dataset**: The algorithm successfully assigns highest importance scores to Bytes 31, 32 and Bytes 36, 37 in every 802.15.4 frame which store the sensing data in question.

These distributions demonstrate that the importance scores calculated by our method successfully identify header fields that are crucial in classifying packets.

14.6.3.2 Impact of Header Fields on Accuracy

The proposed Dynamic Programming algorithm (Algorithm 1) will select as header fields the substrings of the packet bytes that have the highest importance scores. Taking the CICAAGM Android dataset as an example, Figure 14.10a,b show the accuracy and F1 score as we increase the number of header fields we match in the gateway node (K_{max} equal to 1, 2, 3 or 4) and for different header field lengths (L equal to 1, 2 or 4). The byte-to-byte approach corresponds to the packet classifier in Stage 1 of our method described in the previous subsection. Intuitively, the performance improves with the number of header fields. According to the results, it is not necessary to have a large number of header fields. *With three 2-byte-long fields or two 4-byte-long fields, classification is almost as accurate as the byte-to-byte approach.* The difference is around 0.1% in accuracy values.

14.6.3.3 Impact of Header Fields on Costs

Next, we examine the costs associated with the header field definition, measured by the number of flow rules stored in the gateway node. Since more rules lead to a larger memory occupancy and more queries to the control plane, we need to keep their number as low as possible. Figure 14.10c shows that the number of rules required for classification increases with both the length and the number of header fields selected. Therefore, *a tradeoff exists between accuracy and cost.* The balance point can be achieved by adjusting the values of K_{max} and L parameters in our algorithm. Notice

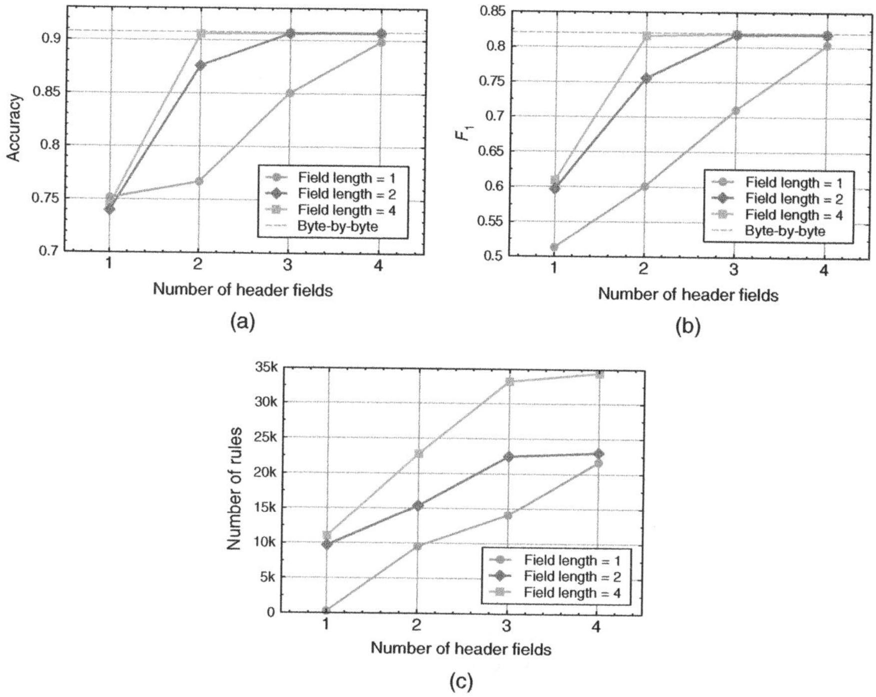

Figure 14.10 Selecting longer or more header fields increases the accuracy and precision, while leading to a higher memory cost. (a) Accuracy. (b) F1 score. (c) Memory cost.

that although the number of all possible values of a header field increases exponentially with its length, the growth is not drastic in practice. In Figure 14.10c, the tendency is closer to a linear growth.

We need to emphasize that the proposed intrusion detection mechanism does not incur much additional costs in other aspects such as network latency and throughput, because it only adds a one-time table lookup in the packet processing procedure. To verify this intuition, we create a virtual network with one BMv2 switch and two hosts using the Mininet emulation platform. We implement several sets of header field definitions and flow tables similar to the results in Figure 14.10. We use this virtual network to measure the maximum throughput achieved by our mechanism for different K and L choices and compare it with the baseline L2 forwarding mechanism that does not perform any intrusion detection. The results are depicted in Figure 14.11. We notice that the maximum throughput is reduced by less than 10% compared with the baseline, i.e. the line speed of packet processing is maintained. In the same scenario, we have another approach that forces packets to go through an application-layer single-thread analyzer based on Scapy [46] before being forwarded, which represents the case of adopting solutions similar to CNN-1 and CNN-2 in the last subsection. In this case, no larger throughput than 1 Mbps is achieved. Therefore, it is extremely beneficial to implement the intrusion detection as a switch function inside the IoT gateway with the help of the programmable data plane feature.

14.6.3.4 Optimal Selection of Header Fields

Last but not least, to demonstrate that the importance scores are proper metrics for the data plane definition, we compare the optimal selection of header fields in our algorithm with random selections. As shown in Table 14.5, with the same number of selected header fields, the performance of

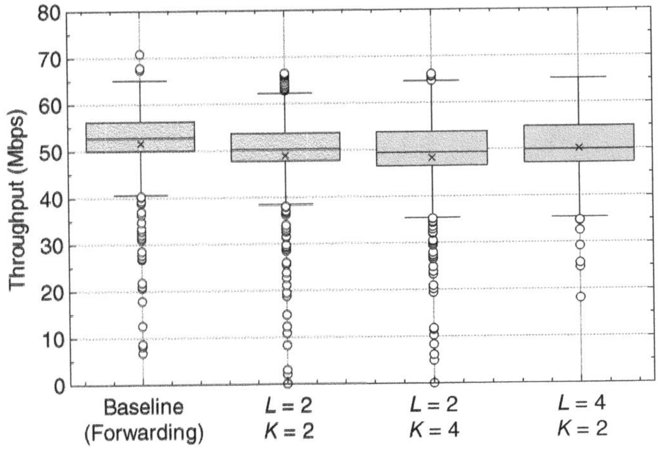

Figure 14.11 With only a small number of header fields used, the throughput decrease caused by performing classification is less than 10%.

Table 14.5 Comparisons between the proposed algorithm and random selected header fields.

Method	Optimal		Random	
# of fields	**Accuracy**	**F_1**	**Accuracy**	**F_1**
1	0.740	0.597	0.689	0.325
2	0.876	0.757	0.765	0.490
3	0.907	0.818	0.775	0.557
4	0.907	0.818	0.802	0.639

(The length of each field is 2 byte in both cases.)

our algorithm is significantly better, with more than 10% accuracy and around 20% more F1 score than the random selection.

14.6.4 Performance of BNN Approach

We then evaluate the performance of the BNN approach in the same environment as described the last subsection, and we concentrate on a similar group of metrics including accuracy, precision and recall rates. Compared with the FRG approach, here we focus more on the lightweight and cost saving features than the capability of handling unknown protocols.

We first investigate the ISCX Botnet 2014 dataset again. Due to the different structure of the model and the binarization requirement of the inputs, the way we process the dataset is slightly different from the FRG evaluations. We choose a very common group of packet-level features, 5-tuple (IP addresses, layer-4 protocol and ports) and IP packet length, and turn them into a 120-bit input vector. Then, although we mainly discuss packet-level features in this chapter, our method can also be easily extended to classify packets based on flow-level features. For this purpose, we consider another dataset, CICIDS2017 [47]. This dataset contains a labeled record of multiple types of attacks and benign flows. Statistics are summarized for each flow. We take two thirds of records for training and the remaining for testing. We convert the layer-4 destination port, bidirectional total amount of packets and bytes into a 144-bit input vector to the BNN. All these statistics can be easily acquired by most programmable switches and SmartNICs.

We implement a BNN in the data plane containing one fully-connected hidden layer with 120 neurons and a single-neuron output. For comparison, we also adopt other state-of-the-art learning algorithms, including the DT and linear SVM methods implemented by scikit-learn [42], as well as the same neural network before binarization (denoted by NN), which has real-valued weights with 32-bit precision. Comparison with this NN will indicate if the binarization leads to performance loss.

Table 14.6 reports our measurement of accuracy, precision and recall rates on CICIDS2017 dataset, where algorithms classify a flow based on several statistics. We observe that the real-valued NN has the same level of performance with DT. Our proposed BNN method has only slightly lower accuracy (0.6%) after the binarization. It also behaves better than SVM. At the same time, the BNN compresses the memory required for weight value storage to 1/32 compared with the real-valued NN and makes it possible to run the algorithm in data plane devices.

While we have shown that our method is valid when performing classification based on flow statistics, we now concentrate on the packet-level features, i.e. matching on header fields, which permits the switch to react to incoming packets in real time. This is the major use case of the proposed method as a switch function. We report performance metrics on the ISCX dataset with such packet-level features as inputs in Table 14.7. As in the previous table, we observe that the binarization incurs minor accuracy loss only (1.05%). Besides, BNN behaves better than both DT and SVM (6% and 7% more accuracy) under this setting.

A high recall rate is especially important for packet classification, since incorrect blockage of non-malicious traffic (false negatives) may hamper normal network functionalities. Therefore we also report precision and recall rates in Table 14.7 and calculate the F1 score, which shows a similar tendency as accuracy.

Table 14.6 Performance metrics of BNN on CICIDS2017 dataset.

Method	Accuracy	Precision	Recall	F_1
BNN	**0.983**	0.966	0.963	**0.965**
NN	0.989	**0.967**	0.987	0.977
DT	0.989	0.962	**0.993**	0.977
SVM	0.957	0.889	0.937	0.913

Table 14.7 Performance metrics of BNN on ISCX dataset.

Method	Accuracy	Precision	Recall	F_1
BNN	**0.945**	0.945	0.766	**0.846**
NN	0.953	**0.992**	0.767	0.865
DT	0.900	0.735	0.767	0.751
SVM	0.890	0.700	0.763	0.730

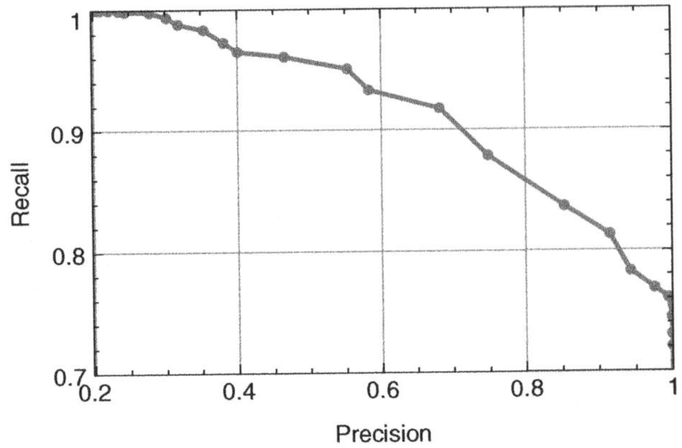

Figure 14.12 To reduce false negatives of BNN inference, a tradeoff can be achieved at a cost of precision.

Moreover, by adjusting the threshold of the Hamming weight calculated in the output layer, a tradeoff can be achieved as depicted in Figure 14.12, which means that a better (higher) recall rate can be acquired at a cost of sacrificing some precision.

14.6.4.1 Main Takeaways

In the FRG approach, a similar level of packet classification accuracy as the byte-to-byte approach can be achieved by merely matching a small number (two or three) of header fields appropriately selected based on the importance scores in the associated neural network. When implemented as a P4 switch function at the IoT gateway, this approach requires low memory and latency cost and incurs small throughput loss for table lookup (less than 10%, i.e. line speed is maintained), while alternative application-layer intrusion detection mechanisms would cause a multi-fold throughput reduction to achieve the same level of functionality.

In the BNN approach, the proposed method performs packet classification with high accuracy based on both flow-level (flow statistics) and packet-level (header fields) features. The BNN method also outperforms several state-of-the-art learning methods in accuracy and F1 score, with only slight performance loss during the binarization.

14.7 Conclusions and Future Challenges

In this chapter, we studied new opportunities for enhancing security in the IoT network brought by the programmable data plane. Namely, we proposed the FRG approach, a two-stage deep learning method based on P4 language that first trains a neural network as the packet classifier and in a later stage selects packet byte substrings as header fields and installs appropriate flow rules to realize intrusion detection functionality inside the IoT gateway. Evaluation results on publicly available and newly developed datasets of IoT scenarios demonstrated the performance benefits and universality of the proposed method compared with state-of-the-art OpenFlow-based methods. We also proposed the BNN approach, a lightweight method executed at the data plane to avoid additional communication latency and memory cost. Evaluation results verified that a more favorable tradeoff between detection accuracy, memory cost, latency and throughput can be achieved by the proposed method.

We believe that this chapter opens exciting directions for future work. First, for the FRG approach, methods that further improve the packet processing efficiency and reducing resource consumption costs remain to be explored. Our current classification model makes decisions based on exact matching of the packet headers. Flow table compression, that properly applies wildcard flow rules, is a promising technology which makes it possible to reduce the amount of flow rules to a large extent. Second, for the BNN approach, we plan to implement prototypes and propose deployment plans for specific hardware including different models of SmartNICs and FPGA boards, with which we are able to perform evaluations on metrics such as network throughput and latency. Last but not the least, for both approaches, we plan to extend the solution to scenarios containing multiple IoT domains and gateways, where proper methods of distributed machine learning model training are required to achieve a better scalability.

Acknowledgment

This research was supported in part by the U.S. Army Research Laboratory and the U.K. Ministry of Defence under Agreement Number W911NF-16-3-0001, and the National Science Foundation under Grants CNS 2132573 and CNS 2112562.

References

1 Kolias, C., Kambourakis, G., Stavrou, A., and Voas, J. (2017). DDoS in the IoT: Mirai and other botnets. *Computer* 50 (7): 80–84.

2 Andrea, I., Chrysostomou, C., and Hadjichristofi, G. (2015). Internet of Things: security vulnerabilities and challenges. *2015 IEEE Symposium on Computers and Communication (ISCC)*, pp. 180–187. IEEE.

3 McKeown, N., Anderson, T., Balakrishnan, H. et al. (2008). OpenFlow: enabling innovation in campus networks. *ACM SIGCOMM Computer Communication Review* 38 (2): 69–74.

4 Lotfollahi, M., Siavoshani, M.J., Zade, R.S.H., and Saberian, M. (2017). Deep packet: a novel approach for encrypted traffic classification using deep learning. *Soft Computing* 24 (3): 1999–2012.

5 Bosshart, P., Daly, D., Gibb, G. et al. (2014). P4: Programming protocol-independent packet processors. *ACM SIGCOMM Computer Communication Review* 44 (3): 87–95.

6 Uddin, M., Mukherjee, S., Chang, H., and Lakshman, T. (2018). SDN-based multi-protocol edge switching for IoT service automation. *IEEE Journal on Selected Areas in Communications* 36 (12): 2775–2786.

7 Courbariaux, M., Hubara, I., Soudry, D. et al. (2016). Binarized neural networks: training deep neural networks with weights and activations constrained to +1 or −1. *arXiv preprint arXiv:1602.02830*.

8 Alaba, F.A., Othman, M., Hashem, I.A.T., and Alotaibi, F. (2017). Internet of Things security: a survey. *Journal of Network and Computer Applications* 88: 10–28.

9 Cao, X., Shila, D.M., Cheng, Y. et al. (2016). Ghost-in-Zigbee: energy depletion attack on Zigbee-based wireless networks. *IEEE Internet of Things Journal* 3 (5): 816–829.

10 Pongle, P. and Chavan, G. (2015). A survey: attacks on RPL and 6LoWPAN in IoT. *2015 International Conference on Pervasive Computing (ICPC)*, pp. 1–6. IEEE.

11 Mayzaud, A., Badonnel, R., and Chrisment, I. (2016). A taxonomy of attacks in RPL-based Internet of Things. *International Journal of Network Security* 18 (3): 459–473.

12 Fu, K. and Xu, W. (2018). Risks of trusting the physics of sensors. *Communications of the ACM* 61 (2): 20–23.

13 Shoukry, Y., Martin, P., Yona, Y. et al. (2015). PyCRA: Physical challenge-response authentication for active sensors under spoofing attacks. *Proceedings of the 22nd ACM SIGSAC Conference on Computer and Communications Security*, pp. 1004–1015. ACM.

14 Antonakakis, M., April, T., Bailey, M. et al. (2017). Understanding the Mirai botnet. *26th USENIX Security Symposium (USENIX Security 17)*, pp. 1093–1110.

15 Li, C., Wu, Y., Yuan, X. et al. (2018). Detection and defense of DDoS attack–based on deep learning in OpenFlow-based SDN. *International Journal of Communication Systems* 31 (5): e3497.

16 Napiah, M.N., Idris, M.Y.I.B., Ramli, R., and Ahmedy, I. (2018). Compression header analyzer intrusion detection system (CHA-IDS) for 6LoWPAN communication protocol. *IEEE Access* 6: 16623–16638.

17 Midi, D., Rullo, A., Mudgerikar, A., and Bertino, E. (2017). Kalis–a system for knowledge-driven adaptable intrusion detection for the Internet of Things. *2017 IEEE 37th International Conference on Distributed Computing Systems (ICDCS)*, pp. 656–666. IEEE.

18 Nguyen, T.D., Marchal, S., Miettinen, M. et al. (2019). DïoT: A crowdsourced self-learning approach for detecting compromised IoT devices. *2019 IEEE 39th International Conference on Distributed Computing Systems (ICDCS)*, pp. 756–767. IEEE.

19 Miettinen, M., Marchal, S., Hafeez, I. et al. (2017). IoT sentinel: automated device-type identification for security enforcement in IoT. *2017 IEEE 37th International Conference on Distributed Computing Systems (ICDCS)*, pp. 2177–2184. IEEE.

20 Wang, W., Zhu, M., Wang, J. et al. (2017). End-to-end encrypted traffic classification with one-dimensional convolution neural networks. *2017 IEEE International Conference on Intelligence and Security Informatics (ISI)*, pp. 43–48. IEEE.

21 Wang, Z. (2015). The applications of deep learning on traffic identification. *BlackHat USA* 24: 1–10.

22 Tokusashi, Y., Dang, H.T., Pedone, F. et al. (2019). The case for in-network computing on demand. *Proceedings of the 14th EuroSys Conference 2019*, ser. EuroSys '19, pp. 211–2116. New York, NY, USA: ACM.

23 McDanel, B., Teerapittayanon, S., and Kung, H. (2017). Embedded binarized neural networks. *arXiv preprint arXiv:1709.02260.*

24 Siracusano, G. and Bifulco, R. (2018). In-network neural networks. *arXiv preprint arXiv:1801.05731.*

25 Siracusano, G., Sanvito, D., Galea, S., and Bifulco, R. (2018). Deep learning inference on commodity network interface cards. *Machine Learning Systems Workshop at the Conference on Neural Information Processing Systems (NeurIPS).*

26 Luo, T., Tan, H.-P., and Quek, T.Q. (2012). Sensor OpenFlow: enabling software-defined wireless sensor networks. *IEEE Communications Letters* 16 (11): 1896–1899.

27 Galluccio, L., Milardo, S., Morabito, G., and Palazzo, S. (2015). SDN-WISE: Design, prototyping and experimentation of a stateful SDN solution for wireless sensor networks. *2015 IEEE Conference on Computer Communications (INFOCOM)*, pp. 513–521. IEEE.

28 Dang, H.T., Wang, H., Jepsen, T. et al. (2017). Whippersnapper: A P4 language benchmark suite. *Proceedings of the Symposium on SDN Research*, pp. 95–101. ACM.

29 Haykin, S. (1994). *Neural Networks: A Comprehensive Foundation.* Prentice Hall PTR.

30 Van Den Oord, A., Dieleman, S., Zen, H. et al. (2016). WaveNet: A generative model for raw audio. *arXiv preprint arXiv:1609.03499*.

31 Yu, R., Li, A., Chen, C.-F. et al. (2018). NISP: Pruning networks using neuron importance score propagation. *Proceedings of the IEEE Conference on Computer Vision and Pattern Recognition*, pp. 9194–9203.

32 Verhelst, M. and Moons, B. (2017). Embedded deep neural network processing: algorithmic and processor techniques bring deep learning to IoT and edge devices. *IEEE Solid-State Circuits Magazine* 9 (4): 55–65.

33 Roffo, G., Melzi, S., and Cristani, M. (2015). Infinite feature selection. *Proceedings of the IEEE International Conference on Computer Vision*, pp. 4202–4210.

34 Bertsekas, D.P. (1995). *Dynamic Programming and Optimal Control*, vol. 1. Belmont, MA: Athena Scientific.

35 Muła, W., Kurz, N., and Lemire, D. (2017). Faster population counts using AVX2 instructions. *The Computer Journal* 61 (1): 111–120.

36 Beigi, E.B., Jazi, H.H., Stakhanova, N., and Ghorbani, A.A. (2014). Towards effective feature selection in machine learning-based botnet detection approaches. *2014 IEEE Conference on Communications and Network Security*, pp. 247–255. IEEE.

37 Lashkari, A.H., Kadir, A.F.A., Gonzalez, H. et al. (2017). Towards a network-based framework for android malware detection and characterization. *2017 15th Annual Conference on Privacy, Security and Trust (PST)*, pp. 23233–23309. IEEE.

38 Le, A., Loo, J., Luo, Y., and Lasebae, A. (2013). The impacts of internal threats towards routing protocol for low power and lossy network performance. *2013 IEEE Symposium on Computers and Communications (ISCC)*, pp. 000789–000794. IEEE.

39 Libelium (2019). Waspmote. http://www.libelium.com/products/waspmote/ (accessed 25 October 2022).

40 Nair, V. and Hinton, G.E. (2010). Rectified linear units improve restricted Boltzmann machines. *Proceedings of the 27th International Conference on International Conference on Machine Learning*, ser. ICML'10, pp. 807–814. USA: Omnipress.

41 Nanda, S., Zafari, F., DeCusatis, C. et al. (2016). Predicting network attack patterns in SDN using machine learning approach. *2016 IEEE Conference on Network Function Virtualization and Software Defined Networks (NFV-SDN)*, pp. 167–172. IEEE.

42 Pedregosa, F., Varoquaux, G., Gramfort, A. et al. (2011). Scikit-learn: machine learning in Python. *Journal of Machine Learning Research* 12: 2825–2830.

43 Abadi, M., Barham, P., Chen, J. et al. (2016). TensorFlow: A system for large-scale machine learning. *12th USENIX Symposium on Operating Systems Design and Implementation (OSDI 16)*, pp. 265–283.

44 Lantz, B., Heller, B., and McKeown, N. (2010). A network in a laptop: rapid prototyping for software-defined networks. *Proceedings of the 9th ACM SIGCOMM Workshop on Hot Topics in Networks*, ser. Hotnets-IX, pp. 191–196. New York, NY, USA: ACM.

45 P.L. Consortium (2018). Behavioral model (BMv2). https://github.com/p4lang/behavioral-model (accessed 25 October 2022).

46 Philippe Biondi (2011) Scapy. https://scapy.net/ (accessed 25 October 2022).

47 Sharafaldin, I., Lashkari, A.H., and Ghorbani, A.A. (2018). Toward generating a new intrusion detection dataset and intrusion traffic characterization. *4th International Conference on Information Systems Security and Privacy (ICISSP)*, pp. 108–116.

15

Distributed Computing for Internet of Things Under Adversarial Environments

Gowri Sankar Ramachandran[1], Luis A. Garcia[2], and Bhaskar Krishnamachari[3]

[1] *School of Computer Science, Queensland University of Technology, Brisbane, Queensland, Australia*
[2] *Information Sciences Institute, University of Southern California, Marina Del Rey, CA, USA*
[3] *Department of Electrical and Computer Engineering, University of Southern California, Los Angeles, CA, USA*

Abstract

IoT applications of increasing complexity for many applications require in-network processing of data from many sources, leveraging hybrid edge and cloud computing resources. In battlefield and defence settings, IoT applications also demand dependable operations, guaranteeing security and resource availability. It is challenging to orchestrate distributed computing for these applications, identifying the optimal placement of computing tasks in the face of dynamics in resource availability and network conditions. It becomes even more challenging to enable distributed computing in an adversarial environment where the data being processed as well as meta-data about resource availability could be manipulated, and the computing and networking resources may be compromised. Therefore, it is essential to develop frameworks to orchestrate distributed computing challenges in battlefield IoT networks. We survey the challenges and the existing state of the art, present some solutions, and identify key directions for future work.

15.1 Introduction

The ubiquity of Internet-of-Things (IoT) devices has led to nascent applications and capabilities for end-users. More specifically, applications in battlefield and defense settings operate in a complex and dynamic environment involving multiple organizations in the form of the military, air force, and contractors providing various services to the government and army. For example, combats and rescue operations in remote areas may require sensing, communication, and real-time processing capabilities while demanding security and reliability from various parties' computing and communication infrastructure. Besides, it is impractical to deploy custom infrastructure in a short span as most of the IoT applications in the battlefield setting are time-sensitive, limiting the response time for the concerned stakeholders. The resulting demand for near-device computation to process massive amounts of data in real-time gave rise to advances in fog and edge computing [1]. Moreover, recent works [2, 3] have proposed orchestration frameworks for optimal computational placement relative to the dynamicity of distributed IoT applications such as location-based services [4].

IoT for Defense and National Security, First Edition. Edited by Robert Douglass, Keith Gremban, Ananthram Swami, and Stephan Gerali.

The inherent dynamics of distributed IoT applications stem from the variability in resource availability and network conditions. For instance, first responders may want to leverage both deployable infrastructure and existing resources available at a response site to enhance situational awareness [5]. The connectivity and resource availability will depend on the highly unpredictable environmental conditions and the cooperation of local resource owners. Recent frameworks such as volunteer cloud computing [6] aim to enable and incentivize device owners to share computation resources. However, introducing untrusted entities into distributed application deployments only expands an already massive attack surface. This chapter aims to taxonomize the security challenges of distributed IoT application deployment in mission-critical environments relative to the existing state-of-the-art distributed computing frameworks.

A large body of work exists to characterize highly-contested, mission-critical environments. In particular, the characterization of the internet of battlefield things (IoBT) [7] encapsulates a superset of the adversarial challenges of emergent distributed computing applications. As a result, IoBT applications represent the next generation of command, control, communications, and intelligent decision-making in mission-critical environments. Agadakos et al. distinguish IoBT from civilian IoT applications based on several factors: device ownership, diverse form factors, dynamic network state, dynamic mission goals, heterogeneous communication channels, and contested operating environments [8]. Specifically, the device ownership characterization of trusted, adversarial, and neutral devices as blue, red, and grey nodes provides a generalized system model basis. Thus, distributed IoBT applications are an exemplar use case for the next generation of command, control, communications, and intelligent decision-making in mission-critical environments. However, initial characterizations of IoBT do not provide a comprehensive overview of adversarial threats relative to emergent distributed computing frameworks (Figure 15.1).

This chapter provides an exhaustive threat model for emergent distributed computing frameworks and their applications. First, we introduce a generalized system model for state-of-the-art distributed computing frameworks and subsequently enumerate adversarial goals and attack vectors in defense settings. We then propose possible solutions and enumerate future directions toward mitigating threats.

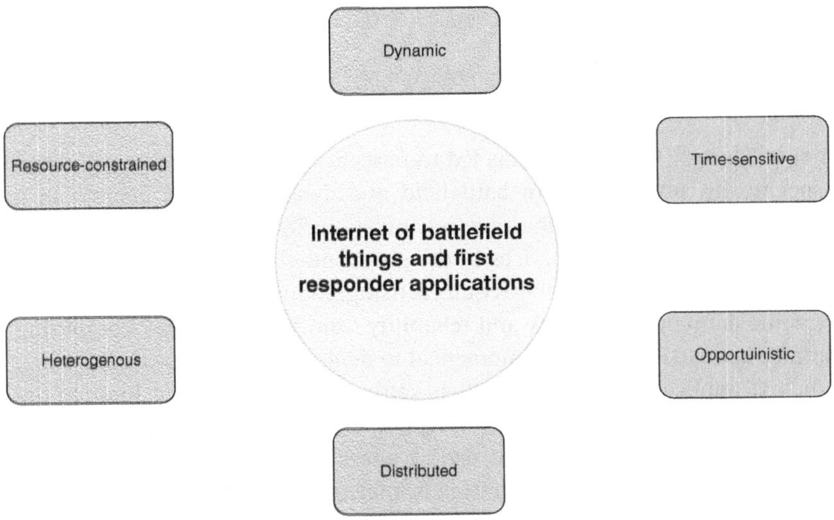

Figure 15.1 Internet of Battlefield and first responder applications operate in dynamic, resource-constrained, and distributed environments with heterogeneous resources while relying on opportunistic computing and networking.

15.2 Distributed Computing for IoT in Defense Applications

15.2.1 Overview of Requirements/Challenges

In defense settings, IoT applications help the field units including the military personnel, battlefield devices, drones, and other equipments gather intelligence from the operational environment in a highly distributed and dynamic setting.

Sensors provide critical information about the hostile forces and their movements. In a battlefield environment, all the sensors may not have a unified view of the operational environment, including hostile forces. Besides, the devices must coordinate with each other not only to gain intelligence but also to process data in the distributed network. Therefore, these applications must collectively fulfil the application requirements in the presence of adversaries, which make distributed computing more challenging in this setting.

In a battlefield setting, the *communication* between devices could also be disrupted by enemies through jamming devices. A lack of communication between devices leads to lack of intelligence, which could potentially put the military forces in danger. Therefore, the communication technology must be able to overcome attacks.

In addition to sensing and communication, devices also *process* information gathered from sensors in-network to reduce latency and avoid long-distance communication. Due to the distributed nature of the defense applications, the processing of sensor data happens in one or more devices. Information from multiple devices are fused and processed at a subset of computation nodes. It is essential to ensure that the computation is scheduled on trusted computation nodes or leverage coded computing nodes to run computations on untrusted nodes. Coded computing approaches aim to ensure confidentiality and integrity. Note that the adversaries could highjack the nodes and start exfiltrating data from the network to project a false view of the battlefield to gain an advantage.

In summary, battlefields represent a highly adversarial setting for distributed computing. Note that the field units may end up making a wrong move if the hardware devices, software components, and communication links are compromised.

15.2.2 Characteristics of Distributed IoBT Applications

- **Dynamic:** IoBT devices constantly move in battlefield settings, introducing a lot of unpredictability. For example, a device cannot off-load a computation to one or more neighboring devices and wait for the result since the devices will move as soldiers navigate the environment to tackle the enemies. Thus, the network that forms the distributed system is not stable, resulting in dynamism.
- **Time-sensitive:** Decisions have to be made rapidly in the battlefield setting since each second is pivotal when the soldiers tackle threats imposed by the enemies. Enabling the devices to communicate and compute promptly is a critical requirement to fulfill the application demands effectively.
- **Resource-constrained:** Devices may have limited resources, including battery, computation power, communication bandwidth, and storage, as the soldiers cannot carry heavy equipment for a battle. Therefore, the distributed computing environment must not consume significant resources when providing intelligence to the field units. Note that too much computation and communication on a single device may deplete its battery, further impacting resource availability.
- **Distributed:** Multiple devices are gathering information from different locations, offering different perspectives in a battlefield setting. Combining data from various devices and their sensors

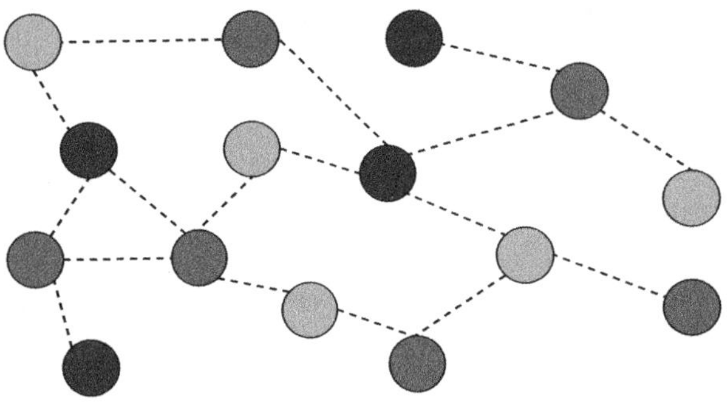

Figure 15.2 IoBT Applications are Opportunistic. Devices come in contact with trusted (Light Gray), yet-to-be-trusted (Dark Gray), and untrusted (Black) devices in the application environment.

is essential to gather insights. This process happens in a distributed network, wherein multiple devices carry out the computation and storage services on behalf of the entire network.

- **Opportunistic:** Due to the unpredictable nature of the battlefield, the devices are not guaranteed to have communication and computation resources at their will when they need them. Hence, the devices may leverage resources available in the vicinity to fulfill their resource demands, making them opportunistic.
- **Heterogenous:** The resource capacity, including the computation capability, communication bandwidth, power, and storage, is not uniform since the devices that are part of mobile vehicles, such as tanks and drones, may have more capacity than the devices carried by soldiers. A distributed system targetting one class of devices may be sub-optimal when heterogeneous devices become part of the network. Figure 15.2 shows that there may even be untrusted devices in the network.

Therefore, the distributed computing environment for IoBT must be versatile to meet the application demands. At the time, it should prevent adversaries from altering the flow of critical information among battlefield devices. In the next section, we will introduce our system model and the threats.

15.3 Threat Model

15.3.1 System Description

We consider a distributed system with a number of stakeholders responsible for the hardware, software, and field operations. In the application environment, devices cooperate and coordinate with each other to achieve the desired application goals. In a vast operational environment, devices establish networks with other devices in the vicinity for communication and computation. Malicious and untrusted devices may sneak into the application network to disrupt the computation and communication processes. We introduce the role of different stakeholders below.

- **Hardware Manufacturers:** The hardware devices responsible for computation, communication, and sensing come from one or more hardware manufacturers. It is important to ensure that the hardware is trusted and reliable.
- **Application Developers:** The software infrastructure that drives the application includes algorithms and protocols for computation and communication, which comes from multiple software developers. A single malicious protocol from a malicious developer could open up an attack window for adversaries, which would alter the behaviour of all the protocols and algorithms.

- **Application Deployers:** Distributed applications are deployed and leveraged by end-users. The placement of the underlying computation may depend on the opportunistically available resources relative to the environmental state, e.g. the user's location, connectivity, or the load distribution in the network. The application deployers are indirectly the tenants of networked resources.
- **Computation Resource Owners:** Entities who own networked resources are incentivized contingent on the success of the deployed applications. Devices may be comprised of computational resources, sensors, or actuators. Moreover, we assume resources may be shared between application components via multi-tenancy, e.g. by using virtualization.
- **Data Sources:** In a distributed setting, devices exchange data between each other to collectively understand the operational environment to make an informed application decision. An adversary may inject malicious data to the devices in the application network to create false interpretations.
- **Sensor and Actuator Device Owners:** Device nodes may have sensors or actuators enabled to drive cyber-physical applications. Attackers may target such devices to impact the physical component of applications, especially in safety-critical contexts.

15.3.2 Threats

When an application operates in a distributed environment, the devices in the network collaborate and coordinate with each other to share and gather intelligence while leveraging the computation power from neighboring devices. Such an application setting provides ample opportunities for adversaries. An adversary could bring down the application by mounting various attacks. In this section, we will discuss the goals of an adversary and attack vectors.

15.3.2.1 Goals of an Adversary

An adversary will aim to disrupt the confidentiality, integrity, and availability.

- **Disrupt Distributed Application Confidentiality:** An adversary could gain access to one or more devices and steal the application data. Such an attack on the battlefield could expose mission-critical data, compromising the security of the field units. Besides, such an attack may also reveal the configuration details of the devices, including the radio frequencies used by the devices and encryption keys.
- **Disrupt Distributed Application Integrity:** An adversary who has access to a networked resource may aim to inject false information in the distributed application pipeline to disrupt the application control flow and to tarnish the reputation of a particular entity. In a deployment scenario where multiple stakeholders offer resources for computations in an ad-hoc fashion, an adversary owning significant computation and storage resources could attack the competitors' infrastructure to attract all the computation and storage jobs. Such an attack would allow an attacker to gain a significant financial advantage. Access to infrastructure owned and managed by a single organization may solve such problems; it may be impractical in a battlefield scenario involving equipment from multiple stakeholders belonging to various organizations. Similarly, an attacker needing computation resources may compromise insecure devices in the network and leverage them for personal gains. Mirai BotNet attack is an example of such an attack [9].
- **Disrupt Distributed Application Availability:** A distributed IoT application gathers intelligence by sensing, computation, and communication. A jamming attack on the communication channels, for example, would cut down the links between the devices, reducing the effectiveness of the applications. Broken or lossy communication links require retransmissions, which demand

energy resources. Therefore, an attack on the communication channel can reduce the lifetime of battery-powered and mobile battlefield IoT devices. An attacker could gain physical access or damage devices from a remote site via weapons to reduce the number of active devices. Such an attack would severely deplete the resources in the distributed network, making battlefield applications less effective. Further, an attacker could also steal the device and reverse-engineer it to cause more harm to the application network.

15.3.2.2 Attack Vectors

- **Malicious Data Source:** A distributed IoT application operate collaboratively in the field by collecting and sharing data through various sensors. When one or more devices get compromised, the data coming from such devices will not provide an accurate representation of the operational environment, thereby reducing the effectiveness of the application while putting the field units in real danger. Securing the devices from unauthorized access would alleviate such attacks, but it may also be helpful to verify the accuracy of the data source before use.
- **Malicious Computation Resources:** Computation resources, including the CPU and memory, may be unreliable. A malicious hardware manufacturer could tamper with the devices to make them erroneous in the operational environment. Note that an occasional bit flip may make the computation inaccurate, causing disruptions. It is important to note that bit-flip errors could also happen due to Cosmic events [10]. Here, we refer to a deliberate attack by malicious hardware manufacturers.
- **Misreporting Resource Usage/Availability:** In a distributed computing environment, computation tasks are often shared with devices in the network due to a lack of local processing capacity. Note that the applications assume that there are computing resources available in the operational environment. In such a setting, each device must honestly report its resource availability and ability to accept new computation tasks. When devices start to report unavailability, citing a lack of resources, application processes may slow down, causing disruptions.
- **Malicious Code:** The dynamic workloads stemming from malicious actors may embed traditional software or network exploits in the code. The attacker may compromise the integrity of the node, perform reconnaissance, or compromise subsequent workloads.
- **Jamming Attack:** An adversary may also jam the communication channels breaking the network. Note that the devices may either be processing data belonging to another device or waiting to accept new results from a neighboring device. Under such circumstances, the jamming attack could hamper the application processes, resulting in retransmissions and rescheduling, which will slow down the computations in the network.
- **False Result Injection by Lazy Nodes:** Computing nodes could report incorrect results or arbitrary results without running computation. Such computing nodes are referred to as *lazy nodes* because of their unwillingness to expend resources for computation. Note that the honest nodes rely on the results produced by neighboring nodes to acquire intelligence in battlefield or mission settings. Incorrect results would provide a false narrative to the honest nodes, jeopardizing the mission. Therefore, it is essential to introduce mechanisms or approaches to mitigate lazy behaviors in the distributed network.
- **Malicious Actuation:** Attackers may target the inherent cyber-physical nature of IoT applications. Generally, the objective of attackers in this context is to maximize the physical impact while maintaining stealthiness–as was done in the Stuxnet attack [11]. Malicious actuation is driven by some adversarial objective function to compromise the safety of the IoT application.

These attack vectors highlight the challenges of distributed computing in battlefield environments involving computing nodes from multiple entities. Existing literature introduces platforms and approaches for distributed and volunteer computing along with strategies to mitigate attacks. The following section reviews the existing literature on these topics.

15.4 Frameworks for Distributed Computing

In a distributed application setting, applications leverage resources belonging to multiple organizations to achieve the desired application functionality. In this section, we review the different distributed computing frameworks and their effectiveness.

15.4.1 Resource and Task Management in Distributed Computing

Distributed IoT applications increasingly rely on remote and third-party computing resources at the edge. Following this model, edge computing frameworks manage the resources and allows clients to leverage neighboring node's computation and storage capabilities [12–14]. Such frameworks enable the resource-constrained end-devices, including sensors, cameras and other data acquisition hardware, to off-load the computation processes to a remote computing infrastructure available at the edge or the cloud to satisfy their processing demands [14]. This model allows the application developers to deploy complex machine learning and AI algorithms for their mission-critical applications since the nodes in the network can fulfill their computation demands collaboratively.

Despite its benefits, contemporary edge computing frameworks essentially assume that the edge computing infrastructure for a given application is deployed and managed by a single organization. Note that in a distributed application environment, nodes in the network typically rely on a centralized orchestrator, which manages nodes and their resources in the network. For instance, when a node requests a computation resource, the orchestrator would look at the resource availability of the nodes and allot the computation to a node with sufficient resources. Under this circumstance, the orchestrator plays a critical role, and it is owned and managed by the same organization. In other words, the end devices and the network/resource manager belong to a single organization, as shown in Figure 15.3a.

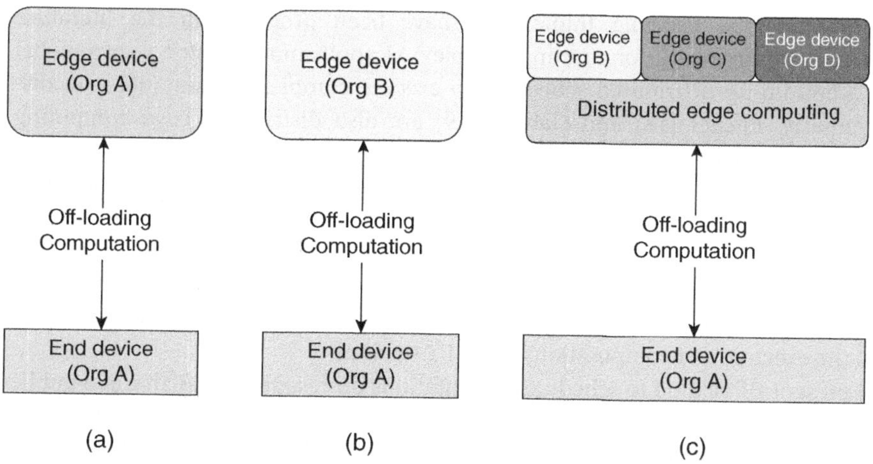

Figure 15.3 The ownership modalities for edge computing. Different colors here represent different organizations. Model A operates in a trusted setting as the devices are owned and managed by a single entity. In contrast, Model B and Model C includes more than one organization demanding trusted solutions.

A single stakeholder deployment model does not scale well while being susceptible to a single point of failure, making them unsuitable for battlefield applications. In addition, it may be challenging to provide edge computing services for mobile end-devices, including UAVs, following this model, because it is practically impossible for a single organization to deploy and manage resources in all possible locations.

Alternatively, multiple providers may voluntarily offer computing services at the edge [15, 16]. In a battlefield setting, these providers may include contractors and coalition partners. Following this model, the devices in the networks can buy or consume computation services from a provider belonging to another organizational domain, as shown in Figure 15.3b. Although this model allows the mobile end-devices to leverage edge computing services, it does not provide trust guarantees to the application developers. Note that the service provider may manipulate the outcome of the computation. Besides, relying on a single untrusted computation platform may prevent application developers from employing computationally-intensive algorithms and applications due to the lack of trust. In a military and defense setting, the edge computing infrastructure may be honest and trustworthy. Still, the communication channel between the edge computing infrastructure and the end-device may be vulnerable, which could impact the application's behavior if the end device relies on a single edge computation platform.

In summary, the mission-critical applications cannot:

- Rely on a centralized orchestrator because such a model is susceptible to single points of failure.
- Execute complex and computationally intensive applications on untrusted edge computing platform.

A distributed edge computing platform involving multiple organizations at the edge allows the end-devices to execute complex computation tasks at the edge, as shown in Figure 15.3c. Frameworks such as Jupiter [2, 17] and Ray [3] are capable of distributing computations over a dispersed network of compute nodes, but such frameworks assume that the compute nodes are honest and trusted. When one or more computation nodes are dishonest in a distributed system and exhibit byzantine faults, the end-devices cannot trust the computation nodes' results and make confident decisions. Therefore, a distributed computing framework without byzantine fault-tolerant (BFT) capabilities may not be effective in a multi-organizational edge computing environment, such as the one shown in Figure 15.3c.

Several distributed machine learning frameworks have been proposed in the literature [2, 3, 18, 19]. Ray [3] is a framework for running complex AI applications, which uses a global control store and a bottom-up distributed scheduler to execute complex computations on distributed nodes. Similarly, Rocket [18] and Gabriel [19] are also distributed edge computing frameworks, and they intelligently schedule computations on edge and the cloud-based on the resource demands of the application and network availability. Another relevant framework in this space is Jupiter [2, 17], which is a dispersed computing framework with support for orchestrating and running complex computations represented in the form of a directed acyclic graph (DAG) to geographically dispersed computing platforms. These frameworks are capable of operating at the edge to provide support for distributed data processing, but none of them explicitly offers support for scheduling and the execution of computations in a BFT manner.

Zhang et al. [20] present BFTCloud to schedule computations on resources available at a public cloud platform, and it starts scheduling computations using a centralized primary node. If the primary node is found to be faulty, it creates replicas to deal with byzantine failures. Similarly, Costa et al. [21] introduce a BFT MapReduce framework to deal with arbitrary faults that could arise in MapReduce jobs, but it does not handle malicious nodes.

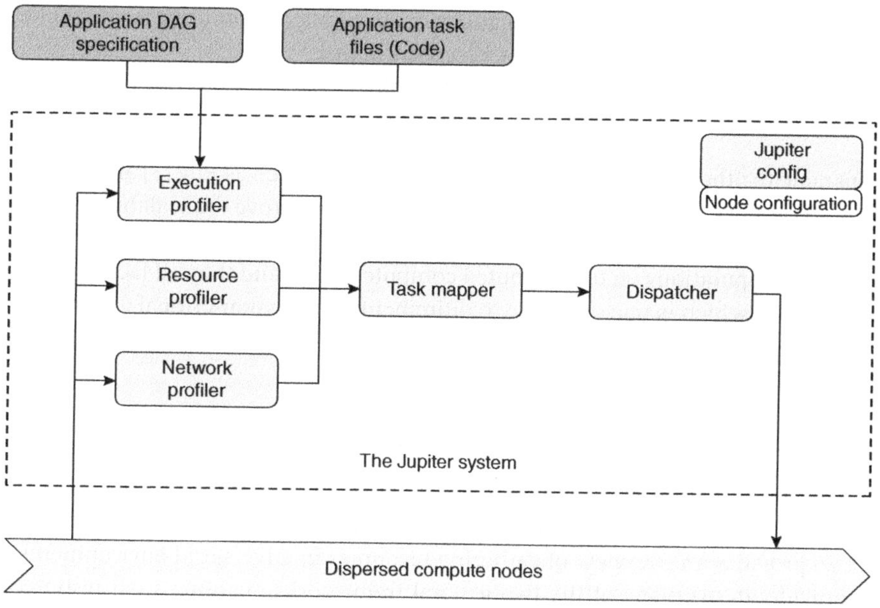

Figure 15.4 The architecture of Jupiter, an orchestration system for dispersed computing. It manages the computation jobs by assessing the resource availability of computing nodes.

Contemporary dispersed computing frameworks such as Jupiter and Ray take the computation tasks, computation nodes, and resource capacity when scheduling computations onto the network of compute nodes. Such frameworks typically provide support for managing the compute nodes and application tasks. This section shows how functionalities such as scheduling and deployment are carried out by Jupiter, one of the dispersed computing frameworks.

Jupiter [2, 17] is an active open-source dispersed computing framework developed at the University of Southern California[1]. Figure 15.4 shows the architecture of the Jupiter framework. The key building blocks of Jupiter are explained below:

- **Jupiter Configuration:** When running distributed computations using Jupiter, the application developer is required to configure the framework by inputting the list of computation nodes and other configuration details. This input must be provided by an user, managing the computation jobs.
- **DAG-based Application:** The application developer is also required to provide the specification of their DAG-based application and the code for the tasks. In dispersed computing frameworks, an application is divided into a set of small jobs, which are connected in the form of DAG. Small jobs are then mapped onto the network of compute nodes for execution. During the execution, the output of one or more jobs is fed as an input to other jobs (or tasks) in the DAG.
- **Execution, Resource, and Network Profilers:** These profilers are responsible for collecting performance statistics from the dispersed computing nodes in the network.
- **Task Mapper (or Scheduler):** This module maps the computation tasks onto compute nodes based on the performance statistics collected from the dispersed computation nodes. Jupiter supports heterogeneous earliest finish time (HEFT) task mapper, which is one of the

1 https://github.com/ANRGUSC/Jupiter

popular algorithms for mapping DAG-based applications on distributed computing nodes while accounting for the communication performance between the nodes.

- **Dispatcher:** This component dispatches the computation tasks to the distributed nodes to begin the execution of tasks.

Jupiter [2] is comparable to other distributed scheduling frameworks such as Ray [3] and Rocket [18] in that they all map computations onto computation platforms to improve the scalability while reducing the makespan. Besides, such existing frameworks, including Jupiter, primarily focus on mapping (or scheduling) computations on to distributed compute nodes, and they all lack support for byzantine fault tolerance, which is one of the key requirements for mission-critical distributed IoT applications.

15.4.2 Gathering Resources in Adversarial Environments

One of the goals of dispersed computing is to enable the usage of opportunistically available resources. Of course, resource sharing in the presence of adversaries poses an additional set of challenges. Figure 15.5 provides an overview of gathering resources in adversarial environments. In this section, we provide an overview of the two general frameworks enabling such resource sharing: volunteer cloud computing and location-based services.

Volunteer Cloud Computing: Volunteer clouds represent a class of cloud computing where systems and applications run over spare resources from volunteer computers [6]. Moreover, Mengistu and Che [6] survey a number of mobile volunteer cloud computation frameworks that have emerged in the past decade [22–26] and have enabled distributed IoT applications such as federated learning [27]. Thus, volunteer cloud frameworks could be ideal for supporting opportunistic sensing and computation from grey assets from citizens or local municipalities. Opportunistic cloud computing can enhance situational awareness where deployed infrastructure is sparse, e.g. a emergency response team leveraging nearby volunteers to provide additional sensing and computational resources. Of course, introducing an untrusted node into the computational framework may compromise the security of a distributed application.

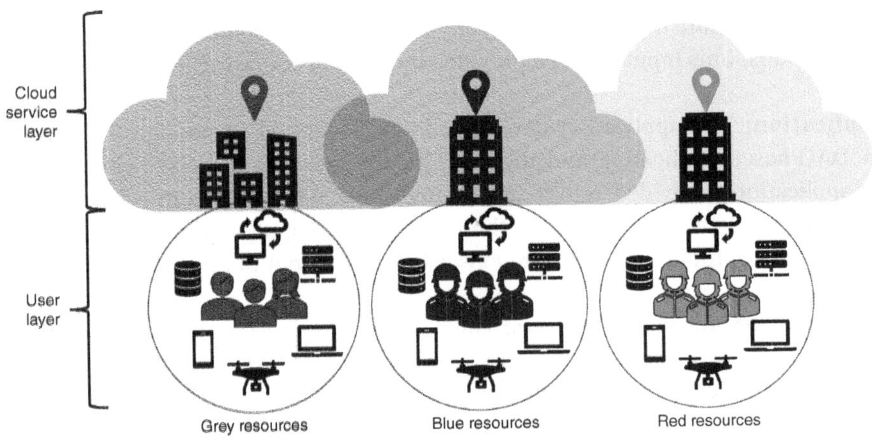

Figure 15.5 Overview of gathering resources in adversarial environments. Resources may be volunteered and provided opportunistically based on location. Here, the trusted, adversarial, and neutral devices are represented with blue, red, and grey colors respectively.

Location-based Services: The ubiquity of IoT has resulted in the widespread use of location-based services, spanning applications in emergency, informational, tracking, and entertainment services [4]. Verifying the location of a network node is vital to the integrity of a service in mission-critical settings, e.g. an emergency service providing information about the nearest hospital. The problem is exacerbated when a service is spread across multiple nodes, e.g. an inference about traffic stemming a variety of city sensors and vehicles. Conversely, the privacy of individual nodes are also at stake when using or contributing to location-based services. In defense settings, exposing location of nodes may have dire consequences.

15.5 Establishing Trust in Adversarial Environments: Solutions and Open Opportunities

In this section, we will review the classes of solutions that could potentially overcome or limit adversarial attacks in distributed computing environment.

15.5.1 Verifiable Computation

Verifiable computation paradigm focuses on crowdsourcing computation platforms. It enables the application or task owner ("submitter") to off-load one or more computation tasks to compute nodes ("worker") to speed up the computation and to exploit the cheap and powerful compute nodes in the vicinity. When the remote compute nodes or workers successfully performed the computation, the results have to be verified by "approvers" or "validators" nodes. To verify the result, the "approvers," "validators" or "submitters" are not required to redo the entire computation since that would minimize the effectiveness of the computation-off-loading approach. The verifiable computation paradigm focuses on ensuring the correctness of the computation without redoing the computation at the "validator" nodes.

To create a trustworthy and reliable verifiable computation platform, it is important to fulfill the following requirements:

- Support for "submitter" or "task owners" to submit the computation task. This process involves selecting the desired computation model and compute nodes.
- Ability to dispatch the compute tasks and associated metadata including inputs and other configuration details to the compute nodes.
- Scheduling of compute tasks on compute tasks while satisfying the demands of the users.
- Verification of the computation result, without redoing the entire computation, to ensure that the compute nodes have performed their work honestly.
- Reputation management framework for rating the compute nodes and the task submitters.
- Payment handler to transfer the computation fee from user's account to the compute node's account.

Anderson et al. [28] contributes SETI @ Home to outsource the analysis of radio signals for a project that searches for extraterrestrial intelligence. Radio signals picked up by telescopes consist of signals from various sources including TV and Satellites. To effectively dissect the signal, a fine-grained frequency analysis is required, and such an analysis requires enormous computation resource. SETI @ Home project allows any machine on the planet to join the SETI network and contribute computation resource in return for credit points. Users receive credits when they complete a task but how the credits are used is not clear.

SETI@Home follows a client-server model, wherein the computation tasks are distributed to the clients from a server. Malicious actors have been known to produce inconsistent or incorrect results to gain high ratings in SETI @ Home project [29]. Result verification through replication is used as a solution to guard the system from malicious workers. Following this approach, the same computation tasks are assigned to multiple clients, and the results from clients are compared to ensure correctness. When a quorum is reached on the results, the worker nodes are rewarded through credit points.

Zhao et al. [29] points out the weaknesses in the replication-based verification schemes. Worker nodes in the network may collude and could return a same incorrect result to gain credit points without performing any computation. Zhao et al. [29] contributes the Quiz mechanism to resolve the issues with the replication-based scheme. The Quiz mechanism inserts quizzes as part of the computation task, and when a client returns the result, the results of the quizzes are verified to ensure correctness since the wrong results to a quiz indicate suspicious behavior. To further strengthen the system, a trust-based scheduling scheme is proposed, wherein the tasks are allocated to trusted clients. In this approach, honest clients are selected based on their past performance and are given high preference during the scheduling phase.

15.5.1.1 Homomorphic Encryption

Encryption schemes are typically used to preserve sensitive information. Systems that use encryption schemes have to decrypt the data before performing any operations on the encrypted data. Homomorphic encryption schemes preserve privacy by performing computation on encrypted data [30]. In other words, the encrypted result of the computations on encrypted data will match the result of the computations performed on plain-text.

Homomorphic encryption schemes have been developed for a number of applications in the last two decades, but the practical use of such systems is still under scrutiny due to their resource demands [31]. Although some recent schemes reduced the overhead of homomorphic encryption schemes, they are not yet ready for resource-constrained applications [47].

15.5.1.2 Proof-based Verification

The secure multi-party computation was introduced in 1982 by Andrew Yao [32] to allow two parties to perform computation without relying on third parties while keeping the local data or input secret [33]. Approaches such as zero-knowledge proof allow the worker and validator to verify the result without exchanging any information except the fact that the verifier knows the correct result [34]. Zero-knowledge proof, one of the interactive proof systems, has the following properties:

Completeness: The worker knows the truth, and he/she will convince the verifier eventually.
Soundness: The worker can convince the verifier only if he/she tells the truth.
Zero-knowledgeness: No information about the truth is exchanged between the worker and the verifier.

ZKSnark [35] is one of the practical implementations of Zero-knowledge proof, and it uses the libsnark library. To use the zero-knowledge proof mechanism, the computation task has to be translated into the right format using the following steps:

- **Computation → Circuit:** The computation task has to be converted into a sequence of logic expressions, which are represented as logic gates. This process is also known as "flattening" process.
- **Circuit → Rank-1 Constraint System (R1CS):** The flattened expressions should be converted into R1CS format. The circuits are converted into a tuple (a,b,c) which results in a solution s. For the R1CS to be satisfiable [36], the s must satisfy the equation: $s . a * s . b - s . c = 0$.

- **R1CS → Quadratic Arithmetic Program (QAP):** R1CS expressions are converted into polynomials, which are then used for verifying the proof of the computation at random points [37]. In other words, the verification of individual outputs from each and every logic gate is cumbersome. Gennaro et al. [37] contributed QAP to negate the verification complexity, wherein the prover can check the constraints at any random point in the polynomial.

The above steps are essential for each computation problem in the ZKSnark system. Besides, the computation currently supports basic arithmetic operations (+, −, *, /), exponent function with constant power, and assignment operator. Loop operations and comparison operators are not currently supported. Although the ZKSnark is promising, it lacks practical tools and frameworks for generic computation problems. LibSTARK is another implementation of the proof-based verification scheme, which is used by ZK-STARK [35].

15.5.1.3 TrueBit

TrueBit [38] presents trustless smart contracts to enable secure computation on the blockchain. In proof-of-work (PoW) based blockchain systems such as Bitcoin and Ethereum, each node in the network executes the smart contract to verify the result, and the nodes are not rewarded for their verification. The node that solves the cryptographic puzzle gets an incentive since it is authorized to create the next block, whereas all the other nodes verify the integrity of the block to make sure they are not attaching themselves to an invalid or malicious chain. TrueBit proposes a novel approach, by which a set of special nodes are elected to perform the verification instead of asking all the validator nodes to expend resources for the computation [38]. TrueBit injects faults to the computation to check the behavior of the validating nodes. Nodes that fail to report the "forced errors" are penalized for their dishonest behavior. This feature enables TrueBit to remove dishonest nodes from the network.

15.5.1.4 Perlin

Perlin [39] is a decentralized compute platform on a DAG ledger technology. In particular, Perlin uses the Avalanche protocol [40], which is a novel DAG-based ledger based on a concept called "metastability." Metastability allows the nodes in the network to come to a consensus in a series of rounds, in which the nodes that are part of the quorum are repeatedly voting to decide on the stable state for a given event. Perlin uses Avalanche protocol to build a novel decentralized compute layer. The key contributions of Perlin include Sybil-resistant identity management through a PoW scheme, whose complexity is selected by the participants in the network through quorum-based voting. Although Perlin announces itself as a decentralized compute layer, it lacks information about the system. Moreover, Perlin's model assumes that the user executes the computation again to ensure correctness, which makes the verifiable computation weaker, as the user can execute the computation on his/her machine in the first place without off-loading to another machine.

15.5.1.5 Open Opportunities

Verifiable computation schemes enable distributed IoT applications to reliably off-load computations to untrusted third parties by leveraging proof-based verifications. However, such an approach introduces additional overhead for the verification while demanding dedicated hardware in some cases. A lightweight and trusted verifiable computation technique for military and defense applications could guarantee trust and security, one of the open challenges.

Computing nodes may join the network and make their resources available for computation. In such a setting, it is essential to ensure that the nodes' identities are verified. For example, blockchain technology introduces the concept of public vs. permissioned networks. Anyone with a decent computation node can join the network in a public blockchain network and start creating and managing the blockchain. On the other hand, the permissioned blockchain includes an extensive admission process, wherein the owner of the computation node must provide some documentation proof to prove their identity. Public blockchain anonymity may introduce vulnerability in the form of Sybil attacks. But, such an attack is difficult in permissioned blockchain, making them a viable solution for distributed battlefield applications. An identity management framework could help the distributed applications validate the identity of the computing nodes before leveraging them for computation and storage.

15.5.2 Byzantine Fault-tolerant Distributed Computing

The distributed systems literature introduced BFT algorithms to tolerate malicious faults [48, 49]. Following the BFT algorithm, application processes get executed on multiple computing nodes. Upon completion of the execution, the computation results get fed into the BFT algorithm for approval. More than two-thirds of the nodes must approve the results to guarantee byzantine fault tolerance. BFT approaches provide safety and liveness guarantees for distributed applications operating in asynchronous settings.

Blue (yet-to-be-trusted) or red (untrusted) nodes may exhibit byzantine behavior in a battlefield setting. The application process should not rely on a single node, primarily when the network consists of untrusted nodes. Therefore, off-loading critical computations to suspicious nodes is risk-prone, compromising the mission's safety while leaving on-field devices in dangerous situations. The BFT algorithm offers a viable solution for such scenarios.

Recall that in Section 15.4.1, we introduced Jupiter, a distributed computing framework, which uses a resource profiler, task scheduler, and dispatcher to orchestrate computation in a distributed environment. In frameworks such as Jupiter, it is assumed that the components responsible for orchestration are centralized, making them susceptible to single-point-of-failure. Distribution of orchestration components following the BFT algorithm would enhance the safety of the applications since the critical functionalities of the applications get executed on multiple nodes as part of the byzantine fault-tolerant consensus process. In this scheme, the network must include many malicious Blue and Red nodes to compromise the application's integrity.

A centralized mapper (or scheduler) may make selfish scheduling decisions, which might not be optimal or dependable for the application developers or the end-users. We recommend that the task mapping process run on N nodes, and the scheduling decision will be considered final *if and only if* more than $N/2$ of the nodes recommend the same identity-aware schedule for the computing tasks (it is easy to show that this can tolerate up to $(N-1)/2$ byzantine faults). As shown in Figure 15.6, a distributed task mapping algorithm has to consider the node configurations, DAG, and BFT scheduling specifications, and it may optionally consider the information coming from the profilers to estimate the mapping for the computation tasks.

While making a scheduling decision, the BFT scheduler is also required to map tasks on computation nodes owned and managed by different organizations based on the criticality levels of the tasks. When the scheduler encounters a critical task in the DAG, all its children tasks must be scheduled on N nodes to provide BFT guarantees. Here, the critical tasks must be scheduled on multiple compute nodes belonging to different organizations. Remember that the critical tasks are replicated to multiple machines; therefore, the byzantine fault-tolerant scheduler introduces additional cost, but this enhances the reliability of the entire DAG-based application.

Byzantine Fault Tolerant Task Dispatching: Our extended framework assumes that the application developer or the end-user submits the DAG and the corresponding code to computing nodes belonging to all the organizations. For the DAG shown in Figure 15.6, each computing node

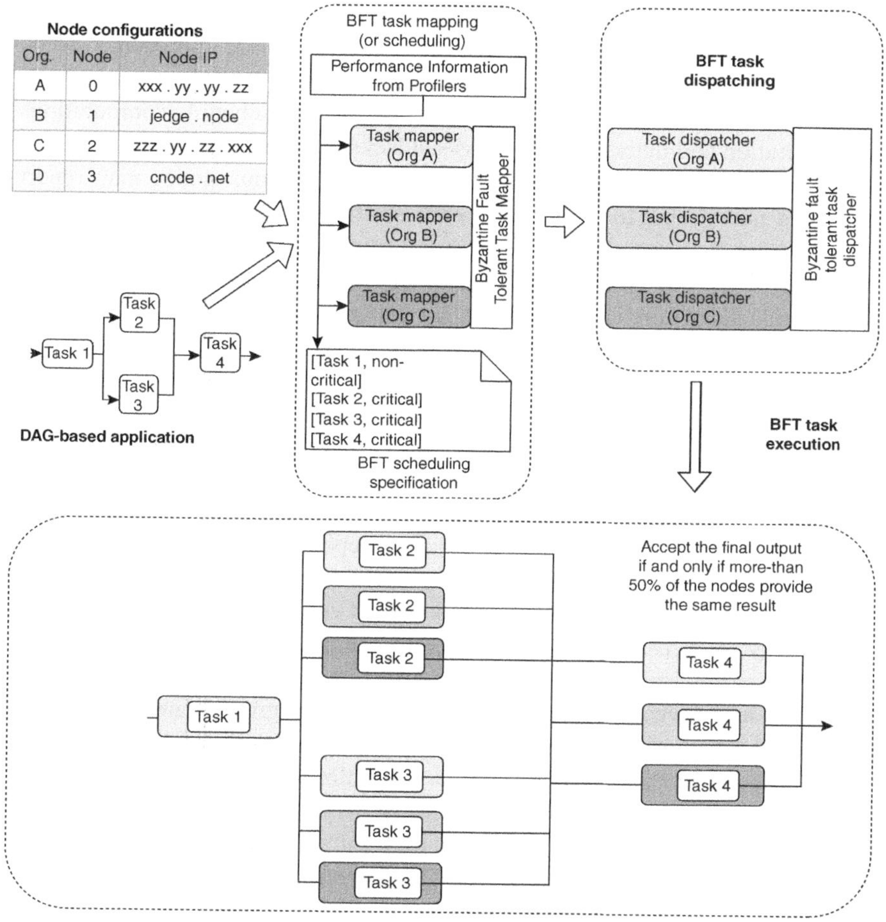

Figure 15.6 A Functional View of the proposed Byzantine Fault Tolerant Jupiter System. The framework schedules critical tasks on multiple nodes to overcome byzantine failures.

belonging to organizations A, B, and C will receive the DAG and the corresponding code. When scheduling is made, each dispatcher will start to map the tasks on computing nodes belonging to their organization following the mapping schedule.

Byzantine Fault Tolerant Task Execution: For the DAG shown in Figure 15.6, the BFT specifications show that tasks 2 and 3 are critical; therefore, the scheduler mapped these tasks on three compute nodes belonging to three different organizations. The subsequent child task in the DAG is also replicated to multiple compute nodes belonging to different organizations. When a task receives results from multiple parents, it has to accept the input as long as more than $N/2$ nodes submit the same result.

15.5.2.1 Open Opportunities

A Byzantine fault-tolerant framework requires multiple nodes to guarantee trust. Distributed applications in battlefield environments involve mobile nodes, which will introduce network partitions. Note that a majority of nodes in the network must approve an operation to provide a BFT guarantee. When there are not enough validator nodes in the network, the operations would not be approved, disrupting the entire application. Therefore, solutions must be developed to deal with network partitions when BFT guarantees are required for operations. SwarmDAG [50] introduces a solution to tackle network partitions in swarm robotics. A similar approach could be used to provide safety and liveness guarantees for distributed applications.

False Advertising of Resources: While frameworks like Jupiter can schedule computations on nodes that offer computation and networking resources, it lacks mechanisms to verify whether the nodes genuinely report their resource capacities. Note that malicious nodes may report incorrect computation and networking to the resource orchestration framework such as Jupiter. In such cases, the orchestration framework may schedule tasks assuming that the nodes can deliver the desired outcome quickly and schedule tasks on malicious nodes. Cryptographic approaches such as verifiable delay function could verify the capacity of the nodes in the network reliably.

15.5.3 Grey Resource Accumulation

Enabling the opportunistic usage of grey resources in distributed settings necessitates the ability to establish a sense of trust between grey resources. Although traditional trust mechanisms are effective in distributed settings, the availability of the resources depends on the incentivization of the resource owners. Incentivization also entails assurances of security, privacy, and confidentiality for the owners. Thus, we first describe two emerging distributed computation abstractions that enable accumulation of grey resources: location-based services and volunteer cloud computing. We describe the challenges of applicability of these frameworks in adversarial settings. We then provide an overview of how incentivization can drive participation from grey resource owners.

Trusted Location-based Services: Gupta and Rao [4] surveyed existing applications, challenges, and possible solutions for location-based services. Centralized approaches with TTP are not feasible in contested or grey node settings. Techniques such as ad-hoc spatial cloaking create anonymized spatial regions with groups of nodes to shroud the accurate location of a node. Such an approach may help obfuscate location in defense settings. Similary, blockchain-based approaches for privacy-preserving, decentralized proof of location have proposed promising approaches for smartphone settings with relatively low latency time (on the order of seconds) [41].

Establishing Trust in Volunteer Clouds: Just as in other distributed computation frameworks, volunteer clouds are susceptible to vulnerabilities stemming from untrusted stakeholders. For example, volunteers may not disclose resource information before joining a network [42], or a volunteer cloud application may be released with bugs that breach the security of a participant host [43]. A variety of solutions have been proposed to address the unique trust challenges of volunteer clouds. For instance, Alsenani et al. [42] proposed an approach to establish host trust in volunteer clouds. They defined a loyalty metric where trustworthiness is based on the priority of a task and is affected by behavioral change in the host. A large body of research has focused on falsifying results from volunteer clouds. The general approach is to pose the same problem or task across multiple resources using replication, and the answer returned by the majority can be taken as the correct answer [28, 44]. Watanabe and Fukushi [45] proposed an attestation-based approach referred to as *spot-checking*, where the server sends a job whose answer is known beforehand to volunteer nodes. If a volunteer node returns an incorrect result for that spotter job, then the host is identified as a malicious node and is blacklisted. Subsequent results from that node are invalidated or no job will be assigned to the node. In distributed settings, an adversary could spawn future instances that circumvent such an attestation. Watanabe et al. [46] proposed an *M-first voting* approach, where the task will be executed repeatedly until the first M matching results are collected by the server. A major drawback of voting techniques is that they waste the scarce donated resources by replicating the tasks. Establishing trust systematically in volunteer clouds will often be ad-hoc and will succumb to exploits in traditional distributed computation settings. Thus, an additional measure in volunteer clouds for establishing trust is to *incentivize* grey assets.

Incentivizing Volunteers: Mengistu and Che [6] categorized the proposed business or incentive models in volunteer computing systems as follows:

- **Volunteer:** a volunteer donates a resource without expecting reciprocation.
- **Trophy:** A non-monetary or non-tangible award is presented to the volunteer.
- **Reputation:** a volunteer builds an individual reputation in a network.
- **Reciprocation:** a volunteer is given higher priority access when they need resources.
- **Auction:** leveraging a market to set a price for a volunteered resource.
- **Posted price:** prices are posted at a set price for volunteered resources.

Incentivization is a fundamental challenge for gathering grey resources. Thus, incentivization in defense settings will provide an abundance of open research opportunities.

15.5.3.1 Open Opportunities

As real-time requirements and heterogeneity increase, we need to develop more robust and efficient techniques to attest staleness of data–especially when nodes are mobile. Additionally, for incentivization in mission-critical settings, volunteer-, trophy-, and reputation-based incentives rely on an informal trust that the local volunteers care about a persistent reputation in the community. In reality, the stakeholder administering a reputation-based system would need a combination of a reputation-based system coupled with the latter three marketing approaches, i.e. reciprocation, auction, and posted price-based incentives. In defense settings, the reciprocation incentives should be tuned to the needs of local communities or municipalities.

15.5.4 Cryptographic Approaches

Distributed applications require and use cryptographic protocols to establish secure communication with remote computing nodes. While protocols such as public-key cryptography help the devices exchange sensitive information securely, many new cryptographic approaches are introduced in the literature to improve security and privacy further.

The zero-knowledge proof [51, 52], discussed in Section 15.5.1.2, allows devices to perform computation on private date to generate insights without revealing any data to remote nodes - a cryptographic proof helps the remote node verify the computation's correctness. Such a scheme would be helpful for distributed computing in adversarial environments. However, existing zero-knowledge schemes require computations and communications, which may limit their application in resource-constrained devices.

The verifiable delay function (VDF) [53] is another cryptographic scheme. VDF introduces provable delays to distributed applications. Computations in VDF require a specific number of sequential steps to evaluate. However, the output of the computations could be verified publicly and efficiently. The problem proposer could adjust the execution time of VDF computations by tuning the computation difficulty. But, the verification process is extremely lightweight and efficient, making VDF elegant. We believe that VDF could be used in a distributed computing environment to validate the resource capacity of remote nodes. Note that a computation's execution time depends on the node's resource capacity - a powerful computation node could run computations faster than resource-constrained nodes. In a distributed computing environment, nodes can assess the resource capacity of the devices in the network by assigning VDF computations.

15.5.4.1 Open Opportunities

Existing cryptography approaches such as Zero-knowledge proof and VDF are widely-used in resource-rich computation environments. Distributed computing in battlefield applications may involve computers with diverse resource capacities, including resource-constrained embedded devices. Cryptography approaches must therefore be tailored for heterogeneous computing environments.

The energy consumption and communication complexity of cryptography approaches must also be assessed to understand its overhead fully. Note that the cryptography scheme should not introduce significant operational overhead.

15.5.5 Secure Computation with Trusted Execution Environments

The aforementioned software-based cryptographic approaches may still leak sensitive data either prior to encryption or when an application uses the data after decryption. Thus, mainstream CPU manufacturers provide hardware-assisted trusted execution environments (TEEs) to protect sensitive data at the time of computation. TEEs are hardware protection mechanisms that isolate the memory into secure memory and unsecure memory. The secure memory can only be accessed by privileged code running inside the TEE while any code can implicitly access the unsecure memory. In ARM TrustZone, the secure memory code resides in secure memory—referred to as the *secure world* (SW), whose high privilege is designated by setting a special ARM instruction *SMC*. The unsecure code resides in unsecure memory—referred to as the *normal world* (NW). The context switch between SW and NW is done through a secure monitor (SM). Intel's secure guard extensions (SGX) enable per-application TEEs, and AMD provides secure execution environments through Secure Encrypted Virtualization. ARM–which is the most common architecture of IoT edge devices–provides the ARM TrustZone that provides data confidentiality and peripheral access protection.

15.5.5.1 Open Opportunities

Recent works [54, 55] have focused on making TEE usage more efficient for IoT applications, including applications that require on-device deep learning. However, there remains a gap between real-time applications and practical usage of TEEs. Moreover, researchers can focus on how to avoid excessive secure memory usage–especially on resource-constrained devices. The recommendation for secure memory usage is on the order of tens of MB [56], whereas the amount of sensitive data used by modern application far exceeds those bounds. More importantly, a larger trusted computing base increases an application's attack surface. Finally, the inherent heterogeneity of device architectures entails the heterogeneity of the TEEs. Future work can explore different TEE mechanisms to interface distributed applications across heterogeneous architectures.

15.6 Summary

This chapter enumerates the challenges, state-of-the-art solutions, and future directions of distributed computing in adversarial environments. In particular, we mapped the large body of recent research in this space to mission-critical settings with the IoBT as a driving application. We provided an exhaustive adversarial model and enumerated existing defense solutions that may mitigate emergent threats in mission-critical settings. More importantly, we formalized open opportunities and challenges centered around verifying a grey nodes' resources, computation, and location.

Acknowledgment

This work was supported in part by Army Research Laboratory under Cooperative Agreement W911NF-17-2-0196.

References

1 Dolui, K. and Datta, S.K. (2017). Comparison of edge computing implementations: fog computing, cloudlet and mobile edge computing. *2017 Global Internet of Things Summit (GIoTS)*, pp. 1–6. IEEE.

2 Ghosh, P., Nguyen, Q., and Krishnamachari, B. (2019). Container orchestration for dispersed computing. *Proceedings of the 5th International Workshop on Container Technologies and Container Clouds*, WOC '19, pp. 19–24, New York, NY, USA: Association for Computing Machinery.

3 Moritz, P., Nishihara, R., Wang, S. et al. (2018). Ray: a distributed framework for emerging AI applications. *13th USENIX Symposium on Operating Systems Design and Implementation (OSDI 18)*, pp. 561–577. October 2018. Carlsbad, CA: USENIX Association.

4 Gupta, R. and Rao, U.P. (2017). An exploration to location based service and its privacy preserving techniques: a survey. *Wireless Personal Communications* 96 (2): 1973–2007.

5 Ventrella, A.V., Esposito, F., and Grieco, L.A. (2018). Load profiling and migration for effective cyber foraging in disaster scenarios with formica. *2018 4th IEEE Conference on Network Softwarization and Workshops (NetSoft)*, pp. 80–87. IEEE.

6 Mengistu, T.M. and Che, D. (2019). Survey and taxonomy of volunteer computing. *ACM Computing Surveys (CSUR)* 52 (3): 1–35.

7 Kott, A., Swami, A., and West, B.J. (2016). The Internet of Battle Things. *Computer* 49 (12): 70–75.

8 Agadakos, I., Ciocarlie, G.F., Copos, B. et al. (2019). Application of trust assessment techniques to IoBT systems. *MILCOM 2019-2019 IEEE Military Communications Conference (MILCOM)*, pp. 833–840. IEEE.

9 Antonakakis, M., April, T., Bailey, M. et al. (2017). Understanding the Mirai BotNet. *26th USENIX Security Symposium (USENIX Security 17)*, pp. 1093–1110.

10 Ziegler, J.F. and Lanford, W.A. (1979). Effect of cosmic rays on computer memories. *Science* 206 (4420): 776–788.

11 Falliere, N., Murchu, L.O., and Chien, E. (2011). W32. Stuxnet dossier. *White paper, Symantec Corp., Security Response* 5 (6): 29.

12 Satyanarayanan, M. (2017). The emergence of edge computing. *Computer* 50 (1): 30–39.

13 Zhou, Z., Chen, X., Li, E. et al. (2019). Edge intelligence: paving the last mile of artificial intelligence with edge computing. *Proceedings of the IEEE* 107 (8): 1738–1762.

14 Porambage, P., Okwuibe, J., Liyanage, M. et al. (2018). Survey on multi-access edge computing for Internet of Things realization. *IEEE Communication Surveys and Tutorials* 20 (4): 2961–2991.

15 Lovén, L., Leppänen, T., Peltonen, E. et al. (2019). Edge AI: A vision for distributed, edge-native artificial intelligence in future 6G networks. *The 1st 6G Wireless Summit*, pp. 1–2.

16 Saad, W., Bennis, M., and Chen, M. (2019). A vision of 6G wireless systems: applications, trends, technologies, and open research problems. *IEEE Network* 34 (3): 134–142.

17 Ghosh, P., Nguyen, Q., Sakulkar, P.K. et al. (2021). Jupiter: A networked computing architecture. *Proceedings of the 14th IEEE/ACM International Conference on Utility and Cloud Computing Companion*, UCC '21. New York, NY, USA: Association for Computing Machinery.

18 Ananthanarayanan, G., Bahl, P., Bodík, P. et al. (2017). Real-time video analytics: the killer app for edge computing. *Computer* 50 (10): 58–67.

19 Wang, J., Feng, Z., George, S. et al. (2019). Towards scalable edge-native applications. *Proceedings of the 4th ACM/IEEE Symposium on Edge Computing*, SEC '19, pp. 152–165. New York, NY, USA: Association for Computing Machinery.

20 Zhang, Y., Zheng, Z., and Lyu, M.R. (2011). BFTCloud: A byzantine fault tolerance framework for voluntary-resource cloud computing. *2011 IEEE 4th International Conference on Cloud Computing*, pp. 444–451, July 2011.

21 Costa, P., Pasin, M., Bessani, A.N., and Correia, M. (2011). Byzantine fault-tolerant mapreduce: faults are not just crashes. *2011 IEEE 3rd International Conference on Cloud Computing Technology and Science*, pp. 32–39. IEEE.

22 Cuervo, E., Gilbert, P., Wu, B., and Cox, L.P. (2011). CrowdLab: An architecture for volunteer mobile testbeds. *2011 3rd International Conference on Communication Systems and Networks (COMSNETS 2011)*, pp. 1–10. IEEE.

23 Ba, H., Heinzelman, W., Janssen, C.-A., and Shi, J. (2013). Mobile computing-a green computing resource. *2013 IEEE Wireless Communications and Networking Conference (WCNC)*, pp. 4451–4456. IEEE.

24 Arslan, M.Y., Singh, I., Singh, S. et al. (2012). Computing while charging: building a distributed computing infrastructure using smartphones. *Proceedings of the 8th International Conference on Emerging Networking Experiments and Technologies*, pp. 193–204.

25 Shi, C., Lakafosis, V., Ammar, M.H., and Zegura, E.W. (2012). Serendipity: enabling remote computing among intermittently connected mobile devices. *Proceedings of the 13th ACM International Symposium on Mobile Ad Hoc Networking and Computing*, pp. 145–154.

26 Habak, K., Zegura, E.W., Ammar, M., and Harras, K.A. (2017). Workload management for dynamic mobile device clusters in edge femtoclouds. *Proceedings of the 2nd ACM/IEEE Symposium on Edge Computing*, pp. 1–14.

27 Lim, W.Y.B., Luong, N.C., Hoang, D.T. et al. (2020). Federated learning in mobile edge networks: a comprehensive survey. *IEEE Communication Surveys and Tutorials* 22 (3): 2031–2063.

28 Anderson, D.P., Cobb, J., Korpela, E. et al. (2002). SETI@Home: An experiment in public-resource computing. *Communications of the ACM* 45 (11): 56–61.

29 Zhao, S., Lo, V., and Dickey, C.G. (2005). Result verification and trust-based scheduling in peer-to-peer grids. *5th IEEE International Conference on Peer-to-Peer Computing (P2P'05)*, pp. 31–38, August 2005.

30 Rivest, R.L., Adleman, L., and Dertouzos, M.L. (1978). On data banks and privacy homomorphisms. *Foundations of Secure Computation* 4 (11): 169–180.

31 Naehrig, M., Lauter, K., and Vaikuntanathan, V. (2011). Can homomorphic encryption be practical? *Proceedings of the 3rd ACM Workshop on Cloud Computing Security Workshop*, CCSW '11, pp. 113–124. New York, NY, USA: ACM.

32 Yao, A.C. (1982). Protocols for secure computations. *23rd Annual Symposium on Foundations of Computer Science, 1982. SFCS'08*, pp. 160–164. IEEE.

33 Goldwasser, S. (1997). Multi party computations: past and present. *Proceedings of the 16th Annual ACM Symposium on Principles of Distributed Computing*, PODC '97, pp. 1–6. New York, NY, USA: ACM.

34 Goldwasser, S., Micali, S., and Rackoff, C. (1989). The knowledge complexity of interactive proof systems. *SIAM Journal on Computing* 18 (1): 186–208.

35 Ben-Sasson, E., Chiesa, A., Tromer, E., and Virza, M. (2014). Succinct non-interactive zero knowledge for a von Neumann architecture. *23rd USENIX Security Symposium (USENIX Security 14)*, pp. 781–796. San Diego, CA: USENIX Association.

36 Ben-Sasson, E., Chiesa, A., Genkin, D. et al. (2013). SNARKs for C: Verifying program executions succinctly and in zero knowledge. In: *Advances in Cryptology – CRYPTO 2013* (ed. R. Canetti and J.A. Garay), 90–108. Berlin, Heidelberg: Springer-Verlag.

37 Gennaro, R., Gentry, C., Parno, B., and Raykova, M. (2013). Quadratic span programs and succinct nizks without PCPs. In: *Advances in Cryptology – EUROCRYPT 2013* (ed. T. Johansson and P.Q. Nguyen), 626–645. Berlin, Heidelberg: Springer-Verlag.

38 Teutsch, J. and Reitwießner, C. (2017). A scalable verification solution for blockchains. url: https://people.cs.uchicago.edu/teutsch/papers/truebit.pdf (accessed 26 October 2022).

39 Iwasaki, K. (2022). Perlinx White Paper. *Technical report*, Perlin. https://drive.google.com/file/d/1vfj1j2OTase4mkj8fj9BmbtUFYgYjwg4/view (accessed July 2022).

40 Team Rocket (2022). Snowflake to Avalanche: a novel metastable consensus protocol family for cryptocurrencies. http://knowen-production.s3.amazonaws.com/uploads/attachment/file/1922/Snowflake%2Bto%2BAvalanche%2B-%2BA%2BNovel%2BMetastable%2BConsensus%2BProtocol%2BFamily.pdf (Accessed July 2022).

41 Nosouhi, M.R., Yu, S., Zhou, W. et al. (2020). Blockchain for secure location verification. *Journal of Parallel and Distributed Computing* 136: 40–51.

42 Alsenani, Y.S., Crosby, G.V., Ahmed, K.R., and Velasco, T. (2020). ProTrust: A probabilistic trust framework for volunteer cloud computing. *IEEE Access* 8: 135059–135074.

43 Cano, P.P. and Vargas-Lombardo, M. (2012). Security threats in volunteer computing environments using the Berkeley open infrastructure for network computing (BOINC). *International Journal of Computer Technology and Applications* 3 (3): 944–948.

44 Costa, F., Veiga, L., and Ferreira, P. (2013). Internet-scale support for map-reduce processing. *Journal of Internet Services and Applications* 4 (1): 1–17.

45 Watanabe, K. and Fukushi, M. (2010). Generalized spot-checking for sabotage-tolerance in volunteer computing systems. *2010 10th IEEE/ACM International Conference on Cluster, Cloud and Grid Computing*, pp. 655–660. IEEE.

46 Watanabe, K., Fukushi, M., and Kameyama, M. (2011). Adaptive group-based job scheduling for high performance and reliable volunteer computing. *Journal of Information Processing* 19: 39–51.

47 Acar, A., Aksu, H., Uluagac, A.S., and Conti, M. (2018). A survey on homomorphic encryption schemes: Theory and implementation. *ACM Computing Surveys (CSUR)* 51 (4): 1–35.

48 Lamport, L., Shostak, R., and Pease, M. (2019). The byzantine generals problem. *Concurrency: The Works of Leslie Lamport*, pp. 203–226.

49 Castro, M. and Liskov, B. (2002). Practical byzantine fault tolerance and proactive recovery. *ACM Transactions on Computer Systems (TOCS)* 20 (4): 398–461.

50 Tran, J.A., Ramachandran, G.S., Shah, P.M. et al. (2019). SwarmDAG: A partition tolerant distributed ledger protocol for swarm robotics. *Ledger* 4 (Suppl 1), 25–31.

51 Grassi, L., Khovratovich, D., Rechberger, C. et al. (2021). Poseidon: A new hash function for zero-knowledge proof systems. *30th USENIX Security Symposium (USENIX Security 21)*.

52 Wu, H., Zheng, W., Chiesa, A. et al. (2018). {DIZK}: A distributed zero knowledge proof system. *27th USENIX Security Symposium (USENIX Security 18)*, pp. 675–692.

53 Boneh, D., Bonneau, J., Bünz, B., and Fisch, B. (2018). Verifiable delay functions. *Annual International Cryptology Conference*, pp. 757–788. Springer.

54 Yun, M.H. and Zhong, L. (2019). Ginseng: keeping secrets in registers when you distrust the operating system. *Network and Distributed System Security (NDSS).*

55 Liu, R., Garcia, L., Liu, Z. et al. (2021). SecDeep: Secure and performant on-device deep learning inference framework for mobile and IoT devices. *Proceedings of the International Conference on Internet-of-Things Design and Implementation*, pp. 67–79.

56 Amacher, J. and Schiavoni, V. (2019). On the performance of ARM TrustZone. In: *Distributed Applications and Interoperable Systems* (ed. J. Pereira and L. Ricci), 133–151. Cham: Springer International Publishing. ISBN 978-3-030-22496-7.

16

Ensuring the Security of Defense IoT Through Automatic Code Generation

M. Douglas Williams[1] *and Robert Douglass*[2]

[1] *Seed Innovations, Colorado Springs, CO, USA*
[2] *Alta Montes, Sandy, UT, USA*

Abstract

IoT devices and networks for military operations face special security challenges not faced by commercial IoT. During military operations it is not possible to guarantee the physical security of all IoT nodes. Hostile agents will attempt to attack, disable, deceive, co-opt, and capture military IoT systems. An IoT network consists of layers of hardware, firmware and software logic produced and programmed by many different organizations residing in multiple nations. Certifying the integrity and vulnerability-free state of all elements and levels of IoT nodes increasingly becomes an intractable challenge for defense use of IoT. Today, if an enemy captures a rifle, one weapon has been lost. However, if an IoT network connects that rifle to all other rifles as well as to an entire logistics train, then the loss of one weapon may allow malware to alter the course of an entire battle. One approach to prevent the compromise of one IoT node from spreading through the network via malware is to ensure that the interface software of all IoT elements contains no exploitable software vulnerabilities. To eliminate software vulnerabilities, interface code can be automatically generated starting from high-level specifications using an automatic code generator that has been verified by formal methods to create vulnerability-free code. This approach has been demonstrated and tested by automatically generating Internet Protocol (IP) routing software that is free from vulnerabilities due to known weaknesses. For defense applications, vulnerability-free interface software can be auto-generated by a trusted authority and augmented with anti-tamper elements to guarantee that it stays secure and unmodified. This approach ensures the security of defense IoT networks by requiring every IoT node to be connected via auto-generated, vulnerability-free software. By so doing, a defense IoT network can limit damage from the capture or compromise of an individual IoT node from spreading beyond the directly affected node.

16.1 The Challenge of IoT in Defense and National Security Applications: The Challenge

Our nation's civilian and military information networks are under constant attack at a time when these networks are increasingly critical to our nation's infrastructure and in particular our military's command and control capabilities. IoT and network element software/firmware is extremely vulnerable, both to direct attack and as the vector for attacks against other infrastructure elements. Mobile networks are particularly problematic: they provide over-the-air access to attack

IoT for Defense and National Security, First Edition. Edited by Robert Douglass, Keith Gremban, Ananthram Swami, and Stephan Gerali.

vectors with no physical connectivity required; they require constant messaging to maintain the changing network topology; and many of the capabilities – especially in military systems – are embedded in firmware, which makes updates more difficult. While millions of new malware attacks are being released each day, they rely on a much smaller number (thousands) of existing vulnerabilities in the software or firmware that is being attacked. These vulnerabilities in turn exist because human software engineers have written code that falls prey to just a few hundred common weaknesses in programming practice.

The goal of automatic code generation is to automatically generate IoT and network element software that is free from vulnerabilities due to known weaknesses, thereby reducing the cyber-attack surface area. Using today's development technologies and processes, it is effectively impossible for humans to write software that accounts for the hundreds of known weaknesses that can lead to vulnerabilities. Additionally, the encouraged practice of code reuse serves to perpetuate and propagate vulnerabilities into new code; when vulnerability patches are applied to the original source, they are unlikely to be propagated to the reused code. However, automatic code generation systems generate vulnerability-free software by using a knowledge base of coding solutions that mitigate for known weaknesses, which will, in turn, reduce software vulnerabilities [1].

This approach is based on the following premises:

- Computers can generate software that is more secure than that written by human programmers by mitigating known weaknesses in the generated code.
- A developer using a high-level protocol description language can generate more code in less time than a programmer using conventional programming tools.
- A developer using auto-generation techniques can generate security updates to existing code faster and with fewer errors than human programmers using conventional tools.
- Auto-generation enables tracing of the provenance of every module of code, preventing attacks that attempt to insert compromised versions of existing modules.

Since automatically generated code will be free from vulnerabilities due to known weaknesses, it will not provide openings for attacks by adversaries' malware. This also reduces the time required to develop IoT and network software modules and reduce the effort required to update software in response to new threats. The result will be more secure software that is developed in less time that can be automatically updated to counter new threats. This process has been demonstrated and proven in principle as part of the Automatic Generation of Network Element Software (AGNES) system that was sponsored by the Office of Naval Research (ONR) and documented in the AGNES Final Report [2].

16.2 Solutions

There are several categories of solutions to the problem of security of defense IoT and network element software. These include:

- Control the interfaces between IoT elements
- Traditional approaches to malware protection
- Auto-code generation for vulnerability-free IoT

We will review these categories of solutions with the realization that a successful system to secure defense of IoT and network element software will likely use elements from all three categories.

16.2.1 Control the Interfaces Between IoT Elements

Perhaps the most obvious approach to the security of defense IoT and network element software is to isolate and control the interfaces between the IoT elements via the network elements and associated protocols. Using signature-based detection, each incoming file is analyzed, assigned a signature or hash (a unique alphanumeric way to identify malware), and then added to the signature database, where it is used for comparison in subsequent malware incidents. When a suspicious file is found on a computer running the antivirus (AV) software, the program looks for patterns that may match with a known malware family. If a match is made with a known variant, it's blocked. This is still a standard security methodology for computers, although not for IoT devices that lack the storage and processing capacity to support it [3].

Static analysis detection techniques are based on machine learning to train computers to recognize and differentiate between benign and malicious files. These techniques take different behaviors (file behaviors, how long the file is open, traffic, everyday behavior, etc.), and classify them to form an estimate on the nature of the file [3]. The current method of controlling the interfaces between IoT elements is by the formal definition of the protocols that implement these interfaces and includes guidance on reporting protocol vulnerabilities [4, 5]. By formal definitions or formal representations, we mean specifications that have undergone such a formal review process.

These formal definitions can themselves be validated by formal proofs of correctness, but this is a time consuming and costly process and is therefore only applied to a small portion of protocol definitions. In this context, formal verification is the act of proving or disproving the correctness of the formal specification using formal methods of mathematics [6].[1] Even if a formal proof of correctness is performed on a protocol definition, the actual implementation of the specification will still be subject to the same software weaknesses to which all such software is prone.

16.2.2 Problems with Traditional Approaches to Malware Protection

There are several traditional approaches to securing IoT and network element software elements that can be divided into a few categories:

- Hardware approaches
- Simulation approaches
- Software approaches

Each of these has their own strengths and weaknesses as discussed in the following section.

16.2.3 Traditional Approaches to Security: Hardware

Many believe that hardware will provide the ultimate solution to securing IoT and network elements by removing software from the critical interface elements in these systems [7].[2] By replacing software running on traditional computational elements for these critical interface elements with Application-Specific Integrated Circuits (ASICs) that are hardwired for a specific purpose and, therefore, cannot be altered by malware, a more secure overall system can be created. This will be most successful if security is embedded in the end-to-end computing and networking infrastructure hardware with no traditional software elements that can be an attack vector for malware.

1 Formal methods are the application of a variety of theoretical computer science fundamentals, in particular logic calculi, formal languages, automata theory, discrete event dynamic system, and program semantics, but also type systems and algebraic data types to problems in software and hardware specification and verification.
2 Note that current firmware is control software deployed with the hardware – usually in a rewritable format and thus still ultimately vulnerable to malware attack.

The adoption of hardware solutions to securing IoT and network elements will be driven by the needs of the IoT domain, as connected devices and vehicles will require it. Many industries, such as the auto industry, are only using the IoT for simple things, like audio-visual connectivity. Increasingly, they will be linking the IoT to more important things and every system in a vehicle will need to communicate. At that point, they will need security that is stronger and more integrated [8].

Another hardware solution to securing IoT and network elements is by using code encryption where the hardware directly executes the encrypted code with on-the-fly decryption of the code and data within the processor. By never exposing the decrypted code to an attacker – even during execution, traditional exploits are neutralized [9].

16.2.4 Traditional Approaches to Security: Simulation

As an adjunct to traditional hardware or software approaches to securing IoT and network element software, simulation approaches may be used to assess and validate IoT and network element security. By performing activities such as intrusion detection and security analysis in a simulated environment, the real IoT and network element software elements can be assessed without risk [10, 11].

16.2.5 Traditional Approaches to Security: Software

Current approaches to securing IoT and network element software is via the software itself. This may be less than optimal because the same computational elements – CPUs and memory – are being used not only for the IoT functionality but also for the security of the network elements themselves. The main point of this chapter is to address the fact that when the software is written by human programmers, it is prone to weaknesses that introduce vulnerabilities that provide points of attack.

16.2.5.1 Coding Weaknesses, Software Vulnerabilities and Malware

A common hierarchy of software security elements includes weaknesses, vulnerabilities, and malware:

- A *weakness* is a flaw, fault, bug, or other error in software or hardware implementation, code, design, or architecture that if left unaddressed could result in systems, networks, or hardware being vulnerable to attack [12].
- A *vulnerability* is a flaw in a software, firmware, hardware, or service component resulting from a weakness that can be exploited, causing a negative impact to the confidentiality, integrity, or availability of an impacted component or components [13].
- *Malware* is software or firmware intended to perform an unauthorized process that will have adverse impact on the confidentiality, integrity, or availability of an information system.

In essence, hundreds of coding weaknesses lead to tens of thousands of software vulnerabilities that can be exploited by hundreds of millions of instances of malware, as shown in Figure 16.1.[3]

This leads us to the conclusion that to produce more secure software we need to concentrate on improving software development methodologies and tools to prevent coding weaknesses from leading to software vulnerabilities in the first place and concentrate less on identifying and mitigating the actual malware instances.

3 As of March 2022, the CWE listed 924 total weaknesses (including duplicates and deprecated entries) [12], the CVE listed 167,047 total vulnerabilities from 1999–2022 [13], and DataProt estimates more than 1 billion malware programs in the wild [14].

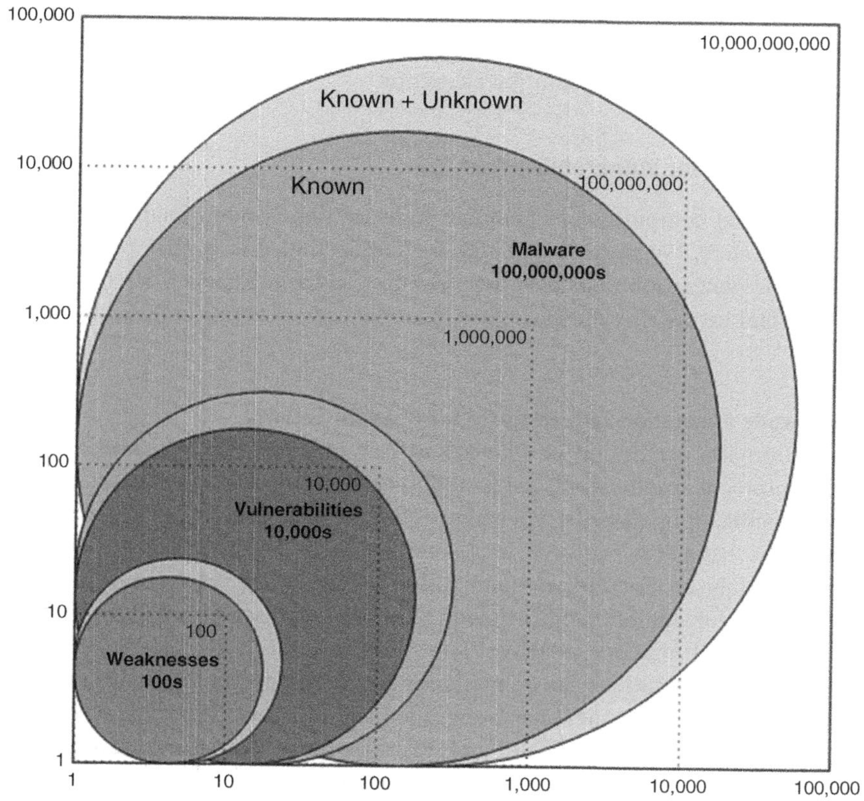

Figure 16.1 The number of weaknesses is two orders of magnitude less than the number of vulnerabilities, which is four orders of magnitude less than the number of malware instances. Source: Adapted from [12–14].

16.2.5.2 Traditional Approaches for Protecting IoT Software

Traditional approaches to protecting IoT software concentrate on identifying and quarantining malware instances before they can adversely impact on the confidentiality, integrity, or availability of the system [3, 14]. This is largely a losing battle, even with almost constant updates to antivirus software, due to the sheer number of malware instances that exist. Secondarily, the approach is a process: vulnerabilities are identified, mitigations developed, patches or new versions released, and systems patched. This, however, takes time and systems are vulnerable until they are patched, and zero-day exploits, which are either unknown to the user community or known and without a patch, may exist which target undisclosed vulnerabilities. Unfortunately, addressing coding weaknesses has not been a priority. Improved software development techniques such as programming language improvements (e.g. using Rust instead of C for network element software) and the use of static code analysis) can improve the security of IoT and network element software, but are not yet widely accepted.

16.2.5.3 Improvements on Traditional Software Approaches

Improvements in the traditional software approaches have included artificial intelligence (AI)/machine learning (ML) techniques to recognize and mitigate malware instances in the wild without requiring constant antivirus software updates. This has led malware developers to use similar techniques to develop and release increasingly stealthy malware [3].

Inverting the current concentration on malware instances, then vulnerabilities, and lastly weaknesses would put the emphasis on the more tractable problem or preventing exploitable vulnerabilities in the first place – in terms of solution elements.

16.2.6 Auto-code Generation for Vulnerability-free IoT

One of our assumptions is that computers can generate software that is more secure than that written by human programmers. We must caveat this by adding that this is only practical in limited domains where code-generation rules have been defined, which is easiest when behaviors have been codified by formal protocol specifications. Automatic code generation is discussed in Section 16.3.

16.2.6.1 Applying Auto-code Generation Selectively for IoT Network Security

As mentioned above, automatic generation of software is only practical when behaviors have been codified by formal protocol specifications using a high-level protocol description language. These formal protocol specifications describe not only packet contents and associated network characteristics but also the behaviors required in response to and producing these external packets. Because networks rely on network element software that by its very nature requires formal protocol specifications for interoperability among IoT devices, they are the obvious initial choice for auto-generation for IoT network security.

Over time, formal protocol specifications can be produced and codified allowing additional IoT software to be auto generated. However, there has been little incentive yet to do so because there are no widely available tools to help create them nor tools to utilize them. Until these are available, we see little reason to expect these formal specifications to emerge.

16.2.6.2 A Practical Approach to Generating Vulnerability-free IoT Networks

Based on the proposed improvements to the traditional approaches to IoT and network element software, we can define a practical approach to generating vulnerability-free IoT networks. Mitigations to the weaknesses from the Common Weakness Enumeration (CWE) [12] would be encoded as part of the coding rules in the code generator. A network developer using a high-level protocol description language can generate more code in less time than a programmer using conventional programming tools.

An additional benefit is that a developer using auto-generation techniques can generate security updates to existing code faster and with fewer errors than human programmers using conventional tools. In the following section (Section 16.3) we detail the key concepts of our approach and the tools that we use to create vulnerability-free software.

16.3 Automatic Code Generation

The automatic code generation approach to building secure IoT and network element software is based on three fundamental capabilities:

(1) High-level representation of building blocks of IoT network protocols.
(2) Machine-readable representations of rules for generating software that is free of vulnerabilities due to known weaknesses.
(3) Auto-generation of executable software.

To measure the utility and usability of automatic code generation, a quantitative evaluation process is defined that enables a head-to-head comparison of manually written open-source network software against auto-generated software.

The knowledge source of software weaknesses is the CWE database, which is an open database maintained by the MITRE Corporation based on input from the larger software development community [12]. There are approximately 1000 weaknesses documented in the CWE – although there are many duplications and deprecated weaknesses in that number and the actual number of unique weaknesses is in the hundreds. No human can keep in mind these hundreds of coding constraints when writing software, and even widely known and understood weaknesses such as buffer overflows are still often introduced into released software through carelessness and oversight. The automatic code generation system builds secure software by storing a domain-specific subset of these weaknesses[4] and coding rules to generate IoT and network element software that is free from the vulnerabilities caused by these weaknesses.

Automatic code generation of IoT and network element software works according to the following steps. These steps rely on several ontologies[5], which are discussed in Section 16.3.2.

(1) Software weaknesses are formally specified in a machine-readable format and maintained in a knowledge base.
(2) For each weakness in the knowledge base, an expert in secure programming develops a set of coding rules that will ensure the code avoids the corresponding weakness. For example, buffer overflow is a common weakness found in code; to prevent a buffer overflow attack, the code should perform bounds checking.[6] The coding rules are also specified in a machine-readable format and stored in the knowledge base.
(3) Based on an ontology of network elements and software patterns, a developer specifies the design of IoT or network element software in a formal representation language.
(4) The code generator interprets the design representation and applies the coding rules to generate executable code.

Figure 16.2 shows a high-level overview of the automatic code generation concept of operation and its main components.

The Common IoT and Network Element Ontology defines the concepts and relationships used to describe and represent IoT and network element components and software, such as classes and subclasses related to IoT and network element components and their characteristics, and relations and sub-relations among concepts and their constraints. Because of the level of formality required, a standard ontological representation language such as the Web Ontology Language (OWL) [15] is needed to define the Common IoT and Network Element Ontology.

The concept of operations starts with a coding expert writing the coding rules based on the CWE descriptions of coding errors that are found in IoT and network element software and storing them in an ontology. An IoT or network software expert then takes IoT or network element specifications and rewrites these so that they can be stored in the ontology. To develop new IoT and network elements, a developer specifies a network element by referencing the element components that

4 Many of the weakness in the CWE are domain-specific and not applicable in other domains. For example, an analysis of the Top 25 weaknesses showed that only ten of them are specifically applicable to network element software.

5 An ontology is a formalization of the properties of a subject area and how they are related by defining a set of concepts and properties that represent the entities in the subject area.

6 Buffer overflow is just one example of the many weaknesses in the CWE, many of which are difficult to detect with static code checkers.

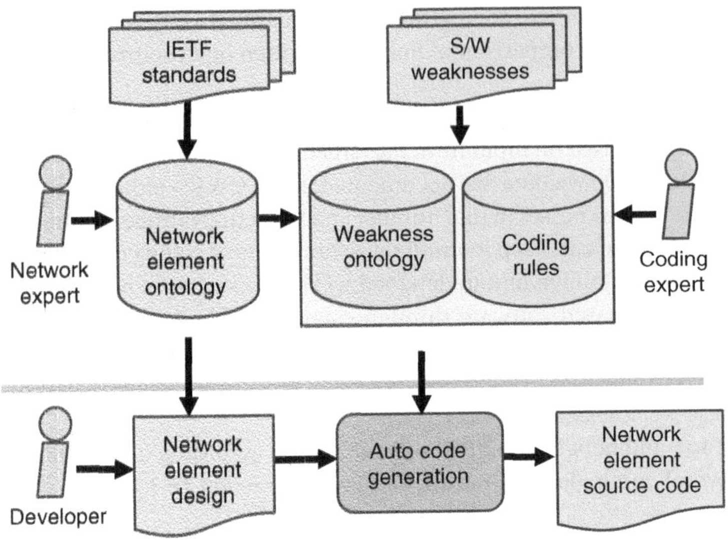

Figure 16.2 Automatic code generation conceptual architecture.

were stored in the ontology by the IoT or network expert. The software for the new elements is generated based on the rules stated in the coding rules ontology. The resulting software is secure and free from security vulnerabilities due to known weaknesses (as defined by the CWE).

16.3.1 Core Auto-generation Engine

Our automatic code generation engine [16, 17], called Content Generation from Templates (Cogent)[7], has been used on several previous projects [16–19] and is a stable piece of software. The original Cogent code generator uses XML to represent the ontology and has been modified to use graph-based Resource Description Framework (RDF) [20]. Cogent use the following packages:

- Cogent incorporates an inference engine that supports both forward-chaining (data-driven) and backward chaining (goal-driven) for developing rule-based systems for automatic code generation [21].
- Cogent uses a standard templating language for generating text – program code in this case. The generated source code could be any language – within reason – but typically C or Rust is used for IoT and network software elements [22, 23].

16.3.2 Semantic Definitions of Software Functions

The fundamental concept in the automatic generation of IoT and network element software is that the representation of domain components (i.e. IoT and network element software/firmware) should be descriptive rather than imperative. This is the same fundamental idea that the Internet Engineering Task Force (IETF) uses in publishing Request for Comment (RFC) documents that describe a network component rather than just publishing source code that implements it [4]. Code generation uses these RFCs as the source of the network element descriptions, encoding the English language description in RDF and OWL.

7 Code generation is no different from generating any other textual content – which is why the term "content generation" is used instead of "code generation" in the name.

An RFC does not always completely describe the behavior required to implement a network component though. For example, the Router Information Protocol (RIP) does not specify how to store router tables or how to efficiently search them, it just describes the message formats and required behaviors of conformant implementations. This is because RIP is applicable to routers that range from small home routers to large Internet backbone routers and, while the functionality is the same, the details may be very different. On the other extreme are some cryptographic standards that are strictly mathematical algorithms and are completely specified. We estimate from experience working with RFCs that they are typically about 80% specified. We do not need to change any RFC. Instead, we produce ancillary descriptions that specify the required additional content, which is exactly what an IoT vendor would have to do.

An important aspect of automatic code generation for IoT and network element software is an ontology for describing protocols to a level required for the automatic generation of IoT and network element software. The ontologies in our approach can be broken down into several interrelated ontologies that cover different aspects of the problem:

- Code Generation Ontology
- Weakness (i.e. CWE) Ontology
- Network Protocol Ontology

The code generation ontology covers basic data structure and algorithm descriptions to the level required to describe structures and processing in network protocols. Some of these elements are inherent in the structure of RDF and RDF Schema. This includes classes/subclasses, relations/sub-relations, collections, and high-level descriptive elements. Also, the data types from XML schema definitions (XSD) are also inherent in RDF.

The code generation ontology must define elements that describe data structures.

- Collections (sets, sequences, and bags) are inherent in RDF and are used to describe many data structures
  - Graphs are based on sets – $G = (V, E)$, where V is a set of vertexes and E is a set of edges between vertexes
  - Stacks, queues, and deques are based on sequences
- Structures
  - Sequences of Fields
  - With Subfields
  - Offsets
- Arrays

It also must describe algorithms at the level of pseudo-code.

- Blocks are sequences of statements
- Statements include
  - Assignments
  - Alternation (if/then/else) – a sequence of alternatives
  - Selection (case) – sequence of selectors
  - Iteration – a sequence of iterators
    - While/Until/Forever
    - Indexed – For
    - Data structure iterators
  - Calls
    - Primitives (e.g. UDPReceiveFrom) are implemented as calls
- Expressions – expression trees (graphs).

Finally, there are high-level structure and control constructs,

- Units – the basic unit for which code can be generated
  - Programs – generated as main programs (i.e. executables)
  - Modules – generated as libraries (header files, etc.)
- High-Level Control Strategies
  - Reactive Control Strategy – events are represented by file descriptors and the control loop implemented by select/poll
    - Timers
    - Kernel interface (e.g. netlink/rtnetlink)
    - Sockets
    - Files

The network ontology includes the primitive elements required for high-level network protocols, such as Internet Protocol (IP), Transmission Control Protocol (TCP), and User Datagram Protocol (UDP). In an ideal world, these protocols would themselves already have been described in RDF and would have their code automatically generated as well.

The Weaknesses Ontology describes the Common Weakness Elements (CWEs). Previously, the weaknesses were recognized and mitigated in the primitives. Each primitive element had been manually evaluated for possible weaknesses through the CWE. Based on the results of the evaluation, the rule set was updated to recognize the potential weaknesses and the code templates are encoded with the mitigations.

For each of the weaknesses applicable to IoT and network element software, rules in structured English are developed to detect when source code is subject to the weakness along with additional rules describing how to avoid the weakness. For example, CWE-805 *Buffer Access with Incorrect Length Value* arises when a buffer access goes outside the bounds of the buffer. To avoid this weakness, it is required that the program track the size of every buffer and check that every access is within bounds. Here are the structured English detection and mitigation rules for the CWE-805 weakness.

IF the program has a statement that accesses a buffer location
AND that statement is not within the scope of a condition guaranteeing that the location(s) accessed are within the buffer start and end bounds
THEN the program is subject to CWE-805 Buffer Access with Incorrect Length Value

(a) Detection rule

IF the program uses pointers (including arrays and strings)
THEN keep every pointer as part of a triple that also contains a pointer to the beginning of the buffer and the buffer size (in bytes)
IF the program has a statement that accesses a buffer location
AND that statement is not within the scope of a condition guaranteeing that the location(s) accessed are within the buffer start and end bounds (as specified by the triple)
THEN replace that statement with a conditional statement that checks whether the location(s) accessed are within the buffer start and end bounds and if so, performs the access, but if not, returns from the current function with an error-indicating return value

(b) Mitigation rules

16.3.3 Formal Methods for Verifying Semantic Definitions

The security of automatically generated IoT and network element software is only as good as the specification that is the basis of the code being generated. Any weaknesses in the protocol specification will necessarily be reflected in the generated code. Formal verification methods can be used to prove (or disprove) the correctness of the protocol [6]. This is recommended for all security related IoT and network element specifications. While formally verifying all IoT software would be time consuming and, in many cases, prohibitively expensive, applying formal verification methods to just network interface specifications is tractable.

16.3.3.1 Static Analysis for Verifying Code Generator Produces Vulnerability-free Code

Weaknesses from the CWE database is our focus along with developing coding rules for the automatic code generator to avoid these weaknesses. In this section, we describe the construction of static analysis tools used to verify that the auto-generated code complies with these coding rules. The rules are strict so that rule compliance can be statically checked.

Potentially Dangerous Function Call Checker: This tool checks that no calls are made to potentially dangerous functions as defined by "Security Development Lifecycle (SDL) Banned Function Calls" (http://msdn.microsoft.com/en-us/library/bb288454.aspx). This checks compliance with the coding rule for CWE-676 *Use of Potentially Dangerous Function.*

Buffer Use Checkers: We developed a buffer implementation to ensure that the generated software would track buffer sizes and check bounds for buffer accesses. These four tools check that the generated software correctly uses the proper buffer implementation. Together they check compliance with the following coding rules:

- CWE-120 *Buffer Copy without Checking Size of Input (Classic Buffer Overflow)*
- CWE-124 *Buffer Underwrite (Buffer Underflow)*
- CWE-127 *Buffer Under-read* and CWE-129 *Improper Validation of Array Index*
- CWE-131 *Incorrect Calculation of Buffer Size*
- CWE-805 *Buffer Access with Incorrect Length Value*

The following is a description of the four tools that check for the above coding rules:

- **Buffer Bounds Checker:** This tool checks that each call to a buffer function either is making a request (buffer read, buffer write, or change to buffer offset) that is guaranteed to be within bounds or has its return value used (so that the caller becomes aware of the bounds error to handle it appropriately).
- **Direct Buffer Access Checker:** This tool checks that there is no direct access to buffers, only access using the validated buffer functions. This includes checking for direct field access, direct initialization of buffers, and casting to or from buffers. (Casting to/from buffers would allow circumventing the other checks.)
- **Pointer Arithmetic Checker:** This tool checks that no pointer arithmetic (including array subscripting) is performed.
- **Buffer Size Calculation Checker:** This tool checks that the `sizeof`[8] function is not applied to a pointer.

Integer Overflow Checker: This tool checks that no arithmetic operations are performed that may result in a value that exceeds the range of values of the underlying numerical representation. This checks compliance with the coding rule for CWE-190 *Integer Overflow or Wraparound.*

Exceptional Condition Handling Checker: This tool checks to make sure that each call to an `int`-returning function either is guaranteed to not return zero or has its return value used. The return value may indicate the status of the function call, so it should either be guaranteed to not indicate an error or checked so that errors may be handled. The checker accepts a whitelist; this should contain `int`-returning functions whose return values may be safely ignored. This tool checks compliance with the coding rules for CWE-754 *Improper Check for Unusual or Exceptional Conditions.*

8 C-functions are used as examples, and they are denoted by the function name with an underscore.

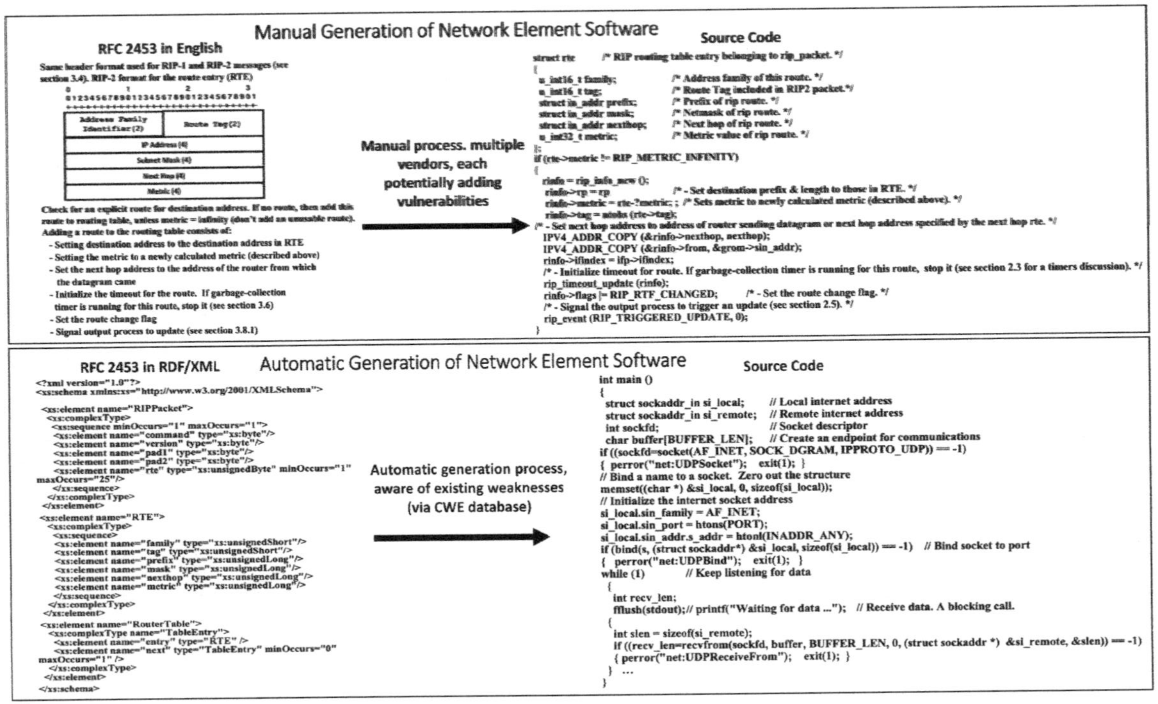

Figure 16.3 Manual vs. automatic generation of network element software.

Variable Initialization Checker: This tool checks that no variable is used before it has been assigned a value. The checker accepts a whitelist; this should contain functions to which it is safe to pass a pointer that points to uninitialized memory (because the function writes to the memory without reading from it). This tool checks compliance with the coding rule for CWE-456 *Missing Initialization of a Variable.*

16.3.4 An Extended Example: Automatic Generation of Router Software

We have used AGNES to generate code for a router that complies with RFC 2453 [24] and is provably vulnerability free due to know weaknesses. We describe that process here as an extended example of auto-code generation. The process of generating the router software and proving it vulnerability free is describe in more detail in AGNES conference paper [16] and the AGNES Final Report [2]. Currently, engineers developing network element software must read and interpret the network standard documents, or RFCs, to determine what code to write. In the AGNES automatic code generation system, these documents are stored using a machine-readable knowledge base along with coding rules based on the CWE. Using the standards and coding rules in the ontology, the code generator generates code for network elements that is free of software vulnerabilities due to known software weaknesses (Figure 16.3).

Since auto-generated code will be free from vulnerabilities due to known weaknesses, it will not provide openings for attacks by adversaries' malware. Automatic code generation will also reduce the time required to develop network software modules and reduce the effort required to update software in response to new threats. The result will be more secure software that is developed in less time and can be automatically updated to counter new threats.

The RIP has been used as a demonstration case for automatic code generation. RIP (currently RIPv2) is in widespread use in existing commercial and military networks, including wireless networks. A complete implementation of RIP, including cryptographic protocols and IPv6, requires understanding ten (10) separate RFCs. Automatic regeneration of this software whenever a new or updated RFC is released will significantly reduce the time to deployment and reduce the number of potential security vulnerabilities.

16.4 IoT Interface-code Issuing Authority

By certifying IoT interface code as vulnerability-free, IoT networks can be created where it is not possible for a compromised element in the network to further compromise other elements in the IoT network. IoT interfaces can be made vulnerability-free by using auto-code generation techniques and a system like AGNES to generate interface code for all IoT elements. IoT vendors and service providers would need to use interface code that is auto generated and certified. To accomplish that, a trusted, independent "IoT Interface-Code Authority" (IICA) is required to ensure that the code generation process itself is not compromised. Logically, this role would be assumed by a country's ministry of defense, such as the US Department of Defense (DOD), or by a regional coalition command, such as NATO. Alternatively, a government agency could contract the code-generating process to an independent, security-cleared contractor.

When an IoT vendor wants to offer a new device for use in an IoT network used for defense, they would be required to submit formal interface specifications for their device to the IICA who then verifies the interface specification and uses a code generator to create secure, trusted interface code. The interface code is then provided back to the vendor to install on their device.

The IoT device provider can then represent this device as having a government-interface-security certification. A service provider wishing to provide IoT services for defense operations, would need to certify that the IoT devices and infrastructure elements, upon which the service depends, operate with certified interface code provided by the IICA. The IICA must also maintain certified vulnerability-free network protocol software (TCP, IP, etc) that it provides to IoT vendors. The IICA would also have the responsibility of reviewing, testing, and approving firmware for all IoT devices as well as any modifications to firmware before installation. Auto-generated interface code that is certified must also protect itself from third-party modifications. Fortunately, software anti-tamper techniques exist and are administered by government agencies such as the US Department of Defense. Integrating anti-tamper software with IICA-issued IoT interface software would ensure that certified vulnerability-free software issued by the IICA would stay vulnerability free after the IoT network was deployed.

16.4.1 Role of IoT Interface-code Authority (IICA)

The IoT Interface-Code Authority or their representative will have the following responsibilities:

- The IICA will maintain a list and ontology of code weaknesses that it will exclude from IoT interface code and rules formally defining them.
- The IICA will obtain and maintain automatic code generation software capable of compiling formal representations and ontologies into a software program in one of a designated set of software languages.
- The IICA will use formal methods to verify that the automatic code generator generates software that excludes all the officially listed code weaknesses.
- A vendor wishing to obtain certified interface software for a device or IoT network element intended for use in defense operations will submit to the IICA formal specifications of the interfaces for the device in a definition language specified by the IICA, for example RDF or OWL.
- The IICA receives and reviews the formal interface specifications from IoT vendors. Any logical errors, inconsistencies, conflicts, or incompatibilities in the definitions will be corrected through interactions with the vendor.
- Once approved, the IICA will input the formal interface representations to the automatic code generator to generate IoT interface code for the device or network element.
- The IICA will integrate anti-tamper software with the interface code.
- The IICA will also auto-generate certified network protocol software with anti-tamper features (e.g. TCP, IP, etc) and provide it to approved defense IoT vendors.
- The IICA will review, test, and approve all firmware and firmware modifications to defense IoT devices and elements.
- The IICA will receive and review all IoT interface code certificates from IoT service providers covering all network elements, devices, and subordinate service providers. The IoT service will be authorized by the IICA for defense use only when it can evidence certifications for all its components.
- The IICA will manage and maintain the required infrastructure to support the foregoing actions, for example, maintaining databases of formal specifications and change histories as well as track what code has been provided to whom and for what purpose as well as track which vendors and IoT service providers are using certified IoT elements.

16.4.2 Precedents and Examples and a Proposed IoT Interface Code Authority

Within the US DOD and other nation's ministries of defense, there are precedents for an agency such as the IICA. Secure communications devices and networks are tightly controlled by specific agencies. For example, in the U.S. secure telephones are provided by and controlled by a specific security agency that is responsible for testing and ensuring the security of these devices. Several different agencies provide secure networking services and are responsible for the secure operation and integrity of the components of the network as well as the network as a whole. Several different DOD entities are responsible for ensuring that anti-tamper software is correctly integrated with applications code and is successfully preventing code tampering.

Incorporation of anti-tamper software is critical to ensuring the integrity of auto-generated vulnerability-free interface code for IoT. Currently, several different organizations in the US DOD and intelligence community hold the responsibility for approving anti-tamper capabilities in application code. Consolidating this authority in one DOD-wide anti-tamper authority would provide uniformity of solutions, as well as simply the process and reduce costs and duplication. Such an anti-tamper authority would be a logical authority to assume the IICA responsibilities as outline in the Section 16.4.1.

16.5 Conclusions

IoT devices and networks for military operations face special security challenges not faced by commercial IoT. During military operations, it is not possible to guarantee the physical security of all IoT nodes. Hostile agents will attempt to attack, disable, deceive, co-opt, and capture military IoT systems. An IoT network consists of layers of hardware and software logic produced and programmed by many different organizations residing in multiple nations. Certifying the integrity and vulnerability-free state of all elements and levels of IoT nodes increasingly becomes an intractable challenge for defense use of IoT. We described one approach to prevent the compromise of one IoT node from spreading through the network. Our approach uses formal specifications of the semantics of the interface to support automatic code generation. The use of auto-code generation from formal specifications has been demonstrated by us and other groups. For example, DARPA's System of Systems Integration Technology and Experimentation (SoSITE) program. SoSITE used their System-of-systems Technology Integration Tool Chain for Heterogeneous Electronic Systems (STITCHES) toolset to link legacy fire-control systems by automatically generating interface software from formal specifications to translate between Army artillery units, fighter aircraft, and shipboard command and control centers [25]. Our work extends auto-code generation by incorporating a software-weakness ontology and rules that prevent the code-generator from creating IoT interface software that contains vulnerabilities due to known weaknesses. By applying formal verification methods to the automatic code generator, it can be proven that the code generator generates software that is vulnerability free. We demonstrated and tested our approach by automatically generating routing software that is free from malware vulnerabilities.

To create a secure IoT defense network, a trusted authority auto-generates weakness-free interface software for IoT elements and augments it with anti-tamper software to guarantee that it stays secure and unmodified. The trusted IoT interface-code authority should reside in a nation's defense ministry or its independent representative. We recommend that for the US, the IICA should be combined with the various anti-tamper authorities into a single, joint service certifying agency. This approach ensures the security of defense IoT networks and IoT services by requiring every IoT node be connected via auto-generated, weakness-free software. By so doing, a defense IoT network can limit the damage from the capture or compromise of an individual IoT node from spreading beyond the directly affected node. An adversary cannot co-opt an entire defense IoT network by co-opting one node.

References

1 Parnas, D.L. (1985). Software aspects of strategic defense systems. *American Scientist* 73 (5): 432–440.

2 Rouff, C., Williams, M.D., and Bennett, D. (2019). Automatic Generation of Network Element Software (AGNES). *Final Report*, March 2019.

3 cynet.com website. A Guide to Malware Detection Techniques: AV, NGAV, AND BEYOND. https://www.cynet.com/blog/a-guide-to-malware-detection-techniques-av-ngav-and-beyond/ (Retrieved 12 March 2022).

4 IETF. Standard Process. https://www.ietf.org/standards/process/ (Retrieved 12 March 2022).

5 IETF. Guidance on Reporting Protocol Vulnerabilities to the IETF. https://www.ietf.org/standards/rfcs/vulnerabilities/ (Retrieved 12 March 2022).

6 Wikipedia Contributors (2021). Formal verification. *Wikipedia, The Free Encyclopedia*. Wikipedia, The Free Encyclopedia. (Retrieved 12 March 2022).

7 Embedded Computing Design. Hardware security in the IoT. https://www.embeddedcomputing.com/technology/security/hardware-security-in-the-iot (Retrieved 12 March 2022).

8 Woods, D. (2018). Cybersecurity's future: powered by hardware. *Forbes*. https://www.forbes.com/sites/danwoods/2018/09/07/cybersecuritys-future-powered-by-hardware/?sh=2c5c04e25ced (Retrieved 12 March 2022).

9 Brooks, R.R., BaeYun, S., and Deng, J. (2012). Cyber-physical security of automotive information technology. *Handbook on Securing Cyber-Physical Critical Infrastructure*, pp. 655–676.

10 U.S. Naval Research Laboratory. Extendable Mobile Ad-hoc Network Emulator (EMANE). https://www.nrl.navy.mil/Our-Work/Areas-of-Research/Information-Technology/NCS/EMANE/ (Retrieved 12 March 2022).

11 gns3.com. The Software that Empowers Network Professionals. https://www.gns3.com/ (Retrieved 12 March 2022).

12 MITRE. CWE: Common Weakness Enumeration. https://cwe.mitre.org/ (accessed 31 October 2022).

13 MITRE. CVE. https://www.cve.org/ (accessed 31 October 2022).

14 DataProt. A Not-So-Common Cold: Malware Statistics in 2022. https://dataprot.net/statistics/malware-statistics/#:~:text=There%20are%20now%20more%20than,fall%20victim%20to%20ransomware%20attacks (Retrieved 12 March 2022).

15 W3C. Web Ontology Language (OWL). https://www.w3.org/OWL/ (accessed 31 October 2022).

16 Rouff, C.A., Williams, M.D., Zhang, Q. et al. (2017). Automatic generation of network element software (AGNES). *Cyber and Information Security Research Conference (CISRC 2017)*. Oak Ridge, TN (4–6 April 2017). https://doi.org/10.1145/3064814.306482.

17 Murray, K., Lowrance, J., Sharpe, K. et al. (2011). Toward culturally informed option awareness for influence operations with S-CAT. *4th International Conference on Social Computing, Behavioral-Cultural Modeling and Prediction (SBP11)*, Springer, volume 6589, pp. 2–9, March 2011.

18 Myers, J., McDowell, R., Rouff, C. et al. (2018). Static analysis of programmatically generated network software: challenges and synergies. *High Confidence Software and Systems Conference* (7–9 May 2018).

19 Murray, K., Lowrance, J., Sharpe, K. et al. (2010). Capturing culture and effects variables using structured argumentation. *1st International Conference on Cross-Cultural Decision Making*.

20 W3C. Resource Description Framework. https://www.w3.org/RDF/ (accessed 31 October 2022).

21 Williams, M.D. Inference Collection: Reference Manual. https://planet.racket-lang.org/package-source/williams/inference.plt/2/0/planet-docs/inference/index.html (accessed 31 October 2022).
22 Flatt, M. and Barzilay, E. Scribble: The Racket Documentation Tool. https://docs.racket-lang.org/scribble/ (accessed 31 October 2022).
23 Flatt, M. and Barzilay, E. Scribble as Preprocessor. https://docs.racket-lang.org/scribble-pp/index.html (accessed 31 October 2022).
24 Malkin, G. *RIP Version 2*. IETF. https://datatracker.ietf.org/doc/html/rfc2453 (accessed 31 October 2022).
25 Jones, J. (2019). Global interoperability without global consensus, a DARPA solution via the STITCHES toolchain. *Proceedings Vol. 11015. Open Architecture/Open Business Model Net-Centric Systems and Defense Transformation*. Baltimore, MD. https://doi.org/10.1117/12.2519443.

Section 4

Introduction: Communications and Networking

Keith Gremban

Ann and HJ. Smead Aerospace Engineering Sciences and Silicon Flatirons Center, University of Colorado Boulder, Boulder, CO, USA

Communications and networking is a fundamental enabler for Internet of Things (IoT). Wikipedia defines the Internet of Things as " … *physical objects (or groups of such objects) that are embedded with sensors, processing ability, software, and other technologies that connect and exchange data with other devices and systems over the Internet or other communications networks.*"[1] Key to this definition is the phrase "… *connect and exchange data with other devices and systems over the Internet or other communications networks.*" Without communications and networking, there is no IoT.

In the commercial world, an IoT developer has access to a host of communications technologies. Typically, an IoT deployment will use some wireless technology (4G, 5G, Wi-Fi, Bluetooth, or others) to connect devices to a local server which is hard-wired to the Internet. From there, data is sent to a processing center or cloud server for data analysis, and instructions for the end devices are sent back the other way. An IoT developer can usually depend on reliable, high-bandwidth connectivity. The only disruptions to the deployment are infrequent, such as power outages, excavations that cut a cable, or similar unpredictable and rare events.

In defense and public safety scenarios, reliable communications are the exception, rather than the rule. Consider a few example use cases:

- In many natural disaster scenarios, such as earthquakes or fires, the commercial communication systems have been destroyed or seriously degraded. Public safety organizations must frequently field their own communications systems. These systems are wireless, and cannot provide the same level of reliability, bandwidth, and quality of service of the fixed infrastructure that they are (temporarily) replacing.
- In a ground combat scenario at the tactical edge, no fixed infrastructure can be utilized. Communications infrastructure is typically among the first targets of combat operations. Hence, tactical edge warfighters must carry their own communications gear. Not only are such deployable systems less reliable and lower bandwidth than systems that make use of fixed infrastructure, but they face unique challenges. Mobile networks are vulnerable to loss of connectivity due to terrain, foliage, and buildings, as well as to deliberate jamming attacks by the enemy. As a result, forward units can lose connectivity with higher echelons to the rear, thereby losing access to critical data servers or computational capabilities.

1 https://en.wikipedia.org/wiki/Internet_of_things

IoT for Defense and National Security, First Edition. Edited by Robert Douglass, Keith Gremban, Ananthram Swami, and Stephan Gerali.

This section discusses the special challenges of communications and networking in the defense and public safety domain, and presents a variety of approaches for dealing with them.

Chapter 17, *Leveraging Commercial Communications for Defense and Homeland Security IoT*, discusses some of the differences in capabilities that have emerged between defense and homeland security communications and commercial communications. Over the past 30 years, wireless technology has enabled consumers to "cut the cord" and go mobile. The growth in investment and resultant increase in capabilities has been staggering. Although the military and commercial operating environments are drastically different, military communications and networking should be able to take advantage of commercial developments. The authors explore the differences in military and commercial capabilities and present opportunities for leveraging commercial developments at the system, sub-system, and technology levels.

Chapter 18, *Military IoT: Tactical Edge Clouds for Content Sharing Across Heterogeneous Networks*, addresses the problem of reliable information dissemination over unreliable wireless networks at the tactical edge. IoT devices, along with human warfighters, are sensors that generate data that contributes to a warfighter's understanding, or *situation awareness*, of what is happening on the battlefield. Because of the unreliability of communications and networking at the tactical edge, reach back to central servers or cloud services is impractical. The authors explore the characteristics of the tactical edge environment, discuss two similar approaches for managing data at the tactical edge, and present an architecture for disseminating critical information across distributed, intermittent, and bandwidth-limited (DIL) connectivity.

Chapter 19, *Spectrum Challenges in the Internet of Things: State of the Art and Next Steps*, addresses the critical, but often overlooked, challenge of reliable access to radio frequency (RF) spectrum. Before the rise of commercial wireless communications in the 1990s, spectrum was a seemingly plentiful resource. Since then, however, commercial communications have been demanding access to more spectrum, making spectrum a rare, and expensive, resource. The authors posit that IoT devices will need to become spectrum agile and energy-efficient to accomplish their missions and for the IoT to continue to grow. The chapter discusses the technical requirements and the enabling technologies for the next generation of spectrum-sharing, energy-efficient IoT wireless devices, and concludes with a discussion of the challenges of spectrum sharing and future directions.

Chapter 20, *Tactical Edge IoT in Defense and National Security*, provides an excellent summary of the communications and networking issues in applying IoT at the tactical edge. The chapter presents a collection of scenarios in which use of IoT at the tactical edge can improve warfighter and public safety officer survivability, decrease cost, and improve effectiveness. The chapter presents a Tactical Edge IoT communications architecture, and closes with a discussion of challenges, and research recommendations.

17

Leveraging Commercial Communications for Defense IoT

Keith Gremban[1] and Paul J. Kolodzy[2]

[1]*Ann and H.J. Smead Aerospace Engineering Sciences and Silicon Flatirons Center, University of Colorado Boulder, Boulder, CO, USA*
[2]*Kolodzy Consulting, LLC, Falls Church, VA, USA*

Abstract

The Internet of Things (IoT) can trace its origins back to the forerunner of the Internet, the ARPANET. The ARPANET was a project conceived of and funded by the Advanced Research Projects Agency (ARPA) within the United States Department of Defense (DoD), so in a very real sense, there would be no IoT without the initial investment from DoD. In recent years, with the advent of Internet-enabled e-commerce, cellular communications, and Wi-Fi, the private sector has invested much more heavily in communications technology than DoD. The IoT is a private sector innovation and thus represents a capability transitioning from the commercial world to the DoD. To leverage the full potential of IoT for defense and homeland security, government agencies should look to the private sector and leverage communications innovations to the greatest extent possible, and target government investment towards meeting unique requirements not fully addressed by the private sector.

17.1 Introduction

The Internet of Things (IoT) has a variety of definitions from different individuals and organizations. Wikipedia [1] defines the IoT as follows:

> The Internet of things (IoT) describes physical objects (or groups of such objects) that are embedded with sensors, processing ability, software, and other technologies, and that connect and exchange data with other devices and systems over the Internet or other communications networks.

The U.S. Department of Defense (DoD) laid the foundation for the IoT through the development of the ARPANET, which was the predecessor to the Internet [2]. The ARPANET was a project funded by the Advanced Research Projects Agency (ARPA) within the U.S. Department of Defense (DoD) to enable computers (originally the command-and-control systems for land-based nuclear missiles) to reliably communicate across a disrupted and fragmented network. The concept of computer networking spread quickly throughout defense, education, and commercial domains, and many separate networks were developed. New technology was developed to interconnect the

IoT for Defense and National Security, First Edition. Edited by Robert Douglass, Keith Gremban, Ananthram Swami, and Stephan Gerali.

profusion of networks into an Internet. In 1990, the ARPANET was officially retired, and academic and commercial entities stepped up to maintain and grow the new Internet. From the initial four nodes of the ARPANET, the Internet is estimated to have 46 billion connected devices by the end of 2021 [3].

One of the most dramatic changes since the founding of the Internet is the means by which devices attach to the Internet. Initially, nearly all the connections were wired. However, this began to change in the mid-1990s, when Wi-Fi was released for consumers [4], along with the first Internet-enabled mobile phone [5]. Since the mid-90s, wireless Internet access has exploded. A 2018 report by the U.S. Census Bureau reported that 68% of households accessed the Internet using mobile broadband [6]. A 2016 report from Bell Labs projected that by 2025, 97% of Internet access will be wireless [7]. It is estimated that 90% of Internet users have wireless access in 2021.

Wireless connectivity is a key enabler for the IoT. Wireless connections, whether cellular, Wi-Fi, or satellite, dramatically reduce the infrastructure and deployment costs of IoT applications. Rather than running a wire to every device – which may require digging trenches, punching through walls, erecting poles, etc. – wireless IoT devices can be installed virtually anywhere and still connect to the Internet. Additionally, wireless access allows IoT devices to be mobile, opening up a vast number of applications.

Throughout most of the twentieth century, wireless communications was dominated by the defense and homeland security domains. Radios were big, bulky, and expensive, but needed to connect military units and first-responders. The development of Wi-Fi and cellular changed the balance between the public and private sector. Today, public sector investment in wireless communications is dwarfed by the private sector. The numbers tell the story. The value of the global military communications market is estimated to be worth $44.87B by 2027 [8]. In contrast, the cellular market in the U.S. alone is estimated to be worth $1268B by 2028 [9]. Adding in Wi-Fi, Bluetooth, and other consumer wireless technologies makes the comparison starker.

A comparison between capabilities also shows a huge divide between defense and private sector capabilities. For example, a typical tactical radio, the AN/PRC-117F, offers data rates up to 56 kbps, which is considered high speed [10]. In contrast, 4G LTE offers data rates up to 1 Gbps, while 5G is specified to offer data rates up to 20 Gbps. Given the incredible developments in commercial wireless technology, defense systems should look to the private sector for technology and systems developments that can be leveraged to improve the performance of defense systems.

The public safety community in the U.S. is adopting commercial cellular communications technology, through building a separate network called FirstNet. The events of 9/11 demonstrated serious problems with multiple non-interoperable public safety agencies each having their own independent communications systems. Additionally, the public safety community is understandably concerned about access to commercial communications in the event of an emergency. The 9/11 commission recommended a nationwide interoperable wireless network for public safety. In 2012, the First Responder Network Authority (FirstNet) was created. FirstNet uses commercial cellular technology, and the network is being built out by a commercial communications provider, AT&T [11, 12]. Given the differences in operational environments, the U.S. DoD has not made any similar commitments to private sector communications technology, although the DoD 5G Initiative is exploring the use of 5G for communications support within the U.S.

The remainder of this chapter will focus on how the defense community can leverage commercial wireless technology, as opposed to developing their own unique solutions. Leveraging has significant potential to both enhance the performance of and reduce the cost of communications systems. Section 17.2 discusses the differences in the defense and commercial requirements.

Section 17.3 discusses the differences between the defense and commercial development processes. Section 17.4 presents specific opportunities for leveraging commercial developments. Section 17.5 summarizes the discussion and presents conclusions.

17.2 Key Differences Between Defense and Commercial Communications Requirements

Defense and commercial communications technologies initially diverged because of the differences in operational requirements. Those differences can be characterized in three major challenge areas: mobility; security; and vulnerability (to environment, electronic warfare, cyber warfare, kinetic warfare, etc.). Furthermore, defense systems must function in every conceivable operating environment: from friendly to hostile; from arctic to rain forest to desert; from land to sea to air to space; from urban to rural to undeveloped. This variety clearly cannot be met by a single system, or even a family of systems – hence one of the reasons for the number of radio systems in the inventory of the U.S. military. (Wikipedia [13] has 92 pages devoted to "Military radio systems of the United States.")

The private sector communications industry has been grappling with many of the same challenges that initially distinguished defense communications. The private sector has therefore been developing solutions that can be adopted for defense purposes. In the sections below, we will examine the initial differences between the defense and commercial requirements and operational environments, identify where the environments have begun to overlap and provide potential for leverage, and discuss the ways in which the two domains will continue to converge.

17.2.1 Interoperability

Warfighters need the ability to communicate with each other in order to share information, plan and execute operations, and coordinate movements, fires, and logistics, among other tasks. However, pervasive communications are difficult to achieve because of the plethora of different communications systems in the defense world. Different systems, different combat arms, different echelon levels, different services, and different nations may all have unique communications systems. These systems often operate at different frequencies, use different waveforms, and use different protocols. Making these systems fully interoperable has yet to be fully accomplished.

Defense communications are typically developed to meet the needs of a particular system or mission. That is, the system defines the communications infrastructure. Each system has a unique collection of requirements for parameters such as range, bandwidth, anti-jam (AJ), low probability of detection (LPD), low probability of interception (LPI), low probability of exploitation (LPE), and anti-geolocation. No single system is able to meet all the requirements across the entire parameter space. Interoperability then requires a gateway between systems, although new technologies like software-defined radios (SDRs) potentially enable customizing systems to meet mission needs while meeting interoperability requirements [14].

In contrast, the commercial world starts with interoperability as the driving requirement. Consequently, the commercial world has settled on a few well-defined communications standards, such as 4G, 5G, Bluetooth, and Wi-Fi. Commercial systems are built on top of standard communications infrastructure. Adherence to these standards has lowered the cost of communications devices and enabled worldwide interoperability. A traveler with a laptop can access the Internet through Wi-Fi at airports, hotels, and coffee shops around the world.

While commercial communications may not meet all the requirements of defense communications, it provides a robust starting point. The defense community should look to leverage commercial standards, and extend them as necessary to meet unique operational requirements.

17.2.2 Mobility

Mobility is a fundamental requirement of defense communications. Commanders and subordinates, whether on land, sea, or in the air, must be able to communicate throughout an operation to provide status and intelligence, and coordinate distributed units. Mobile communications in the naval domain was a reality early in the twentieth century. In the United States, the Wireless Ship Act of 1910 required radio equipment on most ships. During World War I, the first radio communication between ground and air took place. Adoption of radio communications took place rapidly in the naval and air domains in both the defense and private sector.

However, mobile terrestrial communications was largely the province of the defense and public safety sectors until late in the twentieth century. Radio equipment was bulky and expensive, and no civilian networks existed to support mobility. Through much of the twentieth century, private sector communications were wireline connections and were physically made by human operators, establishing circuits between fixed locations.

Cellular systems, along with Wi-Fi, allowed users to "cut-the-cord" and go mobile. This was done through monumental investment in infrastructure in the form of cell towers for wireless connectivity and fiber for network connectivity. Cellular infrastructure has required hundreds of billions of U.S. dollars in investment worldwide. As of 2021, cell coverage is nearly ubiquitous across the world, with over 90% of the world's population having access to some form of cellular service [15].

In some sense, the private sector has conquered the problem of communications support for mobility. However, the private sector accomplished this through investment in fixed infrastructure and high-capacity interconnections (e.g. fiber and microwave backbone networks). And because of the cost of infrastructure, mobility is only supported when return on investment exists for infrastructure and fiber installation. In remote areas, or rural areas with low population density, the cost of infrastructure investment is prohibitive given the low revenue. Thus, while an individual can drive across the United States on an interstate and experience continuous cell service, no service is available when hiking on remote mountain trails, or in many rural areas in which residents are separated by miles.

Military operations, however, cannot depend on fixed infrastructure. Whether in mountains, jungle, desert, or urban terrain, operational units need to communicate to coordinate. Moreover, militaries cannot depend on indigenous infrastructure – these may be operated by or been manufactured by hostile entities, subject to sabotage, or targeted for destruction by the enemy. Hence, military forces either bring mobile infrastructure with them, or, at the tactical edge, rely on technologies like mobile ad hoc networking.

While the defense domain should look to the private sector for opportunities to leverage technology, those opportunities will be tempered by the need for defense systems to work over a range of operational environments. Later in this section, we will explore opportunities for leveraging in different environments.

17.2.3 Security

Security is always a primary concern for military communications. Even the Romans, over two thousand years ago, made use of a variety of secret codes to transmit military orders. Today's defense communications systems make use of multiple technologies, ranging from LPD/LPI waveforms to advanced encryption to provide security.

Private sector communications have slowly been increasing security. The first cellular networks, 1G, did not encrypt the transmissions. Anyone with sufficient skill and the right equipment could eavesdrop on wireless phone calls. 2G introduced encrypted calls, but was still full of vulnerabilities due to lack of authentication, protection from replay attacks, vulnerability to rogue cell towers (stingrays), and lack of encryption in the core network. 3G further improved security by improving encryption, extending it deeper into the core, and improving authentication. However, 3G was still vulnerable to rogue cell towers that would request a downgrade of the user device from 3G to 2G. 4G added on more security improvements but remains vulnerable to denial-of-service attacks and user tracking, among other issues [16]. 5G continues to improve security, but vulnerabilities have already been reported [17]. The Third Generation Partnership Project (3GPP) has a working group, Service and Systems Aspects Working Group 3 (SA3), dedicated to defining upgraded security mechanisms for cellular communications, including 5G.

Wi-Fi encountered – and continues to encounter – security issues. Wi-Fi security went through much of the same evolution as cellular. Initial Wi-Fi networks were open with no encryption or authentication. Then Wired Equivalent Protocol (WEP) became the standard security mechanism for Wi-Fi. This was found to be insecure and replaced by Wi-Fi Protected Access (WPA), which has since seen two more versions, WPA2 and WPA3 [18].

The private sector is very aware of and concerned over the security vulnerabilities of communications systems. Every generation of system is more secure than the last. In 4G and more so in 5G, as well as Wi-Fi, enhanced security has become a driving requirement.

Above and beyond the security of the underlying communications substrate, the private sector has developed mechanisms above the network layer to provide security. Without adequate security guarantees, applications like e-commerce and e-banking would not be possible. Consumers and businesses would not allow transactions over wireless networks without some confidence in the security of the operation.

It should also be pointed out that cellular systems and Wi-Fi are used by billions of users, and that hackers and researchers have access to these systems. In contrast, defense systems are used by perhaps tens of thousands of users and access is restricted, in part yielding security by obscurity.

Security is a major concern in the private sector, with everyone from private citizens to major corporations demanding secure communications. It is not unlikely that in a few generations, commercial communications will be more secure than custom defense systems. With untold number of hackers and researchers investigating wireless security – both attacks and defense – the defense community potentially can leverage commercial technology and augment security mechanisms to overcome certain security weaknesses and meet defense requirements.

An additional security concern, for both commercial and defense systems, is supply chain reliability. Communications equipment, and the underlying components, may be sourced from many different countries. Moreover, opportunities to tamper with components exist all along the supply chain. Mechanisms are needed to ensure the integrity of communications hardware and software.

17.2.4 Vulnerability

Military communications systems face threats that private sector systems do not, such as intentional jamming and physical destruction of equipment, whether targeted (such as in explosive destruction of a cell tower) or unintentional (such as dropping a piece of equipment onto rocks or into water. Indeed, the ruggedness required of military systems is a major cost factor, driven by the fact that a military operation cannot be put "on hold" or temporarily suspended until replacement equipment can be obtained. In combat, you can't run down to the local cellular provider's store to get a replacement device.

In contrast, private sector communications are not designed with an expectation that they will encounter intentional jamming. In the U.S., intentional jamming (except by an authorized Federal agent) is illegal. Similarly, private sector equipment such as cell towers do not face the threat of targeted destruction, and are rarely protected by more than a fence and warning signs. End-user equipment must meet certain standards of ruggedness that represent expected real-world consumer experience. In the case of an event that exceeds the physical requirements of a smart phone, the consumer is expected to head out to the nearest store to purchase a replacement.

Vulnerability is an area in which little, if any, technology from the private sector can be leveraged.

17.3 Key Differences Between Defense and Commercial Technology Development

The pace of advancement of radio communications systems and technology in the private sector has far outstripped that of the defense sector. Since the first commercial cellular call in 1983, the commercial sector has gone through five generations of network architectures (waveforms and core) and many more generations of technology advancements. Even though the standards existed much earlier, the large-scale deployment which started in the early 1990s as shown below (standard complete/major network deployments) and in Figure 17.1:

- 1G Advanced Mobile Phone Systems (AMPS) – (1976/1993) analog voice, unencrypted
- 2G Code Division Multiple Access/Global System for Mobile Communications (CDMA/GSM) – (1991/1999) digital, encrypted signals, text and email
- 3G (CDMA2000/UMTS) – (1999/2003) higher data rates, Internet connectivity
- 4G (LTE) – (2008/2011) higher data rates, integrated voice, video, data
- 5G New Radio (NR) – (2018/2021) enhanced mobile broadband, ultra-reliable low-latency, massive machine type communications

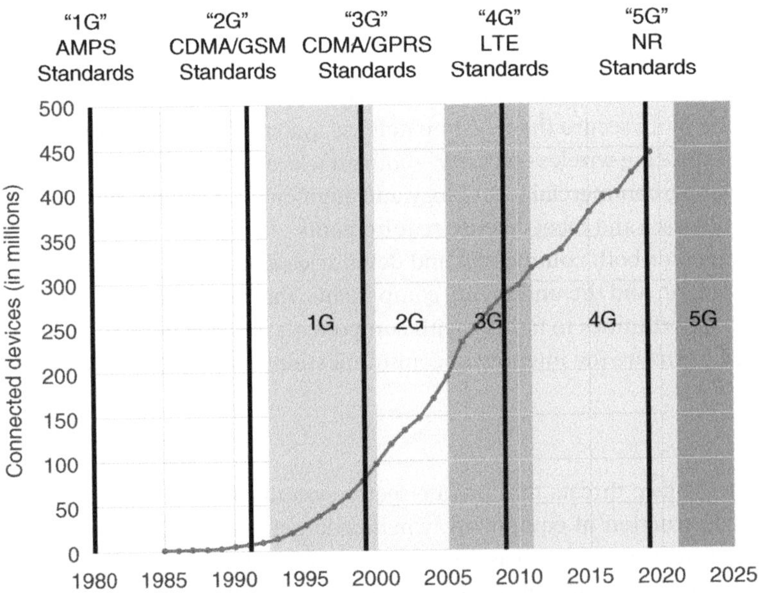

Figure 17.1 The cellular industry has gone through five generations in the 30 years since widespread deployment began. Source: P. Kolodzy.

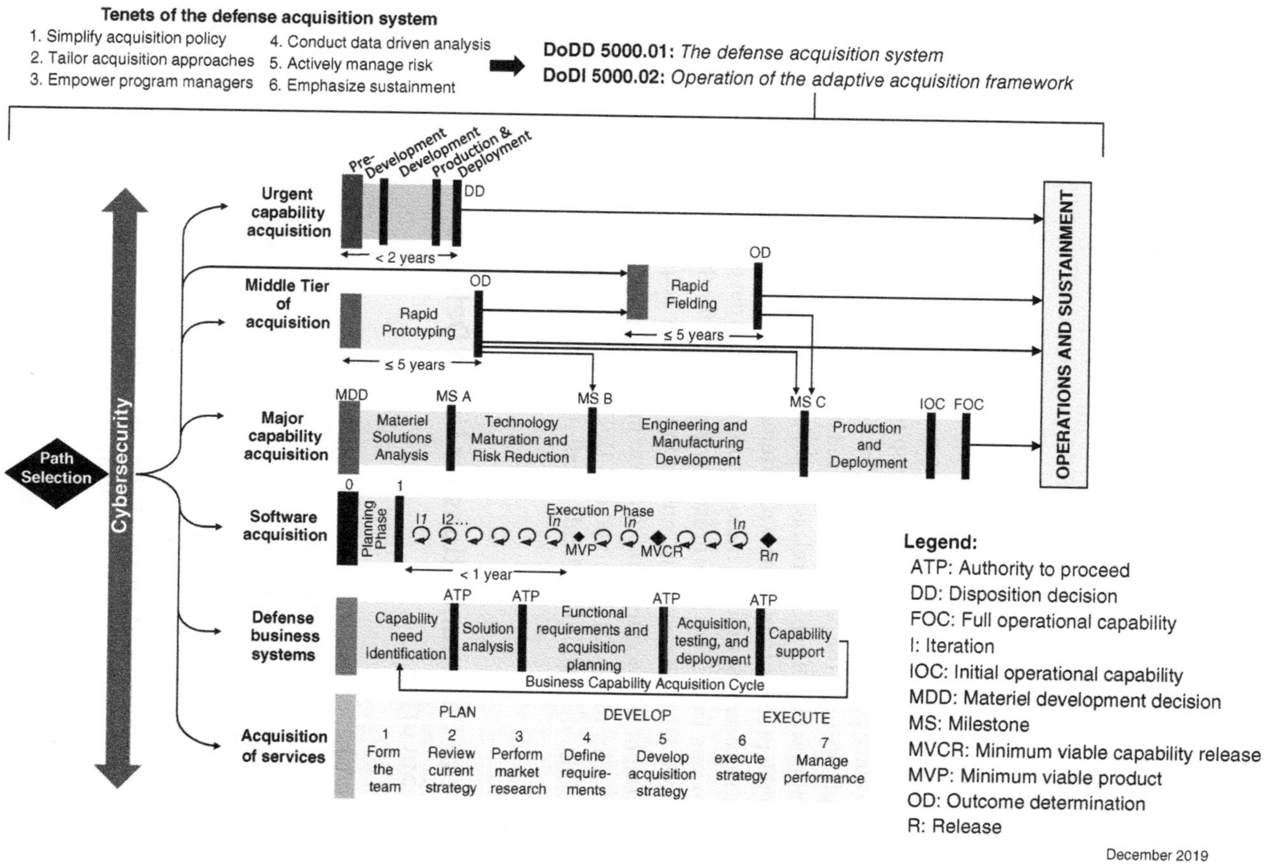

Figure 17.2 DoD 5000 was updated in 2020 to accelerate the acquisition process. Source: [21] /U.S. Department of Defense / Public domain.

That is, five generations of wireless networks and countless more upgrades in technology in roughly 30 years, or a generation every six years, although generations may co-exist. Moreover, end-user equipment goes through a generation every 9–12 months, with improved technology in each generation. Consumers expect upgrades on a regular basis and the majority of users purchase a new, more capable, smartphone every two years.

In comparison, the defense development cycle for communications equipment moves at a snail's pace. Currently, one of the most widely used radios is the Single Channel Ground and Airborne Radio System (SINCGARS). The SINCGARS radio has been in use for nearly 40 years. The program was initiated in the mid-1980s, and has gone through several system upgrades [19]. As another, less successful example, the latest U.S. radio, the Joint Tactical Radio System (JTRS) was initiated in 1997 with a Mission Needs Statement. One facet of the program, the Ground Mobile Radio (GMR) was cancelled in 2012, while the handheld version was not delivered until 2011 and is still being deployed in 2022. That's one radio system developed, deployed, and still used 25 years after the program was initiated.

The story of the JTRS GMR has been explored in a number of articles such as [20]. The principal reason for the failure of the JTRS GMR program, despite the expenditure of billions of dollars, was the constant desire – and effort – to make the radio "better." Making it better meant adding features and increasing complexity until the program collapsed under its own weight. Contrast this with the common commercial practice of rapidly releasing the minimum viable product (MVP), and then routinely upgrading the product to meet new requirements and adapt to customer feedback.

One reason for the slow pace of military radio communications development in the United States is the acquisition process, DoD-5000. This process was designed for the procurement of major systems like a main battle tank or a warship, and didn't consider acquisition of technologies that change virtually overnight. The acquisition policy was updated in 2020 to "deliver warfighting capability at the speed of relevance." [21] The updated DoD 5000 process is illustrated in Figure 17.2. Given the recency of the update, it is difficult to tell if it will move defense acquisition closer to commercial practices.

17.4 Commercial Communications for Use in Defense and Homeland Security

For describing potential opportunities for leveraging commercial communications, we first break the opportunities down into three categories:

- Systems – systems comprise complete communications solutions that may be brought wholesale into the military communications environment.
- Sub-systems – even if commercial communications systems cannot be substituted for military communications systems, there exist non-stand-alone aspects of communications systems that can be integrated into military systems to improve performance.
- Technologies – technologies are basic components of sub-systems and systems.

In previous sections, we discussed the need for military systems to operate in a variety of environments. Some commercial developments are applicable to certain military environments, but not to others. Rather than the typical breakdown into rural, sub-urban, and urban, a more useful way to break down operational environments is based on the demands – and opportunities – for communications:

- Permissive environments, private networks – this category includes government or military facilities that are in United States or allied territory utilizing either local communications or private communications infrastructure.

- Permissive environments, public networks – this category includes government or military facilities that are in the United States or allied territory utilizing telecommunications infrastructure that is not directly controlled by the military.
- Propagation challenged environments – this category includes challenging terrain like dense forest, jungle, or mountains and includes the use of edge devices such as ground or near-ground sensor networks, or edge devices place in defilade.
- Contested environments – this category includes environments with an active opponent able to employ eavesdropping, spoofing, or jamming technologies that either impact assuredness of data delivery or the potential loss or confidentiality of information.

The defense and homeland security sectors can take advantage of commercial systems, subsystems, and technologies. Figure 17.3 summarizes the applicability in terms of the environments. These entries are expanded upon below:

- Systems – defense and homeland security organizations should take advantage of commercial communications systems such as 5G, Wi-Fi, and Wi-Gig when appropriate.
  - *Permissive – Private*: In own or allied territories, defense organizations should consider the option of installing private 5G networks. 5G networks can perform routing at the edge, so that all internal communication is routed internally. 5G provides for encrypting traffic and VPNs can be used as necessary for additional security. Firewalls can be used to secure connections to the outside. Similarly, robust Wi-Fi at endpoints to the private network would provide high-bandwidth connectivity. Wi-Gig can be used inside computer centers to connect devices wirelessly.
  - Wi-Gig can provide near-ethernet speeds, while Wi-Gig frequencies have such poor propagation properties that they can travel only a few 10s of meters of open air, and cannot pass through walls and windows.
  - *Permissive – Public*: In own or allied territories, use of public 5G networks is an option. A defense organization can purchase a 5G network slice, which isolates its traffic. Data is encrypted, but VPNs can be used to enhance security. Wi-Fi security is not up to defense standards, but Wi-Fi could still be used. Wi-Gig is still applicable for use in computer centers that are completely under the control of the defense organization.

	Permissive – Private	Permissive – Public	Propagation Challenged	Contested
Systems	• Private 5G networks • Wi-Fi • Wi-Gig	• Public 4G/5G networks • Wi-Fi • Wi-Gig	• N/A	• Wi-Gig
Sub-Systems	• Massive MIMO • mmWave	• mmWave	• Massive MIMO • Beamforming	• Massive MIMO • Beamforming
Technologies	• Internet of Things • Lo-SWAP devices • Commercial EUDs	• Internet of Things • Lo-SWAP devices • Commercial EUDs	• Lo-SWAP devices	• Internet of Things • Dynamic spectrum access • Lo-SWAP devices

Figure 17.3 Opportunities for leveraging commercial technologies are dependent upon the operating environment. Acronyms: Wi-Gig – 60 GHz Wi-Fi [22]; EUD, End User Device; MIMO, Multiple-Input, Multiple-Output; SWAP, Size, Weight, and Power. Source: K. Gremban.

- *Propagation Challenged*: Most commercial technology does not deal with propagation challenges. The commercial approach is to either install additional infrastructure, or not provide service in propagation challenged areas. Wi-Fi is meant for local area coverage – 10s of meters – and Wi-Fi frequencies are not good for wide area coverage. Wi-Gig is propagation challenged to start with – which is why it is appropriate for indoor, close-range connectivity.
- *Contested*: Commercial technology was not developed to deal with electronic warfare (EW) threats. However, Wi-Gig is still appropriate for use in mobile command centers. Wi-Gig enables close range, high bandwidth connectivity and is immune to medium and long-range EW because of the poor propagation properties of the frequencies used.

- Sub-systems – several commercial technologies are ripe for leveraging by defense communications. In particular, massive Multiple-Input, Multiple-Output (MIMO) arrays, and beamforming.
  - *Permissive – Private*: Even in a private network, the use of massive MIMO can improve end-user experience and increase spectral and energy efficiency. These arrays, which are part of many standards including Wi-Fi and LTE, significantly increase the capacity of radio links. They can also be used for beamforming. mmWave, considered to be frequencies in the 30–100 GHz range, is already part of the 5G standard, and provides dramatic increases in the data rates offered by a system.
  - *Permissive – Public*: If using a public 5G network, defense organizations should take advantage of mmWave, if it is available by the public provider, due to the vastly increased data rates.
  - *Propagation Challenged*: Massive MIMO arrays can improve data rates and spectrum efficiency. Beamforming can extend range and provide increased communications security.
  - *Contested*: Massive MIMO arrays can help overcome attempts at jamming by providing multiple paths and redundant signals. Beamforming significantly reduces the susceptibility to jamming and interception.
- Technologies – commercial devices have made incredible progress in reducing size, weight, and power (SWAP) of end-user devices. Rather than develop defense-specific technologies, defense organizations should take advantage of what is available in the commercial sector. This is increasingly important in the IoT world, in which devices may be deployed and left in place for long periods of time.
  - *Permissive – Private*: On a private network, such as one servicing an installation within the United States, virtually the entire range of commercial communications may be taken advantage of. Particular technologies that can be leveraged, in addition to systems and subsystems above, are low-SWAP devices, and commercial end-user devices (EUDs). A military base with a private network is a perfect environment for establishing a smart installation with IoT technology.
  - *Permissive – Public*: Use of a public network requires additional security features added into all the component technologies. However, low-SWAP devices and commercial EUDs can still be leveraged, and the concept of a smart installation still viable.
  - *Propagation Challenged*: Commercial technologies are not designed to operate in propagation challenged environments. However, given the additional challenges of deploying and servicing IoT devices in operational environments, defense organizations should leverage low-SWAP technologies.
  - *Contested*: Again, given the challenges of deploying and servicing devices, leveraging low-SWAP devices, with the appropriate security added on, would be very useful to defense organizations. Additionally, since spectrum use is unpredictable in a contested environment, dynamic spectrum access technologies would provide significant operational benefit.

17.5 Conclusion

In summary, the defense and homeland security sector has lost the leadership in radio communications due to simple economic drivers. Commercial investment in wireless technology dwarfs that of the defense and homeland security sector. Hence, it is much more efficient to invest in configuring commercial equipment, subsystems, or technologies to meet defense needs than to develop a completely separate technology base.

References

1 https://en.wikipedia.org/wiki/Internet_of_things (accessed 7 August 2022).
2 https://www.cs.utah.edu/birth-of-the-internet (accessed 7 August 2022).
3 https://techjury.net/blog/how-many-iot-devices-are-there (accessed 7 August 2022).
4 https://purple.ai/blogs/history-wifi (accessed 7 August 2022).
5 https://mobilityarena.com/first-mobile-phone-with-internet-access (accessed 7 August 2022).
6 https://www.census.gov/library/stories/2018/08/internet-access.html (accessed 7 August 2022).
7 Weldon, M.K. (2016). *The Future X Network: A Bells Labs Perspective*. CRC Press, Taylor & Francis Group.
8 https://www.emergenresearch.com/industry-report/military-communication-systems-market (accessed 7 August 2022).
9 https://www.zionmarketresearch.com/report/wireless-telecommunication-services-market (accessed 7 August 2022).
10 AN/PRC-117F(C) Multiband Multimission Radio - Applications Handbook. Available at https://manualzz.com/doc/22460132/an-prc-117f-c--multiband-multimission-radio-applications (accessed 7 August 2022).
11 https://firstnet.gov (accessed March 2022).
12 https://www.firstnet.com (accessed March 2022).
13 https://en.wikipedia.org/wiki/Category:Military_radio_systems_of_the _United_States (accessed 7 August 2022).
14 https://modernbattlespace.com/2018/03/14/the-demand-for-military-interoperability-is-resulting-in-modular-comms-approaches/ (accessed 8 August 2022).
15 https://www.statista.com/statistics/1016292/mobile-network-coverage-worldwide-by-regional-type-urban-rural (accessed 7 August 2022).
16 https://www.zdnet.com/article/100-of-4g-networks-vulnerable-to-denial-of-service-attacks-researchers-claim [accessed 7 August 2022]
17 https://techcrunch.com/2019/02/24/new-4g-5g-security-flaws/ (accessed 7 August 2022).
18 https://www.cisco.com/c/en/us/products/wireless/what-is-wi-fi-security.html (accessed 7 August 2022).
19 https://military-history.fandom.com/wiki/SINCGARS (accessed 7 August 2022).
20 https://aida.mitre.org/blog/2020/04/01/jtrs-a-cautionary-tale-for-today/ (accessed 7 August 2022).
21 DoD 5000 Series Acquisition Policy Transformation Handbook, February 5, 2021. https://www.acq.osd.mil/asda/ae/docs/DoD%205000%20Series%20Handbook%20(09%20Feb %2021).pdf (accessed 27 October 2022).
22 https://www.wi-fi.org/discover-wi-fi/wi-fi-certified-wigig (accessed 7 August 2022).

18

Military IoT: Tactical Edge Clouds for Content Sharing Across Heterogeneous Networks

Tim Strayer, Sam Nelson, Dan Coffin, Bishal Thapa, Joud Khoury, Armando Caro, Michael Atighetchi, and Stephane Blais

Raytheon BBN, Cambridge, MA, USA

Abstract

Just as inexpensive and powerful devices are driving ecosystems of connected *Internet of Things* in work and home environments, the battlefield has experienced a proliferation of and reliance on networked devices. However, current networking technologies are ill-suited for content sharing in these emerging military networks where fixed infrastructures and constant connectivity cannot be assumed. The communication paradigm, content-based networking, is proving to be a highly effective solution for operation in mobile infrastructure-less environments where intermittent and disrupted connectivity is expected. We present approaches and tradeoffs for content-centric military IoT architectures that facilitate generation and dissemination of content in challenging *tactical edge* environments. These architectures are designed to address information flow across different underlying tactical data links by managing the dissemination of mission-critical information across an overlay network optimized for disconnected, intermittent, and limited (DIL) connectivity operations.

18.1 Introduction

The proliferation of inexpensive and powerful embedded computational and networking capabilities is driving the explosion of *things* connected to the internet, enabling special purpose sensors and actuators to participate as first class networked devices. We see the internet of things (IoT) most commonly in smart household devices like smart speakers, environmental controls, lighting, and home safety and security. These devices provide users with increased awareness and automated control of work and home environments.

In a similar way, the modern battlefield is turning to sophisticated forward-deployed sensors to build a better, more nuanced view of situational awareness (SA) for the warfighter. This is a modern evolution of the classic Boyd observe, orient, decide, act (OODA) loop [1] from human-centric *observations*, *orientations*, *decisions*, then *actions*, to one where decisions and actions are driven by mission-aware, intelligent data acquisition and dissemination. Such intelligent information sharing is central to visions for fully connected, fully cooperative battlefields such as described by the Department of Defense (DoD's) JADC2 (joint all-domain command and control) vision,

IoT for Defense and National Security, First Edition. Edited by Robert Douglass, Keith Gremban, Ananthram Swami, and Stephan Gerali.

and in initiatives to realize this vision within the service branches with Project Overmatch for the Navy, Project Convergence for the Army, and Advanced Battle Management System (ABMS) for the Air Force.

As military operations become increasingly interconnected, interdependent, and challenged, military information needs are also growing in both scale and complexity. This drives an expanding reliance on richly connected devices – a *military IoT* – at the tactical edge.

Yet, where the civilian ecology of IoT devices benefits from a well-resourced and well-connected infrastructure both at home (Wi-Fi and broadband) and away (LTE and 5G), the military environment is much more challenging. In the civilian model, for example, IoT devices can take advantage of cloud services, where local devices connect to a central data collection site from which users can view the data and to which users can issue commands. Stable, ubiquitous connectivity and always-available access makes this possible. None of these are likely available in combat situations.

Rather, networking at the tactical edge is better characterized as disconnected, intermittent, and limited (DIL) bandwidth [2]. By its nature, the battlefield is a hostile environment, where natural phenomena (landscape, atmospherics, interference, etc.) as well as adversary actions disrupt communications. Communications systems are also often customized for the roles of their users. Each generation of modern aircraft, for example, builds new and very sophisticated radios, so communication across the same platform is secure and efficient and meets the combat needs of that platform, but the very design decisions that enable this sophistication often inhibit interoperability. Further, there are also challenges associated with information protection and access control.

The networking models that serve the internet so well, therefore, are not necessarily appropriate for SA data dissemination at the tactical edge. End-to-end connection-oriented protocols such as TCP are highly optimized for the Internet, yet are entirely ineffective in a DIL environment due to intermittent and low bandwidth links where disrupted connectivity is frequent and unexceptional [3]. While civilian IoT is generally cloud-based today, the underlying paradigm for the cloud still follows a client-server model, which also exposes points of failure in the face of disruption.

Consequently, collection and dissemination of SA data generated and consumed by military IoT devices at the tactical edge require a *new cloud* framework to serve the operational needs for this data. In order to achieve this operating vision, a number of challenges must be addressed including: the ability to store, share and exchange information in DIL environments, mission-aware and network-aware routing to queue and prioritize information and security and trust mechanisms for the information being disseminated. Key to meeting these challenges is novel communications paradigm, called content-based networking (CBN) [4], that considers content, not hosts, as the addressable entity in a network. This replaces the client-server model with one where applications request content from the network directly, via the content name, without necessarily having to know where the content resides.

Raytheon BBN (originally Bolt Beranek and Newman Inc., acquired 2009) has been studying *tactical edge clouds* over the last 10 years, first with ground forces under the DARPA Content-Based Mobile Edge Networking (CBMEN) program [5], then with aerial forces under the DARPA Dynamic Network Adaptation for Mission Optimization (DyNAMO) program [6]. These programs provide key insights into architectural goals and design decisions for constructing a general *tactical edge cloud* framework.

In this chapter, we first motivate the need for a tactical edge cloud framework that builds an ecology of military IoT devices by considering the benefits from wider and more expressive information distribution. In Section 18.3, we introduce two worked models for tactical edge clouds that we will use in Section 18.4 for deriving insights and observations from our experience with those systems.

18.2 The Need for Tactical Edge Clouds

The foundation of successful mission execution is constant and accurate SA information. SA is the accumulation of observations that help in understanding of the area of operation. It aids in achieving the mission by facilitating better decision-making while executing the mission, as well as an increased ability to anticipate and deal with contingencies that may cause deviation from the mission plan.

The critical need for reliable SA information is nicely described in the venerable OODA Loop concept. The OODA Loop suggests that good decisions are made from good observations and understanding of those observations. One way to think about this is to view the battlespace as an infinite time series of battle states. The *observation* and *orientation* steps are essential to understanding the relevant state of the environment and how it is changing over time. The next steps – *decide* and *act* – comprise using that understanding of the environment to transition it from the current state to a more desirable future state.

The steps of the OODA Loop are shown in Figure 18.1, where the OODA Loop is unwound into a coil to show progression of time. The observation step is the act of collecting SA information. In reality, the real-world current state is too complex and much harder to describe, but for the purposes of the OODA Loop, the current state is approximated as the compilation of all collected SA data. This current state collection of SA, weighted for importance, is typically called a common operating picture, or COP. The effectiveness of the OODA Loop in manipulating the current state into a new (more desirable) state is predicated on the availability, timeliness, relevance, authenticity, and accuracy of the observations of the current state. That is, the observer must be able to trust that the information can be relied upon to support making the decisions that lead to taking the action.

Before moving forward with the next steps of the OODA Loop, let's consider the COP concept for a moment. The effectiveness and completion of the COP is subject to four potential issues. First, the COP is only as good as the collected (and possibly inferred) SA data. If there is no access to relevant SA data, either because there is nothing reporting that data or because the collection of the data is delayed or disrupted, the COP cannot be a full picture of the current state. Second, since the COP is a collection of observations from sensors (including humans manually inputting data) and processed data derived from sensors, the COP is essentially a database. So, it has to adhere to the consistency, availability, and partition tolerance (CAP) theorem [7], which essentially states that any distributed system can only achieve two of these three properties at any given time. If the database is centralized, then it is a single point of failure; if it is distributed, then there is a risk that essential parts are not available when needed. Third, the fidelity of the COP has the potential to increase as the amount of data increases, but not all information within the COP is necessarily useful – and is potentially overwhelming – to a particular user with a particular need. In this way, a *relevant* current state may be a subset of the full current state represented by the COP. Finally, the

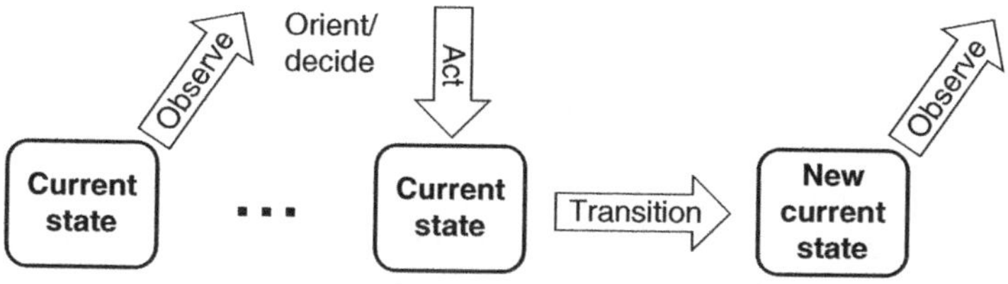

Figure 18.1 The OODA Loop is shown as a coil progressing through time, where actions create new states.

data comprising the COP may not form a consistent representation, particularly if two observations disagree with each other. Either the COP must allow for some uncertainty and, therefore, latitude in the degree to which the COP is accurate, or there must be some authority that adjudicates the contradictions. Inconsistencies and contradictions reduce believability. Note, however, that a contradiction may only be in the eye of the beholder; two observations may be completely true within the context of the observation. Related to this is determining when data is no longer useful, and should be expunged from the COP.

Diving back into the OODA Loop as depicted in Figure 18.1, the *orient* step is where the user applies sense-making to the SA data in the COP. This step is related to the fourth issue above – the believability of the SA data, and how to interpret it. The user's contextual profile, such as the user's role, where the user is physically, the provenance of the data, and what the mission plans are for the user, all factor into how the user orients around the SA data. Examples of orienting include weighing how age or resolution of SA data use, say, for blue force tracking to avoid friendly fire, or images to assess battle damage. When making sense of information, often a user's past experience from similar situations or mental models will influence how the information is viewed; hence this step is where bias can be injected into the system. These factors help shape judgment of the situation, which is a primary input into the decide step.

The *decide* step takes the oriented observations (that is, COP data processed for the user's context) and judgments obtained in the first two steps, along with a set of possible actions, and determines which among those actions is most likely to change the current state into more favorable state, which will then (hopefully) become the new current state. The *act* step executes that decision.

With the advent of a new current state, the COP needs to be updated. This necessitates a new round of observations by sensors and humans, then orienting, then making new decisions that move the state closer to the accomplishing the mission.

Enabling the collection of SA data into a COP that is then used to support decisions and actions is difficult enough in a homogeneous and stable communications environment. At the tactical edge with network heterogeneity, disruptions, varieties of users, and changing mission priorities, the challenges drive toward the need for a novel tactical edge cloud architectural framework.

18.3 Two Architectures

Under DARPA funding, we developed and tested two worked large-scale examples of tactical edge clouds, the first under the CBMEN program with emphasis on communications for ground forces, and the second under the DyNAMO program for solving interoperability issues in aerial networks. The architectures of these two systems are presented here as the experiential basis for drawing design insights and observations in the next section.

18.3.1 Architecture Paradigm 1: DARPA CBMEN

The CBMEN system [8, 9] is an efficient and secure CBN layer developed as part of a research initiative focused on small tactical units operating under harsh communications conditions and no reliable infrastructure, shown in Figure 18.2.

By utilizing recent increases in storage, processing, and communications within mobile devices, CBMEN is able to dynamically create communities, or mobile storage clouds, where devices participating in the network can cooperate to store, serve, and forward content. To do this, the architectural model exploits the tendency for mobile nodes to cluster into small groups. Military

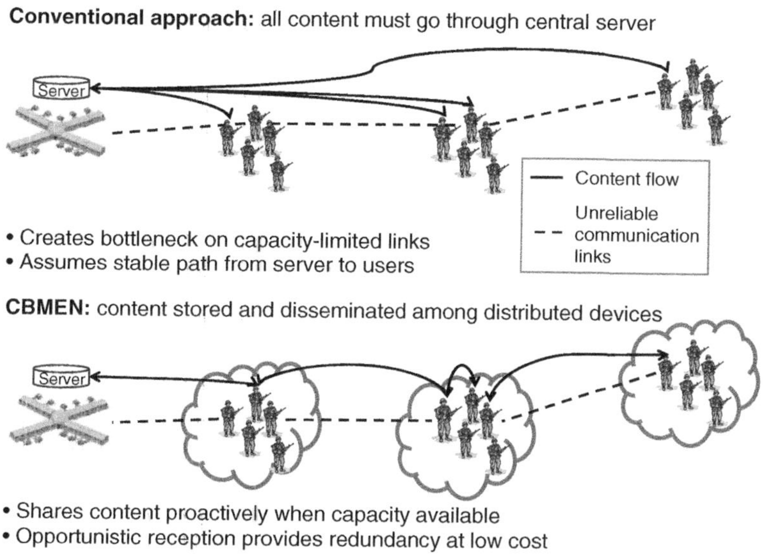

Figure 18.2 CBMEN Program Insight: the conventional approach requires reach-back for all content. CBMEN facilitates local proactive sharing among independent mobile units without reach-back.

squads and first responder teams are good examples of mobile clusters whose common purpose is to accomplish particular missions. The natural cohesiveness of the cluster means that the members will largely remain geographically near each other, even while mobile, which suggests that a mobile ad hoc network (MANET) covering the cluster members will be relatively stable for some time (that is, relative to the harsh environment). A stable network topology further suggests that it is reasonable to use cooperative techniques for storage and retrieval of collected content.

CBMEN also exploits the fact that this clustering creates two distinct types of interactions – the first is within the cluster itself, and the second is between clusters. While individual clusters may form their own MANETs, when clusters approach close enough to establish radio contact with each other, they can form connections that allow the exchange of content. Consequently, we can view the network as having two distinct parts, an *intra-community MANET* and a less stable *intercommunity synchronization*, and can apply appropriate CBN techniques to each.

The CBMEN modular architecture is shown in Figure 18.3. The majority of CBMEN resides between the Application Interface and the Network Interface; these modules manage the content and security functions. Above the Application Interface are the content-aware and legacy applications. Content-aware applications are applications that interact with the network via an interface that allows content to be published, subscribed to, and described. The architecture also provides a shim, called the Mediator, as a translation engine to allow legacy, host-based applications to use the content-aware system. Below the Network Interface are specific MANETs used to transport the content.

The Content Distribution System controls the propagation and routing of content and queries, and deals with network placement and retrieval of uniquely named content objects. Content distribution is agnostic to what is inside a content object; it only requires that the object have a persistent, location-independent, globally unique identifier that allows content distribution to intelligently place the object in the network and efficiently retrieve it later on.

The Content Management System employs an open interface to applications, and implements the content-oriented publish and subscribe functions. The content management interacts with the

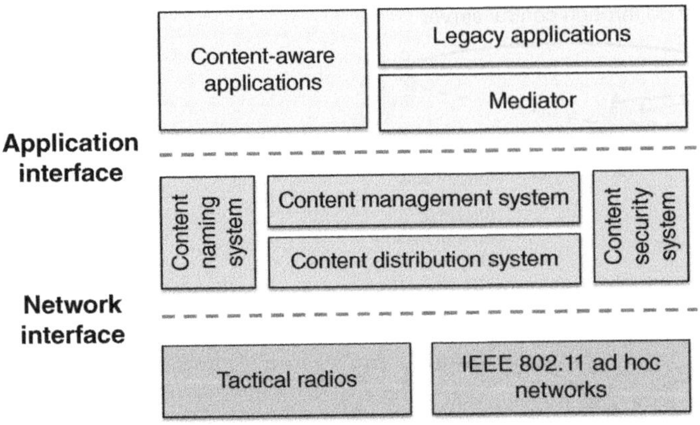

Figure 18.3 The CBMEN Node Architecture supports legacy and new application, and utilizes multiple network technologies. The content and security functions manage the collection and dissemination of the information.

Content Naming System to marshal the metadata used to describe content. Content management uses the Mediator shim to allow legacy applications to operate within the CBMEN system. As a consequence, content management facilitates any application sharing content with any other, even if they were designed to work in a stove-piped fashion and only with content of the same format. Finally, content management prioritizes content based on its relevance to the users of the system.

The Content Security System provides a policy-based access control mechanism for CBMEN using ciphertext-policy attribute-based encryption (CP-ABE) [10, 11]. Since CBMEN uses direct naming via metadata, there is an opportunity for the metadata to reveal information about the content, even when the content is encrypted with an access policy. Our solution is to use CP-ABE over the metadata and its constituent fields as well, requiring the querying node to have access rights to the metadata before being able to access the content [12].

Throughout the development and testing of the CBMEN system, including hundreds of hours of emulation and over-the-air testing using the full code base within Android handheld devices, several important lessons were learned that inform our evolving understanding of IoT in the battlespace.

First, CBN is an effective paradigm for addressing challenges in an infrastructure-less environment due to three important characteristics: (i) CBN replaces the client-server model and breaks the need for the content creator and the content consumer to be present on the network at the same time. (ii) CBN uses content names rather than node addresses, naturally supporting semantically rich query mechanisms based on descriptions of the content. (iii) Content centricity facilitates interoperability between very different applications (this allows the Mediator to work).

Second, the community model, where more stable regions of the network implement distributed storage, is effective in balancing the advantages of a pure *pull* approach (all nodes essentially form one community, and content is served from a central location) with pure *push* (every node is its own community, and all content is epidemically spread through the network). Importantly, a community closely matches small units, such as military squads or groups of first responders, since these units generally stay close together, hence forming stable regions.

Third, the two-part nature of the deployment model supports scalability and opportunistic content dissemination. Small, tightly connected clusters are stable enough to support intra-cluster distributed content sharing. Adding more clusters has little impact on content sharing resources.

Since clusters are mobile, content is spread to other clusters by inter-community synchronization when communities come within network range of each other.

Fourth, access control can be effectively implemented over both the content and the metadata. Since metadata describes the content, unprotected metadata leaks information about the content, even if the content itself is protected. CBMEN's access control method ensures that only those who legitimately have the ability to access the content can also access the metadata, and even then, the individual metadata fields can be protected with different policies.

Next, the Mediator module provides a straightforward mechanism for integrating a wide variety of legacy applications into the CBMEN system, even applications that do not themselves adhere to a content-oriented model. This integration happens without modifying either the application or CBMEN. In this manner, stove-piping is reduced, and even seemingly incompatible applications can be made to share content. This further strengthens the argument that CBN is more appropriate than host-based addressing for supporting application integration.

Finally, prioritizing content with relevance models or even with statically assigned priorities is most helpful during cluster synchronization because there is contention for the communication channels over an unknown time window. Since synchronization is opportunistic, and clusters can separate at any time, the time budget for transfer is unknown, so the most useful content needs to be transferred first.

18.3.2 Architecture Paradigm 2: DARPA DyNAMO

The second architectural paradigm is the DyNAMO system [13]. The goal of DyNAMO was to create an information overlay that allows fighter aircraft to exchange mission-related situational information by including "DyNAMO nodes" within the aircraft, forming a theater-wide network to facilitate cross-platform communication, as shown in Figure 18.4. Specifically, DyNAMO aimed to address three important limitations of current airborne networks: (i) the lack of interoperability among the existing tactical data links, inhibiting information sharing among different aircraft; (ii) the ability to dynamically adapt to conditions in contested environments; and (iii) the ability to manage the information flow to explicitly meet the goals of the mission.

The DyNAMO architecture is viewed as a two-tier stack that enables dynamic and mission-aware sharing of information across multiple security domains in an optimized manner. The *Information Gateway* is the upper part of the stack, providing semantic brokering, mission-centric information shaping, and CBN through a publish–subscribe (pub–sub) interface that supports many different

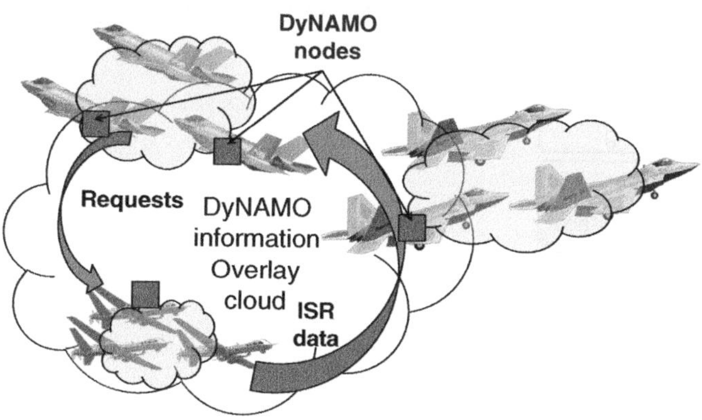

Figure 18.4 DARPA DyNAMO information overlay cloud.

types of applications. It interfaces with applications and manages matching generated data with those remote applications that request that data. The *Network Optimizer* is the lower part of the stack, operating at the network layer to optimizes information dissemination across multiple disparate tactical data links. It interfaces with the tactical data links, understands their capabilities, and forms an overlay network that adapts as the underlying data links change. The interface between these two tiers negotiates service quality so that the Information Gateway can shape the offered data for best fit. Together, the Information Gateway and the Network Optimizer disseminate the right information to the right place at the right time, in a manner that continuously changes based on mission need and dynamic network conditions.

The Information Gateway has five main functional components, as shown in Figure 18.5. The *App Interface* is the point of entry for data from on-board sensors, and the point of delivery for data to devices that consume the data. The interface supports an extensible number of formats through adapter plugins that parse a stream of bytes from the applications into a common structured syntactic representation, then maps the local field names and values into a common ontology capable of describing a rich set of semantic relationships. The *Data Discovery and Matching* function creates a content-based network for managing the data, similar to CBMEN. It is responsible for matching newly generated data published into the system with data needs that are expressed through subscriptions. This module supports rich semantic description and querying. Subscriptions, then, use the expressive SPARQL [14] querying language to request the appropriate data. The subscriptions also specify acceptable maximum and minimum resolutions to facilitate data shaping. The *Data Mission Utility* function allows DyNAMO to incorporate the mission into decisions about how to deliver the data. Typically, the use and structure of military networks are planned well in advance of the mission, and only grossly consider the mission goals. In DyNAMO, mission-centricity is an explicit feature; DyNAMO nodes use mission plans for data prioritization and shaping.

The *Service Negotiation* function works through the Information Gateway-Network Optimizer Interface to meet the data delivery requirements. If the network capacity is less than requested, the *Data Shaping* function reduces the resolution of the data to meet the capacity, if that still fits within the minimum data delivery requirements for that subscription; if not, the data is not sent. If the data can be sent, the *Data Transfer* function engages the Network Optimizer through the shared interface, and the data is placed onto the DyNAMO overlay through the Network Optimizer.

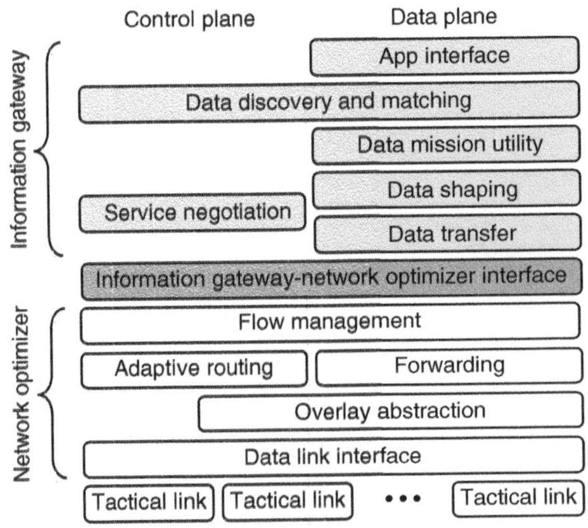

Figure 18.5 DyNAMO's Information Gateway and Network Optimizer work together to create a bridge for data exchange across tactical data links.

The Network Optimizer also has five main functional components as shown in Figure 18.5. *Flow Management* provides a mechanism for the Information Gateway to gain information about the state of the network and the viability of supporting potential traffic flows. Based on the results of the negotiation with the Information Gateway, it sets up flows with the desired services and service quality to the extent possible given the current network conditions. The *Adaptive Routing* and *Forwarding* modules contain a number of innovative techniques for quality of service (QoS)-aware routing state establishment, exploiting configurability at the media access control (MAC) and physical (PHY) layers of an underlying communication system. The *Overlay Abstraction* module creates a virtual link between peer DyNAMO nodes, each virtual link being realized as a (possibly multi-hop) path within a particular tactical data link network. Typical routing control and forwarding can happen over these virtual links in the same way as over direct links, modulo the dramatically different link characteristics. Finally, the *Data Link Interface* provides a generalized mechanism for the Network Optimizer to send and receive packets and information across the different types of tactical data links housed within the DyNAMO node, and provides the ability to query and adjust the data link capabilities.

This architecture provides DyNAMO with the ability to effectively and efficiently exchange information in contested and denied network environments. Like CBMEN, these capabilities have been extensively evaluated in emulated and over-the-air environments, and have been shown to provide significant promise in reducing sharing latencies and in scaling to realistic regimes.

18.4 Tactical Edge Cloud Architectural Insights

Drawing upon our experience with both CBMEN and DyNAMO, we were able to perform a philosophical critique of the effectiveness and suitability of our design choices, and where these systems would benefit from further enhancement.

18.4.1 Information Generation and Discovery

Our experience with CBMEN and DyNAMO argues strongly for using a CBN approach as the base framework for a tactical edge cloud. CBN considers content, not hosts, as the addressable entity in a network. This replaces the client-server model with one where applications request content from the network directly, via the content name, without necessarily having to know where the content resides. The content address, then, is a description or naming of the content. CBN works well with the pub–sub paradigm. Publications describe the newly created information, and subscriptions express the information needs.

CBMEN and DyNAMO are both based on CBN with a pub–sub interface, where one or more pieces of metadata describe the content. The vocabulary used for the metadata comes from an ontology, which defines the terms and inclusive relationships among the terms – MIL-STD-2525B [15] for CBMEN and open mission systems (OMS) universal command and control interface (UCI) [16] for DyNAMO, although both systems are readily extensible to other ontologies.

Augmenting CBN and the pub–sub interface with semantic web technologies [17] creates a powerful information generation and discovery model that can be thought of as a pervasive *knowledge fabric* for the tactical edge cloud, supporting the first steps of the OODA loop. In this way, the knowledge fabric has three roles, as shown in Figure 18.6. It is the underlying *information-centric network* for efficiently collecting and distributing observations; it is a *distributed data store* that

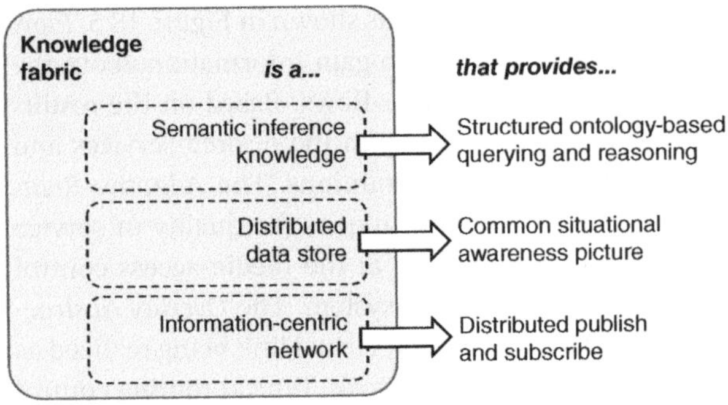

Figure 18.6 Three purposes of a knowledge fabric.

congeals these observations of the current state into the COP; and it is the *semantic inference knowledge* represented by the data contained in it, facilitating higher level reasoning – orienting and deciding – over the collected common picture. As an abstraction, the knowledge fabric is the point of coordination for all system components; no matter where in the battlespace these components reside, all information is available through the same semantically rich pub–sub interface.

There are, broadly speaking, two models for providing information discovery within a content-based network. The first is the advertisement-oriented model. In this model, the availability of newly published information is advertised throughout the network. When this advertisement encounters a subscription for which it is a match, the subscriber seeks the content directly from the publisher. This is the model that CBMEN uses. Because CBMEN is a collection of mobile clusters that only occasionally synchronize with each other, the publishers and subscribers are all within the same relatively small cluster. Also, for resiliency, CBMEN relies on the cluster to create a distributed store for the content, so the content does not necessarily reside at the publisher. This approach is highly disruption tolerant, but can over-utilize network resources, which may make it less enticing for large data streams.

The second model is subscription oriented. Here, when particular information is need, that need is expressed as a subscription, and the subscription is distributed throughout the network. When the subscription encounters a publisher with content that can satisfy the need, that content is sent back to the subscriber. This is the model that DyNAMO uses. Because DyNAMO creates an overlay network, there is less notion of clustering or a distributed store, so content is by default stored at the publisher. Furthermore, when new content is published, there is no guarantee that a subscriber is waiting, while a new subscription does guarantee a need for certain content. Even if there is no current content to match the subscription, each node with the potential to generate such content can note the subscription and answer it as soon as the content is available. This approach can be less disruption tolerant, but is much more friendly to network resources, and very suitable for larger data streams.

Another insight drawn from our experience is the importance of including legacy applications and devices in the overall tactical edge cloud architecture. It is a fairly sure bet that most or all legacy systems will not, at least at first, adhere to the interface of any system that seeks to link them all together. The choice is either to force a refactoring of the legacy interfaces to match the new standard, or to create a mediator shim that is easily customized to natively interface with the legacy applications and devices. At first it may seem a daunting task to customize an interface for every legacy system, but two observations help mitigate the size of the problem. First, there really are only a finite number of military applications and devices in use, and they can generally be grouped so that one mediator approach can serve many similar systems. Second, the tactical edge cloud's pub–sub interface is reasonably straightforward – data comes in, data goes out, and data is tagged with descriptors – so the mediator shim is not very complex.

18.4.2 Information Availability

Information availability largely relies on the ability for the underlying network to heal and restore service. In CBMEN, which is more cluster-oriented, explicit replication and distributed storage of the content across the nodes in the cluster provides robustness against network bifurcation within a cluster. In DyNAMO, the Network Optimizer module uses overlay routing algorithms to adapt to traffic changes and network dynamics while increasing scalability and resilience. In both cases, subscriptions that cannot be satisfied due to disruptions are immediately processed once service has been restored or a new path has been found.

When content is created and published, it must be stored so that requests can be satisfied. An obvious choice is to leave the content with the publisher, but this does not help with balancing storage requirements if one or more publishers are particularly prolific. As content moves through the network to satisfy queries, it may also need to be opportunistically cached at nodes along the route.

Caching is at the heart of disruption tolerance, and efficiently using caches in content-centric networking makes information more available, survivable, and rapidly accessible. Caching is particularly useful in conjunction with the pub–sub paradigm, which provides inherent decoupling of producers and consumers. This allows producers to make information available to currently unreachable consumers, and even consumers who will not be interested in the content until sometime in the future. In other words, caching bridges producers and consumers not only in the space domain, but also the time domain.

Caching can be done reactively or proactively. Reactive caching, which was the primary mechanism of the DyNAMO program, allows nodes (including the publisher) to opportunistically cache information as it flows through the system, en route to an immediate consumer. Since the content is already en route, this results in no extra network resource use, other than storage at the intermediate caching nodes; it is therefore highly efficient. Proactive caching, which was the primary mechanism of the CBMEN program, allows the publisher to proactively distribute information to strategic locations in the network even if those locations were not on the path to a node that requested the information. This is a powerful technique that can significantly increase availability and decrease latency in disruption-prone environment; however, used incorrectly, it can hurt efficiency by unnecessarily consuming network resources.

The insight for a tactical edge cloud is that both proactive and reactive caching are advantageous, but appropriately balancing reactive and proactive caching is an ongoing research challenge, with many factors to take into consideration including information type, network stability, network resources, and interested parties.

18.4.3 Controlling Access

Protecting the confidentiality of published content and meta-data while retaining the benefits of content-based systems is particularly challenging since publishers are often oblivious of subscribers yet still want fine-grained control over who can access their published information. Under CBMEN, we developed a novel application of CP-ABE [10, 11] to support cryptographically enforced access control [12] within content-based networks. The publisher of content encrypts the content and its metadata using a policy defined over the access control attributes. Only users with attributes (and respective secret keys) that satisfy the policy are able to decrypt and access the content; publishers do not need to know exactly which nodes will receive the content in the future. The insight gained here is that the availability of information gained with dissemination and discovery inherent in content-based networks must be tempered with access control mechanisms that work within that paradigm, and CP-ABE provides the right tools to do so.

18.4.4 Information Quality of Service

Our experience with DyNAMO has shown that there are two important aspects to considering QoS in a tactical edge environment. The first is drawn from conventional networks, where QoS mechanisms are employed while setting up a connection signal to the underlying path how the data needs to be handled. In DyNAMO, multidimensional QoS (e.g. capacity, latency, etc.) is enabled by capturing data link characteristics and presenting them to Flow Management in Figure 18.5. In particular, routing state and data link capability is advertised by destinations periodically, and aggregated when possible as it progresses through the network. This allows DyNAMO to establish flows that meet the needs of the subscription.

The second aspect is the use of mission plans for data prioritization and data shaping. By using the capacity information provided by the underlying overlay network as a QoS parameter, the tactical edge cloud can fit as many flows as possible in priority order. Our experience on DyNAMO further suggests that *shaping* the content – that is, reducing its resolution in either frequency or size – will increase the likelihood that the data can be transported on a network (reduce network demand) or that it can be consumed by the target destination (match the destination's capabilities). The mission requirements dictate different shaping levels that are acceptable, ensuring compatibility and usefulness of all shaping levels. Prior to dropping a flow due to lack of capacity, the tactical edge cloud attempts to shape all current flows to make room for the new flow.

18.4.5 Information Importance

Related to QoS is the abstract concept of *importance*. The importance of a piece of data or a task can be measured with respect to accomplishing an objective. This measure allows the importance of two items to be compared to determine that one is "more important" than the other. If there is a way to determine the importance of information, then we can state the following lemma: If, at every decision point, the "most important" information gets preference then, by definition, those conducting the mission plan will get the information most important to the mission objective [18].

Efficient resource utilization and information prioritization are key to delivering the highest value of information to subscribers. There are three primary components to this:

1. Assigning and utilizing globally agreed upon priorities for information objects
2. Purging the least useful information from caches when resources become constrained
3. Dynamically updating the global values that affect prioritization and cache removal policy

The key insight supporting these components in a decentralized tactical edge cloud system is to utilize the centralized information found in the mission. In DyNAMO, the module responsible for understanding the mission is the Data Mission Utility function (recall from Figure 18.5). As with QoS, the mission can force global consensus on information priorities. To address how information should be dropped from caches, the system must contain a cache management policy, which will assign a relative importance value for each information object in that particular cache. When the cache runs out of space, least important objects are dropped first. There are three factors that will account for the importance: (i) the global priority of the content; (ii) how "rare" the object is locally; i.e. how many caches hold the object; and, (iii) how popular the object is locally based on the request trend.

18.5 Summary

The civilian IoT model has a lot of parallels at the military tactical edge, notably networked devices throughout the battlefield providing information for increasing SA and facilitating mission execution. Our experience on two major DARPA programs, CBMEN and DyNAMO, allows us to draw conclusions about how to create tactical edge cloud architectures for military IoT.

A clear conclusion is that one must view these systems from the top down, realizing that information is key, and creating systems that focus on information availability and discovery. In CBMEN, that discovery is advertisement-oriented, where in DyNAMO it is subscription-oriented. We have explored both and have found that the deployment environment drives which is more efficient. Information also comes from many different sources and in many different formats, so a military IoT framework must include flexible and adaptive interfaces to legacy applications and devices in the overall tactical edge cloud architecture. Making the information available also means taking advantage of both proactive and reactive caching opportunities, and balancing the storage and network investment against potential efficiency gains. Availability of information also requires fine-grained access control mechanisms that work within that architecture.

Mission effectiveness is driven by good decisions, and good decisions are driven by good information. The battlefield of the future will rely increasingly on collecting and disseminating this information. Our work shows that content-centric tactical edge cloud architectures support military IoT.

Acknowledgment

The views, opinions, and/or findings expressed are those of the authors and should not be interpreted as representing the official views or policies of the Department of Defense or the U.S. Government. Distribution Statement "A": Approved for Public Release, Distribution Unlimited, Case 35865.

References

1 Boyd, J.R. (1996). The essence of winning and losing. *Unpublished Lecture Notes* 12 (23): 123–125.

2 Scott, K., Refaei, T., Trivedi, N. et al. (2011). Robust communications for disconnected, intermittent, low-bandwidth (DIL) environments. *2011-MILCOM 2011 Military Communications Conference*, pp. 1009–1014.

3 Padhye, J., Firoiu, V., Towsley, D., and Kurose, J. (1998). Modeling TCP throughput: a simple model and its empirical validation. *ACM SIGCOMM Computer Communication Review* 28 (4): 303–314.

4 Carzaniga, A. and Wolf, A.L. (2001). Content-based networking: a new communication infrastructure. In: *Workshop on Infrastructure for Mobile and Wireless Systems* (pp. 59–68). Springer: Berlin, Heidelberg.

5 Gremban, K. (2012). Content-Based Mobile Edge Networking (CBMEN). *DARPA Program Information*. https://www.darpa.mil/program/content-based-mobile-edge-networking (accessed 31 October 2022).

6 Kofford, A. (2016). Dynamic Network Adaptation for Mission Optimization (DyNAMO). *DARPA Program Information*. https://www.darpa.mil/program/dynamic-network-adaptation-for-mission-optimization (accessed 31 October 2022).

7 Gilbert, S. and Lynch, N. (2002). Brewer's conjecture and the feasibility of consistent, available, partition-tolerant web services. *ACM SIGACT News* 33 (2): 51–59.

8 Strayer, T., Nelson, S., Caro, A. et al. (2018). Content sharing with mobility in an infrastructure-less environment. *Computer Networks* https://doi.org/10.1016/j.comnet.2018.07.021.

9 Strayer, T., Kawadia, V., Caro, A. et al. (2013). CASCADE: Content access system for the combat-agile distributed environment. *Military Communications Conference, MILCOM 2013-2013 IEEE*, pp. 1518–1523.

10 Bethencourt, J., Sahai, A., and Waters, B. (2007). Ciphertext-policy attribute-based encryption. *Procedings of the 2007 IEEE Symposium on Security and Privacy*, pp. 321–334.

11 Waters, B. (2011). Ciphertext-policy attribute-based encryption: An expressive, efficient, and provably secure realization. In: *International workshop on public key cryptography* (pp. 53–70). Springer: Berlin, Heidelberg.

12 Khoury, J., Nelson, S., Caro, A. et al. (2014). An efficient and expressive access control architecture for content-based networks. *IEEE Military Communications Conference (MILCOM)*, pp. 1034–1039.

13 Strayer, T., Ramanathan, R., Coffin, D. et al. (2019). Mission-centric content sharing across heterogeneous networks. *2019 International Conference on Computing, Networking and Communications (ICNC)*, pp. 1034–1038.

14 Prud'hommeaux, E. and Seaborne, A. (2008). SPARQL Query Language for RDF. W3C, W3C Recommendation, January 2008.

15 MIL-STD-2525B (1999). Department of Defense Interface Standard: Common Warfighting Symbology, 30 January 1999.

16 U.S. Air Force Virtual Distributed Laboratory. Open Mission Systems (OMS). Universal Command and Control Interface (UCI). https://www.vdl.afrl.af.mil/programs/oam/uci.php.

17 Berners-Lee, T., Hendler, J., and Lassila, O. (2001). The semantic web. *Scientific American* 284 (5): 28–37.

18 Strayer, W.T. (1992). *Function-Driven Scheduling: A General Framework for Expression and Analysis of Scheduling*. University of Virginia.

19

Spectrum Challenges in the Internet of Things: State of the Art and Next Steps

Francesco Restuccia[1], Tommaso Melodia[1], and Jonathan Ashdown[2]

[1]*Department of Electrical and Computer Engineering, Northeastern University, Boston, MA, USA*
[2]*Air Force Research Laboratory, Rome, NY, USA*

Abstract

Thanks to the significant growth of the Internet of Things (IoT) and fifth-generation (5G) networks, the radio frequency (RF) spectrum is poised to become an extremely scarce resource in the next few years. On the other hand, although the IoT will require *timely*, *guaranteed* and *secure* information gathering/delivery, the vast majority of IoT devices are still utilizing protocols, architectures and algorithms that were not designed with reconfigurability and adaptability in mind. To tackle the next-generation spectrum challenges, instead, the IoT must be redesigned to allow spectrum agility "by design", without any manual and/or human intervention. Spectrum agility in the IoT, however, poses significant core challenges that cannot be addressed solely with state-of-the-art technologies. For example, traditional spectrum management systems are *centralized and inflexible*, thus not allowing for fine-grained real-time spectrum management. Moreover, existing spectrum sharing mechanisms do not consider *energy consumption* and *computational constraints* of IoT nodes, which is a fundamental requirement for long-term network survivability. In this book chapter, we will discuss the technical requirements and articulate the enabling technologies for such new generation of spectrum-sharing, energy-efficient IoT wireless devices. We conclude the chapter by discussing existing research challenges and future directions.

19.1 Introduction

The Internet of Things (IoT) has now become a integral part of the technology world's everyday vernacular. In short, the IoT broadly describes the concept of an interconnected network of physical objects, including machines, wearable, buildings, automobiles and anything that we can imagine. These connected "things" will bring new services and deliver new levels of efficiency and safety all around us – in homes, businesses, cities, and across industries. Despite billions of already-connected devices, we are only at the dawn of the IoT era [1], with annual revenues exceeding $470B for IoT vendors [2]. This expansion will be fueled by the rapid growth of exciting new IoT use cases and opportunities all around the world, such as supply chain and logistics, smart home and smart city, health-care, manufacturing, utilities, mining, commerce, surveillance, education, infrastructure management, and transportation, to mention a few [3].

IoT for Defense and National Security, First Edition. Edited by Robert Douglass, Keith Gremban, Ananthram Swami, and Stephan Gerali.

Although the sheer growth of the IoT represents a substantial step forward in humanity's technological advance, it is also a matter of fact that the IoT is quickly saturating industrial, scientific and medical (ISM) spectrum bands [4]. Although the international telecommunication union (ITU) has defined 12 worldwide frequency bands with a total of 4 Gigahertz (GHz) of bandwidth for ISM use, Wi-Fi – the most congested wireless standard – is only using three bands (0.9 GHz, 2.4 GHz and 5.7 GHz) and only 276 Megahertz (MHz) of bandwidth [5]. Perhaps even more importantly, cellular providers usually offload significant amounts of their traffic to Wi-Fi access points. To this end, Cisco recently estimated that 45% of the global smartphone traffic is today offloaded to the unlicensed ISM bands [6]. With the global mobile data traffic projected to reach 164 exabytes per month in 2025 due to the expansion of 5G and beyond (5GB) cellular networks [7], it is only a matter of time before the congestion in the ISM bands will decrease data throughput to intolerable levels, leading to the dreaded *spectrum crunch* issue [4]. Without any action, the IoT's hunger for higher data and processing rates will cause a dramatic decrease in service quality, a sharp increase in prices, and the impossibility of much-needed next-generation IoT applications and services to become reality.

Simply put, *re-engineering IoT devices, protocols and architectures to dynamically self-adapt to different spectrum circumstances has become a compelling necessity.* To this end, researchers are investigating technologies to "squeeze" as many bits as possible into a given portion of the spectrum. For example, dynamic *spectrum sharing* in licensed bands is widely considered as one of the very few options to sustain the IoT growth in the years to come [8–11]. For this reason, in January 2020 the Federal Communications Commission (FCC) opened up 150 MHz around the 3.5 GHz Citizens Broadband Radio Service (CBRS) band [12]. The new spectrum sharing rules will not only open up a plethora of business opportunities, but will also help local operators to provide affordable broadband access to historically underserved areas. In May 2020, the House of Representatives passed the Health and Economic Recovery Omnibus Emergency Solutions (HEROES) Act [13], which includes several provisions to expedite the buildout of rural broadband service in the United States due to the ongoing pandemic. Moreover, new spectrum frontiers in the millimeter-wave (mmWave) [14], terahertz (THz) [15] and visible-light (VL) frequency ranges [16] are being explored by researchers. These bands offer plentiful high-bandwidth spectrum, features that pave the way to the high data rates craved by the IoT.

In addition to moving to less-crowded higher-bandwidth spectrum portions, *researchers are exploring deep learning (DL) and deep reinforcement learning (DRL) techniques to improve how IoT devices access the spectrum* [17]. In other fields, learning-based techniques have been proven to be significantly effective to address a multitude of real-world complex tasks, from playing Atari video games [18] to single-handedly beating world-class Go champions [19]. Beyond offline spectrum analysis, DL can be used for real-time hardware-based spectrum analysis, since different model architectures can be reused to different problems as long as weights and hyper-parameters can be changed through software [20]. Among others, modulation recognition through DL has received significant attention [21–26], while DRL techniques have been leveraged to address handover and power management in cellular networks [27, 28], dynamic spectrum access [29–32], resource allocation/slicing/caching [33–37], video streaming [38–40], and modulation/coding scheme selection [41], just to name a few [42].

While it is true that some significant technological advances in IoT spectrum access have been made, *the harsh reality is that today's commercial IoT devices are still very far from being truly real-time reconfigurable.* For example, when it comes to spectrum sharing, IoT operators that want to utilize the CBRS band must contact a cloud-based spectrum access system (SAS)

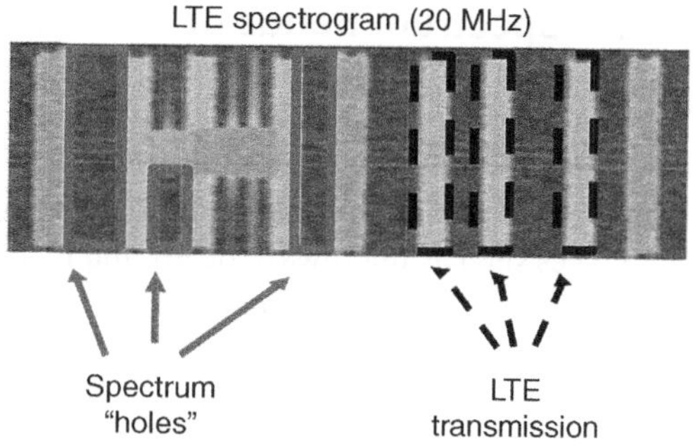

Figure 19.1 Spectrum "holes" during an LTE transmission may be leveraged to increase spectral efficiency.

by indicating precise data about latitude, longitude, and height into the SAS database, and then a SAS administrator determines if the spectrum is available [43]. It is easy to see that this centralized manual approach lacks scalability, and it does not allow for fine-grained real-time spectrum management. Conversely, a scalable and effective solution would be to let IoT devices opportunistically discover which spectrum sub-bands are currently available among ongoing licensed transmissions. This approach would significantly boost spectrum usage without the need of central coordination. To this point, Figure 19.1 shows a 20-MHz spectrogram of a Long Term Evolution (LTE) transmission captured through our experimental testbed, where we distinguish some of the spectrum "holes" left unutilized. *In this simple example, we can clearly see that more than 60% of spectrum could be opportunistically utilized by IoT devices, which is nowadays wasted.* To fully utilize the shared bandwidth without the need of centralized coordination, IoT transmitters should (i) opportunistically determine unutilized spectrum "holes" in the shared bandwidth, also called *spectrum sensing*; (ii) make a decision on which sub-band(s) to utilize based on noise and interference levels. The receiver must (iii) perform spectrum sensing to distinguish the sub-bands occupied by relevant transmissions, and (iv) demodulate the incoming waveforms.

Designing and implementing spectrum sharing and other spectrum management techniques is particularly challenging in the context of the IoT. Among others, *the key challenges are the extreme heterogeneity of spectrum bands, the diverse requirements in terms of energy/performance tradeoff, as well as the unique computational/memory constraints of IoT devices.* In other words, this implies that (i) a spectrum band or spectrum access scheme that is appropriate for one class of IoT devices may not be appropriate for another; (ii) computation-expensive, high-overhead artificial intelligence and machine learning (AI/ML) algorithms and protocols may be applied only to a very limited set of IoT applications and devices, which will require substantial innovation in the AI/ML field. As explained in Section 19.3, existing work has approached the problem of dynamic spectrum access (DSA) in a piecemeal fashion, with techniques that are either too complex to be implemented in real time by tiny, energy-constrained IoT devices, or too simple to be robust to noise and interference. These limitations have so far prevented the enactment of the much-needed "sharing-by-design" vision, where devices "learn to share" in any spectrum condition, at any frequency band, and without human intervention. At the same time, strict real-time constraints, the resource-limited nature of IoT devices, and the unpredictable nature of the wireless spectrum has to be taken into account.

The target of this paper is surveying the state of the art in spectrum management for 5G and IoT technologies, and to propose a roadmap of next-generation research challenges. We first describe the spectrum bands of interest in Section 19.2. We describe related work in Section 19.3, and finally discuss existing challenges in Section 19.4. We draw conclusions in Section 19.5.

19.2 Spectrum Bands of Interest in the Internet of Things

The vast majority of commercially-available IoT devices leverage the spectrum bands reserved for ISM purposes for radio communications. The main reason behind this choice is that a spectrum license is not required in ISM bands. On the other hand, ISM bands are shared, for example, with appliances such as microwave ovens and medical diathermy machines, which may generate a significant amount of interference. As such, IoT devices operating in ISM bands must tolerate any interference generated by ISM applications, and assume that no regulatory protection from ISM device operation in these bands is available. Within the context of ISM bands, a multitude of spectrum bands are currently used or being explored for the IoT. Each portion has peculiar challenges, which are summarized and discussed below.

19.2.1 Low-bands and Mid-bands

A significant number of IoT standards operate in the sub 7 GHz spectrum band, which offers robust and reliable communication with relatively low energy consumption. For example, the IEEE 802.15.4 is a family of radio technologies that operates in the ISM bands 868–868.6 MHz (e.g. Europe) and 902–928 MHz (e.g. North America). Usually, direct-sequence spread spectrum (DSSS) or chirp spread spectrum (CSS) are used for physical layer (PHY) transmissions, although some technologies also use narrowband schemes. With a phenomenal range of a few to tens of kilometers and battery life of ten years and beyond, low-power wide area (LPWA) technologies are the best candidates to realize the Internet of low-power, low-cost, and low-throughput Things [44]. Examples of these technologies working below the 1 GHz band include Sigfox [45], Weightless-N [46], and LoRa [47]. LPWA technologies will face a significant number of challenges. First, cross-technology interference can severely degrade the performance of LPWA technologies. This problem is definitely more severe for LPWA technologies operating in the license-exempt and shared ISM bands. In an anarchy where tens of wireless technologies and massive number of devices are all sharing the same channels, scenarios in which multiple narrowband channels of a LPWA technology are simultaneously interfered with by a single broadband signal will be extremely common. In addition, most LPWA technologies use very simple medium access schemes such as ALOHA, which do not scale well with the number of connected devices. Extensions to the well-know IEEE 802.11 standard include 802.11af [48], which operates between 54 and 790 MHz and employs OFDM technology on TeleVision White Space frequency spectrum (TVWS) bands, and 802.11ah [49], indeed focusing on a long-distance communication with low power consumption and operating in the 900 MHz ISM band. Such slow data rates match the low throughput requirements of most of the applications. Other extensions include IEEE 802.15.4g [50] and IEEE 802.15.4k [51], which provide transmission rates less than 10 Kbps and communication ranges longer than 1 km.

Going up in frequency, Wi-Fi (IEEE 802.11), ZigBee (IEEE 802.15.4) and Bluetooth (IEEE 802.15.1) are the technologies that overwhelmingly support the IoT in the 2.4 GHz and 5 GHz bands [52]. These spectrum bands are ideal for personal area communications (i.e. in the range of tens of meters), which encompass many real-world applications such as smart homes, smart health, and so on. Realizing the compelling need for spectrum sharing technologies, in April 2020, the FCC opened up 1.2 GHz of spectrum in the 6 GHz band to new users [53]. As of last July 2021, lower-band 6 GHz Wi-Fi operation has been allowed in Germany, which has become the first country in Europe to open up the 6 GHz band to Wi-Fi [54]. The German announcement follows the European Commission's announcement on June 30th on releasing 480 MHz of new spectrum in the 6 GHz band. The EU's decision is binding for all member states and local regulation reflecting the EC's decision must be written into law by individual EU members states. Moreover, in January 2020, the FCC opened the 3.5 GHz band for spectrum usage, also known as the "Innovation Band" and widely recognized as the key to unlocking the full 5G potential [12].

19.2.1.1 Millimeter-Wave Bands

The millimeter wave band (mmWave) typically refers to the portion of the spectrum between 30 GHz and 300 GHz [55]. Substantially unexploited and unexplored, mmWave has the potential to support Gigabit-per-second wireless communication. However, compared to conventional spectrum bands, mmWave presents several challenges – most importantly, high propagation loss, high penetration loss, attenuation, and atmospheric/molecular absorption, which ultimately limits the communication range [56]. For this reason, mmWave communications primarily rely on short range an unobstructed communications and directional transmissions.

The characteristics above imply that mmWave may find wide usage in high-speed short-range IoT communications. Indeed, mmWave-based wireless personal area networks (WPAN) and wireless local area networks (WLAN) have already been standardized in IEEE 802.15.3 and 802.11ad, respectively. It is also expected that mmWave will also support the IoT through 5G cellular networks [56]. Specifically, mmWave communications can be used to form small cells, which are short range in nature. Second, if cells are densely deployed, mmWave is also a choice for cellular access. Third, the high speed communication provided by mmWave also envisions a wireless backhaul between densely-deployed small cell base stations.

There are several challenges toward enabling mmWave applications in IoT applications. The high path loss incurred at mmWave frequencies implies higher transmission power, which poses challenges on circuit design and increases energy consumption. Second, medium access control (MAC) remains an area under-exploited. Traditional MAC protocols relying on carrier sensing will not perform well since directional transmissions dominate in mmWave communications. On the other hand, high directivity also provide the opportunity for utilizing spatial diversity. Third, guaranteeing Line of Sight (LoS) communications will also be a major challenge. To this end, reflecting surfaces, spatial diversity – i.e. Multi-Input Multi-Output (MIMO) – and relaying may be employed to mitigate the mmWave path loss.

19.2.1.2 Visible Light and Communications Above 100 GHz

Ultra-high-bandwidth links at 100 GHz and above will enable the multiplexing of thousands of IoT users at the same time, which will enable the next generation IoT [15, 57–59]. Above THz, visible light communications (VLCs) have been envisioned to be one potential RF-complementary solution to alleviate the looming spectrum crisis [60, 61]. First, VLC relies on a substantial portion of unregulated spectrum, ranging from 375 GHz to 750 GHz, providing orders of magnitude (10^4) higher bandwidth than the available radio spectrum. Second, the ubiquity of lighting infrastructure across our private and public spaces make it the perfect "vehicle" to implement ubiquitous IoT devices. Third, VLC can adopt simple modulation and demodulation schemes without complicated complex signal processing, for example, on-off modulation, which is ideal for resource-constrained low-cost IoT devices. Finally, due to the low-penetration property, VLC features itself free from electromagnetic interference and better security.

These above-mentioned factors have combined to launch intelligent lighting, i.e. VLC-IoT. Recent active research focuses on low-power communication technologies for IoT devices. For example, Li et al. [62] applies backscatter [1] in VLC to design low-power uplink for IoT devices.

---

1 Backscatter is a technique to be used to collect the reflections of incoming signals and then harvest energy, where lightweight modulation schemes (e.g. OOK) schemes and MAC protocols are usually adopted.

The authors in [63] proposed DarkVLC, a new VLC primitive that allows the VLC link to be sustained even when the LED lights appear dark or off, to broaden the application scenarios of VLC and to provide a new ultra-low power, always-on connectivity affordable for mobile and IoT devices. As the nodes in the VLC network are required to communicate with nodes in the Internet, a necessary condition for the IoT, therefore, the internet protocol stack must be supported. Schmid et al. [64] presents a software-based VLC PHY layer and a listen-before-talk VLC MAC layer protocol with contention, which only supports LED-to-LED VLC networks. For example, OpenVLC [65] is another low-cost, flexible, open platform for VLC communications, which runs MAC layer, part of PHY layer and offers an interface to Internet Protocols (IPs), thus being a suitable starter for VLC-IoTs.

There are several challenges in implementing IoT solutions with above 100 GHz and VLC communications. For example, transmissions in the THz band are strongly impacted by *time- and distance-dependent factors* such as the molecular absorption loss, which in turn depends on the concentration and the particular mixture of molecules encountered (particularly water vapor) [66]. For this reason, the usable bandwidth in THz channels strongly depends on the distance between the transmitter and the receiver or the ambient humidity, which obviously changes over time. As far as VLC is concerned, we need to integrate VLCs with RF communications to provide backward compatibility with the current IoT. Second, implementing VLC in IoT user space, e.g. implementation of PHY and MAC layers in different communication protocols (e.g. ZigBee, Wi-Fi, etc.) would streamline the testing and development issues related to modifications of existing and new protocols using available software packets.

19.3 Spectrum Management in the Internet of Things: Requirements and Existing Work

Despite the significant advantages of dynamic spectrum access (DSA), today's IoT devices still utilize technologies such as WiFi, Bluetooth and ZigBee, that are not capable of performing DSA. Although DSA techniques have been investigated, the vast majority of existing spectrum sensing algorithms were not designed having the constraints of the IoT in mind. For example, today, IoT operators that want to utilize the CBRS band must contact a cloud-based spectrum access system (SAS) by indicating precise data about latitude, longitude, and height into the SAS database, and then a SAS administrator determines if the spectrum is available [43]. *It is easy to see that this centralized manual approach lacks scalability, and it does not allow for fine-grained real-time spectrum management.* In other words, spectrum sharing is not completely efficient as it is conceived today – for both IoT operators and spectrum owners. Conversely, a scalable and effective solution would be to let IoT devices opportunistically discover which spectrum sub-bands are currently available among ongoing licensed transmissions (i.e. through spectrum sensing). This approach would severely boost spectrum usage without the need of central coordination. In the following, we summarize existing work in the field of spectrum management in the IoT, by highlighting the requirements and the limitations of existing solutions.

Spectrum Sensing. It has been shown that simpler threshold-based algorithms are too simple to perform spectrum sensing in realistic spectrum environments [67]. Nevertheless, effective DSA requires large spectrum chunks (i.e. several tens of MHz) to be sensed almost instantaneously. The receivers have to clearly distinguish transmissions from noise and interference in *dynamic, heavily congested spectrum landscapes where different wireless technologies coexist*, which requires the design of general-purpose, fine-tuneable algorithms. Previous work on spectrum sensing has been divided between narrowband and wideband approaches [68]. Narrowband approaches such as energy detection and matched filtering are hardly adaptable to the diverse and constantly changing spectrum, requiring prior knowledge of the transmission to effectively detect holes [69]. Other narrowband methods such as adaptive thresholds, cyclo-stationary feature detection and even some AI/ML techniques experience good performance in low SNR but are inefficient as they need to sequentially scan the spectrum to get a sense of occupancy for a wider band, occupying time and potentially missing holes [70–72]. Nyquist-based wideband spectrum sensing methods such as multi-band joint detection, wavelets, or utilizing filter banks present high complexity and thus, incur high latency [73]. Sub-Nyquist methods have also being proposed, which do not reach high accuracy [74]. Existing DL-based approaches for DSA are computationally heavy or have not attempted wideband sensing. Liu et al. [75] proposed a technique requiring a signal covariance matrix as input to the CNN, whereas we use unprocessed I/Q signals at the PHY. Naparstek and Cohen [31] leverage DL to formulate a DSA strategy to maximize a given network utility function in a distributed manner without online coordination or message exchanges between users. Similarly, the authors in [76] use DRL to learn an access policy from the observed states of all channels when SUs have no knowledge about the channel model. However, these works operate offline, meaning that the learning algorithms are trained and tested with a static dataset before deployment.

Spectrum Security. Real-world DSA scenarios will require real-time PHY authentication to identify relevant devices in large spectrum bands and protect the spectrum from misuse. Most approaches utilize spectrum permits embedded into different portions of the waveform, for example, in the cyclic prefix (CP) or in PHY symbols by dynamic power control of the secondary users SUs [77]. However, these techniques are not standards-compliant and are limited by the maximum transmission power on each spectrum band imposed by the FCC. Another approach is to use radio fingerprinting (RFP) to identify SUs. RFP has the key advantage that *it does not assume any wireless technology nor does it require any waveform modification or additional overhead*, but leverages the usage of a transmitter's circuitry imperfections. The main focus of early work on RFP has been devising hand-tailored feature extraction techniques [78–84]. Nguyen et al. [79] propose a non-parametric Bayesian method to detect the number of devices through device-dependent channel-invariant radio-metrics, however, the effectiveness of the methodology is tested on 4 ZigBee nodes only. Brik et al. [78] consider a large (i.e. 130 devices) Wi-Fi testbed, and through carefully-tailored transients and offset-based features show that 99% accuracy can be achieved. Conversely, Vo-Huu et al. [80] evaluates, on an in-the-wild Wi-Fi testbed, several feature-based algorithms based on the Wi-Fi scrambling seed, frequency offset and transients, achieving accuracy up to 50% on about 100 devices. Peng et al. [81] devise features based on the ZigBee's PSK constellation to fingerprint 54 radios with about 95% accuracy. Recently, Zheng et al. [85] proposed a function to model a device's modulation and timing errors, frequency offsets and power amplifier noise, and show that high-accuracy can be achieved on a series of 33 devices. Recent work has demonstrated that DL can be successfully used to fingerprint wireless devices with high accuracy [86–90]. Merchant et al. [89] and Das et al. [90] leverage CNN and recurrent neural networks (RNNs) to achieve respectively 92% accuracy on a testbed of 7 ZigBee devices and 90% on 30 LoRa devices. However, the effect of the channel on the performance is not studied. The works in [87, 88] are the first to explicitly evaluate the impact of impairments on the performance of CNN-based fingerprinting algorithms, and propose the introduction of artificial impairments to improve the accuracy. Gopalakrishnan et al. [91] explore the use of complex-valued CNNs for RFP. However, the approaches in [87, 88] do not show how the receiver can accurately compensate the introduced impairments, and it is not clear how to connect the increase in accuracy and the introduction of the hardware impairments.

AI/ML in Spectrum Management. Traditionally used in the computer vision (CV) domain, ML-based techniques are becoming more and more popular in the wireless community [92]. Such AI/ML-based techniques are particularly compelling for at least three reasons: **(i)** By construction, convolutional neural networks CNNs avoid manual feature extraction, which can be cumbersome (or impossible) for the complex spectrum classification problems we will need to address. Moreover, prior work has demonstrated that CNNs can improve modulation classification accuracy by up to 20% with respect to traditional ML techniques [21]. By having significantly more parameters than traditional models, CNNs are more robust to noise and most importantly, do not need to assume *any* underlying wireless technology [22, 24, 93]. This is key for our classification problems, since where different wireless technologies and devices will necessarily co-exist in the same spectrum bands and noise/interference levels will be mostly unpredictable; **(ii)** DRL algorithms can be trained to optimally choose among a set of known network actions (e.g. modulation, coding, medium access) according to the current wireless environment and optimization objective [29–32, 94]. In our context, DRL will prove to be instrumental to design robust and highly-adaptive channel access scheme for IoT devices. Prior work has shown that DRL can successfully be integrated into constrained IoT devices and address practical wireless problems without direct intervention [95]; **(iii)** Not only have ML-based techniques been proven to be significantly effective in the wireless domain [21], but also to be very efficient with respect to latency and energy consumption. Specifically, the work in [20] shows that when implemented in the hardware portion of the IoT platform and integrated with the radio processing chain, these algorithms can deliver 17x and 15x latency and energy reduction with respect to a software-based implementation [20].

19.4 Spectrum Management in the Internet of Things: The Way Ahead

Our vision is simple: the spectrum challenges mentioned above cannot be addressed without a radically different approach to the design of IoT platforms and systems. The significantly dynamic IoT requirements and conditions cannot be addressed with generalized, one-size-fits-all, bolted-on mechanisms. For this reason, the IoT technical designs will need to use context-aware, adaptive software and hardware solutions able to sense the environment and swiftly respond to a range of dynamic and unpredictable network conditions.

19.4.1 Protecting Passive and Incumbent Users from IoT Interference in Shared Bands

It is unquestionable that the decision to open up spectrum bands for shared use in the sub-6-GHz frequencies will fuel wireless growth and innovation for many years to come. In terms of business opportunities, for example, we will see a proliferation of technologies such as small-cell networks and industrial IoT application, while in terms of research we will see groundbreaking advances in fields such as AI/ML, circuit and antenna design, and many others. However, without mechanisms to protect incumbent users – i.e. previous spectrum owners – and passive users – i.e. entities listening to the spectrum for sensing purposes, spectrum sharing could cause severe disruption of key functionalities. For example, based on a study from the Radio Technical Committee for Aeronautics (RTCA), a trade organization that works with the Federal Aviation Administration (FAA) to develop safety standards, some aviation groups worry that 5G networking in the C-band (3700–3980 MHz) could cause disruptive interference to radar altimeters, which measure the distance between an aerial vehicle and the ground [96]. In short, harmful interference could potentially lead to airplane crashes over the United States.

We envision at least three possible sources of harmful interference in spectrum sharing scenarios. The first one regards unintentional interference caused by mismanagement of the spectrum administrators, for example, time synchronization and localization errors that could authorize transmitters when other communications are taking place. The second one is adversarial in nature, and regards the possibility of network operators or individuals abusing the spectrum – either intentionally or unintentionally – by transmitting in forbidden time or frequency slots. Notice that this is different from traditional jamming, where the target of the adversarial action is to disrupt ongoing communications. The third source of harmful interference can be caused by devices transmitting in time-frequency slots that are acknowledged by the spectrum administration entity, yet "spilling over" into other frequencies, for example, due to timing issues or imperfections in the transmitter's hardware circuitry. Each source of harmful interference needs to be addressed "by design" at the system level, with spectrum detection and control mechanisms able to identify spectrum misuse at very fine-grained granularity levels. For example, RFP can be used for spectrum authentication. The core intuition behind RFP is to leverage small-scale hardware-level imperfections such as phase noise, in phase/quadrature (I/Q) imbalance, frequency and sampling offset, and harmonic distortions [97] to obtain a "fingerprint" of a wireless device [86, 98, 99]. Although RFP has been already investigated, there are still key challenges to be addressed. To the best of our knowledge no system has ever used RFP to drive DSP-level decisions in real time. Second, CNN based RFP has been studied in scenarios where the signal being analyzed belongs to a single transmitter, i.e. no other radios are transmitting in the same spectrum band. On the other hand, it is a much more challenging problem identifying waveforms in a much wider spectrum band, with possibly more than one transmitting device (i.e. multi-label classification problem). These issues are further exacerbated by the need to only look at short-spanned "spectrum slices" (i.e. a limited number of I/Q samples coming from the radio interface), that can only partially describe the current spectrum status (see Figure 19.2). The longer these slices are, however, the greater the delay in the decision process, which may further compromise the receiver's performance. To make the problem tractable and scalable with the number of devices, one can carefully "inject" artificial imperfections in the transmitted waveform to help the CNN classifier distinguish among

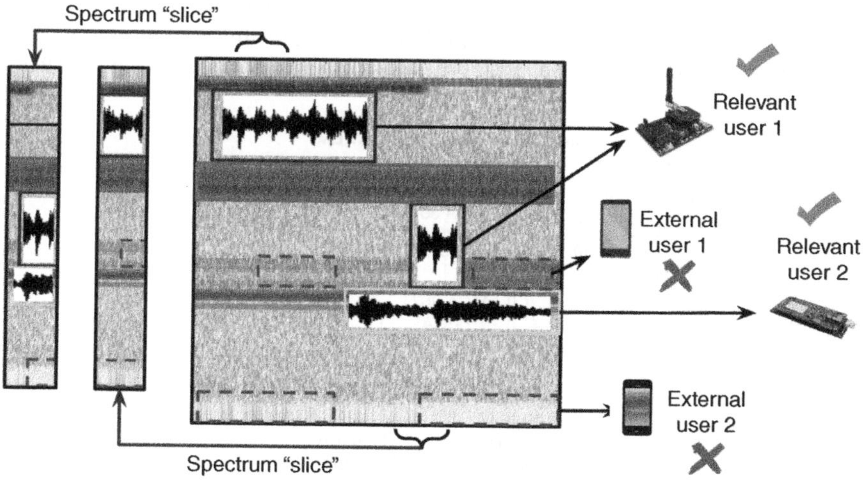

Figure 19.2 Radio fingerprinting in wideband spectrum portions. Source: (a) Libelium Comunicaciones Distribuidas S.L.; (b) Multi-Tech Systems, Inc.; (c) Free SVG.

devices [87]. For example, one could take the I/Q samples from the original waveform and insert a customized I/Q displacement to the given device. We point out that after detection, if the receiver needs to demodulate waveforms from more than one transmitter, more receiver chains should support the demodulation of multiple signals through different digital signal processing (DSP).

It is still unclear whether these techniques can function in practical IoT spectrum sharing systems. Specifically, a crucial issue will be to define mechanisms that will consider interference as a key parameter when accessing the spectrum. For example, when performing spectrum sensing, false positives – i.e. affirming a sub-band is free when it is not – should be a penalizing factor when determining the reward associated to that action, since wireless nodes will incur in higher number of collisions even if from the single node's perspective, the spectrum is utilized more. One possible direction is to design reward functions that consider the throughput of the entire network, instead of considering the single node's.

19.4.2 Experimental Spectrum Sharing at Scale Through the Colosseum and NSF PAWR Testbeds

Addressing the current and future challenges of spectrum sharing in the IoT will require a radical departure from how wireless research and experimentation is done today. The first key issue is that small-scale, individual experimental setups will not be able to analyze a wireless ecosystem in which a massive number of transmitters dynamically utilize different IoT technologies and coexist with each other. Another critical point is that reproducibility of experiments – hardly achievable with "homemade" testbeds – is also fundamental to persuade the major spectrum regulators that a wireless technology provides given constraints in terms of maximum harmful interference. Finally, existing software-defined radio (SDR) testbeds rely on host-based DSP capabilities, which are not able to perform real-time, in-the-loop operations such as spectrum sensing and dynamic spectrum access. Lately, the wireless community has seen a proliferation of large-scale testbeds, which was in part propelled by the National Science Foundation (NSF)'s desire to nurture research in experimental wireless research with the Platforms for Advanced Wireless Research (PAWR) program [100]. These platforms, united with Colosseum [101], the world's largest network emulator, can provide the right tools for both researchers and regulators to define and evaluate spectrum sharing technologies at an unprecedented scale.

Interestingly, in the near future Colosseum could provide the opportunity to augment the current processing capabilities with customized prototypes [101]. As illustrated in Figure 19.3,

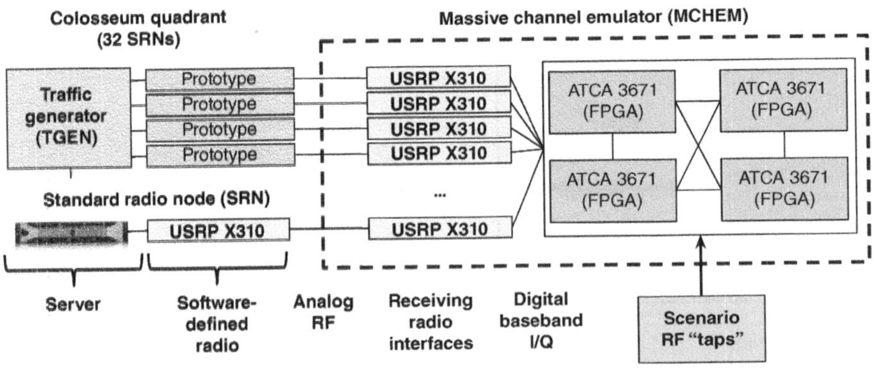

Figure 19.3 Spectrum sharing research in Colosseum. Source: Walmart.

a subset of standard radio nodes (SRNs) can be substituted with customized prototypes with real-time DSP capabilities, and connected to the traffic generator (TGEN) and the massive channel emulator (MCHEM). This heterogeneous setup will allow thorough testing of IoT spectrum sharing algorithms with different co-existing technologies such as LTE and a wide variety of channel conditions, impossible to fully experience in an over-the-air testbed. Another opportunity is given by the Platform for Open Wireless Data-driven Experimental Research (POWDER) [102]. Differently from Colosseum, POWDER provides a large-scale testbed where eight rooftop base stations each with up to four general purpose SDRs are connected to either a broadband or a banded antenna, with and over-the-air operation using srsRAN 4G/5G stack. Importantly, POWDER allows experimenters to bring their own devices to the testbed. This will allow researchers to try out their platforms in real-world cellular network environments. Notice that the purpose of the PAWR platforms is not to evaluate commercially-ready, finalized IoT product. Conversely, the software-defined radios SDRs deployed as part of PAWR could be used to conduct advanced experimental evaluation, as they provide significant flexibility at the wireless communications level. A finalized IoT commercial product would require the usage of application-specific integrated circuits (ASICs) to meet power and cost constraints.

19.4.3 Robust Machine Learning for Effective, Reliable and Efficient Spectrum Management

Spectrum sharing systems will be subjected to environment dynamics (i.e. noise/interference) and rewards (e.g. link or network throughput) that will necessarily change over time. We can also expect that the objective function will change over time – for example, consider a system where IoT platforms prefer wireless performance at the beginning of their lifetime and then prefer energy consumption reduction when their batteries are getting exhausted. Moreover, recent research [86, 98] has exposed the unavoidable effect of the wireless channel, which makes the accuracy of learning models plummet over time. To address this problem, adaptive ML algorithms have been investigated [103–105] in different contexts than wireless systems, such as the one of *lifelong learning* [106]. The key issue is that utilizing new samples by performing stochastic gradient descent [107, 108] may incur in what is known as *catastrophic forgetting* [109–113]. Many continual and lifelong learning aim to learn a variety of tasks without forgetting previous tasks [114–125]. More relevant to the context of RL, *meta-reinforcement learning* aims to rapidly adapt to new settings with small amounts of new experience [126, 127]. Within meta-reinforcement learning, we have optimization-based [128–131] and context-based, which includes both recurrent architectures [132–134] and architectures based on latent variable inference [126, 135–137]. However, the issue with these approaches is that they were not thought applicable for a resource-constrained embedded environment such as the IoT.

Besides the impact of the wireless channel, another key issue is that neural networks are prone to be "hacked" by carefully crafting small-scale perturbations to the input – which keep the input similar to the original one, but are ultimately able to "steer" the neural network away from the ground truth. This activity is known [99, 138–141] as adversarial machine learning (AML). The degree to which adversarial examples can be found is strongly correlated to the applicability of neural networks to the wireless domain [3]. Figure 19.4 shows AML in a wireless context. Specifically, an adversarial agent may decide to carefully jam a legitimate signal in order to cause misclassification of that waveform from a target neural network (TNN) running in the receiver's device. For example, a TNN performing modulation recognition may cause demodulation errors if the inferred modulation class is not correct. According to the threat model, the adversary may use the TNN's

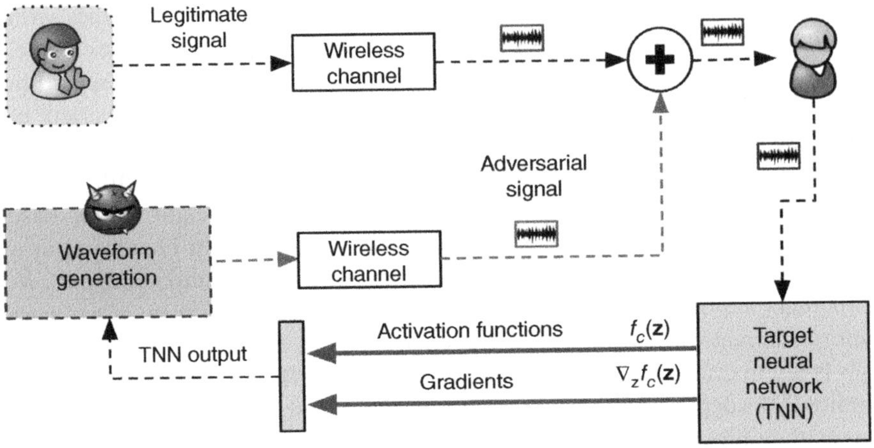

Figure 19.4 Adversarial Machine Learning (AML) in a Wireless Context.

activation functions and gradients to compute an attack waveform (whitebox attack), or simply bruteforce the waveform parameter space to try to compromise the TNN (blackbox attack). As far as literature is concerned, only very recently has AML been investigated by the wireless community. Bair et al. [142] propose to apply a variation of the MI-FGSM attack [143] to create adversarial examples to modulation classification systems. Shi et al. [144] propose the usage of a generative adversarial network (GAN) to spoof a targeted device. However, the proposed solution generates the adversarial example from a noise input. Instead, we modify an existing (decodable) waveform to hack the classifier. Moreover, the evaluation is only conducted through simulation without real datasets. Sadeghi and Larsson [145] proposed two AML algorithms based on a variation of the fast gradient methods (FGMs) [138] and tested on the 11-class RadioML *2016.10A* dataset [146] and with the architecture in [93]. Similarly, Flowers et al. [147] utilize a variation of FSGM to craft adversarial attacks. Sadeghi and Larrson [148] consider a system where the transmitter is an encoder and the channel is considered as random noise. Conversely from prior work, our research will address a more realistic problem: Given a set of target waveforms, design a modification strategy that (i) maximizes the probability to "steer" the classifier away from the ground truth over the set of chosen waveforms, while guaranteeing that the waveforms are still decodable at the receiver's side. Importantly, FSGM or similar algorithms cannot be used, as they can only compute adversarial examples tailored for a specific input and a specific channel condition.

One approach to achieving ML robustness is to utilize techniques based on the "power of the crowd," for example, federated machine learning (FML), to achieve lifelong distributed improvement [149–156]. From its inception by Google in 2017 [157], FML has evolved as a power tool at the intersection of AI and edge computing. Different from centralized approaches, FML allows the periodical fusion of locally-trained models by sharing with a centralized server only the trained model. The core idea is to globally aggregate local learning updates (such as parameters or gradients) of the DRL policy functions trained on the individual IoT platforms in a centralized node, while keeping the raw learning data (i.e. state-action-reward tuples) locally. Such FML-based approach will (i) eliminate the need to stream entire waveforms to the edge; and (ii) improve the global model robustness. For example, consider a number of platforms running spectrum sensing algorithms deployed over a geographic area. The nodes will periodically share their ML parameters with an edge server, which will take care of aggregating the local ML policies into a global ML model, which is then transmitted to the platforms.

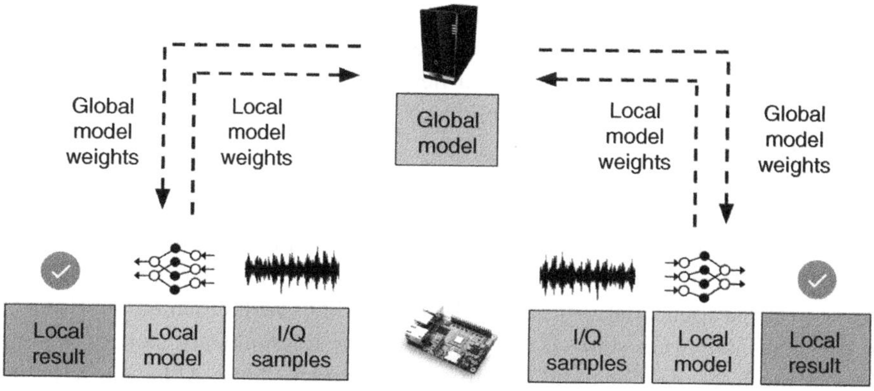

Figure 19.5 Federated Machine Learning (FML) for IoT ML Robustness Sensing.

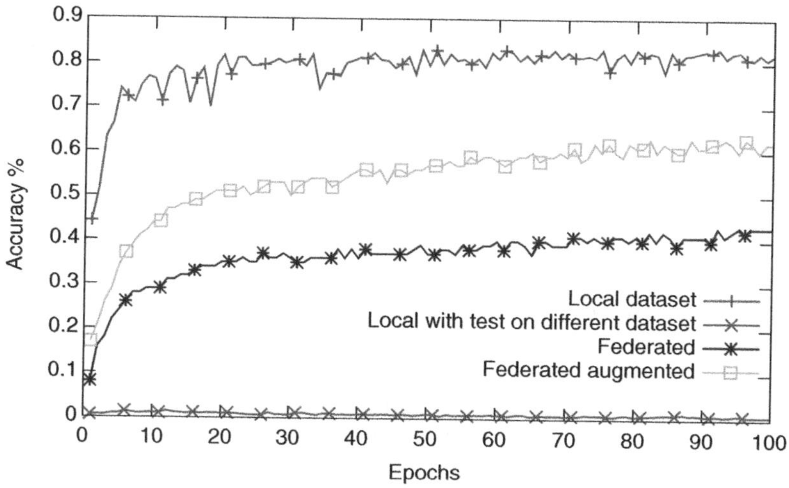

Figure 19.6 Test set accuracy of a CNN bootstrapped with noise-based data augmentation.

To show the improvement given by FML, Figure 19.5 shows the training curves (accuracy as function of the training epoch) of a CNN trained with FML, where the CNN needs to distinguish 200 wireless devices by using RFP. In this example, we consider two IoT nodes, where half of the dataset is located at the first node and the other half at the second node. The first and second curves show the accuracy of the locally-trained CNNs, which are not able to generalize to the dataset trained by the other node, since the accuracy plummets to almost zero percent. The third curve shows the result obtained with FML applied to the two datasets, which improves the accuracy by 40%. To show the effectiveness of data augmentation (curve 4), we further train the CNN by perturbing the data with 5 Gaussian distributions, each with mean 0 and standard deviation 0.1 to 0.5, with increments of 0.1. Figure 19.6 shows that the performance increases to 60%, with respect to the initial 40%.

19.4.4 The Role of O-RAN in Spectrum Sharing

The O-RAN Alliance [158] is an industry consortium that promotes the definition of an open standard for disaggregated, possibly virtualized cellular Radio Access Networks (RAN). O-RAN has two main goals [159]. First, it is in the process of defining an open RAN architecture, with well-defined interfaces between the different elements of the RAN enabling interoperability between network elements manufactured by different vendors. Second, it is promoting the introduction of the RAN intelligent controller (RIC), a new network element that provides an abstraction of analytics and control tools thus enabling programmatic control of the network through ML or other AI techniques. The RAN intelligent controller will provide fine-grained control of spectral interactions, thus potentially enabling new applications in spectrum sharing.

O-RAN introduces two flavors of RICs, the so-called non-real-time and a near-real-time RIC. The service management and orchestration framework operates a non-real-time RIC, which performs control decisions on time scales higher than one second. The near-real-time RIC, instead, is in charge of running applications that control functionalities with much tighter timing requirement (with decision intervals as short as 10 ms), relying on different start, stop, override, or control primitives in the RAN, e.g. for radio resource management. These APIs can be used by different applications running on the near-real-time RIC (referred to as xApps), which can be developed by third-party entities and pulled from a common marketplace. For example, through the near-real-time RIC and its xApps, an operator can control user mobility (e.g. handovers), allocate networking resources according to predicted paths for connected vehicles and UAVs [160], perform load balancing and traffic steering, and optimize scheduling policies [161]. The near-real-time RIC can also leverage ML algorithms trained in the non-real-time RIC.

It is clear that the presence of the RIC and the O-RAN architecture enable new opportunities in spectrum sharing for the Internet of Things. In this context, recent work [162] has proposed for example the Channel-Aware Reactive Mechanism (ChARM), a data-driven O-RAN-compliant framework that allows (i) sensing the spectrum to infer the presence of interference and (ii) reacting in real time by controlling the configuration of O-RAN compliant distributed (DU) and radio (RU) units according to a specified spectrum access policy. ChARM is based on deep convolutional neural networks operating directly on unprocessed I/Q samples to determine the current spectrum context. ChARM does not require any modification to the existing 3GPP standards. It is designed to operate within the O-RAN specifications, and can be used in conjunction with other spectrum sharing mechanisms (e.g. LTE-U, LTE-LAA or MulteFire). The performance of ChARM was demonstrated to enable spectrum sharing between LTE and Wi-Fi in unlicensed bands, where a controller operating over a RIC senses the spectrum and switches cell frequency to avoid Wi-Fi. A full-fledged standard-compliant prototype of ChARM was developed using srsRAN and leveraging the Colosseum channel emulator to collect a large-scale waveform dataset to train the neural networks. Experimental results showed that ChARM achieves accuracy of up to 96% on Colosseum and 85% on Arena [163], demonstrating the ability of ChARM to seamlessly enable coexistence between heterogeneous wireless systems.

19.5 Conclusions

It is now clear that the RF spectrum will become an extremely scarce resource in the next few years. To tackle the next-generation spectrum challenges, the IoT must be redesigned to allow spectrum agility "by design", without any manual and/or human intervention. To achieve spectrum agility in the IoT, we argue that we need to move toward decentralized, flexible, fine-grained real-time spectrum management systems, which will also consider *energy consumption* and *computational constraints* as key fundamental requirements. In this chapter, we have discussed the technical requirements and articulated the enabling technologies for such new generation of spectrum-sharing, energy-efficient IoT wireless devices, as well as discussed existing research challenges and future directions. We hope that this chapter will stimulate further research in the field.

References

1 Cisco Systems (2017). Cisco Visual Networking Index: Global Mobile Data Traffic Forecast Update, 2016–2021 White Paper. http://tinyurl.com/zzo6766 (accessed 31 October 2022).

2 Columbus, L. (2016). Roundup of Internet of Things Forecasts and Market Estimates. (Forbes). http://tinyurl.com/yar5llet (accessed 31 October 2022).

3 Restuccia, F., D'Oro, S., and Melodia, T. (2018). Securing the Internet of Things in the age of machine learning and software-defined networking. *IEEE Internet of Things Journal* 5 (6): 4829–4842.

4 National Institute of Standards and Technology (NIST) (2016). Spectrum Crunch. https://www.nist.gov/advanced-communications/spectrum-crunch (accessed 31 October 2022).

5 IEEE (2012). IEEE Standard for Information technology–Telecommunications and information exchange between systems Local and metropolitan area networks–Specific requirements Part 11: Wireless LAN Medium Access Control (MAC) and Physical Layer (PHY) Specifications - Redline. *IEEE Std 802.11-2012 (Revision of IEEE Std 802.11-2007) - Redline*, pp. 1–5229, March 2012.

6 Cisco Systems (2017). Cisco Annual Internet Report (2018-2023) White Paper. https://www.cisco.com/c/en/us/solutions/collateral/executive-perspectives/annual-internet-report/white-paper-c11-741490.html (accessed 31 October 2022).

7 Ericsson Incorporated (2020). Ericsson Interim Mobility Report, June 2020. https://www.ericsson.com/49da93/assets/local/mobility-report/documents/2020/june2020-ericsson-mobility-report.pdf (accessed 31 October 2022).

8 Zhang, L., Xiao, M., Wu, G. et al. (2017). A survey of advanced techniques for spectrum sharing in 5G networks. *IEEE Wireless Communications* 24 (5): 44–51.

9 Hu, F., Chen, B., and Zhu, K. (2018). Full spectrum sharing in cognitive radio networks toward 5G: a survey. *IEEE Access* 6: 15754–15776.

10 Shokri-Ghadikolaei, H., Boccardi, F., Fischione, C. et al. (2016). Spectrum sharing in mmWave cellular networks via cell association, coordination, and beamforming. *IEEE Journal on Selected Areas in Communications* 34 (11): 2902–2917.

11 Lv, L., Chen, J., Ni, Q. et al. (2018). Cognitive non-orthogonal multiple access with cooperative relaying: a new wireless frontier for 5G spectrum sharing. *IEEE Communications Magazine* 56 (4): 188–195.

12 Davies, J. and Telecoms.com (2020). FCC finally opens up 3.5 GHz for US telcos. https://telecoms.com/502070/fcc-finally-opens-up-3-5-ghz-for-us-telcos/ (accessed 31 October 2022).

13 United States House of Representatives (2020). H.R.6800 - The Heroes Act. https://www.congress.gov/bill/116th-congress/house-bill/6800 (accessed 31 October 2022).

14 Lu, X., Petrov, V., Moltchanov, D. et al. (2019). 5G-U: Conceptualizing integrated utilization of licensed and unlicensed spectrum for future IoT. *IEEE Communications Magazine* 57 (7): 92–98.

15 Polese, M., Cantos-Roman, X., Singh, A. et al. (2021). Coexistence and spectrum sharing above 100 GHz. https://arxiv.org/abs/2110.15187.

16 Kadam, K. and Dhage, M.R. (2016). Visible light communication for IoT. *2016 2nd International Conference on Applied and Theoretical Computing and Communication Technology (iCATccT)*. IEEE, pp. 275–278.

17 Mao, Q., Hu, F., and Hao, Q. (2018). Deep learning for intelligent wireless networks: a comprehensive survey. *IEEE Communication Surveys and Tutorials* 20 (4): 2595–2621.

18 Mnih, V., Kavukcuoglu, K., Silver, D. et al. (2013). Playing Atari with deep reinforcement learning. *arXiv preprint arXiv:1312.5602.*

19 Silver, D., Schrittwieser, J., Simonyan, K. et al. (2017). Mastering the game of go without human knowledge. *Nature* 550 (7676): 354.

20 Restuccia, F. and Melodia, T. (2019). Big data goes small: real-time spectrum-driven embedded wireless networking through deep learning in the RF loop. *Proceedings of the IEEE Conference on Computer Communications (INFOCOM).*

21 O'Shea, T.J., Roy, T., and Clancy, T.C. (2018). Over-the-air deep learning based radio signal classification. *IEEE Journal of Selected Topics in Signal Processing* 12 (1): 168–179.

22 O'Shea, T.J. and Hoydis, J. (2017). An introduction to deep learning for the physical layer. *IEEE Transactions on Cognitive Communications and Networking* 3 (4): 563–575.

23 Wang, T., Wen, C.-K., Wang, H. et al. (2017). Deep learning for wireless physical layer: opportunities and challenges. *China Communications* 14 (11): 92–111.

24 West, N.E. and O'Shea, T. (2017). Deep architectures for modulation recognition. *Proceedings of the IEEE International Symposium on Dynamic Spectrum Access Networks (DySPAN),* Baltimore, MD, USA, March 2017, pp. 1–6.

25 Kulin, M., Kazaz, T., Moerman, I., and Poorter, E.D. (2018). End-to-end learning from spectrum data: a deep learning approach for wireless signal identification in spectrum monitoring applications. *IEEE Access* 6: 18484–18501.

26 Karra, K., Kuzdeba, S., and Petersen, J. (2017). Modulation recognition using hierarchical deep neural networks. *Proceedings of the IEEE International Symposium on Dynamic Spectrum Access Networks (DySPAN),* Baltimore, MD, USA, March 2017, pp. 1–3.

27 Liu, J., Krishnamachari, B., Zhou, S., and Niu, Z. (2018). DeepNap: data-driven base station sleeping operations through deep reinforcement learning. *IEEE Internet of Things Journal* 5 (6): 4273–4282.

28 Wang, Z., Li, L., Xu, Y. et al. (2018). Handover control in wireless systems via asynchronous multiuser deep reinforcement learning. *IEEE Internet of Things Journal* 5 (6): 4296–4307.

29 Yu, Y., Wang, T., and Liew, S.C. (2019). Deep-reinforcement learning multiple access for heterogeneous wireless networks. *IEEE Journal on Selected Areas in Communications* 37 (6): 1277–1290.

30 Wang, S., Liu, H., Gomes, P.H., and Krishnamachari, B. (2018). Deep reinforcement learning for dynamic multichannel access in wireless networks. *IEEE Transactions on Cognitive Communications and Networking* 4 (2): 257–265.

31 Naparstek, O. and Cohen, K. (2019). Deep multi-user reinforcement learning for distributed dynamic spectrum access. *IEEE Transactions on Wireless Communications* 18 (1): 310–323.

32 Chang, H.-H., Song, H., Yi, Y. et al. (2018). Distributive dynamic spectrum access through deep reinforcement learning: a reservoir computing based approach. *IEEE Internet of Things Journal* 6 (2): 1938–1948.

33 Feng, M. and Mao, S. (2019). Dealing with limited backhaul capacity in millimeter-wave systems: a deep reinforcement learning approach. *IEEE Communications Magazine* 57 (3): 50–55.

34 He, Y., Zhao, N., and Yin, H. (2018). Integrated networking, caching, and computing for connected vehicles: a deep reinforcement learning approach. *IEEE Transactions on Vehicular Technology* 67 (1): 44–55.

35 Sun, Y., Peng, M., and Mao, S. (2019). Deep reinforcement learning-based mode selection and resource management for green fog radio access networks. *IEEE Internet of Things Journal* 6 (2): 1960–1971.

36 Li, R., Zhao, Z., Sun, Q. et al. (2018). Deep reinforcement learning for resource management in network slicing. *IEEE Access* 6: 74429–74441.

37 Zhang, H., Li, W., Gao, S. et al. (2019). ReLeS: A neural adaptive multipath scheduler based on deep reinforcement learning. *Proceedings of the IEEE Conference on Computer Communications (INFOCOM).*

38 Pang, H., Zhang, C., Wang, F. et al. (2019). Towards low latency multi-viewpoint 360° interactive video: a multimodal deep reinforcement learning approach. *Proceedings of the IEEE Conference on Computer Communications (INFOCOM).*

39 Wang, F., Zhang, C., Wang, F. et al. (2019). Intelligent edge-assisted crowdcast with deep reinforcement learning for personalized QoE. *Proceedings of the IEEE Conference on Computer Communications (INFOCOM).*

40 Zhang, Y., Zhao, P., Bian, K. et al. (2019). DRL360: 360-degree video streaming with deep reinforcement learning. *Proceedings of the IEEE Conference on Computer Communications (INFOCOM).*

41 Zhang, L., Tan, J., Liang, Y. et al. (2019). Deep reinforcement learning based modulation and coding scheme selection in cognitive heterogeneous networks. *IEEE Transactions on Wireless Communications* 18 (6): 3281–3294.

42 Jagannath, J., Polosky, N., Jagannath, A. et al. (2019). Machine learning for wireless communications in the Internet of Things: a comprehensive survey. *Ad Hoc Networks* 93: 101913.

43 Linda Hardesty (Fierce Wireless) (2020). What is a CBRS spectrum access system? https://www.fiercewireless.com/private-wireless/what-a-cbrs-spectrum-access-system (accessed 31 October 2022).

44 Raza, U., Kulkarni, P., and Sooriyabandara, M. (2017). Low power wide area networks: an overview. *IEEE Communication Surveys and Tutorials* 19 (2): 855–873.

45 Vejlgaard, B., Lauridsen, M., Nguyen, H. et al. (2017). Coverage and capacity analysis of Sigfox, LoRa, GPRS, and NB-IoT. *2017 IEEE 85th Vehicular Technology Conference (VTC Spring).* IEEE, pp. 1–5.

46 Adelantado, F., Vilajosana, X., Tuset-Peiro, P. et al. (2017). Understanding the limits of LoRaWAN. *IEEE Communications Magazine* 55 (9): 34–40.

47 Bor, M., Vidler, J.E., and Roedig, U. (2016). LoRa for the Internet of Things.

48 Flores, A.B., Guerra, R.E., Knightly, E.W. et al. (2013). IEEE 802.11af: A standard for TV white space spectrum sharing. *IEEE Communications Magazine* 51 (10): 92–100.

49 Adame, T., Bel, A., Bellalta, B. et al. (2014). IEEE 802.11ah: the WiFi approach for M2M communications. *IEEE Wireless Communications* 21 (6): 144–152.

50 Chang, K.-H. and Mason, B. (2012). The IEEE 802.15. 4G standard for smart metering utility networks. *2012 IEEE 3rd International Conference on Smart Grid Communications (SmartGridComm).* IEEE, pp. 476–480.

51 Gebremedhin, B.G., Haapola, J., and Iinatti, J. (2015). Performance evaluation of IEEE 802.15.4k priority channel access with DSSS PHY. *Proceedings of European Wireless 2015; 21th European Wireless Conference.* VDE, pp. 1–6.

52 Baronti, P., Pillai, P., Chook, V.W. et al. (2007). Wireless sensor networks: a survey on the state of the art and the 802.15.4 and ZigBee standards. *Computer Communications* 30 (7): 1655–1695.

53 Federal Communications Commission (FCC) (2020). FCC Opens 6 GHz Band to Wi-Fi and Other Unlicensed Uses. https://www.fcc.gov/document/fcc-opens-6-ghz-band-wi-fi-and-other-unlicensed-uses (accessed 31 October 2022).

54 Hetting, C. and Wi-Fi NOW CEO & Chairman (2021). Germany First Country in Europe with 6 GHz Wi-Fi. https://wifinowglobal.com/news-and-blog/germany-becomes-first-country-in-europe-to-open-up-for-6-ghz-wi-fi/ (accessed 31 October 2022).

55 Khan, F. and Pi, Z. (Z.). mmWave Mobile Broadband (MMB): unleashing the 3 - 300GHz Spectrum. *Proceedings of the IEEE Sarnoff Symposium*, Princeton, NJ, USA, May 2011, pp. 1–6.

56 Niu, Y., Li, Y., Jin, D. et al. (2015). A survey of millimeter wave communications (mmWave) for 5G: opportunities and challenges. *Wireless Networks* 21 (8): 2657–2676.

57 Rappaport, T.S., Xing, Y., Kanhere, O. et al. (2019). Wireless communications and applications above 100 GHz: opportunities and challenges for 6G and beyond. *IEEE Access* 7: 78729–78757.

58 Giordani, M., Polese, M., Mezzavilla, M. et al. (2020). Toward 6G networks: use cases and technologies. *IEEE Communications Magazine* 58 (3): 55–61.

59 Polese, M., Melodia, T., and Zorzi, M. (2020). Toward end-to-end, full-stack 6G terahertz networks. *IEEE Communications Magazine* 58 (11): 48–54.

60 Dimitrov, S. and Haas, H. (2015). *Principles of LED Light Communications: Towards Networked Li-Fi.* Cambridge University Press.

61 Ishikawa, N., Sugiura, S., and Hanzo, L. (2018). 50 years of permutation, spatial and index modulation: from classic RF to visible light communications and data storage. *IEEE Communication Surveys and Tutorials* 20 (3): 1905–1938.

62 Li, J., Liu, A., Shen, G. et al. (2015). Retro-VLC: Enabling battery-free duplex visible light communication for mobile and IoT applications. *Proceedings of the 16th International Workshop on Mobile Computing Systems and Applications (HotMobile)*, February 2015.

63 Tian, Z., Wright, K., and Zhou, X. (2016). Lighting up the Internet of Things with Dark-VLC. *in Proceedings of the 17th International Workshop on Mobile Computing Systems and Applications (HotMobile)*, February 2016.

64 Schmid, S., Bourchas, T., Mangold, S., and Gross, T.R. (2015). Linux light bulbs: enabling internet protocol connectivity for light bulb networks. *in Proceedings of the 2nd International Workshop on Visible Light Communications Systems (VLCS)*, September 2015.

65 Wang, Q., Giustiniano, D., and Puccinelli, D. (2015). An open source research platform for embedded visible light networking. *IEEE Wireless Communications* 22 (2): 94–100.

66 Jornet, J.M. and Akyildiz, I.F. (2011). Channel modeling and capacity analysis for electromagnetic wireless nanonetworks in the terahertz band. *IEEE Transactions on Wireless Communications* 10 (10): 3211–3221.

67 Uvaydov, D., D'Oro, S., Restuccia, F., and Melodia, T. (2021). DeepSense: fast wideband spectrum sensing through real-time in-the-loop deep learning. *Proceedings of IEEE INFOCOM)*. https://tinyurl.com/Uvay2021 (accessed 31 October 2022).

68 Arjoune, Y. and Kaabouch, N. (2019). A comprehensive survey on spectrum sensing in cognitive radio networks: recent advances, new challenges, and future research directions. *Sensors* 19 (1): 126.

69 Ranjan, A., Anurag, Singh, B. (2016). Design and analysis of spectrum sensing in cognitive radio based on energy detection. *2016 International Conference on Signal and Information Processing (IConSIP)*. IEEE, pp. 1–5.

70 Alom, M.Z., Godder, T.K., Morshed, M.N., and Maali, A. (2017). Enhanced spectrum sensing based on energy detection in cognitive radio network using adaptive threshold. *2017 International Conference on Networking, Systems and Security (NSysS)*. IEEE, pp. 138–143.

71 Yawada, P.S. and Wei, A.J. (2016). Cyclostationary detection based on non-cooperative spectrum sensing in cognitive radio network. *2016 IEEE International Conference on Cyber Technology in Automation, Control, and Intelligent Systems (CYBER)*. IEEE, pp. 184–187.

72 Gao, J., Yi, X., Zhong, C. et al. (2019). Deep learning for spectrum sensing. *IEEE Wireless Communications Letters* 8 (6): 1727–1730.

73 Hamdaoui, B., Khalfi, B., and Guizani, M. (2018). Compressed wideband spectrum sensing: Concept, challenges, and enablers. *IEEE Communications Magazine* 56 (4): 136–141.

74 Fang, J., Wang, B., Li, H., and Liang, Y.-C. (2021). Recent advances on sub-nyquist sampling-based wideband spectrum sensing. *IEEE Wireless Communications* 28 (3): 115–121.

75 Liu, C., Wang, J., Liu, X., and Liang, Y.-C. (2019). Deep CM-CNN for spectrum sensing in cognitive radio. *IEEE Journal on Selected Areas in Communications* 37 (10): 2306–2321.

76 Nguyen, H.Q., Nguyen, B.T., Dong, T.Q. et al. (2018). Deep Q-learning with multiband sensing for dynamic spectrum access. *IEEE DySPAN),* October 2018, pp. 1–5.

77 Jin, X., Sun, J., Zhang, R. et al. (2018). SpecGuard: spectrum misuse detection in dynamic spectrum access systems. *IEEE Transactions on Mobile Computing* 17 (12): 2925–2938.

78 Brik, V., Banerjee, S., Gruteser, M., and Oh, S. (2008). Wireless device identification with radiometric signatures. *Proceedings of the 14th ACM International Conference on Mobile Computing and Networking (MobiCom).* ACM, pp. 116–127.

79 Nguyen, N.T., Zheng, G., Han, Z., and Zheng, R. (2011). Device fingerprinting to enhance wireless security using nonparametric bayesian method. *Proceedings of the IEEE Conference on Computer Communications (INFOCOM).* IEEE, pp. 1404–1412.

80 Vo-Huu, T.D., Vo-Huu, T.D., and Noubir, G. (2016). Fingerprinting Wi-Fi devices using software defined radios. *Proceedings of the 9th ACM Conference on Security & Privacy in Wireless and Mobile Networks.* ACM, pp. 3–14.

81 Peng, L., Hu, A., Zhang, J. et al. (2019). Design of a hybrid RF fingerprint extraction and device classification scheme. *IEEE Internet of Things Journal* 6 (1): 349–360.

82 Xie, F., Wen, H., Li, Y. et al. (2018). Optimized coherent integration-based radio frequency fingerprinting in Internet of Things. *IEEE Internet of Things Journal* 5 (5): 3967–3977.

83 Xing, Y., Hu, A., Zhang, J. et al. (2018). On radio frequency fingerprint identification for DSSS systems in low SNR scenarios. *IEEE Communications Letters* 22 (11): 2326–2329.

84 Xu, Q., Zheng, R., Saad, W., and Han, Z. (2016). Device fingerprinting in wireless networks: challenges and opportunities. *IEEE Communication Surveys and Tutorials* 18 (1): 94–104.

85 Zheng, T., Sun, Z., and Ren, K. (2019). FID: Function modeling-based data-independent and channel-robust physical-layer identification. *Proceedings of the IEEE Conference on Computer Communications (INFOCOM).* IEEE, pp. 199–207.

86 Restuccia, F., D'Oro, S., Al-Shawabka, A. et al. (2019). DeepRadioID: Real-time channel-resilient optimization of deep learning-based radio fingerprinting algorithms. *Proc. of ACM International Symposium on Mobile Ad Hoc Networking and Computing (MobiHoc).*

87 Sankhe, K., Belgiovine, M., Zhou, F. et al. (2019). ORACLE: Optimized radio classification through convolutional neural networks. *Proceedings of the IEEE Conference on Computer Communications (INFOCOM).* IEEE, pp. 370–378.

88 Riyaz, S., Sankhe, K., Ioannidis, S., and Chowdhury, K. (2018). Deep learning convolutional neural networks for radio identification. *IEEE Communications Magazine* 56 (9): 146–152.

89 Merchant, K., Revay, S., Stantchev, G., and Nousain, B. (2018). Deep learning for RF device fingerprinting in cognitive communication networks. *IEEE Journal of Selected Topics in Signal Processing* 12 (1): 160–167.

90 Das, R., Gadre, A., Zhang, S. et al. (2018). A deep learning approach to IoT authentication. *Proceedings of the IEEE International Conference on Communications (ICC).* IEEE, pp. 1–6.

91 Gopalakrishnan, S., Cekic, M., and Madhow, U. (2019). Robust wireless fingerprinting via complex-valued neural networks. *arXiv preprint arXiv:1905.09388.*

92 Jagannath, J., Polosky, N., Jagannath, A. et al. (2019). Machine learning for wireless communicationsin the Internet of Things: a comprehensive survey. *Ad Hoc Networks (Elsevier)* 93: 101913.

93 O'Shea, T.J., Corgan, J., and Clancy, T.C. (2016). Convolutional radio modulation recognition networks. *International Conference on Engineering Applications of Neural Networks*. Springer, pp. 213–226.

94 Luong, N.C., Hoang, D.T., Gong, S. et al. (2018). Applications of deep reinforcement learning in communications and networking: a survey. *arXiv preprint arXiv:1810.07862*.

95 Restuccia, F. and Melodia, T. (2020). DeepWiERL: Bringing deep reinforcement learning to the internet of self-adaptive things. *Proc. of IEEE Conference on Computer Communications (INFOCOM)*.

96 Radio Technical Committee for Aeronautics (RTCA) (2020). Assessment of C-Band Mobile Telecommunications Interference Impact on Low Range Radar Altimeter Operations. https://tinyurl.com/RTCA-Stud (accessed 31 October 2022).

97 Johnson, E. (1966). Physical limitations on frequency and power parameters of transistors. *1958 IRE International Convention Record*, volume 13. IEEE, pp. 27–34.

98 Al-Shawabka, A., Restuccia, F., D'Oro, S. et al. (2020). Exposing the fingerprint: dissecting the impact of the wireless channel on radio fingerprinting. *Proceedings of the IEEE Conference on Computer Communications (INFOCOM)*.

99 Restuccia, F., D'Oro, S., Al-Shawabka, A. et al. (2020). Generalized wireless adversarial deep learning. *Proceedings of the 2nd ACM Workshop on Wireless Security and Machine Learning*, ser. WiseML '20. New York, NY, USA: Association for Computing Machinery, pp. 49–54. https://doi.org/10.1145/3395352.3402625.

100 PAWR (2018). Platforms for Advanced Wireless Research (PAWR). https://advancedwireless.org (accessed 31 October 2022).

101 Bonati, L., Johari, P., Polese, M. et al. (2021). Colosseum: Large-scale wireless experimentation through hardware-in-the-loop network emulation. *arXiv preprint arXiv:2110.10617*.

102 POWDER (2018). Powder (the Platform for Open Wireless Data-driven Experimental Research). https://powderwireless.net (accessed 31 October 2022).

103 Hassabis, D., Kumaran, D., Summerfield, C., and Botvinick, M. (2017). Neuroscience-inspired artificial intelligence. *Neuron* 95 (2): 245–258.

104 Thrun, S. and Mitchell, T.M. (1995). Lifelong robot learning. *Robotics and Autonomous Systems* 15 (1–2): 25–46.

105 Rosenblatt, F. (1957). *The Perceptron, a Perceiving and Recognizing Automaton Project Para*. Cornell Aeronautical Laboratory.

106 Parisi, G.I., Kemker, R., Part, J.L. et al. (2019). Continual lifelong learning with neural networks: a review. *Neural Networks* 113: 54–71.

107 Kushner, H.J. and Yin, G. (2003). *Stochastic Approximation and Recursive Algorithms and Applications*, vol. 35. Springer Science & Business Media.

108 Benveniste, A., Métivier, M., and Priouret, P. (2012). *Adaptive Algorithms and Stochastic Approximations*, vol. 22. Springer Science & Business Media.

109 McClelland, J.L., McNaughton, B.L., and O'reilly, R.C. (1995). Why there are complementary learning systems in the hippocampus and neocortex: insights from the successes and failures of connectionist models of learning and memory. *Psychological Review* 102 (3): 419.

110 McCloskey, M. and Cohen, N.J. (1989). Catastrophic interference in connectionist networks: the sequential learning problem. In: *Psychology of Learning and Motivation*, vol. 24, 109–165. Elsevier.

111 Ratcliff, R. (1990). Connectionist models of recognition memory: constraints imposed by learning and forgetting functions. *Psychological Review* 97 (2): 285.

112 Lewandowsky, S. and Li, S. (1994). Catastrophic interference in neural networks: causes, solutions, and data, dempster. In: *New Perspectives on Interference and Inhibition in Cognition* (ed. Frank N. Dempster and Charles J. Brainerd).

113 Burgess, N., Shapiro, J., and Moore, M. (1991). Neural network models of list learning. *Network: Computation in Neural Systems* 2 (4): 399–422.

114 Li, Z. and Hoiem, D. (2018). Learning without forgetting. *IEEE Transactions on Pattern Analysis and Machine Intelligence* 40 (12): 2935–2947.

115 Hinton, G.E. and Plaut, D.C. (1987). Using fast weights to deblur old memories. *Proceedings of the 9th Annual Conference of the Cognitive Science Society*. Hillsdale, NJ: Erlbaum, pp. 177–186.

116 Jung, H., Ju, J., Jung, M., and Kim, J. (2016). Less-forgetting learning in deep neural networks. *arXiv preprint arXiv:1607.00122*.

117 Razavian, A.S., Azizpour, H., Sullivan, J., and Carlsson, S. (2014). CNN features off-the-shelf: an astounding baseline for recognition. *Proceedings of the IEEE Conference on Computer Vision and Pattern Recognition Workshops*, pp. 806–813.

118 Donahue, J., Jia, Y., Vinyals, O. et al. (2014). DeCAF: A deep convolutional activation feature for generic visual recognition. *International Conference on Machine Learning*, pp. 647–655.

119 Kirkpatrick, J., Pascanu, R., Rabinowitz, N. et al. (2017). Overcoming catastrophic forgetting in neural networks. *Proceedings of the National Academy of Sciences of the United States of America* 114 (13): 3521–3526.

120 Zenke, F., Gerstner, W., and Ganguli, S. (2017). The temporal paradox of hebbian learning and homeostatic plasticity. *Current Opinion in Neurobiology* 43: 166–176.

121 Polikar, R., Upda, L., Upda, S.S., and Honavar, V. (2001). Learn++: An incremental learning algorithm for supervised neural networks. *IEEE Transactions on Systems, Man, and Cybernetics, Part C (Applications and Reviews)* 31 (4): 497–508.

122 Dai, W., Yang, Q., Xue, G.-R., and Yu, Y. (2007). Boosting for transfer learning. *Proceedings of the 24th International Conference on Machine Learning*.

123 Ren, B., Wang, H., Li, J., and Gao, H. (2017). Life-long learning based on dynamic combination model. *Applied Soft Computing* 56: 398–404.

124 Coop, R., Mishtal, A., and Arel, I. (2013). Ensemble learning in fixed expansion layer networks for mitigating catastrophic forgetting. *IEEE Transactions on Neural Networks and Learning Systems* 24 (10): 1623–1634.

125 Fernando, C., Banarse, D., Blundell, C. et al. (2017). PathNet: Evolution channels gradient descent in super neural networks. https://arxiv.org/abs/1701.08734.

126 Rakelly, K., Zhou, A., Finn, C. et al. (2019). Efficient off-policy meta-reinforcement learning via probabilistic context variables. *International Conference on Machine Learning*, pp. 5331–5340.

127 Nagabandi, A., Clavera, I., Liu, S. et al. (2018). Learning to adapt in dynamic, real-world environments through meta-reinforcement learning. *arXiv preprint arXiv:1803.11347*.

128 Finn, C., Abbeel, P., and Levine, S. (2017). Model-agnostic meta-learning for fast adaptation of deep networks. *arXiv preprint arXiv:1703.03400*.

129 Rothfuss, J., Lee, D., Clavera, I. et al. (2018). ProMP: Proximal meta-policy search. *International Conference on Learning Representations*.

130 Zintgraf, L., Shiarli, K., Kurin, V. et al. (2019). Fast context adaptation via meta-learning. *International Conference on Machine Learning*. PMLR, pp. 7693–7702.

131 Stadie, B.C., Yang, G., Houthooft, R. et al. (2018). Some considerations on learning to explore via meta-reinforcement learning. *arXiv preprint arXiv:1803.01118.*

132 Duan, Y., Schulman, J., Chen, X. et al. (2016). Rl^2: Fast reinforcement learning via slow reinforcement learning. *arXiv preprint arXiv:1611.02779.*

133 Wang, J.X., Kurth-Nelson, Z., Tirumala, D. et al. (2016). Learning to reinforcement learn. *arXiv preprint arXiv:1611.05763.*

134 Mishra, N., Rohaninejad, M., Chen, X., and Abbeel, P. (2018). A simple neural attentive meta-learner. *International Conference on Learning Representations.*

135 Lee, A.X., Nagabandi, A., Abbeel, P., and Levine, S. (2019). Stochastic latent actor-critic: deep reinforcement learning with a latent variable model. *arXiv preprint arXiv:1907.00953.*

136 Zintgraf, L., Shiarlis, K., Igl, M. et al. (2019). VariBAD: A very good method for bayes-adaptive deep RL via meta-learning. *International Conference on Learning Representations.*

137 Xie, A., Harrison, J., and Finn, C. (2020). Deep reinforcement learning amidst lifelong non-stationarity. *arXiv preprint arXiv:2006.10701.*

138 Goodfellow, I.J., Shlens, J., and Szegedy, C. (2015). Explaining and harnessing adversarial examples. *Proceedings of the International Conference on Learning Representations (ICLR).*

139 Moosavi-Dezfooli, S.-M., Fawzi, A., Fawzi, O., and Frossard, P. (2017). Universal adversarial perturbations. *Proceedings of IEEE CVPR*, pp. 1765–1773.

140 Papernot, N., McDaniel, P., Goodfellow, I. et al. (2017). Practical black-box attacks against machine learning. *Proceedings of the 2017 ACM on Asia Conference on Computer and Communications Security*, ser. ASIA CCS '17. New York, NY, USA: ACM, pp. 506–519. http://doi.acm.org/10.1145/3052973.3053009.

141 Carlini, N. and Wagner, D. (2017). Towards evaluating the robustness of neural networks. *IEEE Symposium on Security and Privacy (S&P)*, May 2017, pp. 39–57.

142 Bair, S., DelVecchio, M., Flowers, B. et al. (2019). On the limitations of targeted adversarial evasion attacks against deep learning enabled modulation recognition. *Proceedings of the ACM Workshop on Wireless Security and Machine Learning*, ser. WiseML 2019. ACM, pp. 25–30. http://doi.acm.org/10.1145/3324921.3328785.

143 Dong, Y., Liao, F., Pang, T. et al. (2018). Boosting adversarial attacks with momentum. *Proceedings of IEEE CVPR*, pp. 9185–9193.

144 Shi, Y., Davaslioglu, K., and Sagduyu, Y.E. (2019). Generative adversarial network for wireless signal spoofing. *Proceedings of the ACM Workshop on Wireless Security and Machine Learning*, ser. WiseML 2019. ACM, pp. 55–60. http://doi.acm.org/10.1145/3324921.3329695.

145 Sadeghi, M. and Larsson, E.G. (2019). Adversarial attacks on deep-learning based radio signal classification. *IEEE Wireless Communications Letters* 8 (1): 213–216.

146 O'shea, T.J. and West, N. (2016). Radio machine learning dataset generation with GNU radio. *Proceedings of the GNU Radio Conference*, volume 1.

147 Flowers, B., Buehrer, R.M., and Headley, W.C. (2019). Evaluating adversarial evasion attacks in the context of wireless communications. *IEEE Transactions on Information Forensics and Security* 15: 1102–1113.

148 Sadeghi, M. and Larsson, E.G. (2019). Physical adversarial attacks against end-to-end autoencoder communication systems. *IEEE Communications Letters* 23 (5): 847–850.

149 Tran, N.H., Bao, W., Zomaya, A. et al. (2019). Federated learning over wireless networks: optimization model design and analysis. *Proceedings of the IEEE Conference on Computer Communications (INFOCOM)*. IEEE, pp. 1387–1395.

150 Yang, Q., Liu, Y., Chen, T., and Tong, Y. (2019). Federated machine learning: concept and applications. *ACM Transactions on Intelligent Systems and Technology (TIST)* 10 (2): 1–19.

151 Amiri, M.M. and Gündüz, D. (2020). Federated learning over wireless fading channels. *IEEE Transactions on Wireless Communications* 19 (5): 3546–3557.

152 Yang, K., Jiang, T., Shi, Y., and Ding, Z. (2020). Federated learning via over-the-air computation. *IEEE Transactions on Wireless Communications* 19 (3): 2022–2035.

153 Niknam, S., Dhillon, H.S., and Reed, J.H. (2020). Federated learning for wireless communications: motivation, opportunities, and challenges. *IEEE Communications Magazine* 58 (6): 46–51.

154 Mills, J., Hu, J., and Min, G. (2019). Communication-efficient federated learning for wireless edge intelligence in IoT. *IEEE Internet of Things Journal* 7 (7): 5986–5994.

155 Wang, X., Han, Y., Wang, C. et al. (2019). In-edge AI: intelligentizing mobile edge computing, caching and communication by federated learning. *IEEE Network* 33 (5): 156–165.

156 Samarakoon, S., Bennis, M., Saad, W., and Debbah, M. (2019). Distributed federated learning for ultra-reliable low-latency vehicular communications. *IEEE Transactions on Communications* 68 (2): 1146–1159.

157 McMahan, B., Moore, E., Ramage, D. et al. (2017). Communication-efficient learning of deep networks from decentralized data. *Proceedings of the 20th International Conference on Artificial Intelligence and Statistics, PMLR 54*, pp. 1273–1282.

158 Polese, M., Bonati, L., D'Oro, S. et al. (2022). Understanding O-RAN: architecture, interfaces, algorithms, security, and research challenges. *arXiv preprint arXiv:2202.01032*.

159 Bonati, L., Polese, M., D'Oro, S. et al. (2020). Open, programmable, and virtualized 5G networks: state-of-the-art and the road ahead. *Computer Networks* 182: 1–28.

160 Bertizzolo, L., Tran, T.X., Buczek, J. et al. (2021). Streaming from the air: enabling drone-sourced video streaming applications on 5G open-RAN architectures. *IEEE Transactions on Mobile Computing* 1.

161 Bonati, L., D'Oro, S., Polese, M. et al. (2021). Intelligence and learning in O-RAN for data-driven NextG cellular networks. *IEEE Communications Magazine* 59 (10): 21–27.

162 Baldesi, L., Restuccia, F., and Melodia, T. (2022). ChARM: NextG spectrum sharing through data-driven real-time O-RAN dynamic control. *Proceedings of the IEEE Conference on Computer Communications (INFOCOM)*. IEEE.

163 Bertizzolo, L., Bonati, L., Demirors, E. et al. (2020). Arena: A 64-antenna SDR-based ceiling grid testing platform for Sub-6 GHz 5G-and-beyond radio spectrum research. *Computer Networks* 181: 107436.

20

Tactical Edge IoT in Defense and National Security

Paula Fraga-Lamas and Tiago M. Fernández-Caramés

Department of Computer Engineering, CITIC Research Center, Universidade da Coruña, A Coruña, Spain

Abstract

The deployment of internet of things (IoT) systems in Defense and National Security faces some limitations that can be addressed with Edge Computing approaches. The Edge Computing and IoT paradigms combined bring potential benefits, since they confront the limitations of traditional centralized cloud computing approaches, which enable easy scalability, real-time applications, or mobility support, but whose use poses certain risks in aspects like cybersecurity. This chapter identifies scenarios in which Defense and National Security can leverage commercial off-the-shelf (COTS) Edge IoT capabilities to deliver greater survivability to warfighters or first responders, while lowering costs and increasing operational efficiency and effectiveness. In addition, it presents the general design of a Tactical Edge IoT communications architecture, it identifies the open challenges for a widespread adoption and provides research guidelines and some recommendations for enabling cost-effective Edge IoT for Defense and National Security.

20.1 Introduction

The Internet of Things (IoT) is a distributed system that generates value through data by allowing heterogeneous physical objects to share information and coordinate decision-making. In industrial applications, IoT leads to significant improvements in efficiency and transparency in product development and distribution, as well as in supply chain tracking. It also influences how critical infrastructures are managed and maintained in a wide range of sectors [1], including manufacturing, transportation, logistics, human health and productivity, energy and utilities, home and environment management, or autonomous vehicles. Moreover, IoT redefines how human–machine interactions are performed, being capable of improving equipment performance and enhancing workforce safety [2].

According to a report of McKinsey (November 2021) [3], the potential economic value of IoT will grow from \$5.5 trillion to \$12.6 trillion by 2030. With respect to machine-to-machine (M2M) communications, the market is expected to reach \$83.23 billion by 2030, increasing at a compound annual growth rate of 23.2% [4]. Ericsson states that in 2027 [5], broadband IoT connections with 4G will account for 40% of cellular IoT connections (related to use cases with high throughput, low latency, and large data requirements), while massive IoT connections (associated with use cases

IoT for Defense and National Security, First Edition. Edited by Robert Douglass, Keith Gremban, Ananthram Swami, and Stephan Gerali.

that involve a large number of low-complexity, low-throughput, and low-cost devices with extended battery life) are expected to account for 51% of all cellular IoT connections.

Tactical scenarios are mainly characterized by limited resources, high levels of stress, and a lack of stability (e.g. they can be highly dynamic, complex and hostile, or mission-critical environments). Military and Public Safety (PS) agents are increasingly relying on IoT to support their tactical missions. Such IoT systems are lightweight and sufficiently powerful to run a range of applications to assist dismounted soldiers or vehicles on the move. Nevertheless, they are connected through wireless tactical networks with restricted bandwidth, reachability, reliability, and latency. As a result, tactical network nodes cannot just rely on secure access to cloud computing services to work. Instead, they must look into other options for leveraging *in situ* resources. The concept of Edge IoT involves the use of IoT devices with Edge Computing capabilities, where computation takes place at the network edge. Edge Computing has the potential to improve critical aspects of tactical environments: survivability, resilience, network connectivity, trust, and ease of deployment. Such capabilities can lead to an enhanced Situational Awareness (SA) and decision-making at the edge. Therefore, Tactical Edge IoT can help the military and PS agents in adapting to certain environments where adversaries are located in increasingly complex tactical scenarios.

This chapter continues the work presented five years ago in Ref. [6], where the authors reviewed the future role of IoT for Defense and PS. Thus, this chapter departs from such background knowledge, updating it to provide a comprehensive approach to Edge IoT applied to Defense and PS.

The remainder of this chapter is organized as follows. Section 20.2 introduces the essential concepts that will be used in the chapter. Section 20.3 reviews the opportunities created by current commercial off-the-shelf (COTS) Edge IoT applications for tactical environments. Section 20.4 presents some promising scenarios for Tactical Edge IoT. Section 20.5 overviews the general design of a Tactical Edge IoT communications architecture. Section 20.6 outlines challenges that hinder the adoption of Tactical Edge IoT technologies and introduces some recommendations for further research. Finally, Section 20.7 is dedicated to conclusions.

20.2 Background

20.2.1 Tactical Edge IoT drivers

IoT represents the convergence of different disciplines (e.g. electronics, sensors, networks, computing, communications, signal processing, or Artificial Intelligence (AI)). As a consequence, five main drivers related with these disciplines are fostering IoT rapid expansion. The first driver is the ever-increasing miniaturization and lower cost of powerful microelectronics (e.g. transducers, receivers, or processing units [e.g. microcontrollers, microprocessors, System-on-a-Chip (SoCs), Field-Programmable Gate Arrays (FPGAs), Graphics Processing Units (GPUs), Application-Specific Integrated Circuits (ASICs)].

The second driver is the rapid growth and development of wireless communications systems, from auto-identification and traceability technologies [7, 8] to broadband solutions [9], which are currently being linked to the expansion of 5G/6G wireless connectivity [10–13].

The third driver is the increase on available data storage and computational capabilities offered by System on a Chip (SoC) hardware, which keeps on improving at a rapid rate.

The fourth factor is the guarantee of a compute continuum from IoT to the edge and to the cloud [14], which enables the creation of hyper-distributed intelligent IoT applications. Most IoT applications are currently implemented on cloud computing-based platforms that enable centralized processing and data storage. However, the cloud itself can be seen as a point of failure, since it can be disrupted by attacks or maintenance tasks, which prevent the access to the deployed centralized services and thus block the whole system. In addition, reliable communications are needed to reach the cloud and in tactical and Disaster Relief (DR) situations, communications may be disrupted, denying such an access to the cloud. Furthermore, if an IoT system consists of a high number of connected devices, they will probably generate a lot of data exchanges with the cloud, which will lead to the saturation of the cloud if it is not scaled properly. Additionally, these cloud-based solutions have an inherent high latency and energy consumption.

Due to the aforementioned constraints, novel computing paradigms have emerged thanks to the improvement of IoT end-nodes, which are getting more powerful and efficient, and they have the computational capabilities required to implement end-to-end security mechanisms and high-security cryptographic cipher suites [15]. The main goal of some of the most relevant new paradigms is to process most of the data as close as possible to where end devices are located. Thus, the devices of the higher layers are freed from part of their data processing tasks and the amount of data exchanged with the end devices is reduced significantly [16]. For instance, Mist Computing is a paradigm in which data processing capabilities are moved from gateways to end devices. Edge Computing [17], offloads the cloud from tasks that can be handled by devices at the network edge, close to the end IoT nodes. Fog Computing [18] uses low-power devices on the edge for data acquisition, but data processing is performed at gateway devices. Cloudlets [19] make use of high-end computers that perform heavy processing tasks on the edge [20]. A review of the state of the art on the pervasive edge computing paradigm and its applications to industrial IoT (IIoT) can be found in Ref. [21].

The bandwidth, latency, security, and decentralization requirements of today's modern tactical applications are incompatible with centralized clouds directly connected to a large number of IoT devices. As a result, the centralized cloud computing landscape is rapidly becoming distributed and heterogeneous. The compute continuum is created by combining centralized cloud-based coordination and control with edge devices placed near IoT sensors and actuators. In addition, edge devices enable computing to be moved closer to the point where data are generated, thus reducing latency, increasing overall throughput, and improving security.

Finally, the fifth factor is the availability and rapid evolution of enabling technologies like augmented reality/mixed reality (AR/MR) [22], virtual reality (VR) [23], quantum computing [24], Cyber-Physical Systems and digital twins [25], blockchain [26] and distributed ledger technologies (DLTs), unmanned aerial vehicles (UAVs) [27, 28], or AI [29]. Specifically, the latter is related to the upsurge of innovative software solutions that involve large data processing in areas like AI/machine learning (ML) [30], Big Data, or analytics. Particularly, recent AI-enabled applications rely on supervised learning (where models are previously trained and then used), unsupervised learning (where data are fed into a system to extract data patterns), reinforcement learning (where real-time data are used to adjust the model), or federated learning (in which locally trained models are gathered to generate a more complete one while preserving data privacy).

IoT devices usually include transducers to collect large amounts of data on physical parameters. Such information is then processed by using some of the aforementioned technologies, which are mainly focused on data analytics and on extracting data insights. The knowledge gained from such an analysis can be applied for monitoring, automation, control, and prediction.

There is a lot of research in distributed and reliable ML models that take advantage of edge resources. For example, Park et al. [31] describe the transition from cloud-based training and inference towards Edge ML. Zhou et al. [32] combine edge and cloud approaches to guarantee the compute continuum. In Ref. [33] the authors investigate the convergence of IoT and AI from the perspective of a collaboration between edge and cloud. Deng et al. [34] provide insights into two main research directions: AI for the edge (also known as Intelligence-enabled Edge Computing) (which focuses on the use of AI for constrained optimization problems in Edge Computing) and AI on the edge (which studies distributed approaches that run AI models efficiently on the edge). Thus, new types of design trade-offs have to be reached, including factors such as energy efficiency, latency, privacy, and security.

The previously mentioned five drivers, which can be found in the different layers of the IoT technology stack, will lead the next-generation of IoT networks, notably Edge Computing-based IoT networks, or Edge IoT, as it will be called from now on in this chapter.

20.2.2 Defense and Public Safety

Defense and PS entities are essential for preserving security and when it is necessary to react to emergency situations and natural disasters. For these operations, the U.S. Department of Defense (DoD) commonly uses the term humanitarian assistance and disaster relief (HADR). Previous literature has also used the term public protection disaster relief (PPDR) instead of PS. Such literature mentions PPDR radio communications as a combination of two important areas of emergency response [6]:

- **Public Protection (PP).** This kind of communications are used by entities dedicated to maintain law and order, to protect life and property, and to deal with emergency situations.
- **DR.** These types of communications are used by organizations that fight societal problems that threaten human life. Such problems are usually related to health, economic difficulties, or environmental issues, which can be derived from accidents, sudden natural events, or long-term human activities.

PS agents include police officers, firefighters, border and custom guards, coast guards, medical responders, transportation agents, and other organizations that are among the first ones to go to scenarios where critical situations have occurred. The relationships between such organizations depend on the specific situation, context, and legislation. Figure 20.1 gives an outline of their primary function and the relationships among them [6].

Crisis management aims to minimize the impact and damage to people and property, ensuring communication capabilities in difficult circumstances where critical infrastructure is frequently

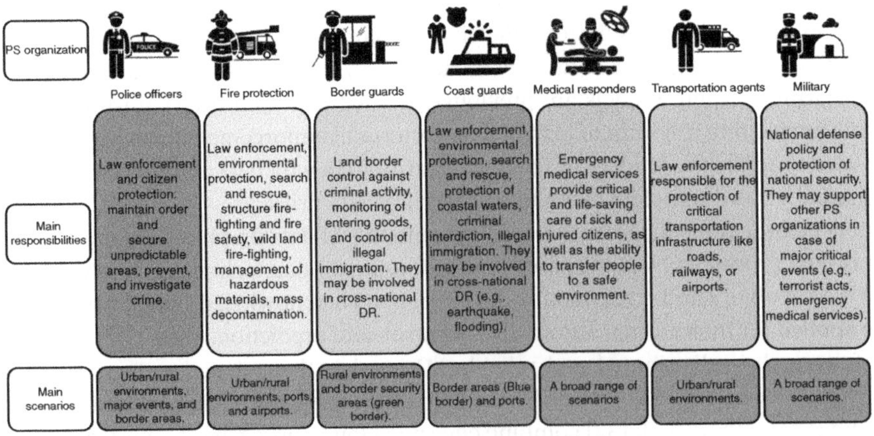

Figure 20.1 Overview of the main PS organizations, their responsibilities, and main scenarios.

degraded or destroyed. It requires a set of capabilities described by standardization bodies (e.g. The Critical Communication Association (TCCA) [35] or European Telecommunications Standards Institute (ETSI) [36]).

For instance, Safety Communications (SAFECOM) Guidance for Fiscal Year 2021 [37] includes a compilation of emergency communications systems and technical standards.

Natural disasters and other emergencies (e.g. COVID-19 pandemic) are frequently unforeseen events that cause panic among civilians and disrupt existing resources. In such cases, various PS organizations may be involved in order to respond to large-scale disasters, when civil communications infrastructures must be operational to send information and warnings to citizens.

In addition, when coordinating DR efforts and establishing SA, PS agents need to exchange data in a timely manner.

Furthermore, interoperability issues may be a problem for fulfilling certain security requirements, like the ones related to communications and data protection. To create and maintain a Common Operational Picture (COP) among PS agencies, as well as between field and central command and control (C2), different types of data coming from heterogeneous devices must be shared.

The remainder of this chapter will be mainly focused on Tactical IoT with Edge Computing capabilities from a military standpoint. This is due to the fact that such a perspective allows for covering some of the most critical situations and challenging scenarios while fulfilling technical and operational requirements of PS organizations.

In the last few years, extensive research has been published focusing on evolving PS communications [38–44]. Some papers focused on the utilization of Long Term Evolution (LTE) and 5G/6G for implementing advanced PS networks [45–49] or specifically, in device-to-device (D2D) communications [50–53]. In addition, extensive research has been published in the optimal deployment of unmanned aircraft systems (UAS) for PS networks [54–59].

There are multiple articles that discuss the diverse parameters that impact the IoT technologies used in Defense and PS applications. For instance, in Ref. [60] the authors describe a fault detection mechanism that makes use of a military network divided into clusters. Other authors have proposed layered architectures and discussed different applications related to, for instance, weapon control solutions [61]. A different research line is presented in Ref. [62], where the authors present a multi-level authentication service that is lightweight, centralized on a cloud, scalable, and that considers the timing constraints of the IoT devices used by PS responders.

Nevertheless, there are not many articles that deal with the merger of Edge Computing approaches and IoT devices in tactical environments to comply with the stringent and critical quality of service (QoS) requirements of the battlefield or C2. For example, Singh et al. [63] present an IoT health platform for the military. Such a platform includes a semantic-edge network model that serves both tactical and non-tactical information. A more sophisticated approach is described in Ref. [64], where the authors propose an IoT framework to monitor troops. The framework is able to enhance decisions that need to be taken during campaigns or battles. The framework uses Mist Computing and ML to further minimize delays. Such a framework is evaluated through simulations of EdgeCloudSim, which show average low network latencies (0.01 s) and failure rates (0.25%) with a high QoS. The main shortcoming of the system includes its inability to perform successfully with an exponential increase in the number of mobile devices. To overcome to a certain extent this constraint, clusters of mobile devices were deployed by the researchers in a specific area and then applied K-nearest neighbors (KNN) to cluster small number of mobile devices. In addition, the system uses distributed data storage and decentralized data analytics. Another relevant publication is a PhD dissertation [65] that demonstrates how federated learning can be adapted to fit in Edge Computing devices that are connected to a tactical data link.

20.3 Compelling COTS Edge IoT Applications

PS operations are currently carried out in complicated, dynamic, and often unpredictable circumstances. In particular, military and warfare scenarios have evolved substantially in recent years.

As a result, there is a pressing need for military technology to evolve at a fast pace [66]. While military contractors and manufacturers develop new and improved technologies, civilian technology advances at a much faster pace. Moreover, the associated expenses (for development and acquisition) in Defense are substantially greater, and the cycle time is much longer than COTS technology.

This section overviews some COTS Edge IoT applications that may be relevant for Defense and PS environments. Two kinds of deployments must be considered: the ones aimed at creating ad hoc IoT for the military field and the ones related to already-developed civil deployments that need to be protected (e.g. smart cities).

- **Transportation.** The authors of Ref. [67] propose a blockchain-enabled Edge IoT framework for maritime transportation systems. Blockchain and smart contracts help in the validation of each block's transactions at edge nodes by estimating their lifetime and trustworthiness, as well as mitigating many forms of security threats. Support Vector Machine (SVM) and Convolutional Neural Network (CNN) are used to predict the number of malicious entities and increase the prediction accuracy of vessel monitoring units.
- **Energy Efficiency.** In Ref. [68] the authors study how to leverage heterogeneous computation resources at the Edge to optimize energy efficiency while satisfying the delay requirements of Mobile Edge Computing (MEC) scenarios. As a result, they come up with an iterative framework that considers two sub-problems: transmission power allocation and computation offloading. The authors indicate as a limitation that their work does not consider heterogeneous edge servers deployed with several types of computation resources (e.g. Central Processing Unit (CPU), GPU). An energy-efficient MEC is also presented in Ref. [69], specifically for the case of UAVs and their specific constraints (e.g. limited flight time, power constraints).
- **Supply Chain/Logistics.** In Ref. [70] the authors present a lightweight authentication protocol for supply chains in a 5G MEC scenario that makes use of blockchain and Radio Frequency Identification (RFID).
- **Smart Cities.** Khan et al. [71] review the state-of-the-art of Edge Computing applications for smart cities. The authors devise a comprehensive taxonomy and discuss the main requirements. Open challenges are outlined to guide further research.

The above-mentioned examples illustrate how the civil sector is leveraging Edge IoT to enable new business models. Concerns, like cybersecurity, cooperative load balancing, collaborative Edge Computing [71], or green IoT [72], still remain as open research.

20.4 Target Scenarios for Tactical Edge IoT

The Network-Centric Warfare (NCW) paradigm [73] connects battlefield assets back to headquarters. Such a concept provides benefits by facilitating the exchange of information among users in a secure and timely way. In addition, the NCW paradigm combines three domains: the physical domain, which generates data where events and actions take place; the information domain, which transmits and stores data; and the cognitive domain, which processes and analyzes data to enable decision-making and mission planning. NCW's three domains correspond to the underpinnings of today's commercial Edge IoT.

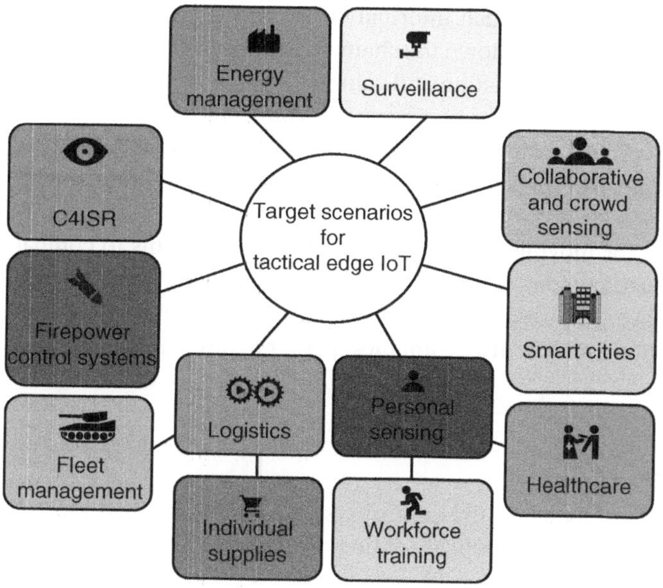

Figure 20.2 Main target scenarios for Tactical Edge IoT in defense and public safety.

In network-centric C2 operations, responsibility is delegated to the battlefield edge [74], creating the so-called internet of battlefield things (IoBT), which can bring together everything on the battlefield that can aid in making informed decisions. However, these dynamics require network paradigms that can ensure network efficiency. In Ref. [74] the authors bring together Information-Centric Networking (ICN) with Software-Defined Networking (SDN) to fulfill such requirements.

This section analyzes some of the most relevant Tactical IoT scenarios with Edge services, which are summarized in Figure 20.2. Applications for command, control, communications, computers, intelligence, surveillance, and reconnaissance (C4ISR) and fire-control systems have dominated the adoption of IoT-related technology for Defense and PS, since sensors are primarily used to collect and communicate data in order to improve C2. Although IoT and Edge Computing technologies have previously been used for applications related to logistics and training, their integration with other systems is often limited.

As it was previously stated, Edge IoT capabilities can be used to provide superior SA in the battlefield. Commanders can make decisions based on real-time analysis derived from the integration of AI/ML extracted data from unmanned/manned sensors and field reports. Ground-based sensors and cameras, as well as human or unmanned devices, vehicles, or soldiers, provide a wide range of information to commanders. The mentioned IoT devices are able to scan the mission environment and then send information to an Edge Computing server, which can be in a forward base. Part of such information may be collected by a Command Center, where it will be processed and fused with information from other sources.

20.4.1 C4ISR

Command, Control, Communications, Computers, Intelligence, Surveillance and Reconnaissance (C4ISR) systems provide advanced SA by deploying millions of sensors across a variety of platforms. Surveillance satellites, aerial platforms, UAVs, ground stations, and soldiers in the battlefield collect different types of data (e.g. radar, infrared, video). Such information is collected, processed, and stored in a platform, which manages the data up and down the chain of command. These platforms provide a COP, allowing for better battlefield coordination and control.

Central operation centers, which receive data from platforms, provide comprehensive SA to high-level military echelons. Lower levels (e.g. platoons and soldiers) have access to data in their specific areas. Combat pilots, for example, receive prioritized data streams that are combined with data from their own sensor systems.

Jiao et al. [75] present a mechanism for moving services from cloud-based systems to combat platforms, thus allowing an effective deployment of C4ISR systems in scenarios where communications bandwidth is limited. Moreover, the researchers propose a two-level search algorithm based on improved quantum evolution, which also combines path planning and pair-wise exchange techniques.

20.4.2 Firepower Control Systems

Sensor networks and advanced AI/ML analytics in fire-control systems enable fully automated reactions to real-time threats and pinpoint accuracy delivery of firepower. Smart weapons can also follow moving targets or be diverted in mid-flight.

20.4.3 Logistics

Multiple low-level sensors are employed in logistics in the Defense sector. Their deployment is mainly focused on safe contexts (e.g. non-combat settings) with infrastructure and human interaction to improve back-end processes. The examples in the following subsections are divided into two main categories: fleet management and individual supplies.

20.4.3.1 Fleet Management

On-board sensors within aircraft and ground vehicle fleets can be used, for instance, to assess the system performance, to monitor part status, or to evaluate the condition of a vehicle and its subsystems. In addition, it is possible to alert users when certain goods (such as fuel or oil) need to be replenished or when a failure is expected. With an Edge Computing approach, sensors would send out real-time alerts, potentially lowering the danger, and the probability of a catastrophic event. Although Edge IoT deployment has upfront expenses, it can result in considerable long-term savings across the different processes.

Defense has an opportunity to leverage the advances in data sharing of the automotive industry (e.g. IIoT-connected vehicles or autonomous vehicles).

There are other relevant parameters that can be useful in real-time fleet management, like the location of the vehicles, their speed, engine status, fuel efficiency, total weight, or the carried load. In addition, in cases when shipments need to be monitored, the location and status of the containers can be tracked in order to detect potential problems.

Regarding aircrafts, current jet engines include sensors that collected huge amounts of data per flight (in the order of terabytes) [76]. The real-time edge analytics information, when paired with in-flight data, can help reduce fuel expenditures, detect small defects, and minimize journey times. It also allows for providing preventive maintenance, which results into longer life cycles (since failures are reduced or totally prevented) and into dedicating less time to repairs.

20.4.3.2 Individual Supplies

Individual supplies can be tracked using standardized barcodes, RFID tags, specific smart labels [77] or a wide range of auto-id and traceability technologies [7]. Edge IoT allows the military to monitor the status of the supply chain in real time, so it enables knowing when goods are delivered, moved, deployed, or consumed.

In addition, real-time tracking can be useful for soldiers, when it is necessary to take a proactive approach to logistics. Critical soldier supplies such as water, food, batteries, or bullets can be tracked and alarms provided if a resupply is required or when an unexpected event happens. Information aggregated from military units (e.g. groups of soldiers, companies, or battalions) can be examined with AI/ML techniques on the edge for further real-time supply improvements for tactical and emergency forces by taking into account context awareness (e.g. characteristics such as the surroundings, body type, daily intake, or weather information).

20.4.4 Smart City Operations

Existing IoT smart city infrastructure can be potentially employed for military operations [78]. For instance, hazardous chemicals can be monitored through environmental sensors, while different sensors have the capacity to track the behavior of people that act suspiciously.

Taking advantage of real-time data offered by current infrastructure could be crucial from some operations. Nevertheless, security risks, such as equipment sabotage or false information, may arise. According to Wang et al. [79], attacks may fall into main four categories: (i) firewalls, software patches, and system design; (ii) malware, security policies, and human factors; (iii) third-party chains and insider threats; and (iv) database schemas and encryption technologies. For instance, an investigation of the different security threats of multi-access MEC from a physical layer perspective is presented in Ref. [80]. Other articles like Ref. [81] have previously analyzed privacy security in Edge Computing and then proposed an efficient intelligent offloading method for smart cities that preserves privacy.

20.4.5 Soldier Healthcare and Workforce Training

Wearable technology embedded within military equipment (e.g. weapon systems, combat suits) allows for the ubiquitous tracking of physical activity and the collection of operational context data as well as biometrics [82]. Inferring and monitoring physical or psychological conditions from context-aware information in real-time and taking preventive actions could be critical. For example, soldiers can be notified of different events (e.g. dehydration, sleep deprivation, elevated heart rate, low blood sugar, speech patterns, likelihood of internal injury based on earlier traumas) and, when needed, notifications can be sent to a base hospital so that the medical team take the necessary actions [63].

In addition, two types of sensing can be distinguished: participatory and opportunistic. The latter could be very useful for undercover troops conducting reconnaissance missions in urban areas.

Also, consider the existence of dismounted soldiers equipped with a mobile device and applications (e.g. Android Team Awareness Kit (ATAK) android app as edge node that can send and receive SA data to and from the tactical operations center [78]).

IoT can also be utilized for achieving an improved experience for training, gaming, and simulation exercises. For example, wearable devices can be used to simulate real-life fighting and to track the workforce [83].

More details on IoT equipment for soldiers can be found in Ref. [6].

20.4.6 Collaborative and Crowd Sensing

Collaborative sensing is a way of sharing the information collected from sensors across mobile devices, often by making use of reliable short-range communications. To supplement their own sensing methods, IoT nodes could use additional sensors. The data fusion information can be made available to soldiers once potential security threats (such as trust and authentication) have been handled.

By pairing sensors with mission assignments, IoT can make ad hoc Intelligence Surveillance and Reconnaissance (ISR) missions easier. As a result, sensors and platforms would not need to be overly equipped to handle missions since they can rely on collaborative Edge Computing-based sensing capabilities to accommodate specific needs on demand.

To create a COP, resource-rich devices may collect data from multiple sources. Much of these data might be stored and processed locally. As a result, Edge Computing functions at a higher level would help in reducing response times and the need for backhaul connections.

Crowdsensing has the potential to be a low-cost tool for flexible real-time monitoring of broad regions, complementing services that may be available in smart cities. Nevertheless, security has to be carefully considered [84]. Data validation is another issue that is further hampered by the great heterogeneity of the devices. Device capabilities and performance may change over time. Low battery levels on a smartphone could result in occasional Global Positioning System (GPS) position updates, resulting in geo-tagging inaccuracies. Oversampling and filtering outlier values are common approaches to ensure data quality. In addition, reputation methods that give trustworthy sensing for PS can be used [85].

Moreover, inherent privacy issues may jeopardize crowdsensing services. For instance, when monitoring certain soldier activities, geo-tagging and timestamping may be needed, which can lead to the revelation of the location of such soldiers. The metadata acquired about devices by performing sensing activities is another privacy concern.

Furthermore, when processing data, the existence of attacks against AI/ML systems shall be carefully considered [86].

20.4.7 Energy Management

The use of real-time IoT data and predictive algorithms can aid in a better understanding of usage trends and drastically reduce military energy expenses.

20.4.8 Smart Surveillance

Real-time remote facility monitoring for security threats is enabled by security cameras and sensors, as well as advanced AI/ML image processing and pattern recognition software. In the case of maritime environments, it is possible to embed different types of sensors into helicopters, airplanes, UAVs, or ships. Thus, IoT solutions allow for monitoring marine activities and ship traffic over large areas, as well as sensing environmental conditions or the status of dangerous oil cargos.

Hazardous environmental parameters can also be monitored by Edge IoT systems, which can alert users fast when certain conditions are met.

20.5 Communications Architecture

After reviewing the current state of the art and the requirements of potential Tactical Edge IoT applications [6], a communications architecture like the one depicted in Figure 20.3 can be devised. As shown in the figure, it is a four-layer Edge Computing architecture that supports data collection and that is composed by the following main components:

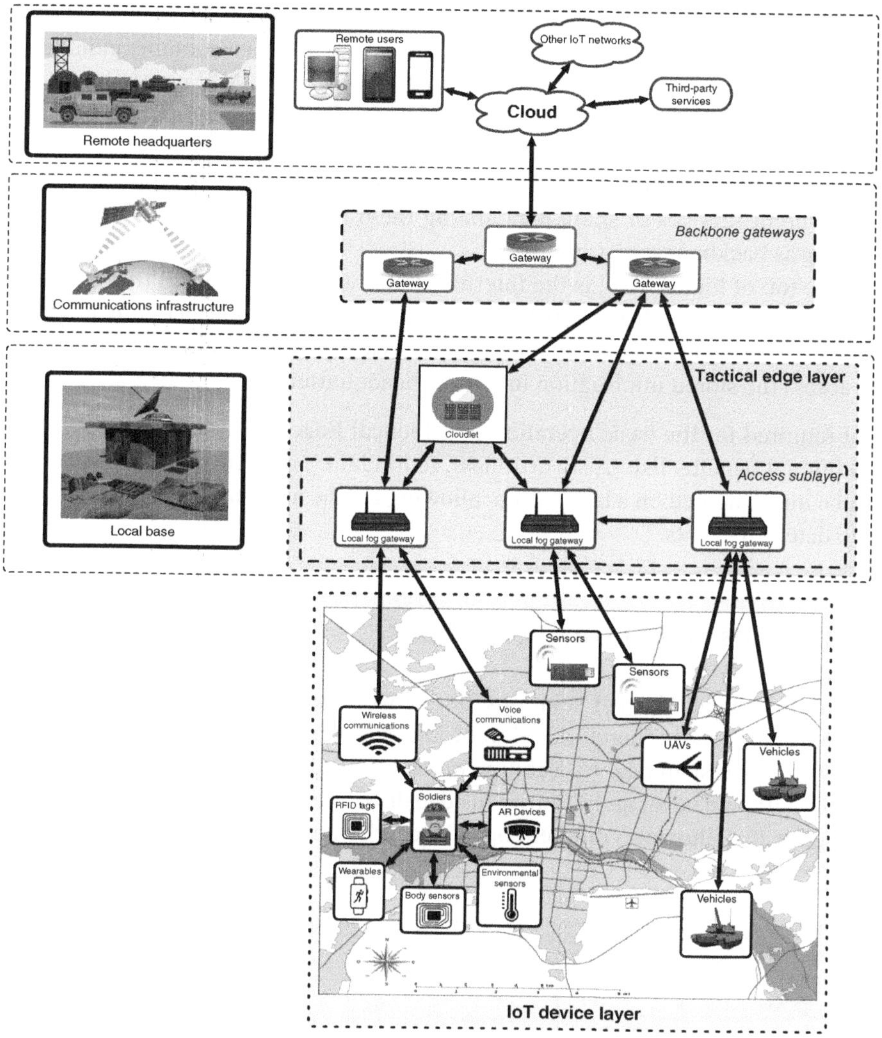

Figure 20.3 Communications architecture of a Tactical Edge IoT system.

- The node layer is at the bottom. Such an IoT Device Layer includes IoT devices that a dismounted soldier might carry (e.g. RFID tags, wearables, body sensors, environmental sensors, AR/VR devices, and handheld radio), as well as other systems like UAVs, vehicles on the move or sensors deployed on the mission scenario.
- Each IoT device exchanges data via a wireless connection with a Tactical Edge Layer gateway, which is usually the one that is physically closest. All local fog gateways are part of a Fog Computing sub-layer (Access Sublayer) that provides services with low latency requirements such as sensor fusion, AI/ML services, position services, or data caching for streaming content to AR/VR devices. Although most of the time only one gateway provides fog services to a single IoT device, the Access Sublayer gateways can cooperate to provide complex and more sophisticated services (e.g. to distribute compute-intensive tasks, to make it easier for the collaboration between remote IoT devices). Local gateways can be Single-Board Computers (SBCs) or similar devices, which are reduced-size low-cost computers that can be quickly deployed in a tactical

scenario. As a result, the architecture assumes that the devices that act as gateways can be easily scattered throughout the battlefield. In addition, the Tactical Edge Layer contains cloudlets, which can perform compute-intensive activities like rendering or AI/ML processing. The cloudlet response latency is significantly lower than that of standard cloud computing systems because it is close to the IoT devices that request its services.

- In the upper layer, the top-level gateway is the point of entry to the lower layers, while the other gateways provide different services or share data among them to reduce the latency response from the cloud, acting as backbone gateways.
- Finally, located at the top of Figure 20.3, is the internal cloud, where the services that demand more processing power are executed. Third-party systems, part of the military IT core, are also connected to the cloud. The cloud makes available certain services to remote users (e.g. commanders that need to access the stored information in remote headquarters).

Although it is not required for the basic operation of a Tactical Edge IoT system, a blockchain or DLT provides additional benefits like trustworthiness, redundancy, or security [87]. Moreover, smart contracts can be implemented on a blockchain, allowing for the automation of certain operations in response to detected events.

20.6 Main Challenges and Recommendations

Although deployments of Tactical Edge IoT solutions for real-world scenarios have already begun, they face major challenges [78, 88]. Therefore, relevant issues still need to be tackled, like trustworthiness (e.g. algorithm transparency, traceability, privacy, and data integrity); capacity (e.g. communications bandwidth and coverage); security in edge distributed architectures; heterogeneity; and scalability. Figure 20.4 illustrates different open research lines.

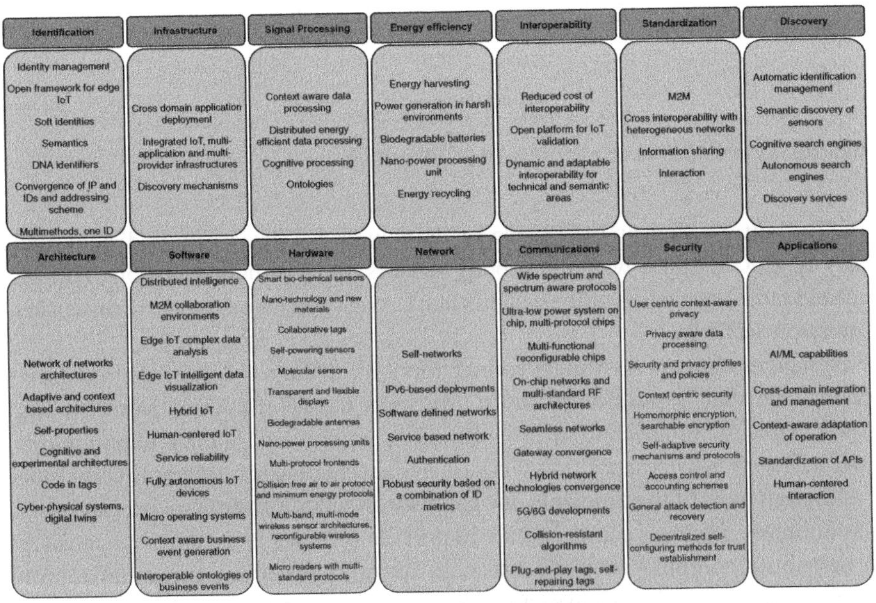

Figure 20.4 Ongoing Tactical Edge IoT research. Adapted from Fraga-Lamas et al. [6].

Some recommendations can be provided to future researchers:

- In order to define metadata across very heterogeneous and different domains, ontologies are required [78].
- With respect to data management, the communications format should include automatic data descriptors allowing for more efficient data management and storage across domains [78].
- Tactical IoT systems must take into account that the architecture's various nodes (e.g. mist nodes, Edge Computing devices, and cloudlets) have varying capacities in terms of communications, computation, storage, and power. The communications format must be supported across platforms, which implies that it should not depend on the used hardware or on underlying software. In addition, developers should consider that IoT devices make use of relatively disruptable networks, which have to tolerate faults and should support requirements related to QoS.
- The use of open standards is highly recommended for military and civilian applications, since it is a way for ensuring long-term interoperability.
- Energy efficiency and low-latency architectures for Tactical Edge IoT systems pose multiple challenges. Therefore, developers should consider aspects like the use of low-power communication technologies (e.g. ZigBee, Wi-Fi Hallow, Long Range [LoRa], Long Range Wide Area Network [LoRaWAN]), the smart management of the radio spectrum or the creation of distributed AI-based solutions able to achieve low inference latencies despite the requirements related to training and learning. In addition, renewable energy sources and energy-harvesting techniques should be considered in order to reduce the dependence on traditional energy source like batteries.
- The protocols used for exchanging messages have to support multi-level security mechanisms, since military information often requires security at various levels [78]. Specifically, security mechanisms are needed to protect Tactical Edge IoT systems from attacks at physical, network, and application levels, as well as from encryption attacks and software vulnerabilities. Moreover, AI learning processes should be protected against adversarial attacks.
- The deployment of 5G/6G networks supposes significant reductions on latency and increases in uplink/downlink data rates. Such changes will enable moving part of the processing power to the edge, where more powerful Edge Computing devices will be demanded. In addition, the popularization of new devices with native global connectivity will require the convergence of Tactical Edge IoT systems with 5G/6G networks and other related technologies (e.g. multi-band radios for low-bandwidth scenarios, mobile ad hoc networks (MANETS) or the use of defensive countermeasures).
- The military should explore forming a specialized technology group made up of military personnel to test new technologies and gather real-world input early on the development process. This evaluation has the potential to create innovative new uses for Edge IoT devices. Its twofold purpose would be to find devices and systems with possible applications, as well as to discover new approaches for completing missions employing COTS.
- The delivery of web-based services can be performed through platform as a service (PaaS) solutions without developing or managing infrastructure, which results in systems that provide more flexibility and scalability for adjustments and updates. The adoption of PaaS in military environments imposes additional challenges, requiring the implementation of security processes by private contractors.
- There is a need for the creation of comprehensive trusted architectures that can fulfill all the military Tactical Edge IoT requirements.

- Governments and Defense may invest in enabling technologies to improve Tactical Edge IoT implementation. Further integration of other digital enabling technologies (e.g. quantum computing, digital twins, blockchain and DLTs, functional electronics, UAVs, AR/VR) is needed.
- Further collaboration with private entities is necessary for updating current Edge IoT systems with the latest technologies. Civil companies are hesitant to collaborate with the military due to cultural gaps, as well as differences when managing intellectual property. For example, private entities may consider as much too challenging the development of certain military Tactical Edge IoT solutions, especially when having to deal with complex and demanding operational requirements. In addition, Tactical Edge IoT implementation will require the compromise of all stakeholders.

20.7 Conclusions

This chapter analyzed some of the ways of how the Defense industry can take advantage of the commercial Edge IoT transformation. Essential issues were discussed concerning the development of Edge IoT applications for the military and PS domains. Specifically, potential Tactical Edge IoT application scenarios were described, like C4ISR, fire-control systems, logistics, smart cities, healthcare, training, crowd sensing, energy management, and smart surveillance. In addition, the design of a generic Tactical Edge IoT communications architecture was presented.

Furthermore, the fact that Defense presents additional challenges to COTS Edge IoT systems was emphasized, mainly posed by tactical surroundings, as well as the inner complex nature of operations and networks. Governments and the Defense sector might gain a competitive advantage by utilizing existing COTS technologies and business methods. As a result, some recommendations were provided for enabling cost-effective Tactical Edge IoT.

Acknowledgments

This work has been funded by the Xunta de Galicia (by grant ED431C 2020/15, and grant ED431G 2019/01 to support the Centro de Investigación de Galicia "CITIC"), the Agencia Estatal de Investigación of Spain, MCIN/AEI/10.13039/501100011033 by grant PID2020-118857RA-100 (ORBALLO), and ERDF funds of the EU (FEDER Galicia 2014-2020 and AEI/FEDER Programs, UE).

References

1 Ferrag, M.A., Friha, O., Maglaras, L. et al. (2021). Federated deep learning for cyber security in the Internet of Things: concepts, applications, and experimental analysis. *IEEE Access* 9: 138509–138542. https://doi.org/10.1109/ACCESS.2021.3118642.

2 Blanco-Novoa, O., Fraga-Lamas, P., Vilar-Montesinos, M.A., and Fernández-Caramés, T.M. (2020). Creating the internet of augmented things: an open-source framework to make IoT devices and augmented and mixed reality systems talk to each other. *Sensors* 20 (3328). https://doi.org/10.3390/s20113328.

3 Chui, M., Collins, M., and Patel, M. (2021). IoT value set to accelerate through 2030: where and how to capture it. November 2021. https://www.mckinsey.com/business-functions/mckinsey-digital/our-insights/iot-value-set-to-accelerate-through-2030-where-and-how-to-capture-it (accessed 29 October 2022).

4 Allied Market Research (2021). Cellular M2M Market by Service: Global Opportunity Analysis and Industry Forecast, 2020–2030. August 2021. https://www.alliedmarketresearch.com/cellular-m2m-market-A13086 (accessed 29 October 2022).

5 Ericsson Mobility Report (2021). https://www.ericsson.com/4ad7e9/assets/local/reports-papers/mobility-report/documents/2021/ericsson-mobility-report-november-2021.pdf (accessed 29 October 2022).

6 Fraga-Lamas, P., Fernández-Caramés, T.M., Suárez-Albela, M. et al. (2017). A review on Internet of Things for defense and public safety. *Sensors* 16 (1644). https://doi.org/10.3390/s16101644.

7 Fraga-Lamas, P., Varela-Barbeito, J., and Fernández-Caramés, T.M. (2021). Next generation auto-identification and traceability technologies for industry 5.0: a methodology and practical use case for the shipbuilding industry. *IEEE Access* 9: 140700–140730. https://doi.org/10.1109/ACCESS.2021.3119775.

8 Fernández-Caramés, T.M., Fraga-Lamas, P., Suárez-Albela, M., and Castedo, L. A methodology for evaluating security in commercial RFID systems. (Crepaldi, P.C. and Pimenta, T.C., eds) *Radio Frequency Identification*. IntechOpen, 37–63.

9 Fraga-Lamas, P., Castedo-Ribas, L., Morales-Méndez, A., and Camas-Albar, J.M. (2016). Evolving military broadband wireless communication systems: WiMAX, LTE and WLAN. *Proceedings of the 2016 International Conference on Military Communications and Information Systems (ICMCIS)*, pp. 1–8. https://doi.org/10.1109/ICMCIS.2016.7496570.

10 Liu, Y., Yuan, X., Xiong, Z. et al. (2020). Federated learning for 6G communications: challenges, methods, and future directions. *China Communications* 17 (9): 105–118. https://doi.org/10.23919/JCC.2020.09.009.

11 Shafique, K., Khawaja, B.A., Sabir, F. et al. (2020). Internet of Things (IoT) for next-generation smart systems: a review of current challenges, future trends and prospects for emerging 5G-IoT scenarios. *IEEE Access* 8: 23022–23040. https://doi.org/10.1109/ACCESS.2020.2970118.

12 Ali, K., Nguyen, H.X., Vien, Q.T. et al. (2021). Review and implementation of resilient public safety networks: 5G, IoT, and emerging technologies. *IEEE Network* 35 (2): 18–25. https://doi.org/10.1109/MNET.011.2000418.

13 Alsamhi, S.H., Afghah, F., Sahal, R. et al. (2021). Green Internet of Things using UAVs in B5G networks: a review of applications and strategies. *Ad Hoc Networks* 117. https://doi.org/10.1016/j.adhoc.2021.102505.

14 Thanh, N.H., Kien, N.T., Hoa, N.V. et al. (2021). Energy-aware service function chain embedding in edge-cloud environments for IoT applications. *IEEE Internet of Things Journal* 8 (17): 13465–13486. https://doi.org/10.1109/JIOT.2021.3064986.

15 Suárez-Albela, M., Fraga-Lamas, P., Castedo, L., and Fernández-Caramés, T.M. (2019). Clock frequency impact on the performance of high-security cryptographic cipher suites for energy-efficient resource-constrained IoT devices. *Sensors* 19 (1). https://doi.org/10.3390/s19010015.

16 Suárez-Albela, M., Fraga-Lamas, P., and Fernández-Caramés, T.M. (2018). A practical evaluation on RSA and ECC-based cipher suites for IoT high-security energy-efficient fog and mist computing devices. *Sensors* 18 (3868). https://doi.org/doi:10.3390/s18113868.

17 De Donno, M., Tange, K., and Dragoni, N. (2019). Foundations and evolution of modern computing paradigms: cloud, IoT, Edge, and Fog. *IEEE Access* 7: 150936–150948. https://doi.org/10.1109/ACCESS.2019.2947652.

18 Bonomi, F., Milito, R., Zhu, J., and Addepalli, S. (2012). Fog computing and its role in the Internet of Things. *Proceedings of the 1st Edition of the MCC Workshop on Mobile Cloud Computing*, Helsinki, Finlad (17 August 2012), pp. 13–16.

19 Dolui, K. and Datta, S.K. (2017). Comparison of edge computing implementations: fog computing, cloudlet and mobile edge computing. *Proceedings of the Global Internet of Things Summit (GIoTS)*, Geneva, Switzerland (6–9 June 2017).

20 Suárez-Albela, M., Fernández-Caramés, T.M., Fraga-Lamas, P., and Castedo, L. (2017). A practical evaluation of a high-security energy-efficient gateway for iot fog computing applications. *Sensors* 17 (9): 1978. https://doi.org/10.3390/s17091978.

21 Narayanan, A., De Sena, A.S., Gutierrez-Rojas, D. et al. (2020). Key advances in pervasive edge computing for industrial Internet of Things in 5G and beyond. *IEEE Access* 8: 206734–206754. https://doi.org/10.1109/ACCESS.2020.3037717.

22 Vidal-Balea, A., Blanco-Novoa, O., Fraga-Lamas, P. et al. (2020). Creating collaborative augmented reality experiences for industry 4.0 training and assistance applications: performance evaluation in the shipyard of the future. *Applied Sciences* 10 (9073). https://doi.org/10.3390/app10249073.

23 Gace, I., Jaksic, L., Murati, I. et al. (2019). Virtual reality serious game prototype for presenting military units. Proceedings of the 2019 15th International Conference on Telecommunications (ConTEL), pp. 1–6. https://doi.org/10.1109/ConTEL.2019.8848505.

24 Fernández-Caramés, T.M. (2020). From pre-quantum to post-quantum IoT security: a survey on quantum-resistant cryptosystems for the Internet of Things. *IEEE Internet of Things Journal* 7 (7): 6457–6480. https://doi.org/10.1109/JIOT.2019.2958788.

25 Fraga-Lamas, P., Lopez-Iturri, P., Celaya-Echarri, M. et al (2020). Design and empirical validation of a bluetooth 5 fog computing based industrial CPS architecture for intelligent industry 4.0 shipyard workshops. *IEEE Access* 8: 45496–45511. https://doi.org/10.1109/ACCESS.2020.2978291.

26 Fraga-Lamas, P. and Fernández-Caramés, T.M. (2020). Fake news, disinformation, and deepfakes: leveraging distributed ledger technologies and blockchain to combat digital deception and counterfeit reality. *IT Professional* 22 (2): 53–59. https://doi.org/10.1109/MITP.2020.2977589.

27 Fraga-Lamas, P., Ramos, L., Mondéjar-Guerra, V., and Fernández-Caramés, T.M. (2019). A review on IoT deep learning UAV systems for autonomous obstacle detection and collision avoidance. *Remote Sensing* 11 (18): 2144. https://doi.org/10.3390/rs11182144.

28 Fernández-Caramés, T.M., Blanco-Novoa, O., Suárez-Albela, M., and Fraga-Lamas, P. (2019). A UAV and blockchain-based system for industry 4.0 inventory and traceability applications. *Proceedings* 4 (1): 26–32.

29 Kibria, M.G., Nguyen, K., Villardi, G.P. et al. (2018). Big data analytics, machine learning, and artificial intelligence in next-generation wireless networks. *IEEE Access* 6: 32328–32338. https://doi.org/10.1109/ACCESS.2018.2837692.

30 Preece, A. (2018). Asking 'Why'in AI: explainability of intelligent systems-perspectives and challenges. *Intelligent Systems in Accounting, Finance and Management* 25 (2): 63–72.

31 Park, J., Samarakoon, S., Bennis, M., and Debbah, M. (2019). Wireless network intelligence at the edge. *Proceedings of the IEEE* 107 (11): 2204–2239. https://doi.org/10.1109/JPROC.2019.2941458.

32 Zhou, Z., Chen, X., Li, E. et al. (2019). Edge intelligence: paving the last mile of artificial intelligence with edge computing. *Proceedings of the IEEE* 107 (8): 1738–1762.

33 Chang, Z., Liu, S., Xiong, X. et al. (2021). A survey of recent advances in edge-computing-powered artificial intelligence of things. *IEEE Internet of Things Journal* 8 (18): 13849–13875. https://doi.org/10.1109/JIOT.2021.3088875.

34 Deng, S., Zhao, H., Fang, W. et al. (2020). Edge intelligence: the confluence of edge computing and artificial intelligence. *IEEE Internet of Things Journal* 7 (8): 7457–7469. https://doi.org/10.1109/JIOT.2020.2984887.

35 TETRA Association (2021). TCCA White Paper Mission-Critical Broadband Device Procurement. *Technical Report*. https://tcca.info/documents/October-21-MC-Device-Procurement-WP.pdf (accessed 29 October 2022).

36 European Telecommunications Standards Institute (ETSI). *Emergency Communications (EMTEL)*. https://www.etsi.org/committee/emtel (accessed 29 October 2022).

37 U.S. Department of Homeland Security, Cybersecurity and Infrastructure Security Agency (2021). *Grants Fiscal Year 2021 SAFECOM Guidance on Emergency Communications. Technical Report*. Washington, DC: Department of Homeland Security.

38 Masood, A., Scazzoli, D., Sharma, N. et al. (2020). Surveying pervasive public safety communication technologies in the context of terrorist attacks. *Physical Communication* 41: 101109.

39 Ferrus, R., Pisz, R., Sallent, O., and Baldini, G. (2013). Public safety mobile broadband: a techno-economic perspective. *IEEE Vehicular Technology Magazine* 8 (2): 28–36. https://doi.org/10.1109/MVT.2013.2252273.

40 Suomalainen, J., Julku, J., Vehkaperä, M., and Posti, H. (2021). Securing public safety communications on commercial and tactical 5G networks: a survey and future research directions. *IEEE Open Journal of the Communications Society* 2: 1590–1615. https://doi.org/10.1109/OJCOMS.2021.3093529.

41 Yu, W., Xu, H., Nguyen, J. et al. (2018). Survey of public safety communications: user-side and network-side solutions and future directions. *IEEE Access* 6: 70397–70425. https://doi.org/10.1109/ACCESS.2018.2879760.

42 Baldini, G., Karanasios, S., Allen, D., and Vergari, F. (2014). Survey of wireless communication technologies for public safety. *IEEE Communications Survey and Tutorials* 16 (2): 619–641. https://doi.org/10.1109/SURV.2013.082713.00034.

43 Marabissi, D. and Fantacci, R. (2017). Heterogeneous public safety network architecture based on RAN slicing. *IEEE Access* 5: 24668–24677. https://doi.org/10.1109/ACCESS.2017.2768800.

44 Cabrero, S., Pañeda, X.G., Melendi, D. et al. (2018). Using firefighter mobility traces to understand Ad-Hoc networks in wildfires. *IEEE Access* 6: 1331–1341. https://doi.org/10.1109/ACCESS.2017.2778347.

45 Jarwan, A., Sabbah, A., Ibnkahla, M., and Issa, O. (2019). LTE-based public safety networks: a survey. *IEEE Communications Survey and Tutorials* 21 (2): 1165–1187. https://doi.org/10.1109/COMST.2019.2895658.

46 Favraud, R., Apostolaras, A., Nikaein, N., and Korakis, T. (2016). Toward moving public safety networks. *IEEE Communications Magazine* 54 (3): 14–20. https://doi.org/10.1109/MCOM.2016.7432142.

47 Qi, Z., Lahuerta-Lavieja, A., Li, J., and Nagalapur, K.K. (2021). Deployable networks for public safety in 5G and beyond: a coverage and interference study. *Proceedings of the 2021 IEEE 4th 5G World Forum (5GWF)*, pp. 346–351. https://doi.org/10.1109/5GWF52925.2021.00067.

48 Fantacci, R., Gei, F., Marabissi, D., and Micciullo, L. (2016). Public safety networks evolution toward broadband: sharing infrastructures and spectrum with commercial systems. *IEEE Communications Magazine* 54 (4): 24–30. https://doi.org/10.1109/MCOM.2016.7452262.

49 Usman, M., Gebremariam, A.A., Raza, U., and Granelli, F. (2015). A software-defined device-to-device communication architecture for public safety applications in 5G networks. *IEEE Access* 3: 1649–1654. https://doi.org/10.1109/ACCESS.2015.2479855.

50 Thomas, A. and Raja, G. (2019). FINDER: A D2D based critical communications framework for disaster management in 5G. *Peer-to-Peer Networking and Applications* 12: 912–923. https://doi.org/10.1007/s12083-018-0689-2.

51 Moghaddam, J.Z., Usman, M., Granelli, F., and Farrokhi, H. (2016). Cognitive radio and device-to-device communication: a cooperative approach for disaster response. *Proceedings of the 2016 IEEE Global Communications Conference (GLOBECOM)*, pp. 1–6. https://doi.org/10.1109/GLOCOM.2016.7841668.

52 Saxena, N., Agiwal, M., Ahmad, H., and Roy, A. (2018). D2D-based survival on sharing: for enhanced disaster time connectivity. *IEEE Technology and Society Magazine* 37 (3): 64–73. https://doi.org/10.1109/MTS.2018.2857640.

53 Ever, E., Gemikonakli, E., Nguyen, H.X. et al. (2020). Performance evaluation of hybrid disaster recovery framework with D2D communications. *Computer Communications* 152: 81–92. https://doi.org/doi:10.1016/j.comcom.2020.01.021.

54 Merwaday, A., Tuncer, A., Kumbhar, A., and Guvenc, I. (2016). Improved throughput coverage in natural disasters: unmanned aerial base stations for public-safety communications. *IEEE Vehicular Technology Magazine* 11 (4): 53–60. https://doi.org/10.1109/MVT.2016.2589970.

55 Tafintsev, N., Moltchanov, D., Gerasimenko, M. et al. (2020). Aerial access and backhaul in mmWave B5G systems: performance dynamics and optimization. *IEEE Communications Magazine* 58 (2): 93–99. https://doi.org/10.1109/MCOM.001.1900318.

56 Duong, T.Q., Nguyen, L.D., Tuan, H.D., and Hanzo, L. (2019). Learning-aided realtime performance optimisation of cognitive UAV-assisted disaster communication. *Proceedings of the 2019 IEEE Global Communications Conference (GLOBECOM)*, pp. 1–6. https://doi.org/10.1109/GLOBECOM38437.2019.9014313.

57 Klaine, P.V., Nadas, J.P.B., Souza, R.D. et al. (2018). Distributed drone base station positioning for emergency cellular networks using reinforcement learning. *Cognitive Computation* 10: 790–804. https://doi.org/10.1007/s12559-018-9559-8.

58 Naqvi, S.A.R., Hassan, S.A., Pervaiz, H., and Ni, Q. (2018). Drone-aided communication as a key enabler for 5G and resilient public safety networks. *IEEE Communications Magazine* 56 (1): 36–42. https://doi.org/10.1109/MCOM.2017.1700451.

59 Lieb, J., Özcan, B., Friedrich, M., and Kippnich, U. (2021). Identifying needs and requirements for an integrated crisis traffic management (iCTM) concept for unmanned aircraft systems supporting first responses in crisis situations. *Proceedings of the 2021 IEEE/AIAA 40th Digital Avionics Systems Conference (DASC)*, pp. 1–8. https://doi.org/10.1109/DASC52595.2021.9594475.

60 Chudzikiewicz, J., Furtak, J., and Zielinski, Z. (2015). Fault-tolerant techniques for the Internet of Military Things. *Proceedings of the 2015 IEEE 2nd World Forum on Internet of Things (WF-IoT)*, Milan, Italy (12–14 December 2015), pp. 496–501. https://doi.org/10.1109/WF-IoT.2015.7389104.

61 Yushi, L., Fei, J., and Hui, Y. (2012). Study on application modes of military Internet of Things (MIOT). *Proceedings of the 2012 IEEE International Conference on Computer Science and Automation Engineering (CSAE)*, Zhangjiajie, China (25–27 May 2012), vol. 3 pp. 630–634, https://doi.org/10.1109/CSAE.2012.6273031.

62 Butun, I., Erol-Kantarci, M., Kantarci, B., and Song, H. (2016). Cloud-centric multi-level authentication as a service for secure public safety device networks. *IEEE Communications Magazine* 54 (4): 47–53. https://doi.org/10.1109/MCOM.2016.7452265.

63 Singh, D., Tripathi, G., Alberti, A.M., and Jara, A. (2017). Semantic edge computing and IoT architecture for military health services in battlefield. *Proceedings of the 2017 14th IEEE Annual*

Consumer Communications & Networking Conference (CCNC), Las Vegas, NV, pp. 185–190. https://doi.org/10.1109/CCNC.2017.7983103.

64 Shahid, H., Shah, M.A., Almogren, A. et al. (2021). Machine learning-based MIST computing enabled internet of battlefield things. *ACM Transactions on Internet Technology (TOIT)* 21 (4): Article 101, 1–26. https://doi.org/10.1145/3418204.

65 Busart, C.E. III (2020). Federated learning architecture to enable continuous learning at the tactical edge for situational awareness. Doctoral dissertation. The George Washington University.

66 Pradhan, M., Gökgöz, F., Bau, N., and Ota, D. (2016). Approach towards application of commercial off-the-shelf Internet of Things devices in the military domain. *Proceedings of the 2016 IEEE 3rd World Forum on Internet of Things (WF-IoT)*, pp. 245–250. https://doi.org/10.1109/WF-IoT.2016.7845516.

67 Munusamy, A., Adhikari, M, Khan, M.A. et al. (2021). Edge-centric secure service provisioning in IoT-enabled maritime transportation systems. *IEEE Transactions on Intelligent Transportation Systems*. https://doi.org/10.1109/TITS.2021.3102957.

68 Li, S., Tao, Y., Qin, X. et al. (2019). Energy-aware mobile edge computation offloading for IoT over heterogenous networks. *IEEE Access* 7: 13092–13105. https://doi.org/10.1109/ACCESS.2019.2893118.

69 Abrar, M., Ajmal, U., Almohaimeed, Z.M. et al. (2021). Energy efficient UAV-enabled mobile edge computing for IoT devices: a review. *IEEE Access* 9: 127779–127798. https://doi.org/10.1109/ACCESS.2021.3112104.

70 Jangirala, S., Das, A.K., and Vasilakos, A.V. (2020). Designing secure lightweight blockchain-enabled RFID-based authentication protocol for supply chains in 5G mobile edge computing environment. *IEEE Transactions on Industrial Informatics* 16 (11): 7081–7093. https://doi.org/10.1109/TII.2019.2942389.

71 Khan, L.U., Yaqoob, I., Tran, N.H. et al. (2020). Edge-computing-enabled smart cities: a comprehensive survey. *IEEE Internet of Things Journal* 7 (10): 10200–10232. https://doi.org/10.1109/JIOT.2020.2987070.

72 Fraga-Lamas, P., Lopes, S.I., and Fernández-Caramés, T.M. (2021). Green IoT and edge AI as key technological enablers for a sustainable digital transition towards a smart circular economy: an industry 5.0 use case. *Sensors* 5745. https://doi.org/10.3390/s21175745.

73 Tunnell, H.D. (2015). The U.S. Army and network-centric warfare a thematic analysis of the literature. *Proceedings of the MILCOM 2015 - 2015 IEEE Military Communications Conference*, Tampa, FL, USA (26–28 October 2015), pp. 889–894. https://doi.org/10.1109/MILCOM.2015.7357558.

74 Leal, G.M., Zacarias, I., Stocchero, J.M., and Freitas, E.Pd. (2019). Empowering command and control through a combination of information-centric networking and software defined networking. *IEEE Communications Magazine* 57 (8): 48–55. https://doi.org/10.1109/MCOM.2019.1800288.

75 Jiao, Z., Zhang, J., Yao, P. et al. (2021). C4ISR service deployment based on an improved quantum evolutionary algorithm. *IEEE Transactions on Network and Service Management* 18 (2): 2405–2419. https://doi.org/10.1109/TNSM.2021.3054752.

76 Oyekanlu, E. (2017). Predictive edge computing for time series of industrial IoT and large scale critical infrastructure based on open-source software analytic of big data. *Proceedings of the 2017 IEEE International Conference on Big Data (Big Data)*, pp. 1663–1669. https://doi.org/10.1109/BigData.2017.8258103.

77 Fernández-Caramés, T.M. and Fraga-Lamas, P. (2018). A review on human-centered IoT-connected smart labels for the industry 4.0. *IEEE Access* 6: 25939–25957. https://doi.org/10.1109/ACCESS.2018.2833501.

78 Pradhan, M., Suri, N., Fuchs, C. et al. (2018). Toward an architecture and data model to enable interoperability between federated mission networks and IoT-enabled smart city environments. *IEEE Communications Magazine* 56 (10): 163–169. https://doi.org/10.1109/MCOM.2018.1800305.

79 Wang, P., Ali, A., and Kelly, W. (2015). Data security and threat modeling for smart city infrastructure. *Proceedings of the 2015 International Conference on Cyber Security of Smart Cities, Industrial Control System and Communications (SSIC)*, Shanghai, China (5–7 August 2015), pp. 1–6. https://doi.org/10.1109/SSIC.2015.7245322.

80 Wang, D., Bai, B., Lei, K. et al. (2019). Enhancing information security via physical layer approaches in heterogeneous IoT with multiple access mobile edge computing in smart city. *IEEE Access* 7: 54508–54521. https://doi.org/10.1109/ACCESS.2019.2913438.

81 Xu, X., Huang, Q., Yin, X. et al. (2020). Intelligent offloading for collaborative smart city services in edge computing. *IEEE Internet of Things Journal* 7 (9): 7919–7927. https://doi.org/10.1109/JIOT.2020.3000871.

82 Castiglione, A., Choo, K.R., Nappi, M., and Ricciardi, S. (2017). Context aware ubiquitous biometrics in edge of military things. *IEEE Cloud Computing* 4 (06): 16–20. https://doi.org/10.1109/MCC.2018.1081072.

83 Cubic. Cubic Awarded Contract to Support US Navy Surface Training Immersive Gaming and Simulations Initiative. https://www.cubic.com/news-events/news/cubic-awarded-contract-support-us-navy-surface-training-immersive-gaming-and (accessed 29 October 2022).

84 Li, J., Su, Z., Guo, D. et al. (2021). Secure data deduplication protocol for edge-assisted mobile CrowdSensing services. *IEEE Transactions on Vehicular Technology* 70 (1): 742–753. https://doi.org/10.1109/TVT.2020.3035588.

85 Kantarci, B. and Mouftah, H.T. (2014). Trustworthy sensing for public safety in cloud-centric Internet of Things. *IEEE Internet of Things Journal* 1 (4): 360–368. https://doi.org/10.1109/JIOT.2014.2337886.

86 Miller, D.J., Xiang, Z., and Kesidis, G. (2020). Adversarial learning targeting deep neural network classification: a comprehensive review of defenses against attacks. *In Proceedings of the IEEE* 108 (3): 402–433. https://doi.org/10.1109/JPROC.2020.2970615.

87 Fernández-Caramés, T.M. and Fraga-Lamas, P. (2020). Towards post-quantum blockchain: a review on blockchain cryptography resistant to quantum computing attacks. *IEEE Access* 8: 21091–21116. https://doi.org/10.1109/ACCESS.2020.2968985.

88 Pradhan, M. (2021). Federation based on MQTT for urban humanitarian assistance and disaster recovery operations. *IEEE Communications Magazine* 59 (2): 43–49. https://doi.org/10.1109/MCOM.001.2000937.

21

Use and Abuse of IoT: Challenges and Recommendations

Robert Douglass

Alta Montes, LLC, Sandy, Utah, USA

Abstract

IoT has tremendous potential to aid national defense and homeland security. It can provide comprehensive surveillance of entire cities and populations as well as implement rapid, distributed measures of response and control. IoT offers power well beyond mass surveillance and information extraction. IoT fuses three elements: sensing/information extraction, intelligent processing, and physical control of the environment through automated actions. By integrating these three elements in real time across dispersed networks of devices, IoT gives us a novel and uniquely powerful new force to physically alter the environment and the people in it. IoT will impact the way nations conduct wars and maintain peace. No nation can ignore the advance of IoT for defense and hope to ensure its national security. The potential benefits are just beginning to be understood. The potential dangers were envisioned at least as early as 1949 by George Orwell. Some nations are already using the power of IoT to help suppress terrorism and enforce the rule of law. Other nations are using it to suppress all opposition to the ruling government. Other chapters in this book have focused on securing IoT from attack by hostile agents. This chapter focuses on the complimentary problem: protecting individuals and societies from attack by hostile agents using IoT. The central theme of this chapter concentrates on the intentional misuse and abuse of IoT by people and organizations who control them. Of most concern is misuse of powerful IoT systems by governments that built them for the purposes of national security. Much research has focused on protecting IoT from attack. Almost nothing has been written on protecting free nations from attack by IoT controlled by hostile agents and governments, including their own. Policy makers must understand both the potential and the dangers that extensive IoT networks pose for democracies. This chapter reviews the three elements of IoT and then focuses on their possible misuse, especially by governments, in the name of national security. It reviews some of the international efforts to regulate aspects of IoT and presents a framework for policy that can mitigate the dangers, while fostering the benefits. This framework provides a starting point for policy discussions among the public, lawmakers, and nations. When misused by individuals, organizations, or governments, IoT threatens the rights, freedoms, property, and lives of individuals and society. Awareness, regulation, and vigilance can ensure that emerging IoT technology realizes its benefits while preventing abuse.

IoT for Defense and National Security, First Edition. Edited by Robert Douglass, Keith Gremban, Ananthram Swami, and Stephan Gerali.

21.1 The Elements of IoT and Their Nature

The internet of things (IoT) represents a technological revolution that will rival the Internet in its impact. Like the Internet, IoT extends beyond technology to impact society. IoT networks will automate control of our world. IoT unites three elements: information extraction, processing, and action (see Figure 21.1). It uses distributed sensors, databases, digital documents, and software applications to extract information from multiple sources spread across time and space. Extracted information can be analyzed locally or by processors dispersed across a cloud of computing resources. The algorithms that process it can be as simple as throwing an on/off switch tied to a single sensor value, such as a temperature on a thermostat. Increasingly though, IoT information is processed by sophisticated artificial intelligence (AI) algorithms that recognize people, places, and events. These algorithms make decisions and plan actions based on what they perceive. The plans may be carried out almost immediately, or they might be synchronized over time by remote devices. Surveillance systems differ from IoT systems because they just observe, while IoT digitally controls actuators that physically alter the environment and the people in it [1].

While the term IoT is relatively new, the concept is not. The concept of digitally uniting sensors with processing and with physical action emerged in the 1960s. All three of IoT's necessary elements exist and have for some time. The infrastructure that it depends on has matured and become powerful and ubiquitous. The elements of IoT and its supporting structure shows itself in networks of video cameras and autonomous drones as well as cloud and edge-based computing and physical actuators, ranging from thermostats to mobile, autonomous weapons. IoT listens to our spoken commands in our homes and turns on the lights. In some situations, it can drive our cars. On the battlefield, IoT systems destroy tanks. Cloud-based intelligent processing exceeds new milestones of human intellectual performance almost monthly. It outperforms human experts in challenging domains, such as chess, poker, predicting protein folding, and air combat. The Internet

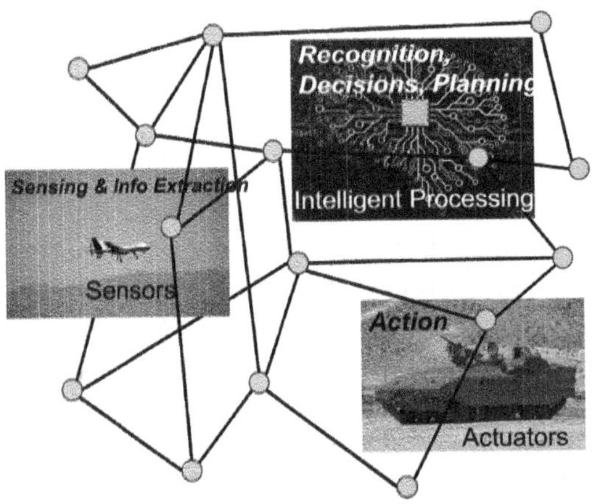

Figure 21.1 IoT represents a revolution in automated control of our world. IoT extracts information from distributed sensors, data bases, mobile applications, and digital documents. It digests information using distant processors running AI algorithms, make decisions, and plans actions. IoT controllers execute actions immediately, autonomously, and remotely through actuators on IoT devices. Source: (a) Mike MacKenzie / Wikimedia Commons / CC BY 2.0; (b) James L. Harper Jr / Wikimedia Commons / Public domain; (c) Ishaiabigail / Wikimedia Commons / CC BY-SA 4.0.

connects these elements together and can control them in synchrony no matter where in the world they physically reside, allowing them to see, think and act in real time both under human supervision and autonomously.

Increasingly, IoT networks alter and control our environment both for defense and in peaceful society. IoT reached an inflection point over the past decade. The technology was largely conceived of in the 1960s and invented by defense ministries but it matured in the commercial arena. IoT now rides high-bandwidth, low-latency wireless connections provided by 5G. Embedded position and timing circuits and on-chip Internet interfaces allow thousands of devices to connect to the Internet and interact with one another. Actuators like switches, motors, and weapons routinely contain embedded Internet connections, allowing them to access and be accessed and controlled by remote processors running AI algorithms. As a norm, mobile devices carry GPS receivers, always knowing their location and the global time. Howard Smith writes: "The internet of things will be the most powerful political tool ever created. By 2020 there will be some thirty billion devices connected to the internet, and political power over the eight billion people on the planet will rest with the people who can control those devices" [2]. IoT's greatest power may lie with how it will be used by nations to defend themselves and ensure their security. These same tools, so powerful for protecting a nation's citizens, can instead suppress individual freedoms and oppress societies when wielded by an unscrupulous government. This chapter focuses primarily on the power of IoT when built by governments in the name of defense and national security, but used instead to secure the government's own power, suppress dissent, and enrich its officials.

Many people consider IoT an obscure topic. Many are unaware of it. Many others question why it is a "thing" different from any other Internet application. In Merriam Webster's definition of "sentient," "a sentient being is one who senses and responds to sensations of whatever kind – sight, hearing, touch, taste, or smell" [3]. In that limited sense, IoT networks are sentient beings. But other definitions of sentience include the notion of "feeling" the things they sense. It is unlikely that any IoT system "feels" its sensations as a living organism does and may never do so. But IoT systems certainly sense, perceive, and react to their environments. Many IoT systems already exceed our own powers in sensing and action. IoT encompasses sensors that surpass the quality and quantity and even the nature of our five senses (see Section 21.1.2). IoT's ability to perceive and formulate a response already exceeds human abilities in some domains (Section 21.1.3). IoT can mobilize actuators that alter their environment at scales both far smaller and far larger than our manual actions (Section 21.1.4). We must understand that IoT is more than an extension of the Internet, more than just another piece of automation. Now is the time to embrace IoT, foster its advance, and harness its benefits. Now is also the time to set boundaries on its use to limit its potential for misuse, abuse, and damage.

IoT endows us with countless beneficial technological advances, many not even yet envisioned. However, if misused or abused, interconnected networks of sensors and actuators that develop plans by themselves present imposing potential threats to society and individual freedoms. Awareness of the dangers of mass technological surveillance grows among the public and lawmakers. But IoT can do much more than just rob your privacy with its surveillance; it can use what it sees to physically alter your world. It can hunt you down and kill you. And you would not be the first (Section 21.1.4.1). Much has been written about how to protect IoT from attacks by hostile agents. Some of our best technologists and brightest policy makers have investigated and strategized about how to secure IoT systems from attack (see Section 3 of this book). This chapter addresses the inverse challenge: how do we protect ourselves from IoT when commanded by hostile agents, whether individuals, organizations, or especially governments?

21.1.1 Use and Abuse of IoT

Numerous books, articles, and websites outline the benefits of IoT in general. In 2015 Philip Howard forecasted a new emerging global society and world order, an emerging "*Pax Technica*" where world peace arises from IoT [1]. Previous chapters of this book outline the potential of IoT to alter the nature of warfare, defense, and national security. By providing comprehensive, survivable battlefield surveillance, IoT can dispel some of the fog of battle. By allowing a commander to step out of the real-time control loop into a supervisory role on top of the loop, IoT makes military systems more autonomous. By increasing the tempo of physical effects, whether robotic weapons or the flow of munitions and food, IoT autonomy can alter the course of wars. By moving soldiers to remote positions of supervisory control, IoT autonomy can save lives. For defense and national security, these benefits are just now beginning to be realized. No nation can afford to advance into the twenty-first century without incorporating IoT into its national defense.

Beyond improved defense, Howard foresaw a world where IoT ubiquitously distributes information and thereby marginalizes extremist groups and drives out false narratives. His 2015 vision has proved premature. Even as some of the benefits of IoT remain unrealized, they already bring some dangers. As early as 1949 George Orwell envisioned these dangers in his book *1984* [2]. A mere three years after the first electronic computer was powered up and decades before the advent of the Internet, Orwell foresaw a possible future enabled by the abuse of IoT. Orwell's dream, or more aptly his nightmare, has not quite yet come to pass. A full blossoming of Orwell's *1984* requires a broader maturation and deployment of IoT. To understand why IoT's benefits are married with a new threat for both individuals and free societies, a summary of the key features of IoT is provided in the next sections.

This chapter focuses primarily on the ***intentional*** misuse and abuse of IoT systems ***by people and organizations who control them***. Its greatest concern centers on intentional misuse of powerful IoT systems by governments that built them for the purposes of national security. To be sure, IoT can be unintentionally misused by anyone or any group. And of course, criminals, unscrupulous organizations, hostile nations, and sociopaths routinely attack and misuse and abuse IoT networks, devices, and systems. Enhancing IoT's security to prevent unintentional or criminal or hostile misuse and abuse is the domain of IoT security. Security is a technical challenge as well as an organizational and process challenge, aspects of which are discussed in many chapters of this book, most notably in Section 3. Improving IoT security sits among the top challenges that hinder the complete realization of IoT's benefits for national security. However, even the most superb security measures cannot protect against intentional misuse and abuse of IoT systems by the people or organizations who control them. When you are the one who defined the Root password, and it is your fingerprint and retina that the system is scanning for, then technical security measures are not the main concern. For these reasons, general IoT security falls outside the scope of this chapter. Security is discussed briefly only in the context of it being a necessary component of successful regulation.

21.1.1.1 What Makes IoT So Powerful?

IoT's ability to perceive and physically change the world is radically new. Control systems are the only similar phenomena, for example Industrial process controls or aircraft autopilots. IoT is a control system for the world. Both the benefits and threats from IoT arise from the fusion of sensing with processing and control of physical actuators. These three elements, supervised by people, will efficiently, rapidly, and effectively monitor, surveille, and control physical aspects of our world. The devices in an IoT system can be all the same, for example video cameras, or they can be heterogenous, combining many different types of devices, for example smoke detectors, cameras, and acoustic sensors. The "Things" in IoT can be widely dispersed across cities or continents, like power control systems, or they may be co-located in a small space associated with a particular system, like the elements in a self-driving car. Because the Internet spans the globe and even reaches into space, an IoT network can potentially reach any device, system, person, or other IoT system connected to the Internet. If IoT can control those entities, IoT can alter the environment anywhere and everywhere.

Sensing may involve a range of information extraction from simply monitoring a single value like a voltage on a power line to 3D imaging with video or radar. It may also involve extraction of information from speech or text or a database or software application. Sensors typically produce raw digital data, processed into information either locally at the sensor or by distributed processors. The term "information extraction" usually refers to the capture of information from digital sources, such as text, databases, location data, multimedia, software applications, email, and social media [4]. With the advent of smartphones and the explosion of mobile apps, information increasingly is collected by software applications as they are being used, and in some cases even when they are not being used but just reside on a smartphone. Both sensing and information extraction involve collection of information that describes some aspect of the world. The term "sensing" will be used here generally to refer to devices that physically transduce some aspect of the environment into digital data. The term "information extraction" will be used here to refer to collecting data from digital sources as diverse as GPS receivers on your phone to extracting information from text on a Facebook page to identifying individuals from video.

Processing converts raw sensed data into information. Intelligent processing, driven by AI algorithms, converts information into knowledge about people, behaviors, events, and activities. Although IoT might use simple algorithms for some applications, this chapter focuses on "intelligent processing" because it enables IoT systems to exercise complex control over our environment either semi-autonomously with human supervision or autonomously. Intelligent processing closes the loop between sensing and action in microseconds as opposed to human decision-making occurring over seconds to minutes to hours. People using IoT typically are said to be "on the loop" rather than "in the loop" meaning that people are not part of the real-time control process, but in a more supervisory position.

The action element of IoT technology refers to devices that can physically alter the environment. Devices alter the environment via actuators, for example switches, motors, or hydraulic pistons. Alterations might involve trivial changes, such as turning on lights as we approach our home. Or they may consist of more vital actions such as releasing a bomb or shooting one of us. But IoT is more than just remote control. It also amplifies our powers more than our current electro-mechanical machines, which already augment our manual strength and dexterity. IoT can coordinate the simultaneous action of multiple devices spread across the globe, ranging in scale from nanometers to devices the size of a battleship. IoT can synchronize actions over microseconds to periods lasting years, adapting its plan of control as it goes. IoT's actuators can be powerful by themselves, but the full force of IoT comes from the integration of many physical actions controlled by intelligent processing fed by massive amounts of sensed and extracted information.

21.1.1.2 Orwell's Vision Has Not Yet Fully Materialized

Orwell, writing in 1949, envisioned a world 35 years in his future where surveillance of every middle- and upper-class individual is pervasive. Surveillance data is universally monitored to detect non-conforming behavior. Such behavior, when observed by ubiquitous cameras and microphones, is discouraged, punished, or eliminated by means of paramilitary police who seize offending individuals and "reeducate" or terminate them. Orwell's world was not possible in

1949 or even by the year 1984. Until recently, detecting "incorrect" behavior required extracting meaning from vast arrays of video feeds and that required human eyes and human brains to convert video data into knowledge about human behavior. For Orwell's surveillance state to exist, almost as many people would be needed to watch video screens as the number of people being watched. Mounting effective, timely actions to control offenders by human enforcers would have been logistically challenging. Controlling the populace required a large body of human enforcers that must organize and then move to the location where an offender can be apprehended. An offender could have escaped in the interim or prepared a resistance, and such a large paramilitary force itself could have posed a threat to ruling powers. Today, IoT is beginning to eliminate such barriers to creating Orwell's 1984. IoT makes it possible to automate analysis of video surveillance and to couple it to mechanical actions without humans required in the loop. IoT can control a population more easily and more efficiently, and it eliminates key risks to the governments deploying it.

Since the actual year 1984, video cameras and many other sensors have dropped at least an order of magnitude in size, weight, power consumption, and cost. Their quality and resolution increased by several orders of magnitude. At the same time, processing power has increased by many orders of magnitude while the cost per unit of processing power has fallen by as much as has its size, weight, and power requirements. Increasing processing power has enabled advances in sensor processing, especially the processing of video. Today, computer vision algorithms can automatically detect complex patterns in still images and video data, such as recognizing an individual face or tracking cars in a cityscape. These advances in sensors and sensor processing allow countries to begin to realize the pervasive world of surveillance envisioned by George Orwell. Now with the integration of the final element of IoT, automated actions, the full suite emerges for Orwell's *1984*.

21.1.1.3 IoT Unites Sensing/Information-Extraction with Intelligent Processing and Action

Widespread surveillance has helped to drive down crime because its existence discourages criminals. It assists law enforcement in apprehending and prosecuting persons observed committing crimes. The UK embraced public video surveillance as early as 1960 and was one of the first nations to install video surveillance on most major roadways, buses, and trains [5]. The country has been quick to add intelligent processing as it has emerged, such as automatic license plate reading and face recognition. In the UK, both government investment and public acceptance of video surveillance has been buoyed by terrorist acts committed first by radical groups like the Irish Republican Army (IRA) and then by Islamic radicals. Similarly, information extraction from digital sources like social media have increased the power of police and the military to identify and prosecute illegal and dangerous actions. While surveillance and sensing continue to be primarily a forensic tool for solving crimes after the fact, increasingly, sophisticated processing is applied to predicting potential crimes and criminals before the fact, although with mixed results [6, 7]. In military affairs, pervasive surveillance enabled by drones has altered the course of military operations and intelligence activities beginning with the Bosnian War of the late 1990s [8].

More recently, the public and governments have realized that comprehensive surveillance and information extraction are a potential threat to privacy and to free societies (Figure 21.2). Public and government concern has begun to surface over the combination of surveillance with intelligent processing. The public now understands that along with the benefits of pervasive information extraction and processing comes the potential destruction of privacy. More slowly the public in many countries has begun to call for regulation of intelligent processing of extracted information. Both citizens' groups and governments are beginning to act to regulate surveillance and information extraction with the protection of privacy as a goal, as discussed in Section 21.3.

Figure 21.2 A piece of the Berlin Wall inscribed in Spanish: "Happy 1984," accompanied by a PlayStation controller denoting mind control through PlayStations or technology in general. Source: Victor Grigas / Wikimedia Commons / Public domain.

By uniting action with information extraction and intelligent processing, IoT can complete the loop from observing the world to changing it. This union yields benefits beyond sensing and processing that represent a discontinuous innovation in the power of connected devices. If abused, the threat posed by IoT extends beyond the loss of privacy to the loss of control by individuals and societies. If not understood and regulated, IoT will complete the last leg of Orwell's vision. Today, we find ourselves reliving the actual year 1984 when visionaries saw the potential of the emerging Internet to freely spread all information everywhere. In 1984, no one was thinking of the dark passenger traveling behind the promise – the unconstrained spread of disinformation and the destruction of everyone's privacy. Pervasive surveillance and information extraction are now ubiquitous threats to privacy. Nevertheless, they form a major segment of our economy [4]. It will now be difficult or perhaps impossible for regulators to rein in the first leg of IoT: sensing/information-extraction, at least in countries like the US that derive great commercial gains from it. However, there may still be a chance to control the misuse of intelligent processing, IoT's second leg. There is an even better chance that we can prevent abuse of IoT by regulating its final leg, control of physical actions. A first step is for the public and their government to become aware of the challenges. With awareness, policy makers can begin to discuss regulatory options and enact protective laws. These laws must protect society while simultaneously fostering the advancement and use of IoT.

Before exploring those options for regulating IoT (Section 21.3), a closer look at each element of IoT is warranted.

Sensing and information extraction, connected via the Internet, is already embedded in all societies across the developed world. While the extent of collection is widely understood, what is less well understood is the diversity of information that can be collected both from individuals and public and private spaces. This diversity of sensor types gives IoT systems a rich perception of our environment and the people in it. It observes their actions more completely than human senses can. Section 21.1.2 provides an overview of sensors and information extraction. Section 21.1.2.1 briefly reviews networked sensing and sensor types as well as information extraction from documents, applications, telecommunications, email, and websites. Because an extensive literature already exists covering these types of information extraction, e.g. see [4], the topic is only briefly reviewed in this chapter in Section 21.1.2.2.

Isolated sensor data and pieces of information extracted by IoT devices by themselves may provide very little threat to privacy and society. However, intelligent processing of sensor data and information from multiple locations over time and space allows recognition of patterns of activity and supports the tracking of individuals. Such patterns of life can be inferred with a detail and accuracy not possible with any single sensor or piece of information. With appropriate processing, seemingly anonymous and innocuous information can be converted into comprehensive life histories of identified individuals, groups, and society. Senor data that is anonymous and noisy can be fused into an accurate and precise picture of individuals identified by name. Section 21.1.3. summarizes the nature of intelligent processing and the sorts of patterns of life deduced. Section 21.1.3 also reviews some of the challenges of understanding the limits and trusting AI algorithms that drive intelligent processing. Because intelligent processing has shown the most dramatic advances in the past decade and because sensing, information extraction, and control of actuators is much better known and understood, additional background information is provided on intelligent processing. Understanding its history and how it works provides an understanding of the challenges for its safe and effective use. Section 21.1.4 discusses how IoT, using intelligent processing, exerts control over the physical world and the potential for its abuse. Actuators close the loop with the physical environment. They execute the plans developed by intelligent processing. Control theory regulates actuators. Both mechanical and electrical actuators are widely known and understood and the control theory that regulates them is well defined and widely studied by engineers. Consequently, Section 21.1.4 on IoT actions limits its discussion to examples of IoT in action.

21.1.2 Pervasive Sensing and Information Extraction

21.1.2.1 Sensors and Sensor Networks

The notion of electronic sensors tied together with a communications network arose at least as early as 1936 with the British Chain Home coastal radar installations designed to detect approaching German aircraft [9]. Early defensive sensor systems used dedicated communications systems that were expensive and provided little flexibility for new types of sensors or new locations. The spread of the Internet created the flexibility and low cost that allowed sensor networks to be created almost anywhere using almost any type of sensor. The Internet Protocol (IP) and Transmission Control Protocol (TCP) network protocols that are the basis for the Internet were developed for the ARPANET, a precursor to the Internet, to explore distributed, redundant, adaptable communication that would also be robust and survivable. Although the early applications of the ARPANET were for data exchange between computer science researchers, the concept was inspired by the goal of creating an IoT network for strategic defense. The funders envisioned a survivable network linking control stations for intercontinental ballistic missiles with early warning sensors, such as the defense early warning (DEW) line [10]. Today, the Internet allows specialized sensor systems to be cheaply interconnected and easily modified in a standard way.

21.1.2.1.1 Video and Visible-Light Imaging

With increasing bandwidth available, standalone cameras began to be connected into small networks covering specific places of business and facilities. Beginning with a couple of temporary video cameras set up by the London metropolitan police to monitor an event in Trafalgar Square in 1960, video cameras now monitor much of our world on a continuing basis. By 1998, the NYC Surveillance Camera Project [11] reported over 2300 cameras operating along the streets of Manhattan. Most of these cameras were standalone or functioned in a small-scale private network. By 2021, a study by Amnesty International found over 15,000 publicly owned surveillance cameras purportedly networked to the New York City police department [12]. Privately owned and installed cameras greatly outnumber the ones operated by the police.

Although estimates vary widely, the UK has perhaps one to two million surveillance cameras, many of them connected to one another and to the Internet. While the majority (70 : 1) are estimated to be privately owned, there is a large government system of networked cameras. For example, HRnews reports that in 2009 there were 23,708 government operated surveillance cameras, 15,516 operated by the Transit Authority alone on buses and underground transit systems [13]. Other estimates claim that 13,900 cameras view public spaces just in London. Estimates indicate 7.5 cameras for every 100 people in the UK [14].

The scale of networked video surveillance in China dwarfs that in the UK. The New York Times reported that there were an estimated 200,000,000 cameras in China, most of which were concentrated in urban areas and networked together as part of the "Skynet" program [15, 16]. China's 2016 five-year plan called for a separate system called "Sharp Eyes" to surveil 100% of China's public spaces by 2021. Surveillance provided by Skynet and Sharp Eyes is augmented by two additional surveillance efforts, Golden Shield and Safe Cities [17]. Key intersections are viewed by multiple cameras (see Figure 21.3). How China processes its surveillance video is of special interest, as discussed below in Section 21.1.3.

The number of installed video surveillance cameras in the world pales in comparison to the number of image and video sensors contained in every one of the most common IoT device: the smartphone. The explosion of smartphones has driven down the cost as well as size, weight, and power required for image and video sensors, making them ubiquitous. Statista estimates that there are over six billion smartphone users in 2021, equivalent to approximately 80% of the world's population [18]. These video sensors are connected to the Internet and to each other for much of the time. The volume of smartphone sales fuels dramatic improvements in camera performance, specifically in resolution, signal-to-noise ratios, and low-light imaging. The smartphone is the ultimate IoT device. It couples image and video sensing with more processing power than super computers possessed just a decade or two ago. The processors make sense of the video, as discussed in Section 21.1.3.2. Smartphones contain far more than just video sensors. The combination of multiple types of information with massive processing power and connectivity makes the smartphone a potent IoT device. Before discussing the implications of processing for IoT, it is worthwhile to review some of the other types of sensors and information that an IoT system can access. Defense departments developed many of these diverse sensor types where they already impact military operations. Many are spreading beyond defense into commercial use.

21.1.2.1.2 A Plethora of IoT Sensor Types Beyond Video

Reviewing the type of sensors on a smartphone provides a good starting place for understanding the diversity of IoT sensors. Besides multiple, excellent-quality visible-light cameras and video,

Figure 21.3 China reportedly has the greatest number of networked video surveillance cameras of any country. Most public spaces in cities are monitored by multiple cameras, such as these six visible cameras on a cell tower in Tianamen Square. Source: Andrey Belenko / Wikimedia Commons / CC BY 2.0.

a typical smartphone contains many other types of sensors, for example, global positioning system (GPS) receivers, six-degrees-of-freedom accelerometers, a high-quality microphone, and the ability to determine location by triangulating with WiFi and cell towers. The ability to provide precise, accurate location, direction, and timestamps tagged to other sensor data greatly increases the value of all smartphone data. In the thousands of businesses outfitted with Bluetooth beacons, smartphones can estimate their location within a meter or less by triangulating among beacons indoors. By 2017 beacons were already being used by 75% of the top 20 retailers in the US [19]. Apple introduced its own version of IoT tracking devices, AirTags, that use Bluetooth as well as proprietary wideband technology to achieve even better tracking accuracy both outdoors and indoors. Apple relies on networks of Apple smartphones, watches, and tablets to triangulate and communicate AirTags locations. By 2021, concerns surfaced regarding corporations using Bluetooth beacons to track individuals without their consent [20]. More ominously, reports of abusive use of AirTags emerged, stating that AirTags are being covertly attached to people's possessions and vehicles to locate and track them with criminal intent [21].

Apple, Inc. released an iPhone in 2021 that contains a lidar sensor. Lidars use laser scanners or bursts of laser light to produce a 3D image where the value of pixels in the image specify the range to a corresponding point in the environment. Lidars have been used on autonomous vehicles or self-driving cars since 1985 [22]. 3D images from lidars identify obstacles in the road and serve as the principal sensor for driving off-road for military autonomous vehicles [23]. iPhone lidars represents a major advance in lower cost, size, weight, and power, but have limited effective range and can be overwhelmed by strong sunlight. Automotive lidars use an electro-mechanically scanned laser beam to provide the necessary range and operation in most lighting conditions, but their precision moving parts are both expensive and prone to failure. Recent research has demonstrated a new lidar technology, which can be fabricated on a chip. It appears to overcome current limitations in range, cost, and lighting conditions while removing the need for electro-mechanical scanning [24]. Some automakers, such as Tesla, have rejected the use of lidar and advocated the use of multiple video frames to compute 3D images, much like the human visual system does both with stereo vision and with a process called motion parallax or shape from motion [25]. While the state of the art in computer-vision algorithms to compute 3D from video sequences has advanced greatly in the past several decades, it can still fail to correctly detect the shape of some types of surfaces. The limitations of Tesla's video-only approach perhaps contributed to more than a dozen deaths in Tesla cars so far when they were driving in "auto-pilot" mode [26, 27]. If solid-state, long-range lidars reach large-scale production, IoT devices will directly perceive their surroundings in 3D. IoT devices with 3D vision can provide accurate targeting information of great value in defense applications.

Beyond smartphone sensors and video cameras, other sensors can sample a much wider range of the electromagnetic spectrum from ultraviolet (UV) light to infrared (IR) light in the near, short-wave, mid and far-IR spectrum. Each spectral band provides a different view into the nature of an IoT device's environment. In mid and far-IR bands, objects both reflect incident IR energy and emit it. Because the thermal energy of objects emits mid and far-IR energy, cameras that can detect those bands can see in total darkness. IR cameras continue to fall in price and size while improving in performance, although not nearly at the rate of visible-light video cameras. The company FLIR, Inc. [28] makes an attachment that clips onto the back of a smartphone and takes IR pictures using thermal emissions to add yet one more sensing modality to smartphone IoT devices. Hyperspectral cameras provide an alternative to using multiple types of cameras to cover the spectrum. They can produce a video stream where pixels have not just three colors, but dozens or hundreds of colors corresponding to spectral slices ranging from UV through visible light down to the IR wavelengths. If the spectral composition of the illuminating light source is known, some objects can be identified by their unique spectral fingerprint.

Going down in frequency below the IR spectrum, sensors can detect, characterize, and track objects at radio frequencies. Automotive collision-avoidance radars have better range and response times than video cameras and lidars. They are not as easily fooled as video cameras by uniform surface colors and textures. Nor do highly reflective surfaces like rain puddles defeat radar collision-avoidance sensors like they can lidars. Radars can do much more than detect and track objects in the open. They can peer through forest canopies to track people and vehicles (see Figure 21.4a). They can see into buildings to recover their layout and detect and follow people as they move around behind walls, for example, the Visibuilding system developed by the Defense Research Projects Agency (DARPA), (see Figure 21.4b) [29, 30]. Radars can even penetrate clothing to detect anomalies like concealed weapons and bombs [31, 32]. They see objects at great distance, even over the horizon of the earth [33]. Radar sensors add a rich layer of data to IoT networks giving them vision into otherwise inaccessible aspects of the environment.

Many sensors can detect signals other than those in the electromagnetic spectrum, such as thermal, magnetic, vibration, chemical, and acoustic sensors (microphones). More exotic sensors beyond these common ones include neutron and ionizing-radiation sensors. Some countries deploy neutron sensors in their transit systems to monitor the movement of radioactive material [34]. Other sensors are designed to detect organic properties such as types of proteins and DNA. DNA sensors can sequence whole-genome DNA strands or detect just specific fragments [35]. A decade ago, DNA analysis required a laboratory's worth of equipment and many manual steps and days to produce a result. In the past decade, DNA sensing has moved to small, on-chip devices and can get

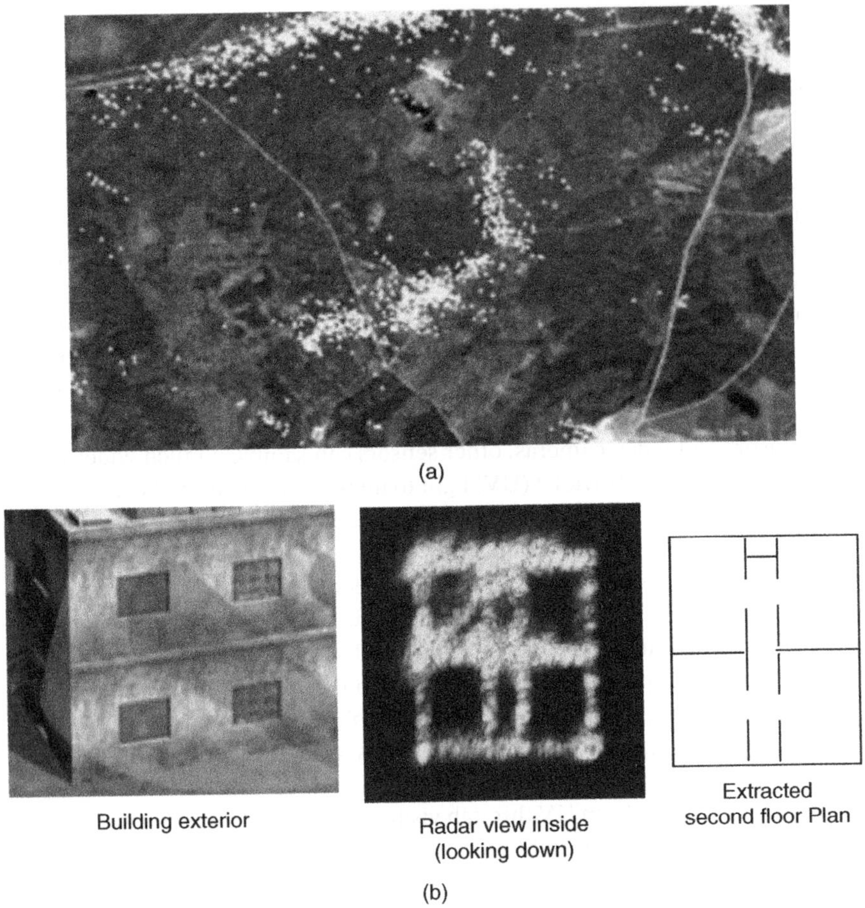

Figure 21.4 Radars can see through forest canopies (a) and inside buildings (b). (a) Low frequency foliage-penetrating radars detect and track people and vehicles moving inside a dense jungle. Source: [29] / from IEEE. (b) DARPA's Visibuilding wall-penetrating radar recovers floor plans and human activity. Source: [30] / from IEEE.

rapid results with minimal human assistance. IoT networks are likely soon to include devices that perform near real-time DNA analysis as part of their suite of sensors. Such IoT networks may one day monitor for pathogens with pandemic potential to detect them before they are discovered in patients with symptoms. US DOD research programs in the mid-2000s explored ways to use *ad hoc* sensor networks to detect mass-terror threats such as biological, chemical, or radiological agents. IoT devices deployed on smartphones, smart watches, and specialized fitness and health devices expand the network of sensors and types of data. These monitor heartrate, cardiac rhythms, temperature, blood oxygen, and detect events such as falls and heart fibrillation. The development of small, low-cost DNA and protein analysis sensors along with existing health and biosensors will add a rich new source of information in the future for detecting and tracking pathogens, but also for detecting and tracking terrorist materials as well as "fingerprinting" and tracking.

IoT sensors, of course, need not be stationary. The trend to decreased size, weight, and power combined with wireless Internet connectivity enables sensors to go mobile. Many types of mobile

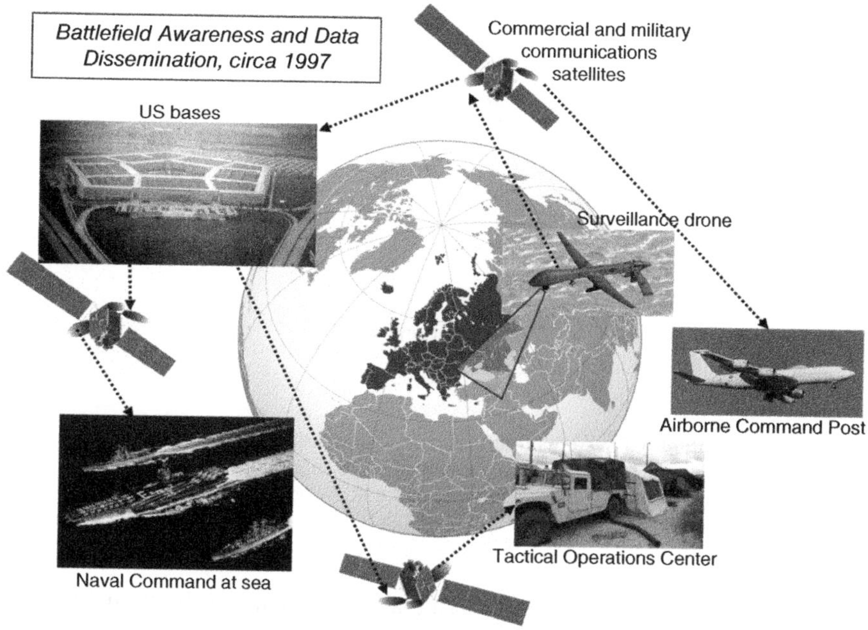

Figure 21.5 By 1997 DARPA, an agency of the US Defense Department, demonstrated how to link mobile sensors on a drone in Europe with fixed and mobile ground, ship, and airborne command posts communicating via space and terrestrial internets across multiple continents. Source: (a) LT Gordon Eugene Martin USNR / Wikimedia Commons / Public domain; (c) U.S. Navy / Wikimedia Commons / Public domain; (d) U.S. Army CERDEC / Wikimedia Commons / Public domain; (f) United States Air Force / Wikimedia Commons / Public domain; (e) Balon Greyjoy / Wikimedia Commons / Public domain.

sensor platforms now exist. DARPA demonstrate the first use of a drone as an Internet-connected video sensor in 1998 in the Battlefield Awareness and Data Dissemination (BADD) program. It linked sensor feeds from a drone over Bosnia via Internet connections to space as well as to command posts on the ground, ships at sea, and airborne command centers spread across several continents and oceans [8] (see Figure 21.5). Since then, drones or unmanned aerial vehicles (UAVs) have proliferated in all shapes and sizes for both defense, commercial, and private use. Many if not most of these aircraft carry at least a video sensor and can connect to the Internet through a smartphone or dedicated ground station.

Airborne and smartphone IoT sensors are not the only mobile platforms. While police vehicles have incorporated video cameras for some time, many if not most recent automobiles and trucks include at least one video camera. Some vehicles like those of Tesla allow for video to be transmitted over the Internet from the vehicle. Tesla cars, like smartphones, are IoT systems and collections of IoT devices. Teslas and similar automobiles not only provide multiple video streams, but also host many other sensors recording specialized data describing the vehicle's operation and state [36] (see Figure 21.6). Specialized sensors, such as GPS, thermometers, accelerometers, pressure gauges, volumetric sensors, and many others round out the suit of sensors in IoT systems on newer cars. They capture the sensor data, exfiltrate it, and integrate it into a comprehensive and ongoing picture of their environment, which can be far richer than that available to human senses. Besides driving and avoiding collisions, automotive IoT send emails to their owners, for example, when their tire pressure is low.

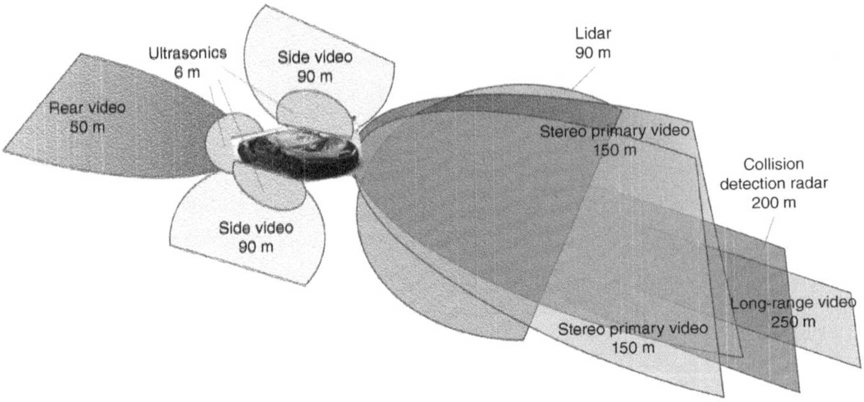

Figure 21.6 Self-driving cars are potent IoT systems that map and characterize their environment, detecting all activity 360° around the vehicle. A single car may have more than a dozen cameras and multiple other sensors including radars, ultrasonics, and lidars. Source: R. Douglass.

21.1.2.2 Information Extraction

Sensors measure physical values of the environment. Digital sensor data can be thought of as a string of bits. To be useful it must be processed into information that supports deductions and inferences about the environment. IoT devices can also obtain information directly from a variety of sources beyond sensors. The process of obtaining and exfiltrating information from a digital source is called "information extraction" [4]. Location data of mobile phone users is one of the most common types of data exfiltrated. Up until 2019, the major telecom companies sold individuals cellphone location data obtained by the phones "pings" to cell towers. Location data included personal information on the phone's user. It was sold to aggregators and most likely government agencies and law enforcement. After a major breach and theft of the data base of one of the largest aggregators in 2018, the telecom companies pledged to quit selling cell-tower based location data by 2019 [37]. Also in 2018, the US Supreme Court ruled that government agencies generally need a court ordered warrant for types of cellphone location data [38]. As cellphone locations from telecoms became unavailable, the location aggregator industry expanded, extracting information from other sources, along with many other types of additional information.

Perhaps the most common source of information extraction is provided by applications running on a computer, tablet, or smartphone. The types of information that can be extracted and fed into the Internet is dizzying. It includes such information as location, time, text communication in SMS and email, audio from phone calls and microphone recordings, images, and video, and cell-tower pings. Metadata accompanies the sensor data indicating the sensor's location, time, owner, etc. One aggregator, Ventel, claims to combine information from 80,000 applications running on millions of computers and mobile devices [39]. IoT systems also extract more abstract information such as addresses associated with phone numbers, email names, personal contact lists, and Facebook friends as well as usernames, government identifying numbers (e.g. driver's license, social security number), date of birth, credit cards, insurance information, residence, place of work, and a myriad of other information. Besides collecting static information, IoT devices can supply information about what applications (apps) are used and on what data. They can exfiltrate how and when apps are used as well as purchases made or considered and what web searches are performed. The popular Venmo app, used to transfer funds and make payments, records and can make public all financial transactions made with the app [40]. Venmo made public visibility of everyone's financial transactions the initial default setting. Some data aggregators claim their extracted information is anonymized by not directly associating the data with an individual's name. However, while initially "anonymous" exfiltrated data can often be easily associated with a specific named individual or organization using intelligent processing, which integrates "anonymous" data with data from other sources, as discussed below.

21.1.2.2.1 How Information Is Extracted

There are several primary ways private information is legally collected over IoT networks.

- Sensors can directly observe and collect information about individuals, their artifacts, and their environment, especially in public spaces.
- Individuals and organizations may willingly sell their information, either knowingly or unwittingly. Frequently, users are paid not with cash, but by getting "free" services, such as free use of a smartphone app in exchange for the app collecting information from the user. The information can then be used or sold by the app's developer or provider.
- Individuals and organizations may intentionally provide their information in response to a marketing effort or for social display, for example by posting information about one's life on social media.
- Data may be covertly but legally extracted by an app or device without the user's knowledge or consent.
- Individuals may grant access to information, knowingly or unknowingly, as part of the necessary process of providing goods or services (for example associating cellphones with specific cell-tower locations is necessary to provide cellphone service).
- Information may be collected legally by government mandate for purposes of taxation and public safety and wellbeing, for example license data, vehicle registrations, and home ownership).
- National and local governments may extract information overtly to increase the safety and well-being of its citizens, such as providing statistics that foster economic growth or public health.
- National governments may overtly or covertly extract information purportedly for intelligence or law enforcement purposes in support of defense, national security, and public safety.

21.1.2.2.2 Abusive Sensing and Information Extraction

Information can be abusively collected in the following ways:

- **Info-1.** Information may be illegally extracted by individual hackers or criminal organizations or by unethical corporations or organizations with hostile intent, normally for financial gain or political advantage.
- **Info-2.** Sensors may legally but unethically collect information about individuals or organizations without explicit knowledge or consent of those being observed.
- **Info-3.** Information may be legally extracted but then used illegally or unethically without benefit or to the detriment of the owners or users of an IoT network. It may be legally extracted but then provided to third parties who may then use it illegally or unethically. Alternatively, information may be willingly provided by IoT users in exchange for goods or services without awareness of how the information will be used or what third parties will have access to it.
- **Info-4.** Governments may legally but unethically extract information covertly, purportedly for intelligence or law enforcement purposes in support of defense, national security, and public safety, but without explicit consent of its citizens or checks and balances on exercising this practice. If there are legitimate concerns for national defense or public safety and public consent and checks and balances on information extraction exist, then this collection would not be considered abusive.

- **Info-5.** Information may be collected by governments, possibly in contravention of their own laws or constitution, for the purposes of maintaining or enhancing their own power or to enrich principals and associates of the government.

Many governments of democratic countries publicly state that they eschew Info-5 collection and have adequate checks and balances, such as court-ordered warrants to prevent type Info-4 abuse. In practice though, many of these governments conduct type Info-4 abusive collections on a large scale with minimal checks and balances and without public awareness or consent. Many countries have mandated laws that require cooperation and special access from communications companies. For example, in the US, the Communications Assistance for Law enforcement Act (CALEA), H.R.4922, allows for extensive collection by law enforcement of communications data, including fixed and mobile phones and broadband transmissions [41]. CALEA also provides some specific limits and exclusions on such collections as well as normally mandating a court-issued warrant authorizing any given collection. Beyond allowing for such collection of information from individuals, the law specifies that communications providers must provide hardware or software mechanisms to access information flows for law enforcement. The law was extended to broadband IP traffic and voice-over-internet communications in 2005 [42]. Most warrants and civil judges place reasonable limits on the scope, place, and person from whom information is collected. Typically, these warrants become public record or at least become available to people prosecuted using the information collected. The US created exceptions to the process in 1978 with the Foreign Intelligence Surveillance Act or FISA, codified as 50 USC. §§ 1801–1885c [43].

After the terrorist attacks of 9/11 and passage of the US Patriot Act [44], FISA rulings allowed massive interception and data collections of individuals within and outside of the US by the National Security Agency (NSA). NSA documents leaked by Edward Snowden showed that in one month in 2013 the US collected more than 97 billion emails and 124 billion phone calls [45, 46]. Secret collections such as these, authorized by the FISA court, are not public record, cannot be seen by any of the accused and are produced by a process that has no transparency, nor public oversight nor consent. It is difficult to understand how such surveillance by the government does not violate national law, for example the fourth Amendment to the US Constitution protecting individuals from unreasonable search and seizure [47]. Or to understand how it does not violate international conventions, such as Article 12 of the United Nations Universal Declaration of Human Rights [48] to which the US is a ratifying signatory. Eventually, the US Court of Appeals for the Second Circuit ruled that such bulk collection is illegal [49].

IoT networks today support massive collection of sensor data and extracted information. But as we will see, additional information can be inferred or deduced well beyond what is directly collected by a sensor or from an information source. That inference combines information gathered over time and space and combined from multiple types and sources of information. Whether obtained from sensor data or extracted directly, much of this information has value in and of itself. However, when different sensors at different locations are merged over time, using intelligent processing, a startling amount of additional information can be deduced or inferred about the activities and associations of an individual, organization, government, and society. Intelligent processing is the second leg of IoT, and the next section describes both the power of intelligent processing and the threat from its misuse.

21.1.3 Intelligent Processing

The most exciting and possibly the most important benefits of IoT for military affairs arises from dramatic advances in intelligent processing. Intelligent processing converts sensed and extracted information into an understanding of a situation. Intelligent processing can then use that understanding to formulate and monitor a plan of action to alter the situation. Intelligent processing is the key ingredient in automating defense operations and moving humans from inside the control loop to a supervisory position. In contrast to the topic of intelligent processing, sensors, and information extraction are much better understood by the public and decision makers. Similarly, control theory for physical systems and actuators are well developed and well understood, at least by engineers. Because the most dramatic gains have come from intelligent processing, this topic receives special attention in this chapter. In particular, understanding how intelligent processing works is necessary for understanding some of the challenges arising from inserting intelligent processing into military operations. As a way of understanding the technology and its challenges, a short history of the development of AI provides insight into the promise and limitations of the technology. It also provides insights into where it may be safely used in IoT as well as misused or abused.

Intelligent processing combines multiple types of sensor data and extracted information over time and space. Any given sensor reading may be noisy or inaccurate or provide a narrow slice of information. A given sensor may be destroyed or disabled, especially if used in military operations. But networked together, multiple sensors and sources provide detailed, highly accurate, and rich information about the environment and individuals in it. Information drawn from across the IoT network is processed into integrated information that is qualitatively better than any one sensor or source. And integrated processing of networked sensors is robust to the loss of any one source. Intelligent processing, however, does much more than just improve the quality of the source information and sensor data. It can filter, abstract, and infer new information. Simply adding more and more data sources may not improve situation awareness or the planning and control of actions. It may create an information fog or information overload, providing more data than a person can digest and analyze. Intelligent processing is essential to convert sensor data to information and information to knowledge and insight into a situation. It distills patterns of life and behavior from thousands of observations and facts. It allows individuals and organizations, such as specific military units, to be located, identified, and tracked. In many cases, intelligent processing can accomplish these actions autonomously with great accuracy and precision.

IoT can generate exquisite situation awareness, but equally importantly or more so, intelligent processing can use its understanding of a situation to formulate a plan of action and to control the execution of those actions. In this way intelligent processing in an IoT system not only understands the world, but it also physically changes it. IoT emerges as such a potent force now to a large degree because of the advances in intelligent processing. Those advances in turn rest on advances in massive, distributed processing and the availability of vast quantities of digital data for training machine-learning and big-data algorithms. In turn, widely available, high-bandwidth networks make it possible to move the data from its source to information processing nodes.

A common type of processing involves tracking individuals using location data from IoT devices. If the IoT devices are cellphones, personal fitness devices, personal navigation devices like GPS receivers, or instrumented automobiles, then it is a reasonable assumption that when the device moves, its owner is moving. Tracking the device is equivalent to tracking the owner. A step up in complexity are intelligent algorithms that can determine an individual's patterns of life and behavior by combining tracking data with other types of extracted information. Patterns of life include routine activities such as commutes to work, shopping trips, and visits to friends or businesses. In national security terms, patterns of life reveal such things as preparations for war, military maneuvers, and an unfolding intelligence operation. Patterns of life can be used to identify an individual, even when sensor data is anonymized. When named individuals cannot be associated with specific tracks, the combination of tracks from multiple individuals or vehicles can provide patterns of life for organizations. Such patterns of life can be deduced by processing even poor quality, low resolution, and noisy sensor data (for example, see Figure 21.4b). For defense purposes, IoT devices observed over time yield unrivalled insights into a hostile power's capabilities, actions, behaviors, and even intentions.

Algorithms that perform these types of processing are usually referred to as AI. Because the definition of "AI" varies, the term "intelligent processing" or just "processing" is preferred here to include AI algorithms but also to encompass other types of information manipulations such as database search and computation of standard statistical measures. Even though IoT networks often include processing that would not be considered "intelligent," the emphasis here is on IoT networks that do include intelligent processing. Such networks can perform automated tasks, previously requiring a human for data interpretation and to control actions. IoT networks with intelligent processing move humans from some decision making or move them to more abstract levels. As a result, IoT networks with intelligent processing hold the greatest benefits but also some of the greatest risks for misuse. The following sections sample some of the tasks IoT systems can accomplish with intelligent processing. A brief summary of AI in the next section (Section 21.1.3.1), illustrates how it creates the potential for IoT's misapplication, abuse, and occasional unsuccessful applications.

21.1.3.1 IoT and the Nature of Intelligent Processing (AI)

While promising tremendous benefits, intelligent processing in IoT systems can become a destructive tool in the hands of hostile agents. Section 21.1.3.3 discusses types of abuse intelligent processing enables. Section 21.3 discusses approaches to regulating intelligent processing so that IoT is free to provide its benefits but constrained from being misused. Beyond being intentionally misused, AI algorithms, especially neural networks trained with machine learning carry special risks and challenges. To understand challenges posed by intelligent processing, it is necessary to understand how it functions. By "intelligent processing" we refer to any algorithm embedded in software and run on a computer processor that converts data to information and information to knowledge about the world and then uses its knowledge to generate a plan and control actions that can impact the state of the world. The concept of intelligent processing is closely aligned with the term AI.

The term AI was coined in 1956 at a conference on computers and thought held at Dartmouth University [51]. The term does not have an agreed upon definition but generally refers to computer solutions and applications that most people would agree require human intelligence. It does not strictly refer to the simulation of human intelligence, but to software or algorithms that can display intelligent behavior. One author defined AI as any mental task on which humans do well and computers do poorly. Consequently, things like long division were excluded from AI in the 1950s, but chess and symbolic solutions of calculus problems were included [52]. This definition has gone by the wayside as AI algorithms have exceeded human mental performance in more and more domains. AI programs are now world champions at chess and GO, translate tolerably well between hundreds of human languages, understand and generate human speech, drive cars, and solve calculus problems. The longest standing definition was given by Alan Turing four years after the first computer was turned on. He defined it, paraphrasing, as machine-generated intelligent behavior that a human would judge to be indistinguishable from behavior exhibited by a human being [53]. While AI can outperform humans on many tasks considered to require intelligence, they do so only when trained to solve specific problems, such as chess or protein folding. A general machine intelligence that can remotely approach a human on a wide range of tasks remains well out of reach of the state-of-the-art in AI, at least at present.

21.1.3.1.1 Brief History of Intelligent Processing/Artificial Intelligence

The field of AI has progressed along two different paths. One path, inspired by the operation of biological neurons, develops ever more powerful networks of artificial neuron-like mathematical entities. One of the most widely used classes of artificial neural networks are computational (or convolution) neural nets or CNNs. Today's CNNs and related algorithms are sometimes referred to as machine-learning algorithms because they are normally automatically generated by feeding training data to a naïve initial neural net. The most successful neural nets today are organized in multiple layers where each layer receives input from the layer below and "innervates" the layer above. For this reason, neural nets are often referred to as deep networks or deep-learning algorithms. The other path pursued by AI researchers explores models and algorithms using symbolic-style reasoning, often using formal logic systems.

These two approaches to AI correspond roughly to the two types of human thinking outlined by the Nobel-Prize winning psychologist Daniel Kahneman. Kahneman calls them System One (S1) or fast thinking, and System Two (S2) or slow thinking [54]. S1 thinking, to simplify, is normally intuitive and produces answers quickly without a conscious awareness of how they were arrived at, for example, knowing that $4/2 = 2$. Normally, S2 thinking involves conscious reasoning through a sequence of deductive or inductive steps or rules, for example, determining that $77/13 = 5.9230769$. Humans can usually explain how they arrived at a decision with S2 thinking. Frequently, we cannot accurately explain how we arrived at a rapid S1 conclusion. Computational neural nets function like the human S1 system while symbolic AI algorithms correspond to S2 slow thinking. Authors also refer to S2-like AI algorithms as modeling "general intelligence" or "rule-based intelligence" or "symbolic intelligence." At present, the most widespread and spectacular AI successes with IoT systems are coming from applications of neural nets. Neural nets and symbolic AI algorithms share key characteristics corresponding with their human S1 and S2 analogs – a fact with important ramifications for IoT, especially IoT used in military applications where lives are at stake. To understand why requires an addition description of how neural network algorithms work and embody knowledge.

One of the first accurate and useful description of how a group of neurons could learn and remember information from input stimuli was developed by Donald Hebb, a neuropsychologist, in 1949 [55]. This model demonstrated mathematically that a network of neurons could learn by modifying the strength of connections (called synapses) between neurons in a network if presented with a repeated set of stimuli or inputs. It took 65 years to finally observe Hebbain learning through synaptic modification in the brain of an animal, specifically a tadpole [56]. In the interim, however, AI researchers seized on computational models of Hebb-type neural nets to demonstrate computer programs that could learn and recognize patterns. One of the earliest was the Perceptron, developed by Rosenblatt [57]. As has occurred throughout the history of AI, the initial success of the Perceptron in solving simple problems inspired an explosion of enthusiasm in the potential of AI to solve complex human problems. The boom was followed by a collapse of interest in AI as limitations of the Perceptron were realized. As a result, funding dried up for a period as did the respectability of AI, an event known as the first AI Winter.

During this first winter, David Hubel and Torsten Weisel published a series of papers beginning in 1962 through the 1970s that outlined the organization of the mammalian visual cortex and resulted in their winning the Nobel Prize in 1981. They showed that the visual cortex is organized in interconnected layers of neural networks (a deep network), each layer of which computes a more abstract description of the visual field and conveys that result to the layer above [58]. Len Uhr and his students at the University of Wisconsin-Madison developed a series of computational models from Hubel and Weisel's neurophysiological findings [59]. They built computer implementations of layered neural nets and demonstrated basic visual recognition tasks [60]. Uhr's deep-layered networks were manually built, but subsequent researchers reformulated and augmented a layered framework with "backpropagation," a technique that enabled practical training of a network by feeding an error term back through the layers [61]. This allowed adjustable-weight connections between nodes to be used to create trainable, multi-layered artificial neural networks. Progress was slow, traveling through a few more AI boom-bust cycles, but by the 2010s such networks were being trained to perform impressive feats of intelligent behavior such as defeating the world champions in a host of board games and understanding and translating spoken language. While there were certainly numerous improvements in machine learning algorithms in the last few decades, among the primary drivers of neural net performance are the exponential expansion of computational power and the explosion of digital data that can be used for training.

While impressive triumphs by neural nets in artificial S1 thinking have caused yet another spike of irrational exuberance for AI, the greatest yet, progress in S2 thinking has advanced steadily but less dramatically. The reality is that many intelligent tasks, such as automatically driving a car, require a combination of S1 and S2 thinking – a combination of neural nets and symbolic reasoning. While neural nets have solved many of the vision problems for self-driving cars, for example enabling them to recognize roadway and lane boundaries and both stationary and moving obstacles, they still need some rule-based S2-type reasoning, for example to embody and apply the rules of the road. The value of S2 thinking has not escaped the notice of neural-net researchers. Recent triumphs of AI processing, such as figuring out the properties of materials from average properties of their electrons, are possible by combining neural nets with S2-type rules or equations that embody the laws of physics [62]. A hybrid neural-net and constraint-based (S1/S2) approach developed by Deep Minds now predicts protein folding for most proteins given the molecular composition; a feat that is not only a triumph for AI; it is a breakthrough for organic chemistry of historic dimensions [63]. The American Association for the Advancement of Science (AAAS) hailed this AI-powered accomplishment as the most important scientific breakthrough in 2021. Mammalian brains may also integrate S2-type constraints to make S1-type learning more efficient, for example see Chung et al. [64]. Researchers investigating IoT for defense applications are also exploring hybrid S1/S2 approaches combining physics models with neural-net-based learning as described by Bastian et al. in Chapter 8.

Many of the most spectacular accomplishments of AI are attributed to neural nets, when in fact these achievements result from systems that combine neural nets with traditional symbolic tree search methods (S1 + S2 thinking). AlphaZero, which learned how to beat the best human players in chess, Go, and Shogi did so by using neural nets to evaluate board positions in an otherwise traditional symbolic tree search [65]. Neural nets get credited though, because they made the difference in powering old algorithms to achieve super-human performance. The neural net portion of these game-playing algorithms acquired their competence by playing thousands and even millions of games against themselves to provide training data. Intelligent processing that can play games at human level or better has great promise for planning and monitoring military operations with IoT. But the need for thousands to millions of trials for training has profound implications for IoT applied to solve problems in the real world, especially when those solutions involve critical military systems like weapons. The next section explores this challenge and others for AI in IoT.

21.1.3.1.2 Challenges for IoT Presented by Neural-Net-Based Artificial Intelligence

Neural nets and S1 thinking, trained by observing large amounts of exemplars, are vulnerable to what Daniel Kahneman calls the "broken-leg case" [66]. He explains with an example: if you are estimating the likelihood of your neighbor going to see a movie during the next week, you might frame your answer based on how often people in general go to movies as well as what you may know about your neighbor's movie-viewing habits. An artificial neural network can learn to make as good or probably better predictions than you if trained on sufficient data. However, if you know that your neighbor broke his leg yesterday, you will immediately drop your estimate to near zero based on S2 type thinking. The neural net could learn about broken legs, given massive amounts of training data, but it may still not know that a range of other singular events can override normal experiences, such as the neighbor having just filed for bankruptcy or having a child who was recently admitted to a hospital. S2-type thinking, whether artificial or human, may be slow but it can efficiently apply rules that completely change how you view past experiences or training exemplars. S1 thinking, whether human or artificial, is not so easily informed by the addition or modification of simple rules, equations, or heuristics. Rules and heuristics can make for far more efficient and less costly learning than translating thousands or millions of training exemplars into internode connection weights in a neural net. As an example, a robot with a hand that could sense damaging heat could learn with a deep learning network not to put its hand on a hot stove in dozens or perhaps thousands of trials. A rat or pigeon might require one or two trials. A human usually requires zero trials – their mother tells them not to put their hand on a hot stove. People file that type of symbolic information away as an S2-type rule without ever scorching flesh.

Financial analysts, economic forecasters, and investors face their own version of "broken-leg" scenarios. Nassim Taleb popularized the notion of a "black swan" event where a totally unexpected and unmodeled event substantially changes the direction of economic markets [67]. He coined the term "black swan" in reference to the belief in Europe that swans were always white, a belief that dissolved when the first black swan was encountered in Australia. He refers to black swans in the financial world as the impact of highly improbable events and maintains that markets will continually encounter black swans given a long enough time. For neural-net based AI, the "highly improbable" can be any significant event that the network has not seen in its training. Deep-learning AI networks already occupy important positions in some investment strategies, and therefore the financial world should be concerned about black-swan or broken-leg vulnerabilities lurking in their AI programs. IoT systems in military operations should fear black swans even more so.

Communities of animals, plants, and microbes form connected networks in ecological systems. An ecological community can be viewed as an IoT network. Although communication does not take place through a digital internet as in man-made IoT systems, the elements of an ecological system connect and communicate in multiple ways. They collectively sense their environment and react to it and in the process alter it. Complex organisms develop and carry out courses of action based on what they sense and how they process the data as well as using models of their environment and fellow living things. They all have effectors or actuators that physically alter their environment. Nodes in an ecological community sense, process, and control actions, just like an artificial IoT network. Organisms modify the nature and the strength of their connections based on feedback, analogous to a neural net. Nodes in this living network learn in the sense that they adjust their behavior during their lifetime and accumulate lasting improvements through DNA mutations that survive and proliferate in proportion to their adaptive utility. This network of living things stores what it learns, not in modified synaptic weights like a neural net, but in the evolving DNA of each species.

In an internet of living things, broken legs and black swans also occur with both beneficial and occasionally devastating consequences. Five major mass-extinction events have occurred in life's history. The best known is the asteroid impact that extinguished the dinosaurs and let mammals radiate out to dominate animal life today. Stephen Gould argued that organisms could not pre-adapt or learn to survive an asteroid impact because ecological communities did not encounter them often enough to incorporate their effects during their evolution [68]. Asteroid impacts are not part of the training set of evolving living communities. They are black-swan events. As artificial IoT networks evolve their level of intelligence and as we further embed them in our lives, we would do well to remember that black swans will always occur, we just don't know when and in what form. IoT that controls networks of weapons and defense systems cannot be expected to react well to events never seen in their training. For now, expecting the unexpected remains a trait exclusive to human intelligence, and a trait that remains rare even among humans.

The foregoing discussion on the differences between S1 and S2 thinking, that is the difference between artificial neural networks and symbolic reasoning, may seem like a digression from IoT, but it is a critical point in terms of challenges presented by IoT systems. IoT decisions relying on neural networks trained with deep learning on occasion make what most humans would consider to be obviously "dumb" decisions. These occur because the algorithm's training did not encompass exemplars of rare events that could have been easily captured with rules.

Less than totally comprehensive training data can also produce more subtle errors or biases in AI neural net decisions. Of special concern are neural networks with biases or poor performance with specific minority groups. Such biases can lead to discrimination in applications ranging from facial recognition for law enforcement to credit histories for deciding who gets a mortgage [69]. Rob Brooks, one of the founding fathers of intelligent devices including the Roomba vacuum cleaner and the military PackBot, states that all AI systems are prone to occasional egregious, unanticipated errors, and therefore, all successful applications of AI occur either where a human is in the loop or where the consequences of failure are low. Examples of the latter include playing games like chess and GO, predicting protein folding, or vacuuming your home [70]. Obviously, the consequences of failure for many applications of IoT for defense and national security are high. Some of the greatest gains for IoT in military operations comes from moving people out of the real-time control loop. In military affairs IoT systems relying on neural nets will always be somewhat prone to "preparing to fight the last war," just like human generals. Human generals find it difficult to anticipate strategy and tactics they've never seen. Having never experienced a Blitzkrieg, the Maginot Line must have appeared an impenetrable defense. Because neural nets distribute their knowledge across thousands of weights on nodes, users of AI-driven IoT systems find it especially difficult to anticipate knowledge gaps and bias in a neural network's training set.

In defense applications, black swan events can lead to fratricide or civilian casualties if IoT has a role in targeting or firing weapons. Not only is it difficult to anticipate and test for such rare events, but it can also be difficult to understand why they occur and therefore difficult to correct for them. In fairness to AI-driven IoT, humans also occasionally commit fratricide and cause civilian casualties in the face of events that they have never experienced nor been trained for.

Challenges due to IoT controlled by AI go beyond rare black swan events. Most neural-net algorithms find it difficult or impossible to provide explanations for why they make the decisions they do. Their knowledge is distributed among the thousands or millions of connections between nodes and layers of the network and is generally impenetrable to humans looking for S2 type rules and reasoning. This opaqueness makes it difficult to trust AI decisions made with neural networks. An AI's decision may seem bizarre and wrong to a human observer but still be correct; however, the human observer has no way to judge that it is correct short of letting the AI proceed with its decision and seeing its consequences. The DARPA Air Combat Evolution (ACE) program is pursuing experiments to see if combat pilots can develop trust in AI-based IoT systems that autonomously control an aircraft in a dogfight. The experiments observe pilots experiencing the consequences of trusting or not trusting IoT [71]. Because IoT networks modify our world, a lack of trust and transparency along with unknown training biases, and occasional serious inaccuracies in judgement due to training limitations should be of great concern to us. If an IoT system plays a part in controlling weapons or even the flow of materiel, such considerations result in life-or-death consequences [72].

Neural net algorithms excel at extracting complex patterns from big data. They exceed human performance in many instances when fed sufficient training data. This property makes intelligent processing an outstanding engine for generating situation awareness with IoT networks. Realizing that promise broadly for IoT assumes that the concerns outline above can be addressed. But intelligent processing holds a second promise of equal or greater importance for enabling IoT: the ability to automatically generate and supervise the execution of plans based on its derived understanding of the situation. When AI can close the loop between IoT sensing and action, IoT then realizes its full benefits by moving humans out of the loop into a supervisory role.

Faith in AI's ability to plan and execute strategies arises from the past decade of success in AI overpowering the best human players in a variety of games, such as chess, Go, and poker. This raises the question of whether game-playing AI algorithms can be transferred to IoT to serve as intelligent planners. If an AI algorithm can best a top human pilot in simulated air combat, why not use AI with an IoT-flown aircraft for real combat? When recalling Rob Brooks' two laws of effective applications of AI, game-playing AI violate Brooks' first rule of successful AI: "a human must be in the loop." To realize the full benefits of IoT, humans must be moved out of the loop into supervisory roles. Brooks' second rule might provide an alternative route to successful AI applications in military IoT without humans in the loop: "the consequences of failure must be low." For some military missions, the consequences of failure may be low, but for many others, the consequences are high and include destruction of property and loss of life. However, even for applications where the cost of the damage resulting from failure is high, the expected-cost of failure may be low. The path to achieve that would be to gain high confidence that an AI algorithm will not fail. The expected cost of failure can then be expressed as:

$$C(f) = D(f) * P(f).$$

Where $C(f)$ represents the expected-cost of failure, $D(f)$ represents the cost of the damage resulting from failure, and $P(f)$ represents the probability that failure will occur.

This formulation implies that the expected-cost of failure may be low even if the damage caused by failure is expensive as long as the likelihood of the failure occurring is low. The failure resulting from the accidental release of a thermonuclear bomb is extremely high – the loss of a city. Therefore, it is hoped that the people who control thermonuclear bombs have ensured that the probability of such an accident is extremely low. On a smaller scale, it may be expensive to lose a surveillance drone on an autonomous mission, but if the mission is constrained so that the drone is operating in uncontested air space, then the probability of that drone being lost maybe very low, leading to a small expected-cost of failure. In such a case AI can be trusted to operate the drone. Assigning the cost of damage, $D(f)$, is relative. A Javelin flying its route autonomously may fail to connect with its target and cause collateral damage. The cost of that damage depends on the context of the strike. In a peacetime law-enforcement action, the cost of collateral damage may be considered unacceptably high. In a combat zone, in the heat of a fight, the same destructive damage may be deemed to be an acceptable cost.

There is one more consideration in determining the expected-cost of failure. The potential cost of failure must be weighed against the expected-cost of not deploying a weapon. Representing the cost of not deploying a weapon as $C(nd)$, the expected-cost of failure can be represented in simple terms as:

$$C(f) = D(f) * P(f) - C(nd).$$

If your enemies have thermonuclear weapons and you do not, then $C(nd)$ is large. That means the expected-cost of a thermonuclear accident may be low even if safeguards do not lower the $D(f) * P(f)$ term close to zero. As a smaller scale example: if a light infantry unit is facing an approaching enemy tank column, the cost of not deploying Javelin missiles, $C(nd)$ is high and therefore $C(f)$ becomes low, even if the probability of failure and the consequences of failure are not low. By Brooks' second law, using AI planners to remove humans from inside the control loop demands that the consequences of failure be low. Making that assessment requires commanders and policy makers to assess the three terms: cost of damage resulting from failure, $D(f)$, the potential cost of not using the IoT system, $C(nd)$, and the probability that the system will fail, $P(f)$.

Commanders routinely must assess the potential cost of collateral damage caused by failure of some tactic or weapon. They also routinely assess the potential damage if they don't use a particular tactic or weapon. Now commanders and decision makers face a new challenge: estimating the probability of failure when using AI controlled IoT. Making and trusting this estimate is tied to the issues of opaqueness of AI decisions, training bias, and black swans as discussed above. But there is one additional issue determining the probability of failure for AI planners in military IoT systems: the complexity and predictability of a mission and its environment. A review of planning and problem solving in AI research provides the context for understanding this last challenge.

AI research focused on automated planning and problem solving since the beginning. By 1959, The general problem solver, developed by Allen Newell and the Noble Prize winner Herbert Simon, demonstrated automatic problem solving for simple, well-defined problems set in highly constrained simulated "world" [73]. They used symbolic reasoning techniques; neural networks were not yet powerful enough to be a competing approach. A similar symbolic approach was used for Shakey, a mobile, indoor robot developed by SRI International under DARPA funding from 1966 to 1972 [74]. Shakey operated in a real set of rooms, but they were carefully painted and sparsely furnished. By the early 1960s advancing into the 1970s, AI algorithms could also play adversarial games – checkers at a human master's level and chess at a mid-range level [51]. AI gradually improved its chess play to grand master level over 50 years as computers got more powerful. Automated intelligent pilots demonstrated success in air combat simulations by 1999 [75]. More complex games, such as Go, eluded machine mastery until enough processing power became available to train neural nets to predict the best set of possible moves for a given board position. Over the last decade, machine learning algorithms coupled with traditional search algorithms quickly overcame human world champions in Go and other board games [76]. Initially, board games, mastered by AI, shared three features: they had just one opponent, the "world" consisted of a simple board with a few types of well-defined and discrete moves, and both players had complete knowledge of the state of the "world" at every step of the game. By 2017, machine-learning algorithms began to master multi-player games where every player had incomplete information about the state of the game and its opponent's full capabilities, for example, StarCraft II, DOTA, and a type of Texas Hold'em poker game [50].

Despite these advances, automated planning in adversarial games or missions still finds most success when planning and executing adversarial games in simulated worlds. These worlds are far simpler environments than the real world and their actions are few, discrete, and well defined. Simulated environments allow AI planners and game-players to succeed because they reduce the size of the search space of alternatives considered while planning. Just as importantly, by abstracting out real-world complexity, simulated worlds can be hosted inside of a computer allowing thousands or even millions of trials to be played to train machine-learning algorithms. For example, it would be cost and time prohibitive to train a machine-learning algorithm for air combat by instrumenting real aircraft with IoT and conducting thousands of real instances of combat. It is cost prohibitive to train human pilots exclusively in the real-world situations. Their training relies on heavy doses of experience gained in simulations. Machine learning algorithms require far more training data than human pilots. Of even more concern, AI algorithms cannot be feasibly trained on multiple failing missions in real-world settings. They must be trained to a high degree in simulated environments and missions. Military commanders must then trust that the AI algorithm, trained in simulation, will transition successfully to real-world missions. Real-world missions will always have unexpected and unpredictable elements not captured in the training simulation. Using AI to plan and execute IoT military missions requires trust that the abstractions and simplifications made in the simulated training environment do not leave holes in IoT's performance when transitioned to real-world missions. These simplifications lead to biases and black swan vulnerabilities in the AI's knowledge. But these vulnerabilities do not come from insufficient training data, but from a lack of fidelity in the training environment.

The real world is overwhelmingly complex and unpredictable. The most complex and unpredictable parts arise from human behavior. Besides simulating the key properties and physics of objects, human behavior needs to be modeled. Humans predict each other's behavior with remarkable accuracy, enabling social cohesion. We do this by building a predictive model of other people. This cornerstone skill of human intelligence is called a "theory of mind" [77]. Few, perhaps no, driverless cars have been trained to develop an extensive theory of mind. A theory of mind is essential to defensive driving – an essential skill for safe driving. It is especially important for driving in urban areas where pedestrians and bicyclists often ignore the rules of the road. But far more trivial elements than human behavior may be abstracted out for training AI algorithms. The first driverless vehicle, the Autonomous Land Vehicle (ALV), was driven off the road by an insect. A review of the recorded video from the main navigation camera after a trial run revealed the reason that the ALV departed the road: a grasshopper was seen flying into the lens of the navigation camera. It grew from a speck in the distance to a giant form that blotted out the vehicle's vision, causing it to lose sight of the road and swerve off. No driverless car simulations at the time, and probably still not at the present time, included grasshoppers as a feature of the environment. Commercial driverless cars are trained both in simulation and on thousands of hours of driving on real roads. It is difficult, however, to train in real situations that are likely to lead to real failures, such as testing a vehicle's reaction to a pedestrian suddenly jay walking in front of it. For driverless vehicles, grasshoppers perhaps remain black swans.

On the ACE program, DARPA is currently conducting a set of experiments using intelligent planning for automated air combat. The automated combatants have demonstrated superiority over a top human pilot in simulated dog fights. The next step is seeing if that training in simulation transitions to real aircraft and real dog fights. The ACE experiments will gain trust in intelligent IoT where AI trained in a simulated world transitions to the real world. ACE is an important experiment, although it may be hard to generalize beyond that specific mission and environment.

For complex, lightly constrained missions, neural net planners must be trained in simulated worlds and then transitioned to real-world missions. Commanders will not know if an essential element of an actual mission was left out of the training world. Testing the AI in real-world trials instills some confidence. But testing is expensive, time consuming, and therefore limited. Testing usually doesn't include unexpected events and by definition it does not include unknown events. Research may invent new technology that engenders trust in AI planners. For example, perhaps formal methods may be invented that can prove the range of possible behavior from a neural net. But short of relying on trust through testing or waiting for some innovation to certify AI-commanded systems, IoT with intelligent processing may best be deployed in missions that are highly constrained and have a constrained environment. Such missions and environments will have a better match to simulations used for training. Selecting such constrained missions lowers the probability of failure and can lower the cost of consequences of failure.

Transitioning AI trained in simulation to real-world missions presents possibilities for intentional misuse, abuse, and defeat. An adversary might exploit elements of the real world or a real mission that were abstracted out of the training simulation. A hostile agent might misuse or abuse AI in an IoT system by inserting elements into a real situation that are known to be absent in the simulation environment. These elements could be used to induce behavior by IoT that harms people or benefits people whom the developers did not intend to harm or benefit. These problems are open research challenges. Experiments are likely to be expensive and challenging themselves because they cannot be performed solely in simulated worlds.

How to gain trust in IoT systems driven by AI algorithms is a topic of great interest to both policy makers and AI technologists. At the research level, the US DOD has funded efforts to create "explainable AI" systems [78, 79]. At the policy level, the US government funded a commission, the National Security Commission on AI, to examine these issues [80]. The European Union has also declared its support for research into explainable AI [81]. A focus of the US Commission has been on how to sufficiently test, examine or analyze an AI system to trust it in a national defense and security role. Researchers on the US Army's Internet of Battle Things program are also investigating mechanisms for explainable AI for IoT (see Chapter 8). Solving the formidable challenges of gaining trust in IoT's AI performance is a critical technical issue as much as a policy or regulatory issue. As such it lies beyond the scope of this chapter. Section 21.3 reviews regulatory and policy measures that might be imposed on IoT that can limit damage where intelligent processing is misused or abused.

21.1.3.2 Intelligent Processing of IoT Sensor Data and Extracted Information

As stated, intelligent processing powers IoT in three ways: converting data and extracted information into situation awareness, using knowledge of the situation to plan actions to evolve the situation, and controlling actions to execute the plan. In this section, we examine some of the ways intelligent processing converts sensor data and extracted information into knowledge of individuals' and organizations' activities, and behaviors.

21.1.3.2.1 Tracking Individuals

Smartphones are the most powerful IoT devices and with every new generation they have gotten better at providing accurate, high-precision location information. As previously described, there are several ways a smartphone knows its location. It may report its location to facilitate services, such as optimizing cellphone connectivity, map-based navigation or finding local retail stores. Third-party companies aggregate location data into individual tracks by linking location reports over time to compile data bases detailing the continuous movements of millions of mobile phones. The track aggregators sell the tracks to yet other companies to use as they desire. The New York Times purchased a week's worth of smartphone tracking data from an aggregator and analyzed it [82]. What they found (Figure 21.7) was that individual phone locations could easily be processing into detailed records of the movement of those phones showing every place thousands of individual smartphones and their owners traveled during the past week. This data is anonymous – named individuals are not associated with specific tracks. But as the Times' reporters showed, it need not remain anonymous.

By fusing anonymous phone locations for specific phones, reported over time, anonymous tracks are formed for each phone. Many if not most of these tracks can be associated with specific individuals by combining the times and locations of a phone with collateral information, such as a

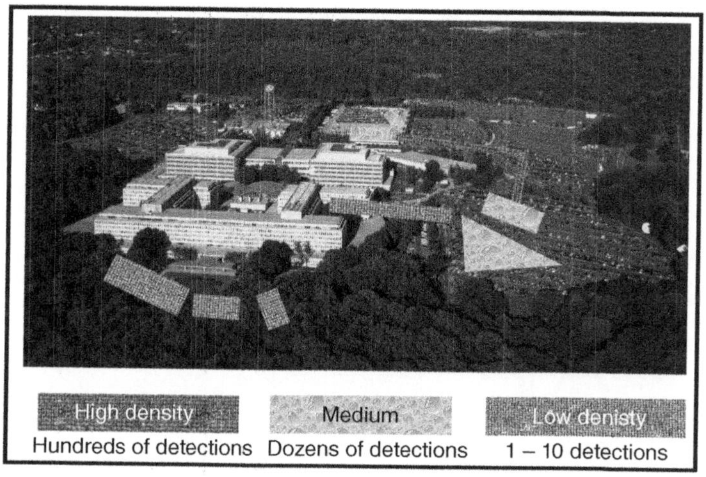

Figure 21.7 Snapshot of densities of mobile phones detected at the headquarters of the US Central Intelligence Agency (CIA). The density map is based on a data base of individual "anonymous" phone-location reports as recorded by a phone-location aggregator using information extracted from applications on the phone. Individual phone detections are purchased and tabulated by the New York Times [82]. Using additional information sources, named individuals can be associated with many of the tracks. Source: Carol M. Highsmith / Wikimedia Commons / Public domain.

phonebook. For example, by following a phone's track back to a street address at the end of the day, a name can be assigned to the track using the name associated with the address in a phonebook. In this manner, tracks can be identified with specific individuals in an automated manner. The New York Times calls it "A Diary of your every movement" [82]. The Times was able to track specific individuals such as celebrities and a known senior Pentagon official taking part in the 2018 Woman's March. Inferences could be made not only about who and where the individual was during the day, but also about their behavior and political views.

The New York Times claims that it took only minutes to find and track the US President given a dataset of 50 billion location reports from the phones of more than 12 million people [83]. Figure 21.8 shows a track that the New York Times proports represents Trump's movements around Palm Beach on 9 February 2019. The newspaper also claimed to have located the Vice President in the D.C. area and tracked him around town. The newspaper discusses the legality and ethics of such tracking and concludes: "The companies that collect all this information on your movements justify their business on the basis of three claims: people consent to be tracked, the data is anonymous, and the data is secure. None of those claims hold up, based on the file we've obtained and our review of company practices." [82]. It should also be noted that an information aggregator obtains far more information than location from apps on someone's phone. In addition to location, information can be extracted such as an individual's search history, contacts, and activity on social media, etc. All this information is associated with an individual's location by time of day. When intelligent processing integrates this information, it reveals a picture of a person's activities, behavior, and associates. As noted below, some AI algorithms go on to infer intent.

Information extraction and intelligent processing of smartphone data has already been used to detect troop build-ups along the Russian and Ukrainian border. That data clearly advertised the pending invasion that occurred afterwards. It was reported that "in 2016 Israeli personnel were banned from playing the mobile game *Pokémon Go* over similar fears of disclosing secret tactical maneuvers" [84]. As another example, the company Strava is an "opt in" social networking

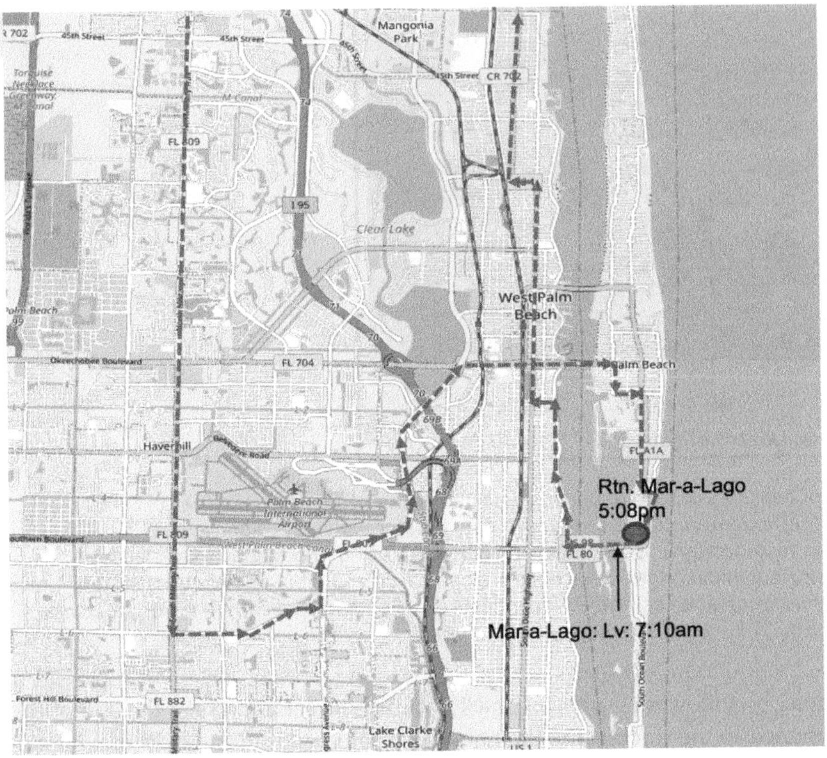

Figure 21.8 The New York Times compiled a "diary" of the US President's movements on 9 February 2019 in the Palm Beach, Florida area. The Times logged the President's whereabouts using supposedly anonymous mobile-phone location data purchased from an information aggregator. The Times associated the "anonymous" tracks with the President by comparing public information on his scheduled meetings and their locations [83]. Source: R. Douglass based on data from New York Times. Base map of Palm Beach: Open Data Commons Open Database License (ODbL) by the OpenStreetMap Foundation (OSMF).

service that combines activity data associated with specific times and places. The information is willing provided by subscribers to Strava that then uses this data to compute a "heat-map" showing activity around the globe [85]. In 2018 it became apparent that observing Strava activity on a heat-map allowed a number of deductions to be made about military and intelligence operations. For example, the locations of US forward-operating bases in Afghanistan and even Central Intelligence Agency (CIA) "black sites" could be deduced. By observing changes in activity at such sites, it becomes possible to develop a pattern of life for the site and detect changes that indicate troop build ups and battle preparations. When Strava was questioned about the risk to military operations, they suggested that military users should "opt out" of sharing information with Strava's heat map. The US military issued an order banning the use of fitness trackers in sensitive areas and made their use subject to the discretion of local commanders in all areas [86]. IoT's intelligent processing can infer far more about defense and national security than could be perceived by any given sensor or intelligence source. Such inferences traditionally have been the domain of human intelligence analysts, but as the power of AI algorithms grows, IoT networks will perform analysis automatically in many instances and in some cases more accurately than a human analysist.

The threat and utility of intelligent tracking is not limited to fitness trackers and smartphones. The noisy, imprecise detections and locations of people by radars that can penetrate building walls and dense foliage can also be fused into precise locations of paths, roads, and facilities as well as maps of building interiors [29, 30]. Activity patterns can be deduced over time even when obscured from human eyes and conventional cameras as shown in Figure 21.4. As self-driving and driver-assist technology becomes pervasive in vehicles along with pervasive Internet connections, vehicles become IoT devices, almost the equal of smartphones. They allow the same sort of tracking, patterns of life, and patterns of society that smartphones do. They observe much more of the surrounding environment and activity than smartphones. Research demonstrates that using only a car's onboard sensors it is possible to "fingerprint" specific drivers by their driving behavior and thus distinguish who is driving [87].

21.1.3.2.2 Comprehensive Mass Surveillance of the Public

Perhaps the most advanced processing of IoT sensor data and extracted information can be found in China's Skynet, Sharp Eyes and Golden Shield projects. Together, these efforts process video and other data to identify, track and attempt to identify improper behavior from jaywalking to national security threats. Intelligent processing is used to identify people and activities through face recognition, gait recognition, license plate readers, smartphone apps, databases on travel, criminal history, texts, and inputs from human observers as well as other emerging information extraction methods. The Sharp Eyes surveillance system aims for 100% surveillance of public spaces and makes video streams of public spaces available to citizens so that they can detect threatening behavior. These technologies are being applied to all citizens, but reportedly especially intensely to the Uighur minorities [15–17]. China is also leading in combining IoT's ability to collect and process multi-sensor/information with direct control of actions – uniting IoT's three elements of sensing, processing, and action [88].

Mass surveillance aims to identify and track large groups of people. It observes activity and actions. It attempts to recognize behavior. The next step is to infer intent and predict intended behavior. Reading media reports indicates some AI firms are already attempting to accomplish this, but with mixed results [89]. The recent Chinese Five-Year Plan seems to indicate that they are also pursuing enhancements to mass surveillance aimed at deducing intent from sensed data. In the case of domestic surveillance in peacetime, this is an ominous development. Inferring intended actions potentially infringes upon our private acts even before we take them. But if IoT systems are to use AI to plan defense and military actions, then it must at least implicitly infer the likely future actions of the adversary, the topic of the next section.

21.1.3.2.3 Intelligent Processing Automates Planning

Intelligent processing, particularly neural networks, are outstanding at recognizing complicated patterns in complex and noisy data. They can convert sensor and extracted information into an understanding of the situation and how it is changing. Equally or more importantly, intelligent processing can observe a situation and formulate plans to alter it in a beneficial way. Intelligent processing also provides top-level control over the actions that execute the plan, modifying them as events unfold. These functions are what allow IoT with intelligent processing to move humans out of the control loop into a supervisory position. Doing so realizes some of the key benefits of IoT for military operations. The actual algorithms that control the mechanical or electrical actuators form part of the action portion of IoT discussed Section 21.1.4. Control theory for controlling actuators is well defined, generally well understood, and implemented with equation-based solutions. It thus is not dependent on intelligent processing, except to the extent intelligent processing rides on top of the control loop and provides supervisory control and adjustments to plans as they unfold. In contrast to controlling actuators, Intelligent processing is the key to automated planning. As explained in Section 21.1.3.1, neural-net planners need to be trained in simulated environments or in very constrained real environments and missions. AI planners trained in simulation are most likely to succeed when they transition to real missions where the real missions are constrained in complexity so that they are not too different from the simulated mission. Several examples of AI for IoT in military operations are discussed here.

Logistics Logistics is essential to winning battles and wars, as Omar Bradley, the first US Joint Chief of Staff, reportedly emphasized. One aspect of logistics that has already gotten a boost from IoT is management of supplies. Radio frequency identification (RFID) tags, stick-on barcodes, and geo-locating tiles have revolutionized the flow of materiel, packages, and baggage in much of the world for both commerce and defense. Chapter 6 reviews some of these applications and the implications of IoT for logistics. Supply management involves tracking thousands of things and routing them to destinations and temporary storage. For military operations, the magnitude and types of materiel is huge. Supply chains are also dynamic. Supplies can be required anywhere in the world on almost no notice. Volumes can escalate quickly from small to vast as events unfold. Intelligent processing is particularly adept at managing supply tasks requiring tracking, scheduling, and navigation planning. It is unfazed by the magnitude and dynamic nature of military supply chains. Using AI and IoT to automate supply chains requires an estimate of the probability of failure and the associated cost of failure. The cost of failure of specific supply chains runs the gamete from low to high. In conflict and emergency relief situations, the cost of failure is often high. Until high trust develops for a particular IoT system, supply chains of lower criticality should be selected for a high degree of automation. The cost of failure can also be lowered with traditional strategies, for example, using multiple supply chains, some of which are manually managed. These concerns regarding AI and IoT for automated supply management also apply to other logistics functions, such as management of maintenance and repair. In terms of misuse and abuse, the features of an automated IoT supply chain that make it effective and efficient also make it rewarding for hostile entities to coopt IoT management through cyber-attack or intentional unethical exploitation.

Aerial Surveillance and Reconnaissance Other than the smallest and simplest drones, UAVs constitute IoT systems with components and sensors networked together. They are networked back to one or more command posts on the ground. Additionally, they may be networked with other drones, piloted aircraft, satellites, ships, and IoT controlled weapons. A surveillance mission frequently meets the AI requirement discussed in Section 21.1.3.1 of having a clear straightforward mission executed in a constrained environment. Surveillance missions often consist of a sensor platform flown to a designated area of interest and once there flying a prescribed track over or around it, for example, flying a racetrack pattern. While flying a commanded surveillance pattern, the drone points its sensors at specified areas or activities. The environment is usually not complex because navigating in the air, while not trivial, is significantly less complex than driving over complex terrain on the ground. The environment is further constrained, especially if using costly aircraft, by locating missions in uncontested airspace. Given these simplifications of environments and missions, AI planners have demonstrated success in autonomous surveillance where remote human pilots operate in supervisory roles. On some drones a human sensor operator manually controls the sensors, even if piloting is largely autonomous. Decades of research using AI techniques for automatic target recognition, moving-vehicle tracking, and video-based geolocation have succeeded in partially automating these functions, but AI algorithms still fail frequently. As a result, most surveillance and reconnaissance missions still rely on human perception for these tasks. On reconnaissance missions, drones may not know what they are specifically looking for,

Figure 21.9 The Global Hawk (US RQ-4) flies 30-h surveillance missions. IoT technology navigates and coordinates with ground stations, other aircraft, and space network nodes. It represents the type of drone aircraft and missions that could be conducted largely autonomously with AI algorithms. Source: U.S. Air Force photo/Bobbi Zapka.

making it difficult to train neural nets to exploit the data. As an example, for several decades now, the Global Hawk has been capable of flying multi-day surveillance missions autonomously, although it is not publicly known how often it does so (see Figure 21.9) [90].

Non-Line-of-Sight Air-to-Ground Weapons: Loitering Drones and Precisions Missiles Loitering air weapons are drones that carry an explosive payload. They are launched by a human operator and fly to their target area. Once there, they communicate images or video back to the operator who designates a target and authorizes a strike. The drone then attacks by diving down from the air onto the target. Alternatively, location coordinates can be provided by the operator or other nodes in the IoT network, such as other drones. If no target is in the area of interest when the weapon arrives, it can loiter for a time looking around the area for a target or waiting until one arrives. Precision non-line-of-sight missiles, exemplified by the Javelin anti-tank missile are fired by a soldier after he has designated a target using the launcher's sensors. The missile then uses its own sensors to fly autonomously to the target. The missile typically flies a non-line-of-sight path, rising above a designated tank and then descends onto the top of the tank. Loitering and precision strike weapons exemplify Intelligent processing used to operate an IoT weapon in a largely autonomous mode, keeping humans in a supervisory role – on the loop instead of in it. The concept was pioneered by the US Army/DARPA NetFires program with its loitering attack missile (LAM) and precision attack missile (PAM) [91]. In the winter and spring of 2022, the Ukrainian infantry destroyed several thousand Russian tanks and other vehicles with several types of these weapons. IoT missiles and loitering drones share some of the credit for victory in the battle for Kyiv in the winter of 2022 [92]. As [92] describes, the Ukrainians are employing numerous types of these weapons against Russian armored vehicles in the 2022 war, including Javelins, NLAWs, Switchblades, Warmates, and the Phoenix Ghost. The Javelin anti-tank missile is a predecessor IoT weapon, while the Switchblade is a quintessential IoT loitering drone (Figure 21.10). These weapons satisfy Brooks' laws for the successful use of AI because the navigation algorithms have a low probability of failure, soldiers can move from a supervisory role to take in the loop control if the AI fails, and target designation is not accomplished by AI. Target designation relies on either furnished coordinates or a soldier's perception.

Figure 21.10 Soldier launching a Switchblade 300 loitering attack drone. These IoT systems can take targeting coordinates from ground controllers or other drones over a network. Alternatively, they can fly to an area of interest to strike a target designated by the soldier in supervisory control or loiter over the area waiting for a target to appear. They represent a new breed of IoT weapons using intelligent processing. Source: U.S. Army AMRDEC Public Affairs / Wikimedia Commons / Public domain.

Self-driving Ground Vehicles Self-driving ground vehicles provide a final example of intelligent processing powering an IoT system. Driverless cars have the ability to drive under autonomous control in a constrained environment in certain conditions. Given a detailed and accurate digital map and sensor input, a driverless car can plan out and navigate well defined roads, such as limited-access highways. Less constrained environments, such as urban streets, and unconstrained environments, such as combat zones, are beyond the state-of-the-art for safe autonomous driving. The US Army is experimenting with truck convoying where only the lead truck has a human driver, and the rest of the trucks follow it under autonomous control. While this is a constrained mission, it would seem to require an uncontested roadway in order to trust AI in planning and control of such an IoT system. The technical and organizational barriers to adopting automated convoys serves as a case study in adopting and employing AI-driven IoT in general [93]. It is nearly four decades since DARPA and the US Army demonstrated the first autonomously driven vehicle, yet such vehicles still struggle to find their way into the US inventory of operational systems. This results in part from the extreme difficulty of driving in complex terrain, especially terrain in a combat zone. The technology, specifically intelligent processing, is not quite mature enough to deserve full trust. The military needs to continue to find missions and environments where the technology can be applied with trust. IoT-driven vehicles, besides making their own occasional mistakes, are open to misuse and abuse as is described in Section 21.1.4.

In summary, intelligent processing or AI adds greatly to IoT's potential. On the positive side, AI can recognize complex, even non-linear patterns buried in millions of pieces of information. It can associate those patterns with effective actions to accomplish specific missions or applications. For defense applications, intelligent processing can recognize and track targets and direct fire onto them in constrained missions. It can automatically provide early indications and warnings of hostile actions. It can optimize logistics and the flow of troops and materiel. The present state-of-the-art allows IoT systems to plan and execute certain constrained combat and surveillance missions and enables weapons such as loitering drone munitions. DARPA's ACE program seeks to create IoT for combat aircraft that can fight autonomously in support of human pilots using AI algorithms. ACE learns how to maneuver a plane and fire its weapons as well as how to recognize tactics of hostile aircraft [71]. The objective of ACE, however, is to understand how to establish trust in AI-powered IoT systems – a critical question on the path to realizing IoT's full benefits for military operations.

21.1.3.3 Abuses of IoT Arising from Problems with Intelligent Processing

Abuse and misuse of IoT intelligent processing can be summarized as one of the following five types:

- **IP-1.** An IoT network can be intentionally programed by its creators or owners through intelligent processing to abuse its users, most often for financial gain.
- **IP-2.** Creators of Intelligent processing in an IoT system can unintentionally abuse its users. For example, a lack of breadth in a training set may introduce racial or ethnic biases [94].
- **IP-3.** Users of an IoT network can knowingly or unknowingly consent to abusive behavior by the owner or operator in exchange for desired goods or services provided by the network.
- **IP-4.** Processing in an IoT network can be co-opted by a hostile third-party, usually for financial or political gain. IoT devices have been the fodder for the largest botnet attacks ever assembled. The fact that IoT devices are usually low-end with little security makes them excellent targets.
- **IP-5.** A government can covertly or overtly co-opt the processing in an IoT system and intentionally or unintentionally abuse its users for legal and ethical reasons, most often for defense and security, but do so without the publics' consent and due process to apply checks and balances.
- **IP-6.** Governments can abuse IoT processing, unethically or illegally, most often for purposes of increasing the government's power and control or enriching government officials and their associates.

21.1.4 Control of Actions by IoT Devices

21.1.4.1 Control of Action

The loss of privacy through massive surveillance and intelligent processing are sources of both IoT's benefits and its challenges to society. However, the potential threat that IoT poses to individuals and society far exceeds the loss of privacy. IoT combines loss of privacy with automated action to control things, people, and society. What do we mean by actions? IoT devices control the environment and

Figure 21.11 An Israeli human-on-the-loop IoT system, like this Samson Remote Controlled Weapon Station, is believed to have combined sensors and a robot to create a remotely fired machine gun to assassinate Iran's top nuclear scientist. Source: MathKnight / Wikimedia Commons / CC BY-SA 4.0.

the people in it by controlling actuators that physically alter the environment. Actuators are devices that physically act on and alter the world. For example, in the author's house IoT devices sense and control sprinklers, furnaces, air conditioners, doorbell, an entrance gate, security cameras, lights, hot tubs, smoke alarms, Echo assistants, automobiles, tire pressures, TVs, playback of music and speakers, and robotic vacuums as well as phones, watches, tablets, and computers. IoT networks can open and close doors, start and stop cars and activate appliances. IoT devices sense the state of these systems, connect to one another and to the Internet, and control electrical and mechanical switches to activate or deactivate systems based on their decisions or on rules or on human command. When Amazon's cloud services, Amazon Web Services (AWS), crashed in December 2021, it stopped a wide variety of devices from running. These ranged from Roomba vacuum cleaners to alerts from Ring security cameras to automated kitty-litter boxes [95] – a small but ample demonstration that IoT already controls many physical interactions in our environment. A human being is in a sense the ultimate actuator and a human can be used as an actuator in an IoT network by engaging with their mental state. Internetworked devices can do more than affect individuals. They have been used to control power grids [96, 97], shut off gasoline pumps across a nation [98], alter the movement of goods, turn off water or power supplies, destroy equipment, and operate armed robots to assassinate people [72, 99] (see Figure 21.11). In Chapter 2, Broadway explains how battlefield IoT networks can control the flow of ammunition and other materiel to the battlefield. IoT can control more than physical entities. Its actions control purchases and other financial transactions. Its actions control human reputations, mental states, and the course of events.

The ability for humans to remotely control specific physical devices was demonstrated at least 8 decades ago or as far back as 17 decades if one counts the telegraph. The ability for a computer to remotely control physical devices is six decades old. What is new is IoT's ability to exercise control over physical actions after automatically assimilating and fusing information from a variety of sources. Absorption of information is followed by intelligent planning with almost instantaneously synchronized action over a wide area.

The power of IoT to act under its own control is not lost on nations bent on controlling their populations and forcing them to the government's will. Significantly, China's 14th Five-Year Plan (2021–2025) includes a call to "strengthen construction of the prevention and ***control*** system for public security" [100] (emphasis this author). China's work on the massive Skynet, Golden Shield, and Sharp Eyes projects concentrates on surveillance and intelligent processing to detect behaviors that are prohibited or a cause of concern. Controlled actions today come mostly in manual form with police interventions. China's emerging IoT network is expanding automated actions to control access to buildings and a host of social and economic services. Access to services is modulated by the "social score" of individuals computed using observations and information about their activities [15]. China's intention to introduce a cyber currency will add a rich new source of information and enable control of all financial transactions using the digital yuan. The government can exercise control over all expenditures and savings because the Chinese cryptocurrency is neither anonymous nor decentralized, the government can extract whatever information it may desire and exert any level of financial control it wants on any individual or organization using it. In this sense it violates the fundamental motivations for early cryptocurrencies like Bitcoin [101]. To induce the use of the digital yuan, China's government has recently outlawed other cryptocurrencies [102] drawing both financial surveillance and control to the government.

Control over all financial operations is sobering. But IoT will soon provide fine grained control over many actions in our physical world. Experiments have produced "research results demonstrating the ability to remotely compromise a vehicle over cellular, Bluetooth, and other non-contact means" [103]. Hacking into a car's systems proved that brakes can be activated or prevented from activating, even on individual wheels. A car's engine can be controlled remotely and independently of the driver. Virtually every other drive-by-wire system in the automobiles tested could be controlled remotely while being driven. Since that study was published, some measures have been instituted by automakers and auto-parts suppliers to help prevent external control overriding the driver. However as more and more new automobiles rely increasingly on computer and software control via IoT, the opportunity increases for control of vehicles by abusing an IoT network.

Beyond remotely stopping a vehicle, IoT systems enable seizing control of self-driving cars and delivering the car's occupants to destinations not of their choosing. IoT networks have already demonstrated that an individual can be identified and tracked, prevented from using an automobile, locked out of or into specific buildings, and targeted by a software-controlled armed robot. Unintended software control errors have inflicted damage in the medical arena. For example, between 1985 and 1987, the Therac-25 radiation therapy machine had a software bug that caused it to grossly over radiate patients on occasion, seriously injuring and killing some [104]. If today an entire hospital's radiation equipment is networked to an IoT system that is compromised, then patients could be killed at the will of someone abusing IoT. These capabilities for controlling the physical world hold great promise for improving traffic flow, securing people and property, improving energy efficiency, healthcare, and generally enhancing the quality of human life. IoT control can also dictate the actions and thereby the freedoms of both individuals and entire societies. All are increasingly subject to IoT control.

Controlling aspects of individuals lives and behavior are not the most sobering consequences of misused IoT. One of the great utilities of IoT for the military is that it cannot just find targets, it can control the weapons that annihilate them. A nation desiring a strong defense needs to develop significant destructive capacity. A destructive capacity wins battles or better yet discourages enemies from engaging in them. On a small scale, IoT can assassinate a single individual of its choosing, as occurred in Iran as described above. However, IoT's destructive capability reaches far beyond taking lives of combatants. IoT, through the control of weapons, can destroy factories, warehouses, vineyards, intelligent highways, hospitals, and smart cities. To be sure, the Russian invasion of Ukraine demonstrates that IoT is not needed to release such destructive wrath on its victims. But IoT can greatly amplify it, make it more efficient, and less costly for the aggressor. And IoT does not have to control weapons to wreak destruction. It can start fires by overloading power grides, halt essential production by altering factory control, disable emergency response, release toxic substances, turn off hospital equipment, and disable water supplies. As we hook up our environments to IoT, it gains the power to disable and destroy at electronic speeds on vast scales with little to no human intervention. Policy makers, lawmakers, and the public need an appreciation of the benefits of IoT. They need to avoid unreasonable and uninformed fear of "killer robots" – a fear played on by movies like Slaughterbots [105]. They do need an informed understanding of the magnitude of the dark side of IoT if it is abused or misused. Policy and regulations can be successfully developed to protect society from misuse. Because it is in an early stage of deployment, IoT's control of actions might be more easily regulated. Regulations must start with an accurate, informed understanding of the issues.

21.1.4.2 Abuse of Action by IoT

IoT can act across a wide area and coordinate its actions in real-time, making IoT control fundamentally different from manual control and action, which occur over human-timeframes and on a case-by-case basis. From an individual's viewpoint, IoT can improve or destroy their life. On a societal scale when wielded by a government, IoT makes George Orwell's *1984* possible. Some nations are already implementing IoT to exercise fine-grained actions to control their populations. While the loss of privacy via information and sensor surveillance is widely understood and is beginning to be addressed, the potential large-scale loss of control of actions that is coming is largely unseen and unappreciated by the public and most governments. Few books or mainstream articles address the issue as of 2022. No organizations nor governments are seeking regulations and guidelines to keep actions by IoT devices in a channel for good. Many governments have woken up to security concerns about IoT, but their focus is almost entirely on how to protect IoT networks from attack and cooption. Government action focuses primarily or exclusively on privacy and security for IoT networks. For examples see the US President's executive order on cybersecurity [106] for IoT, or the US Department of Defense (DoD) policy recommendations on IoT [107]. These documents concentrate on protecting IoT devices from people and not protecting us from IoT devices. The same focus on privacy and security of IoT networks occurs in most policy recommendations, for example, that of Korea [108] and the US Congress [109]. The focus needs to expand from protecting the IoT network from attack to also protecting individuals, the public, and its institutions from attack by IoT. Policy and regulations need to address more than just the harm from invasion of privacy and encompass the potential for harm through the abuse of IoT's actions.

The opportunity for the misuse of actions taken by IoT devices arises in four ways:

- **Act-1.** Individuals and organizations with hostile intent can coopt IoT actions either legally or illegally and either covertly or openly, for example, ransomware attacks where the attacker seizes the power to physically destroy or disable critical systems, such as hospital equipment or oil pipelines [110]. IoT networks may be used by psychotic individuals or terrorists to create fear, for example, terrorizing children by coopting a Ring camera to yell threats at them [111].

- **Act-2.** Organizations or individuals can legally but unethically control devices for the benefit of the device manufacturer or seller or some other third party where such control does not necessarily benefit and can harm the owner or user of the device or IoT network.
- **Act-3.** Governments can abuse IoT actions to carry out defense or national security or law enforcement actions without receiving public consent or providing for due process with independent checks and balances.
- **Act-4.** Governments can misuse IoT actions, intentionally or unintentionally, legally or illegally, to increase or maintain their power or otherwise benefit from control of society and the individuals within it.

If governments use IoT devices to co-opt or violate the rights or lives of individuals, organizations, or other governments for the purposes of its own defense or public security with public consent, then such use would not be construed as misuse or abuse by most nations and societies. A prudent nation or society, however, would wisely apply independent checks and balances to such violations because a government can easily use the pretext of defense or security to commit an Act-3 abuse.

Unintentional abuse of IoT actions can occur given the complexity and scale of IoT networks. Many complex systems suffer from unintended consequences. IoT systems that modify their physical environment are especially prone to unintended consequences and race-conditions due to the impossibility of modeling the world in detail and predicting all external forces acting upon it and their possible timing. Unintentional abuse of the actions of IoT devices can also easily arise because actions taken by an IoT network will increasingly rely on decisions made by intelligent algorithms that are dependent on both the quality and breadth of their training data. Unintentional abuse of IoT networks resulting from technical limitations can be viewed as a technological problem that can be solved through technological means. This chapter focuses instead on policy and regulatory solutions to intentional abuse or misuse of IoT.

21.2 Preventing the Abuse of IoT While Enabling Its Benefits

21.2.1 A General Framework

The framework presented here lays out broad guidelines for developing regulations for the use of IoT and prevention of abuse. Existing efforts by governments to regulate data privacy can be adapted to intelligent processing and action by IoT systems. The framework is intended as a guide for public and legislative discussions of policy. Its objective is to foster the creation of regulations that encourage continued advancement of IoT while imposing some constraints against misuse of IoT. Actions exercised by networked devices deserve special emphasis because they are the new and least entrenched element introduced by IoT. They may be the most amenable to regulation.

21.2.1.1 The Need and Basis for an IoT Framework to Protect Human Rights

Ensuring that a nation exploits IoT's benefits while limiting the potential for damage requires a framework for the use of IoT. Figure 21.12 shows a framework that relies on the consent of the people who use or are subjected to IoT. Consent builds upon and is not possible without the levels underlying it, including transparency, accountability, and security (Figure 21.12). This framework is inspired by and derived from information privacy and processing laws and frameworks developed by the European Union and the 1980 privacy guidelines of the Organization of Economic Cooperation and Development (OECD) [112]. The US Health Insurance Portability and Accountability Act (HIPPA) of 1996 also serves as a reference for this framework [113]. The laws currently enacted address only information privacy. Regulations preventing misuse of intelligent processing are at best in the development stage. Regulations do not exist for regulating IoT actions that alter the physical and mental world, for example control of autonomous vehicles, drones, power grids, or home heating systems.

Passing laws that successfully limit information extraction will be difficult in such countries as the US and China. A large segment of the Internet economy depends on extraction of information as a cornerstone of its revenue strategy, particularly in the US. In China, information extraction appears to be a cornerstone of government control of the environment including its citizens' lives and minds. Intelligent processing is less engrained in the Internet economy, although rapidly becoming so. Regulations on intelligent processing may face less entrenched opposition. On the other hand, because obtaining insight into AI algorithms presents formidable technical problems, it may be resistant to legal constraints that are effective. In contrast, automated or semi-automated actions by IoT systems are in their infancy and are far more transparent in their application than abuses through intelligent processing or invasion of privacy. If IoT causes physical or financial damage, it falls under current laws. But existing laws apply only once damage has occurred and apply only to the specific damage that does occur. No regulations prevent the creation and deployment of IoT systems that have the inherent power and ability to be abused to cause large-scale damage but have not yet done so. Determining who is responsible for such damage after it occurs can be a challenge as well. The author witnessed what was likely to have been the first crash between a driverless vehicle and a human-piloted car in 1986. Software and electronics were directly responsible, but the ultimate responsible human party has yet to be determined. Nevertheless, legislating constrains on potential abuse of the actuation function of IoT, whether physical damage or more subtle injury, is a matter not only of urgency but also feasibility. Fewer vested interests will attempt to stop legitimate regulation of IoT control if enacted soon.

21.2.1.2 Consent by the Public and the Governed

Public consent to the use of IoT to observe and control individuals and their environment, whether by corporations or governments, rests on the public's awareness and understanding of both

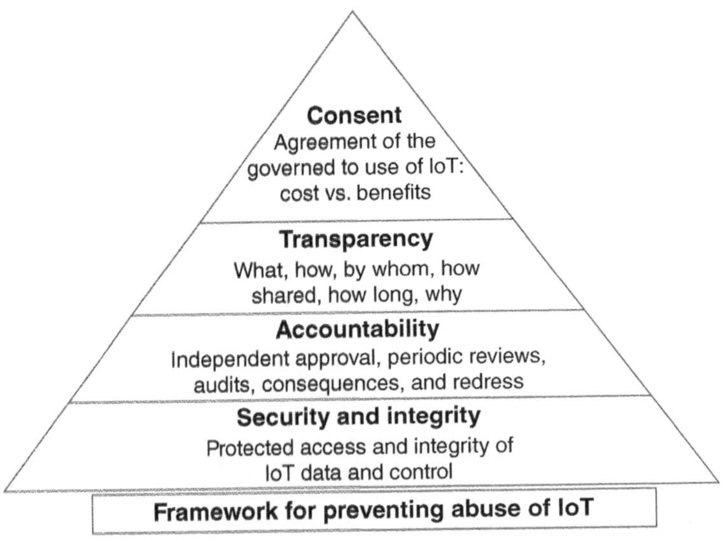

Figure 21.12 A framework for channeling IoT for the greater good while preventing misuse and abuse. Source: Robert Douglass.

the benefits of IoT in a particular application and its potential cost to human rights to achieve those benefits. Consent can only be given, consciously and unambiguously, if the public clearly understands to what it is consenting and how it benefits them and what it costs them. At present, a minority of the public is aware of the concept of IoT and almost none are aware of its potential for misuse and the magnitude of the potential consequences. As discussed below (Section 21.3.1.2), many commercial companies try to achieve the illusion of legal consent by forcing a user to check a box saying they have read and understood a multi-page privacy statement written as a legal document. Such statements frequently contain vague statements on what data is collected and who and how it will be used. Few users read or have the legal expertise to understand such privacy documents. They check the box to obtain the product or service that they are trying to purchase.

Government collection of data is often even more opaque than corporate privacy statements. Consent is assumed based on the passage of laws by the public's representatives but the details of what is being consented to is normally buried in specific regulations that implement the law. These are written by government agencies. Their application and meaning are further defined and refined by the courts as case law. This information, from laws to regulations to court rulings, are available to the public for the most part, but the public often cannot easily obtain them and cannot understand them without study and special expertise. Consequently, the pubic often cannot consciously consent to government practices of information extraction. Government regulations should prevent this practice by both corporations and the government. Laws should ensure that the public is presented with enough transparency to create true informed consent. Moreover, in most cases the public is being asked to consent only to collection of their private information. Consent is not sought as to what intelligent processing may be applied or what actions an IoT device might take. Laws enacted by the European Union are attempting to address just the data collection concern, as discussed in Section 21.3.2 of this chapter. Consent for automated decision making by AI algorithms based on private data is not sought nor is consent obtained for actions taken autonomously by IoT systems. In other countries such as the UK, public consent is not requested nor explicitly given for government collection of private information in many instances or in any instances in the case of China.

Consent can be given only when there is transparency into the benefits received and when the costs to individual freedoms and rights are understood. In free societies, consent occurs indirectly through elected government officials. Consent to control by IoT is not easily nor often achieved in totalitarian societies. Laws and regulations enacted by elected officials that cause IoT to be applied to the public need to have an explicit statement that such officials are acknowledging consent to its use on behalf of those they govern. The use of IoT to control aspects of public life dictated by government officials, whether authorized by specific laws or not, also needs to be accompanied by a clear IoT disclosure statement that is complete and accurate enough to assume public consent. Such statements and the process for producing them need to be open and publicly available. These statements would be analogous to those of the Paperwork Reduction Act in the US [114], but one would hope more concise and understandable in its execution than the 24-page Paperwork Reduction Act. The content of a disclosure statement needs to be publicly accessible and must provide transparency into the use of IoT.

The burden of developing independent IoT disclosure statements to enable consent will bear a cost in time and effort beyond the deployment of an IoT system itself. To control this cost or help spread it over time, applications of IoT could be graded by how critical they are and their potential to cause damage if misused or permitted to malfunction. For example, IoT autonomously controlling weapons or critical infrastructure, or public health would have the highest risk and should have an extensive disclosure process with a detailed statement, even though it adds cost to an IoT deployment. The EU has developed such a risk hierarchy for AI applications, as described below in Section 21.3.2.2. The EU's hierarchy also serves as a starting point for ranking the risk of IoT applications. Grading IoT technology on the criticality of its application is more complex than grading an AI application. IoT networks may be involved in multiple applications that range in criticality. The devices in the network may have been designed for a far different application than that for which the IoT system designer has adapted them.

This chapter has largely ignored the topic of IoT security. Providing security is in part a technical challenge, and many of the chapters of this book discuss technical solutions to security. But security is also a matter for policy, regulation, and enforcement. If IoT carries security and safety vulnerabilities, the public should consent to tolerating them before they are subjected to the IoT system. If you buy a new automobile, you assume that security and safety measures are included. They are included in part because policy dictates regulations that mandate specific safety and security features are incorporated in all new cars. In contrast, IoT products are often the least secure products on the market. IoT devices and systems are often inexpensive and made by early-stage companies rushing new products to market. The public is not aware of the risks of IoT failure or misuse, nor aware of what security and safety measures should be incorporated in the IoT products that they buy. Because the public and the government lack insight into security and safety issues, IoT providers can increase their profit margins by ignoring safety and security, and they frequently do so. This situation is the opposite of what it should be. Because IoT systems interact with the real world and can kill people, security, and safety is much more important than on your laptop and in some cases more important than on your automobile. The public cannot consent to accepting lax security and safety in IoT because the public cannot consent to something that it is unaware of. Lawmakers and the public certainly need education. But in the meantime, the public can be informed by regulations that force IoT product and service providers to include detailed descriptions of security and safety measures in the IoT Disclosure Statement mentioned above. The Disclosure Statement needs to state specifically what security and safety measures are incorporated in a product or service. It needs to detail what threats and risks remain. Such a Statement is like the health side-effect disclosures by drug companies. The risks posed by misused, abused, and failing IoT is no less critical.

21.2.1.3 Transparency: The Foundation of Consent

Transparency in the use of IoT requires that the recipients of IoT's actions understand three things:

(1) Specifically what information is being extracted.
(2) How it is being processed and stored, by whom, and what are the risks.
(3) What specific actions can the IoT system take, including how they are performed and the significance and risks of such actions.

Transparency in the use of IoT needs to come as part of the IoT disclosure statement described above. This disclosure statement goes to users of an IoT device or system as well as to any others who would be affected by the IoT system's actions. Specificity is essential. Also essential is an understanding of who or what entities are extracting information, processing it, and applying actions. Vague statements like "we share your information only with our business partners" does not create transparency. Transparency requires knowing specifically and exhaustively who is applying IoT to whom. The "Who" encompasses parties who access or control or create IoT devices and networks as well as any third parties involved in access or control. Even more than protection of private information, IoT needs to state clearly, simply, and precisely what actions can be taken to affect an individual or their environment physically, mentally, or emotionally. Many IoT vendors today provide a disclosure statement saying what information the IoT extracts and how it will be used and by whom, but few provide that information with any specificity. Almost no IoT vendor clearly states specifically what actions the IoT can take, when and with what consequences. Few or no vendors or operators of IoT today state what sort of processing is performed to arrive at actions taken. Transparency cannot exist without disclosing these specific details for all three elements of an IoT system. The disclosure statements described above should include these kinds of details, analogous to financial and mortgage disclosures required in the US and many other countries.

Providing transparency is not a one-time event. If additional parties, actions, or information extractions are added to an IoT disclosure statement, an update to the disclosure should be generated before a new third party is added, and new consent should be obtained. An email stating that a disclosure statement has been changed is not sufficient to maintain transparency. How IoT usage is being changed needs to be explained in simple, understandable terms, and it needs to be specific. Once transparency is thus reestablished after a change, then consent needs to be reobtained after the updated disclosure has been provided and before use of the IoT system is altered.

In addition to knowing what, who and how IoT is used, creating transparency requires knowing how long IoT will continue to perform, whether passively storing information or applying actions. In addition to disclosing the duration, the extent or scope of actions needs to be disclosed. To achieve transparency requires that people providing consent must be able to trust that the IoT disclosure statement is accurate, complete, and up to date. Such trust comes only if providers and operators of IoT systems are held accountable to provide accurate, complete, and current transparency. Regulations regarding transparency are successful only if there are consequences where transparency information is inaccurate or lacking.

21.2.1.4 Accountability and Consequences

True consent by those affected by IoT systems rests on transparency into how they function. Achieving transparency requires trust in the IoT disclosure process and accuracy and completeness of the disclosure statement. Trust comes from knowing that producers, vendors, and operators of IoT systems are subject to independent review and approval and to ongoing monitoring and periodic auditing of the continuing accuracy of the disclosure statement. To be effective, reviews must be backed by consequences that follow failures in completeness, accuracy, currency, and specificity of IoT disclosure statement. A pre-deployment review and approval of IoT and their associated disclosure statements need to be performed. The review and approval should be independent of the company or government entity deploying the IoT. Once deployed, the use of an IoT system should be routinely re-reviewed and reapproved with updates to disclosure statements if any inaccuracies are uncovered. Practices that do not match the IoT disclosure statement need to be subject to fines and punishments as warranted. If an IoT system is modified or its usage changes, an independent review and approval process needs to be completed along with an updated disclosure statement before the modified IoT system is activated or used. An example of such a process is the court issuance of a warrant that is required before wiretapping a communication network in the US and many other countries. For the highest risk uses of IoT, the reviewing entity would be most effective if it was an independent commission or agency subject to review by an elected body and consisting of appointees that span the spectrum of political parties, expertise, and vested interests. Appointees for high-risk IoT reviews should go through an independent approval process to identify potential conflicts of interest. The commission should function much the way commissions do for the use of radio spectrum or financial markets or trade. High-risk and high-impact IoT should be similarly monitored, controlled, and enforced. Lesser risk activities could be reviewed by an independent group within the agency deploying or using the IoT system, for example the Inspector General offices that reside within many US agencies and departments.

The disclosure statements, the results of a review, and the review process need to be open to public scrutiny.

For national security and defense uses, the details of what devices are used, how they are deployed and exactly what type of intelligence is collected and what actions are taken may need to be classified. In such cases, portions of the disclosure and review reports may also need to be classified. The reviewing body needs to have full access at the appropriate level of clearance and have the power to adjudicate classification assignments. While some technical details and details of usage may be classified, the bulk of the disclosure statement and review report and none of the process of review and approval should be. Where details are obscured by classification, they should have a specific time limit after which they are declassified. A specific reason for classification should be specified. The reasons for classification need to be updated if the classification period is extended beyond the declassification date. Similarly, the intended application of an IoT system should be bounded in scope and time, just like warrants for law enforcement searches.

For classified IoT applications, achieving consent through transparency depends on the independence of the oversight and review entities. They must be committed to the public interest and their rights and freedoms. They must insist on limited classification, permitting it only where there is a true danger of compromise if certain aspects of the IoT system is disclosed. For example, as previously cited, the US National Security Agency justifiably concealed via the FISA court what specific communications from which specific human targets they were intercepting as well as how they were intercepting them. However, it should not be possible to conceal through secret classification that their intercepts involved the incidental collection of billions of pieces of communication on an ongoing basis from citizens who had no meaningful connection to a national security risk. A truly independent approval and review agency would have either disclosed this fact to the public or prevented it from occurring. By minimizing what is secret to what can be specifically justified as needing to be secret and why and for how long, IoT can still be fully effective in defense and national security applications without inviting the types of abuse defined in Section 21.3.1.

Independent approval and periodic reviews are useless unless the reviewing agency can enforce consequences for faulty disclosure statements or IoT usage that violates its disclosure statement. Laws must allow for termination of IoT usage and fines and other punishments if IoT usage is found to be inconsistent with its disclosure statement. For low-risk IoT applications, administrative consequences may suffice. Misuse and malfunctions need to be publicly disclosed as do consequences assigned, if any.

Today, companies in many countries have little incentive to disclose abuses of data privacy, especially when they believe it will not be discovered by the public. Companies can expect no consequences if their malfeasance or inadequate security is not discovered. The US Internet of Things Cybersecurity Improvement Act [115] mandates disclosures and penalties for IoT breaches and known but undisclosed vulnerabilities for IoT, but only if the IoT is owned or used by the government. It does not mandate disclosure by any entity who owns or uses IoT other than the US government. Nor does it mandate the disclosure of secure use of IoT by the government to commit IoT abuses by the government itself. This law needs to be extended to all IoT systems owned or used by any individual or organizations to compel disclosure of known breaches and vulnerabilities. Disclosure mandates must be backed by consequences for non-disclosure. It should also cover any use of IoT that falls outside of the disclosure statements provided by the IoT provider. Similar laws are needed now in all countries as IoT usage expands.

One final type of consequence needed for IoT is redress for wrongs committed through the misuse or malfunction of IoT. Individuals or organizations or governments that are damaged by malfunctioning or misused IoT should have a legal remedy to correct the damage, both by cease-and-desist orders and financial compensation. Courts need to recognize the potential for damage by IoT systems, whether committed knowingly or unknowingly by corporations or intentionally by hostile individuals or governments. They need to recognize the right for an individual to seek redress for the damage. New laws may be required to establish this right. Redress, like any consequence, is difficult or impossible without transparency into the functioning of an IoT system and disclosure of malfunctions, misuse, and vulnerabilities.

It will most likely be much easier to establish the legal right for redress from damage by IoT systems than it will be to allocate responsibility for the damage. IoT systems are complex with potentially dozens of companies providing pieces of hardware, software, and communication and networking services. Beyond problems with individual components, the system as a whole may display emergent behaviors that result from the unique interaction of the disparate components. Such emergent behaviors are difficult to predict and present an even more difficult challenge when assigning blame or credit. The vendor or installer may be at fault. On the other hand, the IoT system itself may be correctly functioning, but the operator of the system may be using it to engage in abuse. Finding responsible parties may be aided by well-crafted laws but will probably emerge only over time through court action and the case law that results. Allocating responsibility may remain as the one of the most difficult challenges for using IoT safely.

21.2.1.5 Security and Integrity

Volumes have been written on technical issues regarding security and integrity for digital data and digital communication, including IoT systems. For defense IoT, Section 3 of this book discusses some of these technical challenges and potential solutions. Beyond technical issues of IoT security, from a regulatory standpoint, IoT providers must be held accountable for misuse or abuse of their products and services when they fail to incorporate sufficient security to protect from known attacks and allow damage due to faulty design or implementation. The security and integrity of both the IoT system and its collateral data must also be maintained. IoT data includes not only extracted information, but also records of its own processing, actions, and observed consequences of those actions. Collateral information includes records of disclosure statements, reviews and audits, and a detailed change history. Maintaining the integrity of such information means not only keeping it secure, but also ensuring that data is captured and stored and archived and retrievable over time. Responsibility for the security and integrity of an IoT system and its use resides with all involved parties from designers to developers to vendors to operators and end users. As mentioned above, the current legal climate in many nations does not incentivize companies or individuals in any of these roles to spend money on strong measures for IoT security and integrity. Laws need to be enacted that add an incentive to IoT producers and operators to incorporate strong security measures and ensure system integrity. Government mandated disclosure of breaches, failures, and vulnerabilities provides one step toward incentivizing companies to incorporate additional security and integrity.

Requiring IoT disclosure statements on the use of IoT systems, as described above, will incentivize producers and operators to increase security and integrity. IoT users or recipients of IoT actions will better understand the risks they run and the costs they may bear based on disclosure statements. That understanding may increase the premium that customers are willing to pay for added security. In the case of government use of IoT, it can empower the public to demand better, more secure systems for government use of IoT. Governments should explore applying to IoT the

EU's approach for AI. That approach incrementally increases the cost of incorporating and enforcing security based on the criticality of an IoT application. While this approach may work for AI, it will be more challenging for IoT. For example, the Mirai botnet took over more than 100,000 IoT devices [116]. The exploitation was able to shut down the Internet on the East Coast for hours. Many captured devices were simple consumer products like "nannycams." For an IoT network, the specific device or application may not be critical, but if it is incorporated in a critical IoT network, then it can lead to the misuse of the entire network. This presents a more complex regulatory challenge than assessing one AI application.

In summary, IoT systems function today in the US and elsewhere like a drug industry operating without the regulations, oversight, and monitoring of a Federal Drug Administration (FDA) and Center for Disease Control and Prevention (CDC). Misused and abused IoT can have at least as great a negative effect as unsafe drugs. It needs to be regulated with similar diligence. The framework described here presents a starting point.

21.3 Types of Abuse and Misuse, and Prevention Through Regulation

21.3.1 Types of Abuse of IoT

There are four general categories of abuse, each of which may be applied to any element of IoT: sensing, processing, action. They correspond with the various types of abuse for each element of IoT as enumerated above. They are:

- **Abuse 1.** Illegal or unethical attacks against any IoT element (sensing, processing, control) by individuals or organizations that benefits the attacker at the expense of the owners or users of IoT. This category includes types of abuse of IoT elements listed above, including Info-1, IP-1, IP-4, Cntl-1.
- **Abuse 2.** Legal abuse of IoT without explicit consent or effective cognizance of users or owners and without benefit or with potential detriment to the users or owners. This category comprises Info-2, Info-3, IP-2, IP-3, and Cntl-2.
- **Abuse 3.** Government abuse of IoT for legitimate reasons of public defense, health, safety, and wellbeing. This category refers specifically to abuse of IoT by law enforcement, public health agencies, and national defense and security organizations in the absence of public consent and without due process providing checks, balances, and redress. Forms of this type of IoT abuse include Info-4, Ip-5, Cntl-3.
- **Abuse 4.** Government abuse of IoT for its own ends, specifically to illegally increase its power and control or to enrich government officials and their associates: Info-5, Ip-6, Cntl-4.

Each of these abuse categories are discussed below in terms of what is being done to regulate it, what more might be done, and how likely any of it is to succeed. Clearly, the answers to these three questions depend on which nation or international body is under consideration. A comprehensive global review is beyond the scope of this discussion; instead, each item will be discussed in terms of a few examples.

21.3.1.1 Type 1 Abuse: Illegal or Unethical Abuse by Individuals or Organizations

There is increasingly broad public and government awareness of this type of abuse of IoT. Awareness is strongest when it concerns sensing and information extraction, for example to enable identity theft or financial fraud. But awareness of the action aspect of IoT is also growing, thanks primarily to ransomware attacks that shut down hospital equipment and energy distribution pipelines because such actions can cost lives and raise the cost and availability of energy. Abuse of IoT is illegal when it contravenes existing laws, and it is unethical if it allows parties to extract benefit from individuals, organizations, or governments without benefiting those being exploited and without their knowledge and consent.

Countering Type 1 Abuse takes a combination of technical solutions, participation of commercial IoT producers and operators, and government regulatory action. In terms of technical solutions for security, no area of IoT use is more challenging than national defense where foreign agents include well-funded, coordinated attacks by hostile nations. Defense and national security sectors are both aware of the need for IoT security, protection, and integrity and are willing to fund research leading to solutions. They increasingly are willing to demand that security be incorporated into IoT networks by vendors and operators. Examples of technical solutions for these most stressing of IoT applications are covered in the chapters of this book in Sections 3 and 4. Protecting IoT from abuse in the commercial and civil domain is covered by numerous books, journal articles, and conference submissions (see for example, [117]). Technical issues and solutions will not be discussed here. However, Type 1 abuse is not just a technical problem, it is also a policy and legal issue.

Obviously, perpetrators of Type 1 Abuse need to suffer legal consequences, and laws need to cover such abusive actions. But limiting laws to just attempting to apprehend and punish wrong doers is not sufficient. In the cost competitive IoT marketplace, IoT designers, manufacturers, and vendors are incentivized to lower the complexity and cost of their devices and one way to do that is to forgo inclusion of strong security measures. Moreover, both users and producers of IoT products and services are often reluctant to report abuse or security flaws publicly or to law enforcement because public disclosures of breaches can damage reputations, reduce sales, and engender lawsuits. IoT producers frequently will not address known vulnerabilities until they become public, usually as a result of a successful exploitation. Mandatory disclosure of vulnerabilities, at the time they become known by the producer, forces the producer to immediately correct a flaw before it is announced or shortly thereafter. Successfully distributing patches to fix disclosed vulnerabilities in millions of devices is a technical challenge, but one that will add cost and cause resistance to regulations. Policy makers should anticipate this cost and seek regulations to mitigate it. Laws are needed to force public disclosure of security vulnerabilities and material security breaches in a timely manner. Some governments and agencies have taken action to enact these sorts of disclosure laws, for example in the previously cited executive orders, laws, and agency guidelines like those of the US DoD [106, 107]. The Internet of Things Cybersecurity Improvement Act of 2020 focuses specifically on disclosure of vulnerabilities and attacks against IoT devices owned or used by the US government agencies:

> "This important legislation provides for the creation of IoT security guidelines for devices sold to federal agencies and their management by agencies as well as the establishment of guidelines for vulnerability disclosure. Though focused on IoT devices purchased by the federal government, this legislation is likely to have significant consequences for IoT manufacturers across the economy. It also likely will have important implications for enterprise cybersecurity through its vulnerability disclosure provisions" [115].

It remains to be seen if this legislation will indeed have an impact on IoT abuse beyond government owned or operated systems. It seems unlikely, given that defense use of IoT constitutes a small portion of the total market. The act should be broadened to cover a wider class of IoT products and services to ensure that it does have an impact on the wider economy and the public in general. In the US a new law is needed to force disclosure of known vulnerabilities and security breaches of commercial devices regardless of who owns or operates or produces them. The entire spectrum from producers to end-users needs to be held accountable. When users of IoT systems are held accountable, they are incentivized to select IoT producers and vendors who have demonstrably more secure products. This type of disclosure law is essential to creating transparency for Type 1 Abuse, in keeping with the solution framework presented above in Section 21.2.1. It may be desirable or even necessary that governments go beyond mandating disclosure to also mandating the incorporation of standard security measures. This latter approach is more problematical because IoT security standards are just beginning to emerge, and they cover only portions of IoT vulnerabilities. Such laws will likely continue to have limited effectiveness due to the rapidly evolving nature of IoT technology and evolving threats.

Both a breach/vulnerability-disclosure law and laws mandating a certain level of security would have costs to users of IoT technology. It is likely that the cost is small compared to the cost of the damage from successful attacks on IoT networks. Regardless, the cost could be lowered by applying the law to only higher-risk applications. A starting point for defining a risk-hierarchy is provided by the EU risk scale developed for applications that depend on AI processing [118]. Alternatively, the cost could be spread out over time by gradually applying the law to successively lower-risk applications, but this approach is subject to the same difficulties mentioned above with respect to the complexity of components in IoT vs. a single AI application. Governments could provide monitoring and enforcement of IoT suppliers and users, with an attendant cost to governments. Creating transparency into security risks requires enforcement and fines or sanctions when suppliers fail to provide it. Such fines may help offset the cost of enforcement. Alternatively, laws could be adjusted or enacted as need be to allow aggrieved parties to pursue civil suits and financial damages from producers, vendors, and operators of compromised systems. At present, few companies in the supply chain and operations are held accountable either by governments or by injured members of the public for security breaches due to lax security. Even if regulations made it easier to sue for damages resulting from poor security, without government laws and enforcement of disclosure mandates, some vendors and operators will continue to conceal breaches caused by substandard security to avoid lawsuits. Enforcement by civil suit may save governments the cost of monitoring and enforcement, but it creates enforcement that is less consistent in punishing wrong doers and highly variable in imposed punishments. Frivolous civil suits will also undoubtedly occur and add unnecessary cost to IoT systems. Government legislation, if done correctly, will provide a more effective solution. What is clearly needed is increased public awareness as well as engagement by policy makers, lawmakers, and the public in seeking better solutions to Type 1 IoT abuse.

The global nature of the supply chain for IoT presents a significant hurdle to implementing disclosure laws for IoT producers and operators that hold them accountable for breaches. It is unlikely that any IoT network of any complexity, whether commercial or military, consists of hardware and software components all produced in a single country. IoT components are completely globalized. A company in the US that sells an IoT device will likely be using chips produced in one country but assembled onto a board in another country. Software, operating systems, firmware, middleware, and IoT network components could come from anywhere and probably come from multiple countries. Moreover, it is difficult to tell which country something came from because companies themselves are multinational and use subcontractors in still other countries. Software can be produced by foreign nationals working in the US with some US citizens who develop the software residing outside the US. The company that sells the software may not know which country or nationality produced which piece of software. Holding the end vendor or operator of an IoT system accountable for security will be challenging. They may have no way to validate firmware, software, or hardware buried deep inside a subsystem of an IoT device purchased on the open market. This globalization may present the greatest challenge to regulating increased security for IoT. Policy and regulators should hope for and fund comprehensive technical solutions.

21.3.1.2 Type 2 Abuse: Legal Abuse of IoT Without Consent or Benefit to Users or Owners

If IoT is used by a producer or vendor or supplier of IoT services to legally benefit themselves without benefiting or by harming the users or owners of the IoT, then it can represent legal but potentially unethical use of IoT. This use of IoT is exploitative and becomes unethical and abusive if the users or owners do not understand the true costs vs. the benefits to themselves and if they have not been allowed to withhold consent. Providing consent requires that users/owners are cognizant and truly understand the costs and benefits to them vs. the benefits to providers, producers, and any third parties. Many corporations and governments create the illusion of consent today by forcing users to check a box indicating that they have read, understood, and consent to the terms of usage and privacy statements. These "terms" typically run many pages and are crafted by attorneys as legal documents. As discussed in the previous Section 21.2.1.2, such documents do not enable true informed consent by users and are a form of Type 2 abuse. They are legal by design but reside on the borders of the unethical. Laws and regulations need to be enacted or tightened to eliminate this legal but unethical abuse of IoT.

21.3.1.2.1 Government Protections Against Type 2 Abuse

When it comes to sensing and information extraction by IoT, the public and governments are aware of existing and potential abuse. Although this type of abuse is at the forefront of discussions across the globe, specifically in terms of individual data privacy, the extent to which effective regulations have been enacted varies widely. Data is the life blood of effective IoT, so any attempts to limit data collection must be made with the understanding that the power of IoT will be limited as well. The most aggressive countries regulating information extraction are those whose economies depend the least on tech companies who rely on extracting information. Countries whose economies are intertwined with tech companies who are dependent on wide-spread information extraction express concern over data privacy but generally have done little to limit it. At the far extreme are totalitarian countries who try to control information extraction by entities other than the governments themselves or demand that companies provide the government with all data they collect.

There is also a growing public and government awareness of problems arising from the misuse of intelligent processing using AI algorithms, whether part of an IoT network or not. As reviewed above, this concern focuses on biases, lack of transparency, and lack of explanation, and occasional but surprising failures. There is growing concern regarding poor AI performance and broken-leg/black-swan type errors, which usually arise from inadequate training sets. In contrast, few countries and almost no public concern has focused on misuse of IoT's action function.

In this section we will review efforts to limit IoT type 2 abuse in several regions and in commercial IoT practice. We end with recommendations for regulations that are needed to prevent type 2 abuse with some prognoses of the chance of success. Countries fall into three general categories: countries that already have existing regulations applying to aspects of IoT, countries that are considering laws regulating IoT, and countries whose governments participate in third party abuse.

The European Union is the most advanced in terms of regulating Type 2 abuse, with extensive regulations addressing it. These regulations levying fines for companies that violate them. The European Commission states that their General Data Protection Regulation (GDPR) "is the toughest privacy and security law in the world" [119]. Put into effect in May 2018, the law covers organizations anywhere that collect data on people in the EU and imposes harsh fines for violations. This law explicitly excludes defense and intelligence operations from its provisions. It applies only to data that can be related directly or indirectly to a person who can be identified; however, its description of indirect identification includes some of the techniques described above to combine location data with other types of data to assign identities where they are not explicitly provided. These regulations run hundreds of pages. A review is out of scope of this discussion, but the GDPR serves as a model for data privacy and rights for other governments. It provides the elements of the framework for abuse prevention described in the section above, including elements to ensure consent through transparency, accountability, and security of the extracted information.

The EU GDPR is also unusual in that it addresses processing of extracted and sensed information, not just its collection, whether or not the processing is AI-based. It lays out specific requirements for when personal data can be processed. A key requirement is that unambiguous consent exists. Beyond GDPR, the EU is developing a framework for regulating the use of AI [120–123]. The EU has a balanced approach toward AI in general. It aims to foster the development and use of AI while at the same time acknowledging and seeking to mitigate abuse by AI. The approach focuses on developing trustworthy AI to limit potential damage from AI pattern recognition and decision making. The most widespread AI algorithms used today are neural nets developed with machine learning. By strictly controlling the collection and processing of IoT data, the GDPR also indirectly provides some checks on the use of AI for decision making in IoT systems. However, starving an AI algorithm of sufficient training data can lead to it making poor decisions and exhibiting biased performance. The EU's approach to classifying applications of AI technology in a category as "very high" to "low risk" provides a way to limit the cost of controls on intelligent processing by limiting regulations to higher-risk domains. Alternatively, the cost of implementing restrictions and safeguards on AI could also be spread over time by applying regulations in a phased approach starting with high-risk applications.

The European Commission summaries its GDPR principals of protection and accountability as follows [119]:

- **Lawfulness, Fairness, and Transparency.** Processing must be lawful, fair, and transparent to the data subject.
- **Purpose Limitation.** You must process data for the legitimate purposes specified explicitly to the data subject when you collected it.
- **Data Minimization.** You should collect and process only as much data as necessary for the purposes specified.
- **Accuracy.** You must keep personal data accurate and up to date.
- **Storage Limitation.** You may only store personally identifying data for as long as necessary for the specified purpose.
- **Integrity and Confidentiality.** Processing must be done in such a way as to ensure appropriate security, integrity, and confidentiality (e.g. by using encryption).
- **Accountability.** The data controller is responsible for being able to demonstrate GDPR compliance with all of these principles.

In contrast to the EU's leading role in regulating information extraction and intelligent processing, there are no restrictions on the abuse of IoT's use of physical actuators, at least none that appear to be under consideration at present. Much of what the EU has developed to address data privacy and limit the application of AI could be directly adapted to the actuator element of IoT to both foster its adoption while simultaneously reducing type 2 abuse.

In contrast to the EU, the US represents several countries that, while they are deep into discussions of regulation of the data privacy issues of IoT, they have yet to enact any meaningful laws or guidelines to limit type 2 abuse. With respect to data privacy, whether collected by IoT devices or otherwise, the Washington Post states "Don't hold your breath for a law" [124]. The Post sums up the situation as it stands at the end of 2021 as:

> "But while there's universal agreement that [the US] Congress needs to do more than talking — specifically, setting rules around the collection and use of consumer data — action has remained elusive. Despite a litany of privacy scandals in Silicon Valley, legislation has been stagnating for years."

The US National Security Commission on Artificial Intelligence made recommendations on regulating possible damage from the misuse of intelligent processing in applications such as IoT as noted above. The Commission's report is both noteworthy and rare in its emphasis on possible abuse of AI in charge of IoT actions. They state:

> "Our country should clarify the rights and freedoms we expect data-driven technologies [i.e., AI machine learning] to respect. What exactly those are will require discussion, but here are some possibilities: your right to know when and how AI is influencing a decision that affects your civil rights and civil liberties; your freedom from being subjected to AI that hasn't been carefully audited to ensure that it's accurate, unbiased, and has been trained on sufficiently representative data sets; your freedom from pervasive or discriminatory surveillance and monitoring in your home, community, and workplace; and your right to meaningful recourse if the use of an algorithm harms you." [80]

The Commission's basic rights or freedoms should be applied to IoT systems in general and can be stated as follows:

- Individuals and organizations must know and consent to influence or control by IoT decisions or actions that affects their civil rights and liberties.
- Individuals must not be subjected to IoT information extraction and control that has not been carefully audited to ensure it is accurate, unbiased, and trained on sufficiently representative data sets.
- Individuals and organizations must not be subjected to pervasive or discriminatory surveillance, monitoring, or control in their home, place of operation, community, and workplace.
- Individuals and organizations must have transparency into sensing, intelligent processing, and control exerted by IoT technology and have a means providing informed consent and have meaningful recourse if harmed by the technology.

Despite recommendations such as the Commission and considerable testimony in Congress, the US has done little to impose restrictions on any type 2 abuse associated with the collection of information, how it is processed, or what actions are exercised by IoT systems. The exception are regulations regarding abuse by foreign nations or individuals committed against IoT devices owned or operated by the US government.

Some countries have enacted some regulations or guidelines on type 2 abuse, but with limited to no effectiveness. For example, the UK established guidelines in 2013 on the use of surveillance video called the Surveillance Camera Code of Practice, pursuant to the Protection of Freedoms Act 2012 [125]. Far from providing effective limits on mass surveillance, the Code states that a person can expect to be surveilled in public places: "An individual can expect to be the subject of surveillance in a public place as closed circuit TV (CCTV), for example, is a familiar feature in places that the public frequent. An individual can, however, rightly expect surveillance in public places to be both necessary and proportionate, with appropriate safeguards in place." Covert surveillance by public authorities is specifically excluded from the Code. In effect, the code is saying that the British public should expect mass surveillance by the government and the government is exempt from the code if it is done covertly. The guidelines are mandatory for government authorities if not done covertly, but voluntary for all others. Moreover, the guidelines are vague in critical sections. For example, they state that "Surveillance camera systems operating in public places must always have a clearly defined purpose or purposes in pursuit of a legitimate aim and necessary to address a pressing need (or needs)." While they provide some examples of "specific purposes," the wording leaves a wide margin for an official or organization or any individual to define what is a "specific purpose" with a "legitimate aim." The Code also fails the Enforcement and Consequences part of the counter-abuse framework described above because "A failure on the part of any person to act in accordance with any provision of this code does not of itself make that person liable to criminal or civil proceedings." It is a code purported to help protect human rights under the European Convention on Human Rights (ECHR) [126] (enacted as the UK Human Rights Act of 1998); yet, in fact it legitimizes mass surveillance of the public by the government of the UK. It justifies such a violation of the Convention on Human Rights by stating that "a person's right to respect for their private and family life, home and correspondence, as provided for by Article 8 of the ECHR" are conditional rights that can be superseded and interfered with by a public official if it is necessary for national security, public safety, economic wellbeing, or the prevention of crime and disorder. Such exceptions lie at the core of Abuse Type 3 discussed in Section 21.3.1.3.

21.3.1.2.2 Commercial Constraints on Type 2 Abuse

Governments are not alone in concern and responsibility for the ethical use of IoT. Commercial organizations realize that "personal data is worth billions" [127]. Intelligent processing of that data and control of actions by IoT devices will ultimately be worth even more. Given the value to be realized, it is not surprising that commercial companies by in large have done little to meaningfully stem Type 2 abuse, when not forced to do so. Beyond legal regulations, commercial developers, and providers of IoT technology and services also have an ethical responsibility. Ethical responsibility would argue against providers of IoT products and services benefiting at the expense of users who are not benefiting and not effectively consenting to be so exploited. If they are not already forced to use ethical practices by regulations such as those of the EU, companies can be categorized as falling in one of three categories:

1. Companies may intentionally or unintentionally use IoT unethically.
2. Companies may attempt to comply with accepted ethics for use of IoT, most typically by requesting consent from users by asking them to review and agree to a stated privacy policy.
3. Companies may position themselves as proactively providing products or services that protect users from other companies' unethical use of IoT, specifically unethical information extraction.

There are no protections for the public from category 1 companies if laws are not passed to make unethical practices illegal or at a minimum force the disclosure of how IoT is being used and unethically abused. Most large and established companies fall into Category 2. Much of the Internet economy depends on users trading what passes for consent in exchange for goods and services, such as the use of software apps. What passes for consent usually consists of forcing a potential user to check a box saying they have reviewed and agreed to the company's privacy statement. Many, probably most, potential users have neither the legal skills nor the patience to review a lengthy legal privacy disclosure document and consequently they check the consent box without reading the disclosure document. Those that do read and have the legal skills to understand the document will often find that it is vague and provides no effective limits on what or how much private data is extracted, how it is used, and with whom it is shared. An example will illustrate the *potential* for Type 2 abuse, whether it is occurring of not.

Tesla's vehicles can be viewed as large collections of sophisticated IoT devices operating together and connected with other IoT nodes beyond the vehicle. Tesla's Customer Privacy Notice [128], while 13 pages long, is clearly and simply written (unlike many privacy agreements) and it laudably contains sections covering many of the elements of transparency, audit, and security listed in the framework for preventing abuse presented in Section 21.2. It therefore provides an excellent example for understanding how privacy agreements can generate customer consent when in fact many such agreements provide no guarantees against Type 2 Abuse. It is not maintained here that Tesla or any other company is committing Type 2 Abuse, which would be legal in the US at any rate. What is argued is that even well written agreements like Tesla's are vague in critical places and provide no effective protection or redress against Type 2 abuse. For example, Tesla's statement on protections regarding collection and access to location data is:

> "To protect your privacy, location data is either processed directly without leaving your vehicle, is in a form that does not personally identify you or remains inaccessible to Tesla. For safety purposes, your vehicle's location may be used when you experience a safety event (such as a collision, an airbag deployment, or automatic emergency braking event) to aid response efforts. This means that unless captured as a result of a safety event— Tesla does not link your location to your identity or know where you've been. You may also choose to enable or disable the collection of 'Road Segment Data Analytics' at any time within your vehicle's touch screen by navigating to Safety & Security > Data Sharing. Please note, some advanced features such as real-time traffic and intelligent routing rely on such data" [128].

Note that Tesla can and does collect information on your vehicle's location in the event of a "safety event." Safety events are not precisely defined but are illustrated with examples. Clearly, the agreement leaves Tesla with the authority to define anything as a safety event. Additionally, Tesla does say it collects "Road Segment Data Analytics" which presumably includes location data, unless you opt out of certain navigation services. While Tesla may not explicitly connect your identity to any location data, except in the event of a safety event, as described above, location data can easily be associated with an identity using collateral information such as your home address, which Tesla has per another part of their privacy statement. Tesla carefully lists the uses of data collected from you and your vehicle, but that use includes developing and promoting new products and services as well as "for business purposes." These terms are open ended and sufficiently broad to include use of your IoT data for virtually any purpose. Tesla also carefully lists the categories of third parties with which they may share your data, but the list includes "service providers and business partners as well as affiliates and subsidiaries." While Tesla gives examples of such entities, no entity is specifically excluded. In summary, Tesla's privacy agreement is one of the better ones; yet, it appears to provide no effective limits on what information its IoT systems extract, how it is used, with whom it might be shared, or how a third party might use it or further share or sell it.

Commercial companies, unless compelled by law, cannot be relied upon to protect information extracted by IoT devices. While privacy statements may be well intentioned, they generally do not provide transparency on IoT information extraction, nor do they permit unambiguous consent. Some countries have gone in the opposite direction concerning regulating information extraction. China has enacted laws that not only fail to prevent commercial companies from extracting private data, instead they mandate that companies turn over all such extracted private data to the government for analysis and archiving [129–132].

Because many countries do not provide constraints on private data extraction, some commercial companies have begun to provide products and services aimed at limiting extraction or at least making it more transparent. Apple has begun to brand itself as a company that limits information extraction in its default settings. It also requires app vendors to have their software tested and accompanies lists of available apps with disclosure of the information their application extracts, although only in board terms and it appears to allow exceptions and vagueness. Apple's business model relies on the sale of hardware, and Apple can afford to limit information extraction. Companies like Alphabet (Google's parent) and Meta (formerly Facebook) depend on information extraction as a key part of their revenue strategy. Such companies have been far less willing to limit information extraction in either their own or third-party software. In contrast to both Apple and Alphabet, companies in category 3, such as DuckDuckGo and NordVPN, have emerged. Their principal products and services are tools to assist a user in limiting extraction of their information.

Despite the few companies that work to limit information extraction, much of the Internet economy is powered by extraction and monetization of private data. Given that situation, many countries may lack the will to effectively control information extraction by IoT systems. Beyond information extraction, few if any companies provide transparency or an opportunity for consent to automated intelligent processing of data extracted from you and your IoT devices. Because limiting intelligent processing through legislation entails solving difficult technical problems, the prospect for doing so effectively seems remote. Similarly, few if any companies provide statements that would constrain how their IoT systems act upon and control you and your environment. For example, Tesla explains at length the types of information they extract from your vehicle and how that data might be used, but they lack any statement of how or when they control the actions of your vehicle. It is important to note, however, that few companies and only a small portion of the Internet economy depend on control of actions by IoT. Therefore, governments and the public should focus soon on debating and enacting regulations to limit the abuse of actions exercised by IoT systems. For IoT products, a "statement of control of actions" is at least as important as a privacy statement. Acting now is imperative, before automated, third-party control becomes completely embedded in our devices, economy, and society.

21.3.1.2.3 Recommendations for Limiting Type 2 Abuse

For IoT devices to provide significant benefit to us, they need data that is intelligently processed to arrive at control decisions that improve or protect our lives and environment. But to avoid type 2 abuse, users need to clearly consent to giving up their data and having automated decisions made that then control actions by IoT devices. To give consent, users must know what they are consenting to: what information is extracted, how is it used and by whom, and what actions will be taken based on the information processed. Laws are needed for all three elements of IoT: information extraction, processing, and action. The existing and emerging regulations by the EU are currently the best model available for the first two elements of IoT. A model for regulating the third element of IoT, action, is sorely lacking.

Such laws may not be possible in some countries and may not be successful even where implemented. Because extracted information is worth billions of dollars and much of the Internet economy depends on it, it will be difficult to meaningfully control information extraction, especially in countries that have significant economic dependence on tech companies reliant on information extraction. Controlling intelligent processing to avoid type 2 abuse is also a formidable challenge. Technical advances need to be developed before widespread trust in AI algorithms can be achieved. Even with the technical means to achieve trust in a particular algorithm, processing inside IoT systems can be opaque. Companies may regard its algorithms as key intellectual property and resist disclosure. Processing may be distributed over many devices and through a cloud using many different vendors' products. It may be layered across many architectural levels from hardware to firmware to operating systems to end-applications. Given the difficulty of regulating intelligent processing, new laws are perhaps better aimed at limiting the action component of IoT. Besides being at a much earlier stage of deployment than information extraction, control of actions by IoT devices is generally explicit and easily observable, unlike processing. Actions are not hidden inside a cloud or concealed behind proprietary walls. If an IoT device takes an action altering the physical world, it is visible and measurable. An action that modifies a human's emotional or mental state may be much more subtle than physical action, but it still is more likely to be observed and therefore transparent, disclosed, and regulated. Discussion and debate on regulating actions by IoT is urgently needed by policy and lawmakers and the public. It is probably our best hope to prevent type 2 abuse spreading across IoT systems.

In contrast to the US, UK, and the EU, China limits extraction of data by foreign entities, opposing calls by the US and Japan among others for increased data sharing across national boundaries [133]. However, China has not stopped extraction by its own companies, choosing instead to reserve the right to access all the information that is extracted by its private companies. It augments commercial extraction with an aggressive expansion of its own massive surveillance while investing heavily in AI for automated processing. China also calls for increased control in its most recent five-year plan [100, 132]. China's use of IoT leads to the final two types of abuse: type 3 and 4: abuse of IoT by a government.

21.3.1.3 Type 3 Abuse: Government Abuse While Using IoT for Public Defense, Health, Safety, and Wellbeing

Governments need IoT, especially in the defense and national security domain. As chapters of this book show, IoT enables a shift in how wars are fought, and peace is maintained. That shift can be seen in the Ukraine–Russian war today. If a nation fails to develop and deploy IoT in its own interest, it will be at a disadvantage to its peers and adversaries who do. Governments also have a generally acknowledged right to deploy IoT to protect their citizens and homeland, even when it interferes with individuals and their environment. As noted, nations have clarified or augmented national and international declarations of human rights to make it clear that governments can invade an organization and individual's person, homes, property, and lives to serve a greater societal good, particularly for the health, safety, and security of a nation. The European Council's ECHR, Article 8 explicitly so states, for example [126]. However, there is a fine line between legitimate use of IoT by a government and abuse of IoT by a government. That line should be established through public consent by those governed, not by government dictate. The difference between legitimate government use and type 3 abuse of IoT depends on the public's agreement that any cost to individual rights is outweighed by the collective benefit to society. There is also a narrow gap between legitimate use by a government of IoT and type 4 abuse where a government uses the power of IoT to further its own ends at the expense of those governed. The difference between type 3 and type 4 abuse can come down to a government's intent when using IoT. That intent should be understood and agreed to by those governed.

When a government subjects its populace to monitoring and control by IoT, the public should be the judge of the tradeoff between costs and benefits and the legitimacy of intent, not the government. It is the public that pays the price of abuse. To make that judgement requires transparency. Transparency requires monitoring, enforcement, consequences, and redress as discussed in Section 21.2. The challenge comes because many of the most important government uses of IoT require some level of covertness, specifically for purposes of law enforcement, defense, and intelligence collection on behalf of public safety and national security. Where the use of IoT is fully covert, no direct, specific transparency can exist, and therefore, the public cannot directly and specifically consent to the infringement of its rights, even though the net benefit to society may justify it. Where the use of IoT does not need to be covert, for example, public health and economic wellbeing, laws should clearly follow the framework presented in Section 21.2.1.3 to provide for and enforce transparency so that the public can provide unambiguous consent. The remainder of this section will focus on the three areas where covertness of IoT makes it challenging to follow a framework for abuse-prevention: law enforcement, defense, and intelligence.

21.3.1.3.1 Law Enforcement Use of IoT

Clearly in some instances, obtaining prior consent for surveillance or actions against individuals would alert the individuals and defeat the purpose of surveillance or planned actions. But all free societies have checks and balances on the use of covert law enforcement measures that allow surveillance or police action without obtaining prior consent. These have been successfully extended to new technology as it emerges, for example wiretapping of telephones or implanting video surveillance cameras in private spaces. These measures can and need to be extended beyond surveillance, either through regulations or case law, to actions taken by IoT systems. Law enforcement can use IoT not just to surveille subjects but to kill them in an automated way. The process for covert use of intrusive surveillance in private places in most free countries follows a process that requires a warrant issued by an independent agent, such as a judge or court. A warrant's issuance requires some evidence of probable wrongdoing, and it set limits on the scope and duration of surveillance. Typically, the nature and results of such covert surveillance become either public knowledge or at least open to those accused as a result of the surveillance. This process provides transparency, although delayed, and a chance for redress. Extending this process and protections to law enforcement use of IoT would seem both straightforward and essential to avoiding abuse. IoT's actions need specific protections, for example before using IoT devices to disable someone's vehicle. These checks and balances along with the process of delayed transparency and the possibility of redress, developed for law enforcement in free countries, can also serve as a model for preventing type 3 abuse of IoT in defense and intelligence applications.

Advances in IoT using intelligent processing are already providing law enforcement with new tools that cannot only solve crime but can help to predict where it is likely to happen and who is likely to commit it. For example, Clearview AI, a leader in automated facial recognition, says that 2400 law enforcement agencies were using their facial recognition system as of March 2021. Clearview AI's system is trained on a database of more than three billion photos scrapped from Internet applications like Facebook, Instagram, Google, and Venmo, without explicit, informed consent of the people behind the faces [134]. If law-enforcement agencies use individuals' data without their consent, whether directly or through third parties, they need to amend their practices. Some states have passed laws that broaden the definition of personal information to explicitly include biometric data like images of faces, for example Illinois [135]. Consistent laws are needed that clearly define when private information can be extracted and how it can be used. Such laws need to define due process regarding information extraction for law enforcement without consent.

Regulations and case law define when law-enforcement agencies can exercise physical control over individuals without consent, for example, when a law officer may shoot a fleeing suspect. Do such regulations and case law apply to an IoT system that uses face recognition to identify a subject and applies force to apprehend them? If so, then accommodations and adjustments are needed for AI processing limitations, biases, and failures when used by an IoT system that controls people and their environments. Law enforcement agencies are becoming aware of some of the problems with AI for law enforcement [136], but such awareness is not generally shared across all agencies and police forces. Police and the public need clear laws regulating the exercise of IoT actions by law enforcement. For example, warrants are generally needed to obtain information from Nest about an individual's thermostat settings, but it is not clear if laws specifically address law enforcement's ability to turn the thermostat up or down or seize control of someone's IoT-enabled car. Law enforcement's use of IoT actions should be allowed but constrained to follow processes and limitations parallel to those for seizing of an individual's private information. When IoT exerts control, physically, financially, or otherwise, it should be subject to the same legal constraints that regulate a human agent's actions. Abuse of IoT's actions for law enforcement may be constrained in the US by appeals to the Supreme Court and the fourth Amendment. Nevertheless, for the sake of clarity and consistency, new laws are needed that explicitly govern IoT actions initiated by law enforcement. These laws could parallel those governing seizure of property and information and imposing bodily force on people.

21.3.1.3.2 IoT in National Defense and Intelligence Activities

Five factors make defense and intelligence applications of IoT more vulnerable to abuse than law enforcement because they extend the secrecy around its application by necessity. A research arm of the US Congress defines the military and intelligence circumstances requiring covert action of any type, and presumably including the deployment of IoT, as ones with "a serious risk of exposure, compromise of information, loss of life, and a possible requirement to conceal U.S. sponsorship" [137]. The reasons for extended secrecy in this domain include:

1. The recipients of the actions of IoT are normally foreign nationals or nations that are often not protected by the laws and rights provided to citizens of the country deploying IoT.
2. The need for covertness frequently extends over a greater timeframe than in law enforcement, for example surveillance measures used to maintain ongoing peace through mutual deterrence.
3. The technical aspects of IoT or aspects of its deployment may represent a competitive edge over an adversary, and a nation may not want to reveal its existence or capabilities or means of employment, for example surveillance from space was long a closely guarded secret by governments engaged in it.
4. The action of IoT devices may constitute or border on an act of war or an international crime, for example assassinating a foreign official or shutting off energy supplies, and therefore the nation deploying the IoT may want to deny doing so.
5. Military and intelligence operations often occur in chaotic or dangerous environments, and they demand secrecy around their execution to protect those involved and to ensure success. The immediacy of life-threatening defense situations may not allow for a formal approval process. As a result, monitoring the use of IoT for abuse becomes difficult and makes any possibility for enforcement of regulations or limitations and redress of abuse less likely.

For Cases 3, 4, and 5 above, a nation may use IoT to surveille and control individuals, organizations, and other governments in the interests of protecting itself. Such use may be acceptable and not abusive under a given nation's laws and international covenants. A growing use of IoT in the name of national security is essential for any government attempting to maintain parity with its peers and adversaries. However, IoT raises the stakes beyond surveillance for the abuse of individuals and organizations in the name of national security. While IoT devices have long been used to provide surveillance, in the last decade or two they have been used to kill individuals and destroy property as cited above. IoT greatly raises the potential to skirt the recognized definition of war as armed human conflict, obfuscating what acts can be construed as war. Robotic machine guns, autonomous drone swarms and automated devices that increasingly control energy, health, and critical infrastructure will offer ever more momentous targets for IoT used as tools of war and national competition. Abuse of such power will tempt many governments to covertly use IoT under the guise of defense and security.

Government abuse of foreign nationals and nations is adequately addressed by the UN declaration of human rights as described in Section 21.3.2.1, and it can be applied to IoT. With the notable exception of China, all major nations are signatories to these covenants and therefore these rights apply to all people, regardless of nationality and regardless of whether an individual nation's own laws apply to foreign nationals or just its citizens. Nevertheless, as most national and international laws acknowledge, secrecy and classification of information are essential to many aspects of national defense and most aspects of intelligence operations. However, governments sometimes use secrecy and information restrictions to hide practices and information that may be embarrassing or invoke public backlash because it reveals abusive practices. For example, in the 1970 and 1980s the US withheld all information about US spy satellites from the American people, long after the Soviet Union had obtained detailed design documents for a key system. Purportedly this was due at least in part to concern about potential negative US public reaction to spying from space and not a concern about disclosing capabilities to a prime adversary [138]. While secrecy is necessary, it should be limited in extent and duration and follow a disciplined practice of oversight. Oversight needs to consider limitations on secrecy as well as its enforcement.

21.3.1.3.3 Recommendations for limiting Government Abuse of IoT Used for Defense and National Security

The prospect of autonomous weapons continues to capture the public's imagination and incite discussion regarding limits on such weapons. But autonomous weapons are just one application of IoT that needs to be regulated. Most democratic nations have laws that set limits and constraints on acceptable activities, such as covert and clandestine operations, for example, US Title 10 for the DOD and military and US Title 50 for the Intelligence Community [137]. Such statutes are necessarily general and high-level. Deployment of autonomous IoT networks raise questions of potential abuse that require additional guidance and regulation, specific to IoT. An analogy is the specific doctrine that defines the process for authorizing and executing lethal air strikes. Before subjecting anyone or any nation or organization to potentially harmful or abusive IoT, an independent assessment must be made certifying an evidenced-based, credible threat exists to the security and safety of the nation deploying the IoT. Such an assessment needs to be of the same form as completed before using weapons on an adversary or conducting an air strike. IoT can incorporate weapons directly, but it can also be used as a weapon, even if the IoT network does not explicitly contain explosive devices or firearms. Such a certification should have reasonable limits in scope and duration and have a process for review with possible consequences and redress.

This process of truly independent, in-depth approval and review can also address case 2 above (long term IoT application by governments) by applying a review and reapproval process at regular time intervals. For these reviews to lead to public consent, it is imperative that they be declassified in a timely manner within a standard period. Any extension of the period of secrecy needs to be justified with specific reasons and a limited amount of extended time. The process of review (as opposed to its detailed findings) needs to be transparent and open to all lawmakers and the public. This process could be tailored or streamlined based on a hierarchy of severity of impact of the IoT system being employed. As mentioned, such a hierarchy might be modeled after the EU risk-hierarchy for AI applications described previously.

The US government attempted to put just such a process in place for surveillance with its FISA of 1978 [139]. FISA established a court and a process partially complying with the process outlined in the previous paragraph. It applies to surveillance of both non-US citizens and US citizens in certain circumstances. It compels cooperation from communications providers. The details of the Act, the FISA Court and its shortcomings and successes are beyond the scope of this discussion. What few audits have been done have shown "widespread" problems [140]. Requests for surveillance are rarely turned down, in contrast to warrants for law enforcement in criminal courts. As has been previously stated, the FISA Court has allowed massive surveillance of most Americans and hidden it from the public and most lawmakers [141]. At present lawmakers, policymakers, and the US Supreme Court are engaged in an intense debate about the constitutionality of keeping the FISA court records secret. With Merrick Garland, the current US Attorney General, arguing for keeping them secret after having previously argued to make them public as a US Federal judge. The process needs more transparency and accountability, even if the rulings of the FISA Court are not made public. If the process can be improved to provide more transparency, then a law or specific guidance is still needed to extend it to IoT's processing and action elements as well as surveillance. When the public is instilled with enough fear, they demonstrate a willingness to allow governments to abuse their rights. A process striving for transparency with accountability for government use of IoT needs to be put in place now and not abandoned with the next public terror.

An independent, pre-approval process is not possible in the case of defense against an immediate threat, for example in self-defense against an unfolding attack. Immediate action may be required without time for an independent review and approval. Nevertheless, as with law enforcement treatment of self-defense incidents, a post hoc review is required, for example as would happen with a law enforcement investigation of an officer-involved shooting. The review needs to be independent, timely and have consequences as required. Few democratic nations would disagree with such a process, and many mandate them. But at times nations use self-defense rationales where they are not justified and use reasons 3–5 above to thwart a timely and thorough independent review [142].

21.3.1.4 Type 4 Abuse: Government Use of IoT to Enhance Its Own Power and Enrich Officials

Type 3 abuse occurs from an excess of zeal for protecting and benefiting the public. Type 3 abuse occurs most often when the government, in a vacuum, gets to decide what is necessary for a society's defense and security. Perhaps the greatest threat of type 3 abuse is that it can easily drift into George Orwell's nightmare *1984* scenario of type 4 abuse – the use of IoT to enhance a government's power and enrich its officials at the expense of those governed.

There are many totalitarian and oppressive regimes today, but only a few who are already attempting to exploit IoT to further consolidate their power and control over their citizens. The People's Republic of China stands out as having a combination of mature digital infrastructure, technical sophistication, and a government that has the will to pursue the use of IoT to further consolidate its power. In 2015, Howard stated: "The Chinese internet is already the most expensive and elaborate system ever built for suppressing political expression. The Chinese are trying to extend it by exporting their technologies to authoritarian regimes in Asia and Africa." [2, p. 183]. Since then, China has made even more impressive investments in both deployment of IoT sensors and development of AI-based processing as previously reviewed. Its 2021 five-year plan calls for increased control over actions – one of the few nations to recognize the importance of the third and decisive element of IoT. China reportedly is engaging in a massive exercise of type 4 abuse, most clearly in its campaign of domination of its Uighur minority [143]. This application of IoT appears to lack public consent of those affected. When a nation's media is controlled by the state and foreign information is filtered through a vast Internet firewall, there is no true transparency, and the public cannot effectively consent.

The first line of defense against type 4 abuse is an independent news media and an informed public. Short of armed rebellion, the people of a nation suffering type 4 abuse have little recourse. Their best hope is that other nations recognize and acknowledge violations of UN covenants on human rights and use both moral suasion as well as political and economic sanctions to force an offending government to cease its abuse. A free society must guard against abuses of types 1, 2, and 3, but more importantly they must ensure that they do not incrementally move from type 3 to type 4 abuse. The governed in a free society can force action by their government to regulate types 1, 2, and 3 abuse. Societies rapidly lose the ability to stop or resist type 4 abuse. Once it begins, it self-perpetuates, growing as it goes [144].

21.3.2 Regulating IoT to Prevent Abuse While Advancing Its Benefits

Governments and policy organizations are beginning to address potential misuse of IoT from the standpoint of the loss of individual privacy. If successful, these efforts can help fence in IoT so that it can provide its benefits without letting loose potentially harmful aspects. But these efforts to date address only the sensing/information-extraction side of IoT via regulations on privacy of information. Overregulation of sensing and information extraction starves IoT of the data it needs to make intelligent decisions. Governments needs to consider the impact that regulation of sensing and information extraction will have on IoT performance and balance that impact with the desire to protect individuals. To a much lesser extent, regulations on intelligent processing are being considered. In contrast, governments and the public seem to be almost totally unaware of the need to also address threats from the misuse of IoT's actions. It is the combination of control of actions with sensing and intelligent processing that makes IoT so powerful and so much more of a concern than a mere loss of privacy. Moreover, current attempts to constrain misuse of IoT focus on misuse only by individuals and organizations. They largely neglect to address misuse and abuse by governments themselves. Some constitutions and laws could be interpreted as applying to actions taken by IoT, such as the fourth Amendment of the US Constitution and the UN Universal Declaration of Human Rights (UN UDHR) [47, 48]. Abuse by individuals and organizations can rob the privacy, health, and wellbeing of individuals. Misuse by governments can destroy free societies. The following sections will discuss policy and regulatory efforts with a focus on preventing misuse of IoT by governments while not blocking the primary benefits of IoT.

21.3.2.1 The Right to Limit and Regulate IoT

Before summarizing strategies for regulating IoT, one needs to ask the question of whether regulation is justified or even permissible under existing national and international norms. This section answers the question: "Do governments have the right to regulate the use of IoT?".

There are ample national and international examples that provide moral authority to enact regulations that limit IoT's potential for abusive outcomes. The UN Universal Declaration of Human Rights, Article 12, cited above states:

> "No one shall be subjected to arbitrary interference with his privacy, family, home or correspondence, nor to attacks upon his honour and reputation. Everyone has the right to the protection of the law against such interference or attacks" [145].

The Universal Declaration of Human Rights (UN UDHR) was written in 1948 and adopted as a declaration and as such did not have the force of international law. However, it inspired four UN covenants that are binding for the ratifying nations as multilateral treaties, the basic one being the International Bill of Human Rights (IBHR), General Assembly Resolution 217 (III), that mandates that "No one shall be arbitrarily deprived of his property" Article 17.2 [145]. The IBHR entered into force as a multilateral treaty along with supporting covenants and conventions in 1976, introduced as the International Covenant on Civil and Political Rights (ICCPR) [146]. The US ratified the resolution in 1992, and as of 1 December 2021, it has been adopted as well by 173 members of the UN and accepted by the European Union [147]. Many nations have adopted national laws inspired by ICCPR. Only six nations have so far failed to accept the ICCPR, most notably China, although not all ratifying countries follow all the articles of the Covenant in practice.

Many countries have their own equivalent of Article 12 of the UDHR, many dating a century or two before the U.N.'s declaration. For example, the fourth Amendment of the US Constitution states:

> "The right of the people to be secure in their persons, houses, papers, and effects, against unreasonable searches and seizures, shall not be violated, and no Warrants shall issue, but upon probable cause, supported by Oath or affirmation, and particularly describing the place to be searched, and the persons or things to be seized."

The French "Declaration of Human and Civic Rights" of 1789, Article 17 states:

> "Since the right to Property is inviolable and sacred, no one may be deprived thereof, unless public necessity, legally ascertained, obviously requires it, and just and prior indemnity has been paid" [148].

The European Union defined its own Charter of Fundamental Rights [149] in 2000, Chapter II, Articles 6, 7, 8, and 17. It became legally binding in the EU in 2009 and was strengthened by a Communication released in 2020 [150]. The EU's Charter was proceeded by the European Council of ECHR, 1950 [126]. Article 8 of the Convention specifies that "everyone has the right to respect for his private and family life, his home and his correspondence." However, article 8 leaves a large hole in those rights by also specifying that inference of those rights by a public authority is justified by the interests of a democratic society, such as for national security and public safety. The UK has used Article 8's exception to justify both overt and covert mass surveillance of its population when in public spaces, as discussed above.

Today "Property" is widely, if not universally, understood to include intellectual property and personal data, whether physically manifest or digitally represented; an understanding made explicit in the EU Charter in Articles 8 and 17. With near global proforma acceptance of the UDHR/ICCPR comes general acceptance that the protection from interference of privacy and property includes personal digital information whether collected by sensors or extracted from digital communications or from information stored in computer files or contained inside software applications. More recent laws by some nations provide explicit controls over digital data, for example the Health Insurance Portability and Accountability Act (HIPPA) of 1996 in the US [113]. These numerous precedents demonstrate that national and international governments are well within their rights to establish boundaries around the abuse of IoT. Such regulation links to almost universal recognition of basic human rights.

Table 21.1 Types of potential misuse and abuse of IoT are constrained by varying degrees of existing regulations. Additional regulations are needed to address the emerging technology of IoT, especially the automated control of actions by IoT – largely unregulated today and outside of the public's awareness.

Type of IoT abuse	Existing regulations	Additional protections needed
1. Illegal or unethical abuse by individuals or organizations	Existing laws protecting information extraction vary widely but are expanding. Regulation of intelligent processing is under development in the EU, but awareness is just emerging in countries like the US.	Laws mandating disclosure of weaknesses and reporting of breaches of information privacy are needed. Civil actions against negligent IoT providers become possible with mandated disclosure. Regulations on intelligent processing need to be completed in the EU and adopted in similar form elsewhere. Awareness of misuse of IoT actions is needed. A process to formulate regulations of IoT actions should begin immediately.
2. Legal abuse of IoT without consent or benefit to users or owners	Widespread awareness and some emerging regulations address abuses of personal privacy, varying widely by nation. Little to no regulations protect against abuse of intelligent processing and automated actions by IoT.	EU regulations for protection of information should be used as a model for other nations. Regulations on intelligent processing need to be completed in the EU and adopted in similar form elsewhere. A process to formulate regulation of IoT actions should begin immediately.
3. Government abuse of IoT for reasons of public health, safety, and wellbeing	Many nations have laws governing government extraction of information. Discussion is ongoing but no laws yet governing government use of intelligent processing. Neither awareness nor regulations exist regarding action by IoT devices.	Stronger regulation and oversight are required for covert government information extraction. New legislation is needed by all nations to address government abuse of intelligent processing and actions by IoT systems. Legislation should implement the framework presented in this chapter. Public awareness is needed now.
4. Government use of IoT to enhance its own power and enrich officials	Regulations, if any exist, protecting against type 3 abuse need to be rigorously strengthened and applied to prevent type 3 abuse from progressing to type 4.	Adequate checks on Type 3 abuse will constrain the emergence of Type 4. Creation and preservation of independent media, informed public, citizen political action, and engaged voters. Pressure and sanctions from other nations. Ultimately, the option of armed rebellion.

Source: R. Douglass.

It should be noted that almost all nations recognize the right to interfere with property and persons if they present a threat to the nation or society. Most nations are aware of the potential for abuse of such exceptions and impose special restrictions on when rights can be overridden for security and defense purposes, for example by requiring independent court orders of warrants detailing the threat and what can be seized and why. However, many nations with strong statements of property rights and personal rights assert that those rights apply only to citizens of their own nations. They are not extended to foreign nationals and organizations. For example, the US fourth Amendment has generally been applied only to US citizens, despite the US's ratifying of the U.N.

UDHR, which applies to all individuals in the world, regardless of citizenship. At times of national crisis, the fourth Amendment has not been applied to all citizens, such as the Japanese US citizens in World War II [151]. When the special checks and balances on exceptions to rights in cases of defense and security are not applied, it most frequently is because foreign nationals and foreign organizations are judged to be a national security threat. Nevertheless, the extent that the sensing element of IoT interferes with privacy, it can be justly regulated per national and international declarations. The intelligent processing element of IoT can also be regulated under Article 17 of the UN ICCPR if decisions or actions are used to arbitrarily interfere with persons or property. To the extent that the control of actions by IoT interferes with a person or their property, it can also be justly regulated. It can also be argued that it is just and fair to require checks and balances on governments when they override regulations protecting human rights in the name of defense and security.

21.3.2.2 Regulating IoT: A Summary

Table 21.1 briefly summaries the type of potential misuse or abuse of IoT, the state of existing protections, and what is needed or could be added to further protect individuals, organizations, societies, and nations. Regulations have costs in terms of time and money to enforce them. These financial costs are of most concern to individuals and corporations. Regulations also have hidden costs in that they can stymie the development and spread of new technologies and rob a nation of potential benefits. These indirect costs are of greater concerns to nations and societies than financial costs of regulation because of the strategic importance of IoT technology for national defense and security. Regulations for IoT should follow a hierarchical risk model, such as the European model being developed for AI regulation where applications are tiered by their criticality and potential cost of failure or abuse. The amount of regulation and therefore its cost can be stepped up according to the criticality of the application of IoT and the potential cost of its misuse. As described, IoT presents greater challenges to a risk-level approach than AI applications alone.

Table 21.1 summaries existing regulations and needed additional regulations or protections. It is not a set of prescriptions, but rather a starting place for discussions and debates among lawmakers, policy makers, and the public. Suggested regulatory measures need to be paired with advancing technical solutions for better IoT security and protection as described in Sections 3 and 4 of this book.

21.4 Concluding Remarks: A Call to Action

The power of IoT comes from the fusion of information extraction combined with intelligent processing coupled to devices that can physically alter the environment and the people in it. The control of the environment can occur almost instantaneously across wide geographic areas, over spans of time and with little or no human intervention. The author hopes that this book makes evident the game-changing potential of IoT for defense and national security. IoT's power extends far beyond the power of mass surveillance. For defense and national security, IoT's triumvirate of functions is the automation or semi-automation of logistics, surveillance, weapons, and security operations and much more. No nation can fail to develop IoT for defense and hope to ensure its national security. The public and governments are beginning to see the potential of IoT and the need to protect IoT systems. They are also aware of the potential dangers of IoT's ability to extract vast amounts of information and end privacy. They are aware of the potential of intelligent processing to improve the world through automated decision making. The public is beginning to become aware of the potential limitations and dangers of IoT systems that use intelligent processing to remove human reasoning and compassion from decisions. The public and governments are largely unaware of the power of IoT to control and alter the world by combining actuation with sensing and processing. This new force changes the world for the good but also carries an ominous dark passenger that can change it for the worse. When the Internet was freed from its defense department cradle, many understood its potential; few understood its threat. It is the author's hope that readers of this chapter now understand IoT's potential for the bad as well as the good. When misused by individuals, organizations, or governments, IoT can threaten the rights, freedoms, property, and lives of individuals and societies. Awareness, regulation, and vigilance can ensure that emerging IoT technology realizes its benefits while preventing damage. The time for all three is now. However, no nation seeking security in the twenty-first century can ignore the power of IoT to defend and protect itself. It is already shaping outcomes on battlefields. IoT must be enthusiastically embraced for national security while simultaneously being understood and controlled.

References

1 Howard, P. (2015). *Pax Technica: How the Internet of Things May Set Us Free or Lock Us Up*. New Haven: Yale University Press.

2 Orwell, G. (1949). 1984, London: Secker & Warburg.

3 Merriam Webster (2022). Sentient. https://www.merriam-webster.com/dictionary/sentient (accessed 22 January 2022).

4 Zuboff, S. (2019). *The Age of Surveillance Capitalism: The Fight for a Human Future at the New Frontier of Power*. New York: Public Affairs.

5 NY Surveillance Camera Players (1998). A History of video surveillance in England. http://www.notbored.org/england-history.html (accessed 1 December 2021).

6 Puente, M. (2019). LAPD pioneered predicting crime with data. *LA Times* (3 July 2019).

7 Puente, M. (2019). LAPD to scrap some crime data programs after criticism. *LA Times* (5 April 2019).

8 Douglass, R., Mork, J., and Suresh, R. (1997). Battlefield awareness and data dissemination (BADD) for the warfighter. *Proceedings of SPIE 3080, Dgitization of the Battlefield*, Orlando.

9 Buderi, R. (1996). *The Invention That Changed the World*. New York: Touchstone.

10 Lukasik, S. (2011). Why the arpanet was built. *IEEE Annals of the History of Computing* 33 (3): 4–20.

11 NYC Surveillance Camera Project (1998). Surveillance Camera Project Summary. http://www.mediaeater.com/cameras/summary.html (accessed 4 February 2022).

12 Amnesty International (2021). Surveillance City, Amnesty International News, 3 June 2021. https://www.amnesty.org/en/latest/news/2021/06/scale-new-york-police-facial-recognition-revealed/ (accessed 27 November 2021).

13 HRnews (2020). Number of CCTV Cameras in the UK reaches 5.2 million. number-of-cctv-cameras-in-the-uk-reaches-5-2-million (accessed 1 December 2022). CCTV.co.uk (2021). How Many Cameras are there in the UK?. https://www.cctv.co.uk/how-many-cctv-cameras-are-there-in-london/ (accessed 2 November 2021).

14 CCTV.co.uk (2020). How Many Cameras are there in the UK?. https://www.cctv.co.uk/how-many-cctv-cameras-are-there-in-the-united-kingdom/ (accessed 2 November 2021).

15 Mozur, P. (2018). Inside China's Dystopian Dreams: A.I., Shame and Lots of Cameras. *New York Times* (8 July 2018).

16 Kumar, A. (2021). "Skynet" -- China's Massive Video Surveillance Network. *Advanced Tech World* (8 March 2021).

17 Gershgorn, D. (2020). China's 'Sharp Eyes' Program Aims to Surveil 100% of Public Space. *OneZero* (2 March 2020).

18 Statistica (2021). Number os Smartphone Users from 2016 to 2021. https://www.statista.com/statistics/330695/number-of-smartphone-users-worldwide/ (accessed 5 November 2021).

19 Afaneh, M. (2020). Wireless Connectivity Options for IoT Applications -- Indoor Navigtion, Bluetooth Blog. https://www.bluetooth.com/blog/wireless-connectivity-options-for-iot-applications-indoor-navigation/ (accessed 27 October 2021).

20 Kwet, M. (2019). In Stores, Secret Surveillance Tracks Your Every Move. *New York TImes* (5 July 2019).

21 Mac, R. and Kashmir, H. (2021). Are Apple AirTags Being Used to Track People and Steal Cars? *New York Times* (30 December 2021).

22 Douglass, R. (1987). Second generation architecture of the Autonomous Land Vehicle. *Proceedings of 1987 IEEE International Conference on Robotics and Automation*, San Jose, CA, USA.

23 Lowrie, J., Isler, W., and Douglass, R. (1986). The autonomous land vehicle. *Proceedings of SME Conference on Mobile Robots*, Monterey, CA, USA.

24 Rogers, C., Piggott, A.Y., Thomson, D.J. et al. (2021). A universal 3D imaging sensor on a silicon photonics platform. *Nature* 590: 256–261.

25 Korosec, K. (2021). Tesla is no longer using radar sensors in Model 3 and Model Y. https://techcrunch.com/2021/05/25/tesla-is-no-longer-using-radar-sensors-in-model-3-and-model-y-vehicles-built-in-north-america/ (accessed 23 August 2021).

26 Shepardson, D. (2021). U.S. safety agency probes 10 Tesla crash deaths. *Reuters* (18 June 2021).

27 TeslaDeaths (2021). Tesla Deaths. https://www.tesladeaths.com (accessed 4 February 2021).

28 FLIR, Inc. Mobile Accessories. https://www.flir.com/browse/home-amp-outdoor/mobile-accessories/ (accessed 31 July 2022).

29 Mori, S., Hoang, H., Arambel, P.O. et al. (2014). Group state estimation algorithm using foliage penetration GMTI radar detections. *17th International Conference on Information Fusion*, Salamonica.

30 Baranoski, E. (2006). Visibuilding: sensing through walls. *4th IEEE Workshop on Sensor Array and Multichannel Processing*, Waltham, MA, USA.

31 Robbins, S. (2010). U.S. Army trains troops on new device to detect suicide bombers. *Stars and Stripes* (13 June 2010).

32 Douglass, R., Gorman, J., and Burns, T. (2008). System and method for standoff detection of human carried explosives. US Patent 7900527, 5 June 2008.

33 Zinger, W. and Krill, J. (1997). Mountain Top: beyond-the-horizon cruise missile defense. *Johns Hopkins APL Technical Digest* 18 (4): 501–520.

34 Leidos, Inc. (2021). Exploranium Radiation Portal Monitors. https://www.leidos.com/products/exploranium (accessed 5 February 2022).

35 Dhagat, V. and Faquir, J. (2017). Microfulidics and sensors for DNA analysis. *International Journal of Engineering Research and Science* 3 (3): 23–91.

36 Tesla, Inc. (2022). Autopilot, Advanced Sensor Coverage. https://www.tesla.com/en_GB/autopilot?redirect=no (accessed 4 February 2022).

37 Palmer, A. (2019). Major phone companies including Sprint and Verizon say they've finally stopped selling user location data. *Daily Mail* (17 May 2019).

38 Liptak, A. (2018). In ruling on cellphone location data, supreme court makes statement on digital privacy. *New York Times* (22 June 2018).

39 Electronic Privacy Information Center, epic.org. Location Tracking. https://epic.org/issues/data-protection/location-tracking/ (accessed 15 July 2022).

40 Rayome, A.D. (2022). 9 Venmo Settings You Should Change Right Now to Protect Your Privancy. https://www.cnet.com/personal-finance/9-venmo-settings-you-should-change-right-now-to-protect-your-privacy/ (accessed 31 July 2022).

41 U.S. Congress (1994). Communications Assistance to Law Enforcement Act, Public Law 103-414-OCT.25 1994. https://www.congress.gov/bill/103rd-congress/house-bill/4922 (accessed 19 August 2021).

42 U.S. Federal Trade Commission (2005). First Report and Order and Further Notice of Posted Rule Making, FCC-05-153. https://ndcac.fbi.gov/file-repository/34.pdf (accessed 19 August 2021).

43 U.S. Federal Courts (2021). United States Foreign Intelligence Surveillance Court. https://www.fisc.uscourts.gov/about-foreign-intelligence-surveillance-court (accessed 1 December 2021).

44 U.S. Congress (2001). Uniting and strengthening America by providing appropriate tools required to intercept and obstruct terrorism (USA Patriot Act). *107th Congress Public Law 56* (26 October 2001). https://www.govinfo.gov/content/pkg/PLAW-107publ56/html/PLAW-107publ56.htm (accessed 7 September 2021).

45 Cole, D. (2014). 'No Place to Hide' by Glen Greenwald on NSA's sweeping efforts to 'Know it All'. *Washington Post* (12 May 2014).

46 Macaskill, E. and Dance, G. (2013). NSA Files: Decoded, What the revelations mean for you. *The Guardian* (1 November 2013).

47 U.S. Congress (1791). 4th Amendment, Constitution of the United States, 15 December 1791. https://constitution.congress.gov/constitution/ (accessed 15 October 2021).

48 General Assembly of the United Nations (1948). Universal Declaration of Human Rights, General Assembly Resolution 217A. https://www.un.org/en/about-us/universal-declaration-of-human-rights (accessed 22 September 2021).

49 Savage, C. and Weisman, J. (2015). N.S.A. Collection of Bulk Call Data Is Ruled Illegal. *New York Times* (7 May 2015).

50 Moravčík, M., Schmid, M., Burch, N. et al. (2017). DeepStack: Expert-level artificial intelligence in heads-up no-limit poker. *Science* 356: 508–513.

51 Feigenbaum, E. and Feldman, J. (1963). *Computers and Thought*. New York: McGraw-Hill Book Company.

52 Douglass, R. (1982). *Artificial Intelligence: Definitions and Characteristics*. Los Alamos: Los Alamos National Laboratory Report.

53 Turing, A. (1950). Computing machinery and Intelligence. *Mind* 59: 433–460.

54 Khaneman, D. (2011). *Thinking, Fast and Slow*. New York: Macmillan.

55 Hebb, D. (1949). *The Organization of Behavior*. New York: Wiley.

56 Munz, M. (2014). Rapid Hebbian axonal remodeling mediated by visual stimulation. *Science* 344 (6186): 904–909.

57 Rosenblatt, F. (1957). The Perceptron -- A Perceiving and Recognizing Automaton. *Report 85-460-1*. Ithica: Cornell Aeronautical Laboratory.

58 Hubel, D. and Wiesel, T. (1962). Receptive fields, binocular interaction and functional architecture in the cat's visual cortex. *The Journal of Physiology* 160 (1): 106–154.

59 Uhr, L. (1978). Recognition cones and some test results. In: A. Hanson and E. Riseman, *Computer Vision Systems*, 363–378. New York: Academic Press.

60 Douglass, R. (1977). Recognition and spatial organization of objects in natural scenes. *International Joint Conference on AI-5 (IJCAI-5)*, Cambridge, MA.

61 Rumelhart, D., Hinton, G., and Williams, R. (1986). Learning representations by Back-propogating errors. *Nature* 323: 533–536.

62 Kirkpatrick, J. (2021). Pushing the frontiers of density functionals by solving the fractional electron problem. *Sciences* 374 (6573): 1385–1389.

63 Thorp, H. (2021). Proteins, proteins everywhere. *Science* 374 (6574): 1415–1415.

64 Chung, A., Jou, C., Grau-Perales, A. et al. (2021). Cognitive control persistently enhances hippocampal information processing. *Nature* 600: 484–488.

65 Cambell, M. (2018). Mastering board games. *Science* 362 (6419): 1118.

66 Kahneman, D., Sibony, O., and Sunstein, C. (2021). *Noise: The Flaw in Human Judgement*. New York: Harper Collins.

67 Taleb, N. (2007). *The Black Swan: The Impact of the Highly Improbable*. New York: Random House Trade Paperback.

68 Gould, S.J. (1989). *Wonderful Life: The Burgess Shale and the Nature of History*. New York: W. W. Norton & Company.

69 Lander, E. and Nelson, A. (2021). Americans need a bill of rights for an AI-powered world, wired opinion. *Wired Magazine* (8 October 2021).

70 Brooks, R. (2021). An inconvenient truth about AI. *IEEE Spectrum*. https://spectrum.ieee.org/rodney-brooks-ai (accessed 1 December 2022).

71 DARPA Outreach (2021). Collaborative Air Combat Program Makes Strides. https://www.darpa.mil/news-events/2021-03-18a (accessed 21 August 2021).

72 Bergman, R. (2021). The scientist and the A.I.-asssited remote-control killing machine. *New York Times* (18 September 2021).

73 Newell, A. and Simon, H. (1963). GPS, A program that simulates human thought. In: Feigenbaum, E. and Feldman, J., *Computers and Thought*, 279–309. New York: McGraw-Hill.

74 SRI Interantional. Shakey the Robot. https://www.sri.com/hoi/shakey-the-robot/ (accessed 21 June 2022).

75 Jones, R., Laird, J.E., Nielsen, P.E. et al. (1999). Automated intelligent pilots for combat flight simulation. *AI Magazine* 20 (1): 27–41.

76 Silver, D., Hubert, T., Schrittwieser, J. et al. (2018). A general Reinforcement learning algorithm that masters chess, shogi, and Go through self-play. *Science* 362: 1140–1144.

77 Whiten, A. (1991). *Natural Theories of Mind*. Oxford: Blackwell Publishers.

78 Phillips, P., Hahn, C., Fontana, P. et al. (2020). *The Four Principles of Explainable Artificial Intelligence*, NISTIR 8312. Washington, DC: U.S. National Institue of Standards and Technology.

79 Turek, M. (2018). Explainable Artificial Intelligence. DARPA Whitepaper. https://www.darpa.mil/program/explainable-artificial-intelligence (accessed 3 December 2021).

80 Schmidt, E., Smith, S.A., Sorensen, J.A. et al. (2021). *Final Report*. Washington, DC: National Security Commission on Artificial Intelligence.

81 European Commission (2018, 2018). *Artificial Intelligence for Europe, Communication COM (2018) 237 Final*. Brussels: European Commission to the European Parliament.

82 Tompson, S. and Warzel, C. (2019). Twelve million phones, one dataset, zero privacy. *New York Times* (19 December 2019).

83 Thompson, S. and Warzel, C. (2019). How to track president trump. *New York Times* (20 December 2019).

84 Tait, A. (2019). Fitness app Strava breached US security -- it's time to consider the dangers of data. *The New Statesman* (31 January 2019).

85 Hern, A. (2018). Strava suggests military users 'opt out' of heatmap as row deeps. *The Guardian* (29 January 2018).

86 Baldor, L. (2018). Pentagon restricts use of fitness trackers, other devices. *AP News* (6 August 2018).

87 Enev, M., Takakuwa, A., Koscheer, K., and Kohno, T. (2016). Automobile driver fingerprinting. *Proceedings of Privacy Enhancing Technologies*, Darmstadt.

88 U.S.-China Economic and Security Review Commission (2018). China's Internet of Things, Special Report, 25. https://www.uscc.gov/research/chinas-internet-things (accessed 26 January 2022).

89 Biddle, S. (2016). The Intercept: Troubling Study Says Artificial Intelligence Can Predict Who Will Be Criminals Based On Facial Features. https://theintercept.com/2016/11/18/troubling-study-says-artificial-intelligence-can-predict-who-will-be-criminals-based-on-facial-features/ (accessed 2 May 2022).

90 US Air Force. RQ-4 Global Hawk. https://www.af.mil/About-Us/Fact-Sheets/Display/Article/104516/rq-4-global-hawk/ (accessed 23 July 2022).

91 Designation-Systems.net (2009). Raytheon PAM Lockheed LAM (NetFires NLOS-LS). http://designation-systems.net/dusrm/app4/netfires.html (accessed 15 July 2022).

92 Hambling, D. (2022). Has Elusive 'Phoenix Ghost' Loitering Munition Broken Cover At Last?. *Forbes* (2 August 2022).

93 McKay, S., Boyer, M.E., Beyene, N.M. et al. (2020). Automating Army Convoys: Technical and Tactical Risks and Opportunities. https://www.rand.org/pubs/research_reports/RR2406.html (accessed 7 May 2022).

94 Lohr, S. (2021). Group backed by top companies moves to combat A.I. bias in hiring. *New York Times* (8 December 2021).

95 Palmer, A. (2021). Dead Roombas, stranded packages and delayed exams: how the AWS outage wreaked havoc across the U.S. *CNBC*, 9 December 2021.

96 Sanger, D. and Periroth, N. (2019). U.S. Escalates online attacks on Russia's power grid. *New York Times* (15 June 2019).

97 Sanger, D. (2018). Trump's national security advisor calls russian interference 'incontrovertible'. *New York Times* (17 February 2018).

98 Fassihi, F. and Bergman, R. (2021). Israel and Iran broaden cyberwar to attack civilian targets. *New York Times* (27 November 2021).

99 Doucet, L. (2020). Qasem Soleimani: US kills top Iranian general in Baghdad air strike. *BBC News* (3 January 2020).

100 Murphy, B.e. (2020). The 14th Five-Year Plan for National Economic and Social Development and Long-Range Objectives for 2030. Proposal of the Central Committee of the Chinese Communist Party, 3 November 2020. https://cset.georgetown.edu/wp-content/uploads/t0237_5th_Plenum_Proposal_EN-1.pdf (accessed 7 December 2021).

101 Vrz, L. (2021). 4 Things to know about China's new cryptocurrency. *The Washington Note* (16 April 2021).

102 Deng, S. and Liakos, C. (2021). Bitcoin plummets after China intensifies cryptocurrency crackdown. *CNN Business* (24 September 2021).

103 Koscher, K., Czeskis, A., Roesner, F. et al. (2010). Experimental security analysis of a modern automobile. *2010 IEEE Symposium on Security and Privacy*, Oakland.

104 Wikipedia. Therac-25. https://en.wikipedia.org/wiki/Therac-25 (accessed 31 July 2022).

105 IMDb. Slaughterbots. https://www.imdb.com/title/tt7659054/ (accessed 31 July 2022).

106 Biden, J. (2021). Improving the Nation's Cybersecurity, Executive Order 14028. *Federal Resister* (12 May 2021).

107 Office of U.S. DoD Chief Information Officer (2016). DoD Policy Recommendations for The Internet of Things (IoT), DoD CIO portal, December 2016. https://dodcio.defense.gov/Portals/0/Documents/Announcement/DoD%20Policy%20Recommendations%20for%20Internet%20of%20Things%20-%20White%20Paper.pdf?ver=2017-01-26-152811-440 (accessed 13 November 2021).

108 Kim, K., Kim, I., and Lim, J. (2017). National cyber security enhancement scheme for intelligent surveillance capacity with public IoT environment. *The Journal of Supercomputing* 73: 1140–1151.

109 Kirschbaum, J. (2017, 2017). *Internet of Things: Enhanced Assessments and Guidance Are Needd to Address Security Risks in DoD, GAO Report to Congress, GAO-17-668*. Washington, DC: US General Accounting Office.

110 US Office of Cybersecurity, Energy Security, and Emergency Response. Colonial Pipeline Cyber Incident. https://www.energy.gov/ceser/colonial-pipeline-cyber-incident (accessed 31 July 2022).

111 Lasker, A. (2019). 'Frantic' children terrorized by Ring Camera hacker: 'Your family's going to die'. Yahoo!life, 16 December 2019. https://www.yahoo.com/lifestyle/frantic-children-terrorized-ring-camera-154614911.html (accessed 31 July 2022).

112 Solove, D. (2006). A Brief History of Information Privacy Law, George Washington University Law Scholarly Commons. https://scholarship.law.gwu.edu/cgi/viewcontent.cgi?article=2076&context=faculty_publications (accessed 14 October 2021).

113 104th Congress of U.S. (1996). Public Law 104-191, Health Insurance Portability and Accountability Act of 1996, 21 August 1996. https://www.congress.gov/104/plaws/publ191/PLAW-104publ191.pdf (accessed 7 December 2021).

114 U.S. 104th Congress (1995). Paperwork Reduction Act (44 U.S.C. 3501 et seq.), 22 May 1995. https://www.govinfo.gov/content/pkg/PLAW-104publ13/html/PLAW-104publ13.htm (accessed 22 November 2021).

115 116th Congress of U.S. (2020). Internet of Things Cybersecurity Improvement Act of 2020, Public Law 116-207, 4 December 2020. https://www.congress.gov/116/plaws/publ207/PLAW-116publ207.pdf (accessed 18 December 2021).

116 Center for Internet Security, CIS. The Mirai Botnet -- Threats and Mitigations. https://www.cisecurity.org/insights/blog/the-mirai-botnet-threats-and-mitigations (accessed 31 July 2022).

117 Camp, J., Henry, R., Kohno, T. et al. (2020). Toward a secure Internet of Things: directions for research. *IEEE Security and Privacy* 18 (4): 28–37.

118 European Commission (2021). Regulatory framework proposal on Artificial Intelligence, 1 December 2021. https://digital-strategy.ec.europa.eu/en/policies/regulatory-framework-ai (accessed 1 December 2021).

119 European Commission (2021). What is GDPR, the EU's new data protection law?, 1 December 2021. https://gdpr.eu/what-is-gdpr/ (accessed 1 December 2021).

120 European Commission (2021). Regulation of the European Parliament and the Council Laying Down Harmonised Rules on Artificial Intelligence and Amending Certain Union Legislative Acts, COM(2021) 206 final, 6 January 2021. https://eur-lex.europa.eu/resource.html?uri=cellar:e0649735-a372-11eb-9585-01aa75ed71a1.0001.02/DOC_1&format=PDF (accessed 3 November 2021).

121 European Commission (2021). A European approach to artificial intelligence, 1 December 2021. https://digital-strategy.ec.europa.eu/en/policies/european-approach-artificial-intelligence (accessed 1 December 2021).

122 European Commission (2021). Laying Down Harmonised Rules on Artificial Intelligence and Amending Certain Union Legislative Acts, Proposal for a Regulation of the European Parliament and of the Council, SEC(2021) 167 final, 3 December 2021. https://eur-lex.europa.eu/resource.html?uri=cellar:e0649735-a372-11eb-9585-01aa75ed71a1.0001.02/DOC_1&format=PDF (accessed 4 February 2022).

123 European Commission (2021). Fostering a European approach to Artificial Intelligence, Communication from the Commission to the European Parliament, COM(2021) 205 final, 21 April 2021. https://ec.europa.eu/newsroom/dae/redirection/document/75790 (accessed 20 October 2021).

124 Lima, C. (2021). Congress is reviving the data privacy debate. Don't hold your breath for a law. *Washington Post,* 24 September 2021.

125 UK Home Office (2013). Surveillance Camera Code of Practice, 1 June 2013. https://assets.publishing.service.gov.uk/government/uploads/system/uploads/attachment_data/file/282774/SurveillanceCameraCodePractice.pdf (accessed 16 August 2021).

126 European Council (1950). The European Convention on Human Rights, European Court of Human RIghts, Rome, 4 November 1950. https://www.echr.coe.int/Documents/Convention_ENG.pdf (accessed 2 September 2021).

127 Haggin, P. (2021). Personal data is worth billions. These startups want you to get a cut. *The Wall Street Journal* (4 December 2021).

128 Tesla, Inc. (2021). Customer Privacy Notice, 5 December 2021. https://www.tesla.com/legal/privacy#updates-to-this-notice (accessed 5 December 2021).

129 Rafaelof, E., Creemers, R., Sacks, S. et al. (2021). Translation: Data Security Law of the People's Republic of China (effective September 1, 2021). DigiChina: Stanford University, 12 Novemeber 2021. https://digichina.stanford.edu/work/translation-data-security-law-of-the-peoples-republic-of-china/ (accessed 12 November 2021).

130 Pottinger, M. and Feith, D. (2021). The most powerful data broker in the world is winning the war against the U.S. *New York Times* (30 November 2021).

131 Li, J. (2021). Beijing has a new legal architecture for sweeping control over user data, Quartz, 19 August 2021. https://qz.com/2051268/china-aims-to-control-but-also-unleash-the-economic-power-of-data/ (accessed 28 October 2021).

132 Periroth, N. (2021). How China transformed into a prime cyber threat to the U.S. *New York Times* (19 July 2021).

133 Koizumi, M. (2019). Japan's pitch for free data flows 'with trust' faces uphill battle at G20 amid 'splinternet' fears. *The Japan Times* (27 June 2019).

134 Bhuiyan, J. (2021). Clearview AI uses your online photos to instantly ID you. *Los Angeles Times* (9 March 2021).

135 Illinois General Assembly, State of Illinnois (2008). Biometric Information Privacy Act, 740 Illinois Compiled Statutes 14/1-99, 3 October 2008. https://ilga.gov/legislation/ilcs/ilcs3.asp?ActID=3004 (accessed 15 December 2021).

136 Ramaijmakers, S. (2019). Artificial intelligence for law enforcement: challenges and opportunities. *IEEE Security and Privacy* 17 (5): 74–77.

137 US Congressional Research Service (2019). Covert Action and Clandestine Activities of the Intelligence Community: Selected Definitions in Brief, 14 June 2019. https://sgp.fas.org/crs/intel/R45175.pdf (accessed 15 December 2021).

138 Lindsay, R. (2018). *The Falcon and the Snowman: A True Story of Friendship and Espionage.* New York: Open Road Media.

139 U.S. Bureau of Justice, U.S. Dept. of Justice (2021). The Foreign Intelligence Surveillance Act of 1978 (FISA), 22 December 2021. https://bja.ojp.gov/program/it/privacy-civil-liberties/authorities/statutes/1286 (accessed 22 December 2021).

140 Barrett, D. (2021). Inspector general finds 'widespread' problmes in FBI's FISA applicaitons. *Washington Post* (30 September 2021).

141 Liptak, A. (2021). At the supreme court, a plea to reveal secret surveillance rulings. *New York Times* (3 October 2021).

142 Khan, A. (2021). Hidden Pentagon records reveal patterns of failure in deadly airstrikes. *New York Times* (18 December 2021).

143 Buckley, C., Mozur, P., and Ramzy, A. (2019). How China Turned a City into A Prison: a surveillance state reaches new heights. *New York Times* (4 April 2019).

144 Lucretius, T. (2008). *De Rerum Natura*. Madison, Univ. of Wisconsin Press.

145 United Nations (1948). Resolution 217 (III), International Bill of Human Rights, 10 December 1948. https://www.un.org/ga/search/view_doc.asp?symbol=A/RES/217%28III%29 (accessed 10 October 2021).

146 United Nations (1966). International Covenant on Civil and Political Rights, 16 December 1966. http://www.un.org/ruleoflaw/files/International%20Covenant%20on%20Civil%20and%20Political%20Rights.pdf (accessed 24 August 2018).

147 United Nations (1976). International Covenant on Civil and Political Rights, UN Treaty Collection, 23 March 1976. https://treaties.un.org/Pages/ViewDetails.aspx?src=TREATY&mtdsg_no=IV-4&chapter=4&clang=_en (accessed 24 August 2021).

148 National Assembly of the French People (1789). Declaration of Human and Civic Rights, 26 August 1789. https://www.conseil-constitutionnel.fr/sites/default/files/as/root/bank_mm/anglais/cst2.pdf (accessed 16 August 2021).

149 European Commission (2000). Charter of fundamental rights of the European Union. *Official Journal of the European Commission* 18 (12): 396–407.

150 European Commission (2020). Communication from the Commission to the European Parliament, Strategy to strengthen the application of the Charter of Fundamental Rights in the EU, COM (2020) 711 final, 2 December 2020. https://eur-lex.europa.eu/legal-content/EN/TXT/PDF/?uri=CELEX:52020DC0711&from=EN (accessed 2 September 2021).

151 US National Archives. Japanese-American Incarceration During World War II. https://www.archives.gov/education/lessons/japanese-relocation (accessed 31 July 2022).

Index

IoT for Defense and National Security, First Edition. Edited by Robert Douglass, Keith Gremban, Ananthram Swami, and Stephan Gerali.

i

n

o

p

Printed and bound by CPI Group (UK) Ltd, Croydon, CR0 4YY
17/09/2026
14964259-0006